Elasticity of Transversely Isotropic Materials

SOLID MECHANICS AND ITS APPLICATIONS
Volume 126

Series Editor: G.M.L. GLADWELL
Department of Civil Engineering
University of Waterloo
Waterloo, Ontario, Canada N2L 3GI

Aims and Scope of the Series

The fundamental questions arising in mechanics are: *Why?, How?,* and *How much?* The aim of this series is to provide lucid accounts written by authoritative researchers giving vision and insight in answering these questions on the subject of mechanics as it relates to solids.

The scope of the series covers the entire spectrum of solid mechanics. Thus it includes the foundation of mechanics; variational formulations; computational mechanics; statics, kinematics and dynamics of rigid and elastic bodies: vibrations of solids and structures; dynamical systems and chaos; the theories of elasticity, plasticity and viscoelasticity; composite materials; rods, beams, shells and membranes; structural control and stability; soils, rocks and geomechanics; fracture; tribology; experimental mechanics; biomechanics and machine design.

The median level of presentation is the first year graduate student. Some texts are monographs defining the current state of the field; others are accessible to final year undergraduates; but essentially the emphasis is on readability and clarity.

For a list of related mechanics titles, see final pages.

Elasticity of Transversely Isotropic Materials

by

Haojiang Ding

Zhjiang University,
Hangzhou, China

Weiqiu Chen

Zhjiang University,
Hangzhou, China

and

L. Zhang

The University of Sydney,
NSW, Australia

Springer

A C.I.P. Catalogue record for this book is available from the Library of Congress.

ISBN-10 1-4020-4033-4 (HB)
ISBN-13 978-1-4020-4033-7 (HB)
ISBN-10 1-4020-4034-2 (e-book)
ISBN-13 978-1-4020-4034-4 (e-book)

Published by Springer,
P.O. Box 17, 3300 AA Dordrecht, The Netherlands.

www.springer.com

Printed on acid-free paper

Printed in the Netherlands.

Contents

Preface

This book aims to provide a comprehensive introduction to the theory and applications of the mechanics of transversely isotropic elastic materials. There are many reasons why it should be written. First, the theory of transversely isotropic elastic materials is an important branch of applied mathematics and engineering science; but because of the difficulties caused by anisotropy, the mathematical treatments and descriptions of individual problems have been scattered throughout the technical literature. This often hinders further development and applications. Hence, a text that can present the theory and solution methodology uniformly is necessary.

Secondly, with the rapid development of modern technologies, the theory of transversely isotropic elasticity has become increasingly important. In addition to the fields with which the theory has traditionally been associated, such as civil engineering and materials engineering, many emerging technologies have demanded the development of transversely isotropic elasticity. Some immediate examples are thin film technology, piezoelectric technology, functionally gradient materials technology and those involving transversely isotropic and layered microstructures, such as multi-layer systems and tribology mechanics of magnetic recording devices. Thus a unified mathematical treatment and presentation of solution methods for a wide range of mechanics models are of primary importance to both technological and economic progress.

The authors aim to achieve a systematic structure for this complex subject in a single volume and provide the reader with state-of-the-art solution strategies for transversely isotropic elasticity under a unified umbrella. The subject matter has been organized into ten chapters to incorporate fundamental theories, solution skills and applications into an organic whole.

Chapter 1 begins with a concise summary of the basic equations of anisotropic elasticity used in the book, including thermo-elasticity. The materials presented here construct the framework for the theories and solutions of transversely isotropic problems.

The success of solutions relies largely on the strategies and mathematical treatments. Chapter 2 is therefore arranged to explain the basic methodologies for obtaining the general elastic solutions of transversely isotropic materials. In this way, the reader becomes clearer about the specific approaches for individual mechanics models in the later chapters.

Point force solutions are fundamental in solving various problems. Hence, Chapter 3 is devoted to establishing the relevant basics using a unified method that avoids the existing confusions in the literature. Meanwhile, this chapter focuses on infinite body problems and serves as an introduction to solution skills for more complex cases.

With the understanding gained and theory developed in the previous chapters, Chapters 4 to 10 discuss the solution of complicated engineering problems, including half-spaces, layered media, cones, thermal stress, frictional contact and bending, vibration and stability of plates and shells. These provide the reader not only with specific methods for tackling mathematical systems involving transverse isotropy, but also the fundamental solutions that can be extended to more complex situations.

This book is suitable for engineers, designers, researchers and postgraduates who are interested in the solution of transversely isotropic elastic materials. The authors believe that the reader who takes time to study this book will find ample reward.

The first author is indebted to Professor Hu Haichang, who introduced him to the field of elasticity and has offered invaluable help and guidance for many years. The authors appreciate very much the valuable comments and suggestions made by Professor Graham Gladwell. The financial support from the National Natural Science Foundation of China, Natural Science Foundation of Zhejiang Province as well as the 151 Talent Project of Zhejiang Province are very much appreciated. Finally, the authors wish to thank their families for their assistance; without their encouragement the book would never have been completed.

Haojiang DING, Weiqiu CHEN
Hangzhou, P. R. China

Liangchi ZHANG
Sydney, Australia

March 2005

1

BASIC EQUATIONS OF ANISOTROPIC ELASTICITY

This chapter introduces the basic equations of anisotropic elasticity which are essential for solving transversely isotropic elasticity problems. For simplicity, we ignore the mathematical details of deriving these equations; the reader can find them in the relevant references provided in **Appendix A**.

1.1 TRANSFORMATION OF STRAINS AND STRESSES

Consider an anisotropic and ideally elastic continuous solid subjected to small deformation. Assume that the solid is free of stress before deformation. The *stress-strain relationship* in this case is linear, *i.e.*, it follows the *generalized Hooke's law*. If the solid is homogeneous, the coefficients in the stress-strain relationship are constant, but if it is inhomogeneous, they will vary because the elastic properties at different points in the solid are different; they will be functions of the coordinates.

We can use various coordinate systems when studying the stresses and strains in a solid generated by external loading. In this book, we use a *Cartesian coordinate system*, (x, y, z), a *cylindrical coordinate system*, (r, α, z) or a *spherical coordinate system*, (R, θ, α). There are simple relationships between these coordinates, as listed in Tables 1.1 and 1.2, where angle α $(0 \le \alpha \le 2\pi)$ is measured from the positive direction of the *x*-axis to that of the *y*-axis, and angle θ $(0 \le \theta \le \pi)$ is measured from the positive direction of the *z*-axis to the negative direction of the same axis.

Correspondingly, we use (u, v, w), (u_r, u_α, w) or $(u_R, u_\theta, u_\alpha)$, respectively, to denote the displacements at a point in the solid in Cartesian, cylindrical or spherical coordinate systems. In tensor form, the displacements will be written as u_i $(i = 1, 2, 3)$, but in matrix form, we write them as $\{u\} = [u, v, w]^{\mathrm{T}}$, $[u_r, u_\alpha, w]^{\mathrm{T}}$ or $[u_R, u_\theta, u_\alpha]^{\mathrm{T}}$, respectively, here the superscript T stands for transpose.

Table 1.1 Direction cosines between coordinate axes in Cartesian and cylindrical coordinates.

	x	y	z
r	$\cos\alpha$	$\sin\alpha$	0
α	$-\sin\alpha$	$\cos\alpha$	0
z	0	0	1

Table 1.2 Direction cosines between coordinate axes in Cartesian and spherical coordinates.

	x	y	z
R	$\sin\theta\cos\alpha$	$\sin\theta\sin\alpha$	$\cos\theta$
θ	$\cos\theta\cos\alpha$	$\cos\theta\sin\alpha$	$-\sin\theta$
α	$-\sin\alpha$	$\cos\alpha$	0

In the three coordinate systems, the stresses and strains can be written, respectively, as

$$\begin{bmatrix} \sigma_x & \tau_{xy} & \tau_{zx} \\ \tau_{xy} & \sigma_y & \tau_{yz} \\ \tau_{zx} & \tau_{yz} & \sigma_z \end{bmatrix}, \quad \text{and} \quad \begin{bmatrix} \varepsilon_x & \frac{1}{2}\gamma_{xy} & \frac{1}{2}\gamma_{zx} \\ \frac{1}{2}\gamma_{xy} & \varepsilon_y & \frac{1}{2}\gamma_{yz} \\ \frac{1}{2}\gamma_{zx} & \frac{1}{2}\gamma_{yz} & \varepsilon_z \end{bmatrix},$$

$$\begin{bmatrix} \sigma_r & \tau_{r\alpha} & \tau_{zr} \\ \tau_{r\alpha} & \sigma_\alpha & \tau_{\alpha z} \\ \tau_{zr} & \tau_{\alpha z} & \sigma_z \end{bmatrix}, \quad \text{and} \quad \begin{bmatrix} \varepsilon_r & \frac{1}{2}\gamma_{r\alpha} & \frac{1}{2}\gamma_{zr} \\ \frac{1}{2}\gamma_{r\alpha} & \varepsilon_\alpha & \frac{1}{2}\gamma_{\alpha z} \\ \frac{1}{2}\gamma_{zr} & \frac{1}{2}\gamma_{\alpha z} & \varepsilon_z \end{bmatrix},$$

$$\begin{bmatrix} \sigma_R & \tau_{R\theta} & \tau_{\alpha R} \\ \tau_{R\theta} & \sigma_\theta & \tau_{\theta\alpha} \\ \tau_{\alpha R} & \tau_{\theta\alpha} & \sigma_\alpha \end{bmatrix}, \quad \text{and} \quad \begin{bmatrix} \varepsilon_R & \frac{1}{2}\gamma_{R\theta} & \frac{1}{2}\gamma_{\alpha R} \\ \frac{1}{2}\gamma_{R\theta} & \varepsilon_\theta & \frac{1}{2}\gamma_{\theta\alpha} \\ \frac{1}{2}\gamma_{\alpha R} & \frac{1}{2}\gamma_{\theta\alpha} & \varepsilon_\alpha \end{bmatrix},$$

where γ_{ij} are the *engineering shear strains* and have the relationship, $\gamma_{ij} = 2\varepsilon_{ij}$ $(i \neq j)$, with the *strain tensor*, ε_{ij}. Thus, in Cartesian coordinates $\varepsilon_{12} = \varepsilon_{xy} = \gamma_{xy}/2$, in cylindrical coordinates $\varepsilon_{12} = \varepsilon_{r\alpha} = \gamma_{r\alpha}/2$, and in spherical coordinates $\varepsilon_{12} = \varepsilon_{R\theta} = \gamma_{R\theta}/2$.

In tensor form, the *stress tensor*, σ_{ij}, and the strain tensor, ε_{ij}, are expressed respectively as

$$\begin{bmatrix} \sigma_{11} & \sigma_{12} & \sigma_{13} \\ \sigma_{21} & \sigma_{22} & \sigma_{23} \\ \sigma_{31} & \sigma_{32} & \sigma_{33} \end{bmatrix}, \quad \text{and} \quad \begin{bmatrix} \varepsilon_{11} & \varepsilon_{12} & \varepsilon_{13} \\ \varepsilon_{21} & \varepsilon_{22} & \varepsilon_{23} \\ \varepsilon_{31} & \varepsilon_{32} & \varepsilon_{33} \end{bmatrix}.$$

Both the stress and strain are symmetric tensors of rank two, *i.e.*, $\sigma_{ij} = \sigma_{ji}$ and $\varepsilon_{ij} = \varepsilon_{ji}$ and they follow the following *transformation rule*:

$$\sigma_{p'q'} = l_{p'i} l_{q'j} \sigma_{ij}, \tag{1.1.1}$$

$$\varepsilon_{p'q'} = l_{p'i} l_{q'j} \varepsilon_{ij}, \tag{1.1.2}$$

where σ_{ij} stands for the stresses in Cartesian coordinates, (x, y, z), $\sigma_{p'q'}$ represents the stresses in a new Cartesian coordinate system after rotation, (x', y', z'), and $l_{p'i}$ are the direction cosines between two coordinate axes, as listed in Table 1.3. For example, $l_{1'1} = \cos(x', x)$, $l_{1'2} = \cos(x', y)$, $l_{2'3} = \cos(y', z)$.

Table 1.3 Direction cosines between coordinate axes.

	x	y	z
x'	$l_{1'1}$	$l_{1'2}$	$l_{1'3}$
y'	$l_{2'1}$	$l_{2'2}$	$l_{2'3}$
z'	$l_{3'1}$	$l_{3'2}$	$l_{3'3}$

In Eqs. (1.1.1) and (1.1.2), the repetition of a subscript in a term denotes a summation with respect to the index over its range from 1 to 3. In tensor analysis such an index is called a *dummy index*, while one that is not summed out is called a *free index*. This *summation convention* will apply throughout the book unless otherwise stated.

Using Eq. (1.1.1) and Tables 1.1 and 1.3, we can easily obtain the relationships between the stresses in cylindrical and Cartesian systems, *i.e.*,

$$\begin{aligned}
\sigma_r &= \sigma_x \cos^2\alpha + \sigma_y \sin^2\alpha + 2\tau_{xy} \sin\alpha\cos\alpha, \\
\sigma_\alpha &= \sigma_x \sin^2\alpha + \sigma_y \cos^2\alpha - 2\tau_{xy} \sin\alpha\cos\alpha, \\
\tau_{r\alpha} &= (\sigma_y - \sigma_x)\sin\alpha\cos\alpha + \tau_{xy}(\cos^2\alpha - \sin^2\alpha), \\
\tau_{zr} &= \tau_{zx}\cos\alpha + \tau_{yz}\sin\alpha, \\
\tau_{\alpha z} &= -\tau_{zx}\sin\alpha + \tau_{yz}\cos\alpha, \\
\sigma_z &= \sigma_z.
\end{aligned} \tag{1.1.3}$$

Note that in deriving the above equations, we have used the symmetry of the stress tensor and simplified the formulae because some of the direction cosines are zero. These reduce the nine summation items in Eq. (1.1.1) to six and bring about the constant factor 2 in the first two equations and terms $\cos^2\alpha$ and $-\sin^2\alpha$ in the third.

On any infinitesimal area in a solid with an external normal $\vec{n}$, if the projections of the stress in *x*-, *y*- and *z*-directions in a Cartesian coordinate system are p_x, p_y and p_z, then

$$\begin{aligned} p_x &= \sigma_x \cos(n,x) + \tau_{xy} \cos(n,y) + \tau_{zx} \cos(n,z), \\ p_y &= \tau_{xy} \cos(n,x) + \sigma_y \cos(n,y) + \tau_{yz} \cos(n,z), \\ p_z &= \tau_{zx} \cos(n,x) + \tau_{yz} \cos(n,y) + \sigma_z \cos(n,z). \end{aligned} \tag{1.1.4}$$

We can get similar formulae for stresses in cylindrical and spherical coordinate systems.

1.2 BASIC EQUATIONS

The basic equations of elasticity are *geometric equations* (strain-displacement relations), *equations of motion* and *constitutive equations* (stress-strain relations). Using the coordinate transformation discussed above, we can easily get the basic equations in different coordinate systems, as listed below.

1.2.1 Geometric Equations

In Cartesian coordinates, we have

$$\begin{aligned} \varepsilon_x &= \frac{\partial u}{\partial x}, \quad \gamma_{yz} = \frac{\partial v}{\partial z} + \frac{\partial w}{\partial y}, \\ \varepsilon_y &= \frac{\partial v}{\partial y}, \quad \gamma_{zx} = \frac{\partial w}{\partial x} + \frac{\partial u}{\partial z}, \\ \varepsilon_z &= \frac{\partial w}{\partial z}, \quad \gamma_{xy} = \frac{\partial u}{\partial y} + \frac{\partial v}{\partial x}. \end{aligned} \tag{1.2.1}$$

In cylindrical coordinates, they become

$$\begin{aligned}
\varepsilon_r &= \frac{\partial u_r}{\partial r}, \quad \gamma_{\alpha z} = \frac{\partial u_\alpha}{\partial z} + \frac{\partial w}{r\partial \alpha}, \\
\varepsilon_\alpha &= \frac{1}{r}\frac{\partial u_\alpha}{\partial \alpha} + \frac{u_r}{r}, \quad \gamma_{zr} = \frac{\partial w}{\partial r} + \frac{\partial u_r}{\partial z}, \\
\varepsilon_z &= \frac{\partial w}{\partial z}, \quad \gamma_{r\alpha} = \frac{1}{r}\frac{\partial u_r}{\partial \alpha} + \frac{\partial u_\alpha}{\partial r} - \frac{u_\alpha}{r}.
\end{aligned} \tag{1.2.2}$$

In spherical coordinates, they can be written as

$$\begin{aligned}
\varepsilon_R &= \frac{\partial u_R}{\partial R}, \quad \varepsilon_\theta = \frac{1}{R}\frac{\partial u_\theta}{\partial \theta} + \frac{u_R}{R}, \\
\varepsilon_\alpha &= \frac{1}{R\sin\theta}\frac{\partial u_\alpha}{\partial \alpha} + \frac{u_R}{R} + \frac{u_\theta}{R}\cot\theta, \\
\gamma_{\theta\alpha} &= \frac{1}{R}\left(\frac{\partial u_\alpha}{\partial \theta} - u_\alpha \cot\theta\right) + \frac{1}{R\sin\theta}\frac{\partial u_\theta}{\partial \alpha}, \\
\gamma_{\alpha R} &= \frac{1}{R\sin\theta}\frac{\partial u_R}{\partial \alpha} + \frac{\partial u_\alpha}{\partial R} - \frac{u_\alpha}{R}, \\
\gamma_{R\theta} &= \frac{1}{R}\frac{\partial u_R}{\partial \theta} + \frac{\partial u_\theta}{\partial R} - \frac{u_\theta}{R}.
\end{aligned} \tag{1.2.3}$$

In tensor form, the geometric equations in Cartesian coordinate system can be written concisely as

$$\varepsilon_{ij} = \frac{1}{2}(u_{i,j} + u_{j,i}), \tag{1.2.4}$$

with $2\varepsilon_{ij} = \gamma_{ij}$ when $i \neq j$. Here $u_{i,j}$ means $\partial u_i / \partial x_j$. Sometimes, it is also useful to express the equations in matrix form, *i.e.*,

$$\{\varepsilon\} = E^{\mathrm{T}}(\nabla)\{u\}, \tag{1.2.5}$$

where $\{u\} = [u, v, w]^{\mathrm{T}}$ and $E(\nabla)$ is an operator matrix defined by

$$E(\nabla)=\begin{bmatrix} \dfrac{\partial}{\partial x} & 0 & 0 & 0 & \dfrac{\partial}{\partial z} & \dfrac{\partial}{\partial y} \\ 0 & \dfrac{\partial}{\partial y} & 0 & \dfrac{\partial}{\partial z} & 0 & \dfrac{\partial}{\partial x} \\ 0 & 0 & \dfrac{\partial}{\partial z} & \dfrac{\partial}{\partial y} & \dfrac{\partial}{\partial x} & 0 \end{bmatrix}. \tag{1.2.6}$$

The geometric equations specify strains when displacements are known. They can also be regarded as the first order partial differential equations for solving displacements when strains are known. In this case, however, we will be using six equations to solve for three displacement components. This cannot yield a solution unless the six strain components follow certain conditions, called the *compatibility conditions*, which can be obtained easily as

$$\varepsilon_{ij,kl}+\varepsilon_{kl,ij}-\varepsilon_{lj,ki}-\varepsilon_{ki,lj}=0\,. \tag{1.2.7}$$

Equation (1.2.7) represents six independent equations. In Cartesian coordinate system, they can be written as

$$\begin{aligned}
&\frac{\partial^2\varepsilon_y}{\partial z^2}+\frac{\partial^2\varepsilon_z}{\partial y^2}-\frac{\partial^2\gamma_{yz}}{\partial y\partial z}=0,\\
&\frac{\partial^2\varepsilon_z}{\partial x^2}+\frac{\partial^2\varepsilon_x}{\partial z^2}-\frac{\partial^2\gamma_{zx}}{\partial z\partial x}=0,\\
&\frac{\partial^2\varepsilon_x}{\partial y^2}+\frac{\partial^2\varepsilon_y}{\partial x^2}-\frac{\partial^2\gamma_{xy}}{\partial x\partial y}=0,\\
&\frac{\partial^2\varepsilon_x}{\partial y\partial z}+\frac{1}{2}\frac{\partial}{\partial x}\left(\frac{\partial\gamma_{yz}}{\partial x}-\frac{\partial\gamma_{zx}}{\partial y}-\frac{\partial\gamma_{xy}}{\partial z}\right)=0,\\
&\frac{\partial^2\varepsilon_y}{\partial z\partial x}+\frac{1}{2}\frac{\partial}{\partial y}\left(\frac{\partial\gamma_{zx}}{\partial y}-\frac{\partial\gamma_{xy}}{\partial z}-\frac{\partial\gamma_{yz}}{\partial x}\right)=0,\\
&\frac{\partial^2\varepsilon_z}{\partial x\partial y}+\frac{1}{2}\frac{\partial}{\partial z}\left(\frac{\partial\gamma_{xy}}{\partial z}-\frac{\partial\gamma_{yz}}{\partial x}-\frac{\partial\gamma_{zx}}{\partial y}\right)=0.
\end{aligned} \tag{1.2.8}$$

In other coordinate systems, the explicit form of the compatibility conditions is lengthy. For convenience, we list only those in cylindrical coordinates when deformation is axisymmetric, *i.e.*,

$$\varepsilon_r - \varepsilon_\alpha - r\frac{\partial \varepsilon_\alpha}{\partial r} = 0, \quad \frac{\partial \varepsilon_z}{\partial r} + r\frac{\partial^2 \varepsilon_\alpha}{\partial z^2} - \frac{\partial \gamma_{zr}}{\partial z} = 0. \tag{1.2.9}$$

1.2.2 Equations of Motion

In Cartesian coordinates, the equations of motion can be expressed as

$$\begin{aligned}
&\frac{\partial \sigma_x}{\partial x} + \frac{\partial \tau_{xy}}{\partial y} + \frac{\partial \tau_{zx}}{\partial z} + F_x = \rho\frac{\partial^2 u}{\partial t^2}, \\
&\frac{\partial \tau_{xy}}{\partial x} + \frac{\partial \sigma_y}{\partial y} + \frac{\partial \tau_{yz}}{\partial z} + F_y = \rho\frac{\partial^2 v}{\partial t^2}, \\
&\frac{\partial \tau_{zx}}{\partial x} + \frac{\partial \tau_{yz}}{\partial y} + \frac{\partial \sigma_z}{\partial z} + F_z = \rho\frac{\partial^2 w}{\partial t^2},
\end{aligned} \tag{1.2.10}$$

where ρ is the density of the material and F_i is the component in i-direction of the body force per unit volume.

In cylindrical coordinates, these equations become

$$\begin{aligned}
&\frac{\partial \sigma_r}{\partial r} + \frac{1}{r}\frac{\partial \tau_{r\alpha}}{\partial \alpha} + \frac{\partial \tau_{zr}}{\partial z} + \frac{\sigma_r - \sigma_\alpha}{r} + F_r = \rho\frac{\partial^2 u_r}{\partial t^2}, \\
&\frac{\partial \tau_{r\alpha}}{\partial r} + \frac{1}{r}\frac{\partial \sigma_\alpha}{\partial \alpha} + \frac{\partial \tau_{\alpha z}}{\partial z} + \frac{2\tau_{r\alpha}}{r} + F_\alpha = \rho\frac{\partial^2 u_\alpha}{\partial t^2}, \\
&\frac{\partial \tau_{zr}}{\partial r} + \frac{1}{r}\frac{\partial \tau_{\alpha z}}{\partial \alpha} + \frac{\partial \sigma_z}{\partial z} + \frac{\tau_{zr}}{r} + F_z = \rho\frac{\partial^2 w}{\partial t^2}.
\end{aligned} \tag{1.2.11}$$

In spherical coordinates, these equations can be written as

$$\begin{aligned}
&\frac{\partial \sigma_R}{\partial R} + \frac{1}{R}\frac{\partial \tau_{R\theta}}{\partial \theta} + \frac{1}{R\sin\theta}\frac{\partial \tau_{\alpha R}}{\partial \alpha} + \frac{1}{R}(2\sigma_R - \sigma_\theta - \sigma_\alpha + \tau_{R\theta}\cot\theta) + F_R = \rho\frac{\partial^2 u_R}{\partial t^2}, \\
&\frac{\partial \tau_{R\theta}}{\partial R} + \frac{1}{R}\frac{\partial \sigma_\theta}{\partial \theta} + \frac{1}{R\sin\theta}\frac{\partial \tau_{\theta\alpha}}{\partial \alpha} + \frac{1}{R}[(\sigma_\theta - \sigma_\alpha)\cot\theta + 3\tau_{R\theta})] + F_\theta = \rho\frac{\partial^2 u_\theta}{\partial t^2}, \\
&\frac{\partial \tau_{\alpha R}}{\partial R} + \frac{1}{R}\frac{\partial \tau_{\theta\alpha}}{\partial \theta} + \frac{1}{R\sin\theta}\frac{\partial \sigma_\alpha}{\partial \alpha} + \frac{1}{R}[2\tau_{\theta\alpha}\cot\theta + 3\tau_{\alpha R})] + F_\alpha = \rho\frac{\partial^2 u_\alpha}{\partial t^2}.
\end{aligned} \tag{1.2.12}$$

The equations of motion in Cartesian coordinates can be also written concisely in tensor form as

$$\sigma_{ij,j} + F_i = \rho \ddot{u}_i, \quad (i = 1, 2, 3), \tag{1.2.13}$$

where a dot indicates partial differentiation with respect to time t. If the motion of the solid does not involve acceleration, Eq. (1.2.13) reduces to the *equations of equilibrium*, *i.e.*,

$$\sigma_{ij,j} + F_i = 0, \quad (i = 1, 2, 3). \tag{1.2.14}$$

1.2.3 Constitutive Equations

The constitutive equations in linear elasticity are represented by the generalized Hooke's law. If the state of vanishing strain corresponds to zero stress, then in Cartesian coordinates the generalized Hooke's law can be written as

$$\sigma_{ij} = c_{ijkl}\varepsilon_{kl}, \tag{1.2.15}$$

where c_{ijkl} are components of a fourth-rank tensor, representing the properties of a material, which generally varies from one point to another in the material. If c_{ijkl} do not change across a material, it is called a homogeneous material. This book will only consider homogeneous and elastic materials whose c_{ijkl} are independent of coordinates.

Since σ_{ij} is symmetric, the exchange of indices i and j in Eq. (1.2.15) does not alter the result, which gives rise to

$$c_{ijkl} = c_{jikl},$$

Without losing the generality, c_{ijkl} can be regarded symmetric with respect to the last two indices as detailed below.

First, define

$$c'_{ijkl} = (c_{ijkl} + c_{ijlk})/2, \quad c''_{ijkl} = (c_{ijkl} - c_{ijlk})/2,$$

which shows $c'_{ijkl} = c'_{ijlk}$ and $c''_{ijkl} = -c''_{ijlk}$, *i.e.* c'_{ijkl} are symmetric and c''_{ijkl} are antisymmetric with respect to the last two indices. Then, c_{ijkl} can be expressed as

$$c_{ijkl} = c'_{ijkl} + c''_{ijkl}.$$

Thus, Eq. (1.2.15) can be written as

$$\sigma_{ij} = c'_{ijkl}\varepsilon_{kl} + c''_{ijkl}\varepsilon_{kl}.$$

Noting that the second term of the right-hand side vanishes because $\varepsilon_{kl} = \varepsilon_{lk}$ and $c''_{ijkl} = -c''_{ijlk}$, we have

$$\sigma_{ij} = c'_{ijkl}\varepsilon_{kl},$$

where c'_{ijkl} are symmetric with respect to either the first two indices or the last two indices.

It is therefore reasonable to assume that c_{ijkl} in Eq. (1.2.15) has the following symmetry:

$$c_{ijkl} = c_{jikl} = c_{ijlk}.$$

Thus, among the total 81 components of c_{ijkl}, the maximum number of independent ones is 36.

To avoid the double summation over k and l in Eq. (1.2.15), introduce the following notations

$$\sigma_{11} = \sigma_1, \quad \sigma_{22} = \sigma_2, \quad \sigma_{33} = \sigma_3, \quad \sigma_{23} = \sigma_4, \quad \sigma_{31} = \sigma_5, \quad \sigma_{12} = \sigma_6,$$
$$\varepsilon_{11} = \varepsilon_1, \quad \varepsilon_{22} = \varepsilon_2, \quad \varepsilon_{33} = \varepsilon_3, \quad 2\varepsilon_{23} = \varepsilon_4, \quad 2\varepsilon_{31} = \varepsilon_5, \quad 2\varepsilon_{12} = \varepsilon_6,$$

where ε_4, ε_5 and ε_6 are the engineering shear strains. Equation (1.2.15) can then be rewritten as

$$\sigma_1 = c_{11}\varepsilon_1 + c_{12}\varepsilon_2 + c_{13}\varepsilon_3 + c_{14}\varepsilon_4 + c_{15}\varepsilon_5 + c_{16}\varepsilon_6,$$
$$\sigma_2 = c_{21}\varepsilon_1 + c_{22}\varepsilon_2 + c_{23}\varepsilon_3 + c_{24}\varepsilon_4 + c_{25}\varepsilon_5 + c_{26}\varepsilon_6,$$
$$\cdots\cdots\cdots\cdots\cdots\cdots$$
$$\sigma_6 = c_{61}\varepsilon_1 + c_{62}\varepsilon_2 + c_{63}\varepsilon_3 + c_{64}\varepsilon_4 + c_{65}\varepsilon_5 + c_{66}\varepsilon_6,$$

or in a more concise form,

$$\sigma_i = c_{ij}\varepsilon_j, \quad (i, j = 1, 2, \cdots, 6).$$

The corresponding matrix form is

$$\{\sigma\} = [c]\{\varepsilon\}, \tag{1.2.16}$$

where $\{\sigma\}$ and $\{\varepsilon\}$ are vectors of stress and engineering strain, respectively. In Cartesian coordinates, they become

$$\{\sigma\} = [\sigma_x, \sigma_y, \sigma_z, \tau_{yz}, \tau_{zx}, \tau_{xy}]^{\mathrm{T}},$$
$$\{\varepsilon\} = [\varepsilon_x, \varepsilon_y, \varepsilon_z, \gamma_{yz}, \gamma_{zx}, \gamma_{xy}]^{\mathrm{T}}.$$

In Eq. (1.2.16), $[c] = [c_{ij}]$ should be a nonsingular and reversible matrix, *i.e.* $\det[c] \neq 0$. Hence, Eq. (1.2.16) can also be written as

$$\{\varepsilon\} = [s]\{\sigma\}, \tag{1.2.17}$$

where $[s] = [s_{ij}]$ is the inverse of $[c]$, *i.e.* $[s] = [c]^{-1}$. In the above, c_{ij} are called the *elastic stiffnesses* (or moduli) of a material, having the dimension of stresses (F/L^2) because strains are dimensionless, and s_{ij} are called the *elastic compliances* of the material with the dimension of L^2/F.

If there exists a *strain energy density function*,

$$W = \frac{1}{2}c_{ij}\varepsilon_i\varepsilon_j, \quad (i, j = 1, 2, \cdots, 6) \tag{1.2.18}$$

then

$$\frac{\partial W}{\partial \varepsilon_i} = \sigma_i = c_{ij}\varepsilon_j, \tag{1.2.19}$$

and

$$\frac{\partial^2 W}{\partial \varepsilon_i \partial \varepsilon_j} = c_{ij} .$$

Similarly,

$$\frac{\partial^2 W}{\partial \varepsilon_j \partial \varepsilon_i} = c_{ji} .$$

Therefore $c_{ij} = c_{ji}$, since the order of differentiation is immaterial. This indicates that the number of independent elastic stiffnesses c_{ij} is further reduced from 36 to 21. Similarly, we have $s_{ij} = s_{ji}$. Thus, for a general anisotropic elastic material, there are only 21 independent elastic stiffness constants or elastic compliance coefficients. Because the strain energy density W is always non-negative and becomes zero only when $\varepsilon_i = 0$, $(i = 1, 2, \cdots, 6)$, it is clear that the stiffness matrix $[c]$ and its inverse, the compliance matrix $[s]$, are both positive definite.

In a different coordinate system (x', y', z'), the constitutive equations will have the same form as Eq. (1.2.16) or (1.2.17), *i.e.*,

$$\{\sigma'\} = [c']\{\varepsilon'\}, \tag{1.2.20}$$

or

$$\{\varepsilon'\} = [s']\{\sigma'\}. \tag{1.2.21}$$

Using Eqs. (1.1.1) and (1.1.2), we can transfer Eq. (1.2.20) into the linear relationship between $\{\sigma\}$ and $\{\varepsilon\}$. Then by comparing it with Eq. (1.2.16), we can easily obtain the transformation formula between $[c']$ and $[c]$.

As just mentioned, for a general anisotropic material, $[c]$ or $[s]$ has twenty-one independent elements and hence the application of the constitutive equation (1.2.16) or (1.2.17) will bring about tremendous difficulties in solving a problem. Fortunately, the equation can be much simplified when the elastic properties of a material possess certain symmetries. We will now introduce the simplified constitutive equations for various materials with special properties.

(1) Plane of elastic symmetry

At any point in a solid, if there exists a plane about which the elastic properties are symmetrical, the number of independent elements in $[c]$ will reduce to thirteen. The direction perpendicular to this *plane of elastic symmetry* is often called the *principal elastic direction* or the *principal direction of the material.* Consider a substance elastically symmetric with respect to the xOy coordinate plane. The symmetry is expressed by the statement that $[c]$ is invariant under the transformation $x' = x$, $y' = y$, and $z' = -z$. Thus, according to Eq. (1.2.20), we have

$$\{\sigma'\} = [c]\{\varepsilon'\}. \tag{1.2.22}$$

For this transformation, we have

$$l_{1'1} = l_{2'2} = 1,\ l_{3'3} = -1,\ l_{1'2} = l_{2'1} = l_{2'3} = l_{3'2} = l_{3'1} = l_{1'3} = 0.$$

Substituting these direction cosines into Eqs. (1.1.1) and (1.1.2) yields

$$\sigma_{x'} = \sigma_x,\ \sigma_{y'} = \sigma_y,\ \sigma_{z'} = \sigma_z,\ \tau_{y'z'} = -\tau_{yz},\ \tau_{z'x'} = -\tau_{zx},\ \tau_{x'y'} = \tau_{xy},$$
$$\varepsilon_{x'} = \varepsilon_x,\ \varepsilon_{y'} = \varepsilon_y,\ \varepsilon_{z'} = \varepsilon_z,\ \gamma_{y'z'} = -\gamma_{yz},\ \gamma_{z'x'} = -\gamma_{zx},\ \gamma_{x'y'} = \gamma_{xy}.$$

Using these relations, we get from Eqs. (1.2.22) and (1.2.16)

$$c_{14} = c_{15} = c_{24} = c_{25} = c_{34} = c_{35} = c_{46} = c_{56} = 0. \tag{1.2.23}$$

Thus the generalized Hooke's law of Eq. (1.2.16) becomes

$$\begin{aligned}
\sigma_x &= c_{11}\varepsilon_x + c_{12}\varepsilon_y + c_{13}\varepsilon_z + c_{16}\gamma_{xy},\\
\sigma_y &= c_{12}\varepsilon_x + c_{22}\varepsilon_y + c_{23}\varepsilon_z + c_{26}\gamma_{xy},\\
\sigma_z &= c_{13}\varepsilon_x + c_{23}\varepsilon_y + c_{33}\varepsilon_z + c_{36}\gamma_{xy},\\
\tau_{yz} &= c_{44}\gamma_{yz} + c_{45}\gamma_{zx},\\
\tau_{zx} &= c_{45}\gamma_{yz} + c_{55}\gamma_{zx},\\
\tau_{xy} &= c_{16}\varepsilon_x + c_{26}\varepsilon_y + c_{36}\varepsilon_z + c_{66}\gamma_{xy}.
\end{aligned} \tag{1.2.24}$$

(2) Orthotropic material

If there exist three orthogonal planes of elastic symmetry at any point in a solid, then there are nine independent elements in $[c]$, and the material is said to be orthotropic. Let the three coordinate planes of a Cartesian system, *xOy*, *xOz* and *yOz*, coincide with these planes of symmetry. Performing the similar transformation with respect to each coordinate plane, we will get, in addition to Eq. (1.2.23),

$$\begin{aligned} &c_{14}=c_{16}=c_{24}=c_{26}=c_{34}=c_{36}=c_{45}=c_{56}=0,\\ &c_{15}=c_{16}=c_{25}=c_{26}=c_{35}=c_{36}=c_{45}=c_{46}=0. \end{aligned} \tag{1.2.25}$$

Hence Eq. (1.2.24) further reduces to

$$\begin{aligned} &\sigma_x=c_{11}\varepsilon_x+c_{12}\varepsilon_y+c_{13}\varepsilon_z,\\ &\sigma_y=c_{12}\varepsilon_x+c_{22}\varepsilon_y+c_{23}\varepsilon_z,\\ &\sigma_z=c_{13}\varepsilon_x+c_{23}\varepsilon_y+c_{33}\varepsilon_z,\\ &\tau_{yz}=c_{44}\gamma_{yz},\quad \tau_{zx}=c_{55}\gamma_{zx},\quad \tau_{xy}=c_{66}\gamma_{xy}. \end{aligned} \tag{1.2.26}$$

In this case, all the coordinate axes are in the principal directions of the material. Equation (1.2.26) shows that in an orthotropic material, normal stresses depend only on normal strains; a shear stress on a plane depends only on the shear strain on the same plane. This makes the stress and deformation analysis of an orthotropic solid much easier than that of a general anisotropic material.

We can also express the strains in Eq. (1.2.26) in terms of the stresses using the compliance matrix $[s]=[c]^{-1}$,

$$\begin{aligned} &\varepsilon_x=s_{11}\sigma_x+s_{12}\sigma_y+s_{13}\sigma_z,\\ &\varepsilon_y=s_{12}\sigma_x+s_{22}\sigma_y+s_{23}\sigma_z,\\ &\varepsilon_z=s_{13}\sigma_x+s_{23}\sigma_y+s_{33}\sigma_z,\\ &\gamma_{yz}=s_{44}\tau_{yz},\quad \gamma_{zx}=s_{55}\tau_{zx},\quad \gamma_{xy}=s_{66}\tau_{xy}. \end{aligned} \tag{1.2.27a}$$

In the literature, one often finds engineering constants, E_i, G_{ij} and ν_{ij}, in the stress-strain relationship of Eq. (1.2.27*a*), *i.e.*,

$$\varepsilon_x=\frac{1}{E_1}\sigma_x-\frac{\nu_{21}}{E_2}\sigma_y-\frac{\nu_{31}}{E_3}\sigma_z,$$

$$
\begin{aligned}
\varepsilon_y &= -\frac{\nu_{12}}{E_1}\sigma_x + \frac{1}{E_2}\sigma_y - \frac{\nu_{32}}{E_3}\sigma_z , \\
\varepsilon_z &= -\frac{\nu_{13}}{E_1}\sigma_x - \frac{\nu_{23}}{E_2}\sigma_y + \frac{1}{E_3}\sigma_z , \\
\gamma_{yz} &= \frac{1}{G_{23}}\tau_{yz} , \quad \gamma_{zx} = \frac{1}{G_{31}}\tau_{zx} , \quad \gamma_{xy} = \frac{1}{G_{12}}\tau_{xy} ,
\end{aligned}
\tag{1.2.27b}
$$

where

$$
\nu_{21} / E_2 = \nu_{12} / E_1 , \quad \nu_{31} / E_3 = \nu_{13} / E_1 , \quad \nu_{32} / E_3 = \nu_{23} / E_2 .
$$

The most important and frequently encountered type of curvilinear anisotropy is *cylindrical anisotropy*. Natural wood is a representative example of such materials. In this case, if cylindrical coordinates (r, α, z) are adopted with the z-axis coincident with the axis of anisotropy, then the generalized Hooke's law still holds its form of Eq. (1.2.16) or (1.2.17), while the stress and strain vectors should now be written as $\{\sigma\} = [\sigma_r, \sigma_\alpha, \sigma_z, \tau_{\alpha z}, \tau_{zr}, \tau_{r\alpha}]^{\mathrm{T}}$ and $\{\varepsilon\} = [\varepsilon_r, \varepsilon_\alpha, \varepsilon_z, \gamma_{\alpha z}, \gamma_{zr}, \gamma_{r\alpha}]^{\mathrm{T}}$. Accordingly, the number of independent elastic stiffness constants c_{ij} or compliance coefficients s_{ij} is still 21. If any plane perpendicular to the z-axis is a plane of elastic symmetry, then Eq. (1.2.23) also holds, and the number of independent elastic stiffness constants c_{ij} (or compliance coefficients s_{ij}) reduces to 13. Moreover, if, at an arbitrary point in the material, there exist three planes of elastic symmetry that are perpendicular to r-, α- and z-directions, respectively, then Eqs. (1.2.23) and (1.2.25) keep unchanged, and the number of independent elastic stiffness constants c_{ij} (or compliance coefficients s_{ij}) further reduces to 9. Materials having this type of anisotropy are cylindrically orthotropic. Similar to Eqs. (1.2.26) and (1.2.27), for a *cylindrically orthotropic material*, we have

$$
\begin{aligned}
\sigma_r &= c_{11}\varepsilon_r + c_{12}\varepsilon_\alpha + c_{13}\varepsilon_z , \\
\sigma_\alpha &= c_{12}\varepsilon_r + c_{22}\varepsilon_\alpha + c_{23}\varepsilon_z , \\
\sigma_z &= c_{13}\varepsilon_r + c_{23}\varepsilon_\alpha + c_{33}\varepsilon_z , \\
\tau_{\alpha z} &= c_{44}\gamma_{\alpha z} , \quad \tau_{zr} = c_{55}\gamma_{zr} , \quad \tau_{r\alpha} = c_{66}\gamma_{r\alpha} ,
\end{aligned}
\tag{1.2.28}
$$

or in an inverse form

$$\begin{aligned}
\varepsilon_r &= s_{11}\sigma_r + s_{12}\sigma_\alpha + s_{13}\sigma_z, \\
\varepsilon_\alpha &= s_{12}\sigma_r + s_{22}\sigma_\alpha + s_{23}\sigma_z, \\
\varepsilon_z &= s_{13}\sigma_r + s_{23}\sigma_\alpha + s_{33}\sigma_z, \\
\gamma_{\alpha z} &= s_{44}\tau_{\alpha z}, \quad \gamma_{zr} = s_{55}\tau_{zr}, \quad \gamma_{r\alpha} = s_{66}\tau_{r\alpha}.
\end{aligned} \tag{1.2.29a}$$

Using engineering symbols, we can rewrite Eq. (1.2.29*a*) as

$$\begin{aligned}
\varepsilon_r &= \frac{1}{E_r}\sigma_r - \frac{\nu_{\alpha r}}{E_\alpha}\sigma_\alpha - \frac{\nu_{zr}}{E_z}\sigma_z, \quad \gamma_{\alpha z} = \frac{1}{G_{\alpha z}}\tau_{\alpha z}, \\
\varepsilon_\alpha &= -\frac{\nu_{r\alpha}}{E_r}\sigma_r + \frac{1}{E_\alpha}\sigma_\alpha - \frac{\nu_{z\alpha}}{E_z}\sigma_z, \quad \gamma_{zr} = \frac{1}{G_{zr}}\tau_{zr}, \\
\varepsilon_z &= -\frac{\nu_{rz}}{E_r}\sigma_r - \frac{\nu_{\alpha z}}{E_\alpha}\sigma_\alpha + \frac{1}{E_z}\sigma_z, \quad \gamma_{r\alpha} = \frac{1}{G_{r\alpha}}\tau_{r\alpha},
\end{aligned} \tag{1.2.29b}$$

where

$$E_i \nu_{ji} = E_j \nu_{ij}, \qquad (i,\ j = r,\ \alpha,\ z). \tag{1.2.30}$$

Another frequently encountered type of curvilinear anisotropy is *spherical anisotropy,* of which a typical example is the model of Earth considering the effect of curvature in its constitutive description[1]. In this case, if spherical coordinates (R, θ, α) are used, the generalized Hooke's law is still in the form of Eq. (1.2.16) or (1.2.17), but the stress and strain vectors need to be replaced by $\{\sigma\} = [\sigma_R, \sigma_\theta, \sigma_\alpha, \tau_{\theta\alpha}, \tau_{\alpha R}, \tau_{R\theta}]^{\mathrm{T}}$ and $\{\varepsilon\} = [\varepsilon_R, \varepsilon_\theta, \varepsilon_\alpha, \gamma_{\theta\alpha}, \gamma_{\alpha R}, \gamma_{R\theta}]^{\mathrm{T}}$. For a general case, there are only 21 independent elastic stiffness constants c_{ij} (or compliance coefficients s_{ij}). If, at any point in the material, there are three planes of elastic symmetry that are perpendicular to R-, θ- and α-directions (Fig. 1.1), respectively, then Eqs. (1.2.23) and (1.2.25) still hold and the number of independent elastic stiffness constants c_{ij} (or compliance coefficients s_{ij}) becomes 9. Materials with this type of anisotropy are spherically orthotropic.

The generalized Hooke's law for a *spherically orthotropic material* takes a similar form to Eq. (1.2.26), *i.e.*

---

[1] See the ACY400 Earth model in Montagner and Anderson (1989). The analytical determination of stress fields in the interior of the Earth using this model was presented and discussed by Ding, Zou and Ding (1996).

$$\begin{aligned}
\sigma_R &= c_{11}\varepsilon_R + c_{12}\varepsilon_\theta + c_{13}\varepsilon_\alpha, \\
\sigma_\theta &= c_{12}\varepsilon_R + c_{22}\varepsilon_\theta + c_{23}\varepsilon_\alpha, \\
\sigma_\alpha &= c_{13}\varepsilon_R + c_{23}\varepsilon_\theta + c_{33}\varepsilon_\alpha, \\
\tau_{\theta\alpha} &= c_{44}\gamma_{\theta\alpha}, \quad \tau_{\alpha R} = c_{55}\gamma_{\alpha R}, \quad \tau_{R\theta} = c_{66}\gamma_{R\theta}.
\end{aligned} \tag{1.2.31}$$

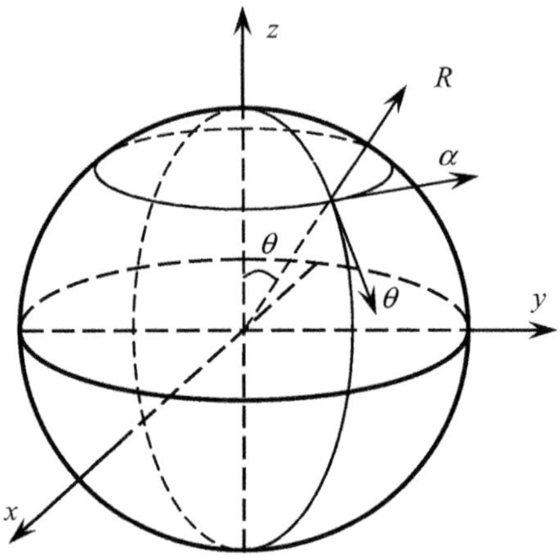

Fig. 1.1 The directions of coordinates at a point on the spherical surface.

(3) Transversely isotropic material

If at any point there is an axis of symmetry such that the elastic properties in any direction within a plane perpendicular to the axis are all the same, the total number of independent elements in $[c]$ will reduce to five. The plane is called an *isotropic plane* and the material is called a *transversely isotropic material.* The hexagonal crystals, like Cadmium and Zinc, are transversely isotropic.

If we take the coordinate plane xOy to coincide with the isotropic plane, then the z-axis is the axis of symmetry. Taking a new Cartesian system such that $x' = y$, $y' = -x$, and $z' = z$, then we have

$$l_{1'2} = l_{3'3} = 1, \quad l_{2'1} = -1, \quad l_{1'1} = l_{1'3} = l_{2'2} = l_{2'3} = l_{3'1} = l_{3'2} = 0.$$

Substituting into Eqs. (1.1.1) and (1.1.2), yields

$$\begin{aligned}
&\sigma_{x'} = \sigma_y, \quad \sigma_{y'} = \sigma_x, \quad \sigma_{z'} = \sigma_z, \quad \tau_{y'z'} = -\tau_{zx}, \quad \tau_{z'x'} = -\tau_{yz}, \quad \tau_{x'y'} = -\tau_{xy}, \\
&\varepsilon_{x'} = \varepsilon_y, \quad \varepsilon_{y'} = \varepsilon_x, \quad \varepsilon_{z'} = \varepsilon_z, \quad \gamma_{y'z'} = -\gamma_{zx}, \quad \gamma_{z'x'} = -\gamma_{yz}, \quad \gamma_{x'y'} = -\gamma_{xy}.
\end{aligned}$$

Using these relations as well as Eqs. (1.2.23) and (1.2.25), we get from Eqs. (1.2.22) and (1.2.26)

$$c_{11} = c_{22}, \quad c_{13} = c_{23}, \quad c_{44} = c_{55}. \tag{1.2.32a}$$

In view of this equation, there are only six independent constants in the constitutive relations in Eq. (1.2.26). Now taking another transformation by rotating the original coordinate system 45 degree about the z-axis, we have the relations

$$x' = (x+y)/\sqrt{2}, \quad y' = (y-x)/\sqrt{2}, \quad z' = z,$$

and

$$l_{1'1} = l_{1'2} = l_{2'2} = 1/\sqrt{2}, \quad l_{2'1} = -1/\sqrt{2}, \quad l_{3'3} = 1, \quad l_{1'3} = l_{2'3} = l_{3'1} = l_{3'2} = 0.$$

Substituting these direction cosines into Eqs. (1.1.1) and (1.1.2), yields

$$\sigma_{x'} = (\sigma_x + \sigma_y + 2\tau_{xy})/2, \quad \sigma_{y'} = (\sigma_x + \sigma_y - 2\tau_{xy})/2, \quad \sigma_{z'} = \sigma_z,$$
$$\tau_{y'z'} = (\tau_{yz} - \tau_{zx})/\sqrt{2}, \quad \tau_{z'x'} = (\tau_{yz} + \tau_{zx})/\sqrt{2}, \quad \tau_{x'y'} = (\sigma_y - \sigma_x)/2,$$
$$\varepsilon_{x'} = (\varepsilon_x + \varepsilon_y + \gamma_{xy})/2, \quad \varepsilon_{y'} = (\varepsilon_x + \varepsilon_y - \gamma_{xy})/2, \quad \varepsilon_{z'} = \varepsilon_z,$$
$$\gamma_{y'z'} = (\gamma_{yz} - \gamma_{zx})/\sqrt{2}, \quad \gamma_{z'x'} = (\gamma_{yz} + \gamma_{zx})/\sqrt{2}, \quad \gamma_{x'y'} = \varepsilon_y - \varepsilon_x.$$

Using these relations as well as Eqs. (1.2.23), (1.2.25) and (1.2.32a), we get from Eqs. (1.2.22) and (1.2.26)

$$2c_{66} = c_{11} - c_{12}. \tag{1.2.32b}$$

Thus, Eq. (1.2.26) becomes

$$\begin{aligned}
\sigma_x &= c_{11}\varepsilon_x + c_{12}\varepsilon_y + c_{13}\varepsilon_z, \\
\sigma_y &= c_{12}\varepsilon_x + c_{11}\varepsilon_y + c_{13}\varepsilon_z, \\
\sigma_z &= c_{13}\varepsilon_x + c_{13}\varepsilon_y + c_{33}\varepsilon_z, \\
\tau_{yz} &= c_{44}\gamma_{yz}, \quad \tau_{zx} = c_{44}\gamma_{zx}, \quad \tau_{xy} = c_{66}\gamma_{xy},
\end{aligned} \tag{1.2.33}$$

where $c_{66} = (c_{11} - c_{12})/2$.

Similarly, the expressions for strains in terms of stresses, given by Eq. (1.2.27*a*), become

$$\begin{aligned}
\varepsilon_x &= s_{11}\sigma_x + s_{12}\sigma_y + s_{13}\sigma_z, \\
\varepsilon_y &= s_{12}\sigma_x + s_{11}\sigma_y + s_{13}\sigma_z, \\
\varepsilon_z &= s_{13}\sigma_x + s_{13}\sigma_y + s_{33}\sigma_z, \\
\gamma_{yz} &= s_{44}\tau_{yz}, \quad \gamma_{zx} = s_{44}\tau_{zx}, \quad \gamma_{xy} = s_{66}\tau_{xy},
\end{aligned} \tag{1.2.34}$$

where $s_{66} = 2(s_{11} - s_{12})$. We can also express this equation in terms of engineering constants, *i.e.*,

$$\begin{aligned}
\varepsilon_x &= \frac{1}{E}(\sigma_x - \nu\sigma_y) - \frac{\nu'}{E'}\sigma_z, \quad \gamma_{yz} = \frac{1}{G'}\tau_{yz}, \\
\varepsilon_y &= \frac{1}{E}(-\nu\sigma_x + \sigma_y) - \frac{\nu'}{E'}\sigma_z, \quad \gamma_{zx} = \frac{1}{G'}\tau_{zx}, \\
\varepsilon_z &= -\frac{\nu'}{E'}(\sigma_x + \sigma_y) + \frac{1}{E'}\sigma_z, \quad \gamma_{xy} = \frac{1}{G}\tau_{xy},
\end{aligned} \tag{1.2.35}$$

where

$$2G = E/(1+\nu). \tag{1.2.36}$$

In cylindrical coordinates, Eq. (1.2.33) becomes

$$\begin{aligned}
\sigma_r &= c_{11}\varepsilon_r + c_{12}\varepsilon_\alpha + c_{13}\varepsilon_z, \\
\sigma_\alpha &= c_{12}\varepsilon_r + c_{11}\varepsilon_\alpha + c_{13}\varepsilon_z, \\
\sigma_z &= c_{13}\varepsilon_r + c_{13}\varepsilon_\alpha + c_{33}\varepsilon_z, \\
\tau_{\alpha z} &= c_{44}\gamma_{\alpha z}, \quad \tau_{zr} = c_{44}\gamma_{zr}, \quad \tau_{r\alpha} = c_{66}\gamma_{r\alpha},
\end{aligned} \tag{1.2.37}$$

where $c_{66} = (c_{11} - c_{12})/2$, and Eq. (1.2.34) becomes

$$\begin{aligned}
\varepsilon_r &= s_{11}\sigma_r + s_{12}\sigma_\alpha + s_{13}\sigma_z, \\
\varepsilon_\alpha &= s_{12}\sigma_r + s_{11}\sigma_\alpha + s_{13}\sigma_z, \\
\varepsilon_z &= s_{13}\sigma_r + s_{13}\sigma_\alpha + s_{33}\sigma_z, \\
\gamma_{\alpha z} &= s_{44}\tau_{\alpha z}, \quad \gamma_{zr} = s_{44}\tau_{zr}, \quad \gamma_{r\alpha} = s_{66}\tau_{r\alpha},
\end{aligned} \tag{1.2.38}$$

where $s_{66} = 2(s_{11} - s_{12})$.

The constitutive relations for a spherically orthotropic material are given by Eq. (1.2.31). Further, if the elasticity property in any direction is the same at the point of intersection in a plane perpendicularly intersecting the radial R, the material is said to be spherically isotropic[2]. By a derivation similar to that of Eq. (1.2.33), we get the generalized Hooke's law of a *spherically isotropic material* in spherical coordinates

$$\begin{aligned}
\sigma_R &= c_{11}\varepsilon_R + c_{12}\varepsilon_\theta + c_{12}\varepsilon_\alpha, \\
\sigma_\theta &= c_{12}\varepsilon_R + c_{22}\varepsilon_\theta + c_{23}\varepsilon_\alpha, \\
\sigma_\alpha &= c_{12}\varepsilon_R + c_{23}\varepsilon_\theta + c_{22}\varepsilon_\alpha, \\
\tau_{\theta\alpha} &= c_{44}\gamma_{\theta\alpha}, \tau_{\alpha R} = c_{55}\gamma_{\alpha R}, \tau_{R\theta} = c_{55}\gamma_{R\theta},
\end{aligned} \tag{1.2.39a}$$

where

$$2c_{44} = c_{22} - c_{23}. \tag{1.2.39b}$$

If Eq. (1.2.39) is used to describe a small portion of material where R is very large, the material can be treated approximately as transversely isotropic since the effect of curvature may be ignored. For this approximation, it will be convenient to rewrite Eq. (1.2.39) by rearrangement of the subscripts of elastic constants as

$$\begin{aligned}
\sigma_\theta &= c_{11}\varepsilon_\theta + c_{12}\varepsilon_\alpha + c_{13}\varepsilon_R, \\
\sigma_\alpha &= c_{12}\varepsilon_\theta + c_{11}\varepsilon_\alpha + c_{13}\varepsilon_R, \\
\sigma_R &= c_{13}\varepsilon_\theta + c_{13}\varepsilon_\alpha + c_{33}\varepsilon_R, \\
\tau_{\alpha R} &= c_{44}\gamma_{\alpha R}, \tau_{R\theta} = c_{44}\gamma_{R\theta}, \tau_{\theta\alpha} = c_{66}\gamma_{\theta\alpha},
\end{aligned} \tag{1.2.40a}$$

where

$$2c_{66} = c_{11} - c_{12}. \tag{1.2.40b}$$

The materials described by Eq. (1.2.40) are also called transversely isotropic materials in some references, see Shenderov (1985), Bufler (1998), Khoma (1998), among others.

Using engineering symbols, we have

---

[2] A classical solution of spherically isotropic material was obtained by Saint-Venant for a spherical shell subjected to uniform internal and external pressures, see Love (1927) and Lekhnitskii (1981).

$$
\begin{aligned}
\varepsilon_\theta &= \frac{1}{E}(\sigma_\theta - \nu\sigma_\alpha) - \frac{\nu'}{E'}\sigma_R, \quad \gamma_{\alpha R} = \frac{1}{G'}\tau_{\alpha R},\\
\varepsilon_\alpha &= \frac{1}{E}(-\nu\sigma_\theta + \sigma_\alpha) - \frac{\nu'}{E'}\sigma_R, \quad \gamma_{R\theta} = \frac{1}{G'}\tau_{R\theta},\\
\varepsilon_R &= -\frac{\nu'}{E'}(\sigma_\theta + \sigma_\alpha) + \frac{1}{E'}\sigma_R, \quad \gamma_{\theta\alpha} = \frac{1}{G}\tau_{\theta\alpha},
\end{aligned}
\tag{1.2.41}
$$

and the coefficient relationship described by Eq. (1.2.36) still holds.

(4) Isotropic material

If any plane in the material is a plane of elastic symmetry, then the material is isotropic, and has only two independent elastic constants, which can be shown by a new coordinate transform: $x' = z$, $y' = x$ and $z' = y$ with

$$
l_{1'3} = l_{2'1} = l_{3'2} = 1, \; l_{1'1} = l_{1'2} = l_{2'2} = l_{2'3} = l_{3'1} = l_{3'3} = 0 .
$$

Substituting these direction cosines into Eqs. (1.1.1) and (1.1.2), yields

$$
\begin{aligned}
&\sigma_{x'} = \sigma_z, \; \sigma_{y'} = \sigma_x, \; \sigma_{z'} = \sigma_y, \; \tau_{y'z'} = \tau_{xy}, \; \tau_{z'x'} = \tau_{yz}, \; \tau_{x'y'} = \tau_{zx},\\
&\varepsilon_{x'} = \varepsilon_z, \; \varepsilon_{y'} = \varepsilon_x, \; \varepsilon_{z'} = \varepsilon_y, \; \gamma_{y'z'} = \gamma_{xy}, \; \gamma_{z'x'} = \gamma_{yz}, \; \gamma_{x'y'} = \gamma_{zx} .
\end{aligned}
$$

Using these relations as well as Eqs. (1.2.23), Eq. (1.2.25) and (1.2.32), we get from Eqs. (1.2.22) and (1.2.33)

$$
c_{12} = c_{13}, \quad c_{11} = c_{33}, \quad c_{44} = c_{66} . \tag{1.2.42}
$$

Thus, only two independent elastic constants are involved in an isotropic material. By the introduction of *Lamé constants* λ and μ, the generalized Hooke's law is written as

$$
\begin{aligned}
\sigma_x &= (\lambda + 2\mu)\varepsilon_x + \lambda\varepsilon_y + \lambda\varepsilon_z,\\
\sigma_y &= \lambda\varepsilon_x + (\lambda + 2\mu)\varepsilon_y + \lambda\varepsilon_z,\\
\sigma_z &= \lambda\varepsilon_x + \lambda\varepsilon_y + (\lambda + 2\mu)\varepsilon_z,\\
\tau_{yz} &= \mu\gamma_{yz}, \quad \tau_{zx} = \mu\gamma_{zx}, \quad \tau_{xy} = \mu\gamma_{xy},
\end{aligned}
\tag{1.2.43}
$$

where we have used the relations $c_{12} = c_{13} = \lambda$, $c_{44} = c_{66} = \mu = G$ and $c_{33} = c_{11} = \lambda + 2\mu$. In engineering, one prefers to use the following form of Eq. (1.2.43):

$$\begin{aligned} \varepsilon_x &= \frac{1}{E}[\sigma_x - \nu(\sigma_y + \sigma_z)], \quad \gamma_{yz} = \frac{2(1+\nu)}{E}\tau_{yz} = \frac{1}{G}\tau_{yz}, \\ \varepsilon_y &= \frac{1}{E}[\sigma_y - \nu(\sigma_z + \sigma_x)], \quad \gamma_{zx} = \frac{2(1+\nu)}{E}\tau_{zx} = \frac{1}{G}\tau_{zx}, \\ \varepsilon_z &= \frac{1}{E}[\sigma_z - \nu(\sigma_x + \sigma_y)], \quad \gamma_{xy} = \frac{2(1+\nu)}{E}\tau_{xy} = \frac{1}{G}\tau_{xy}, \end{aligned} \tag{1.2.44}$$

where E, ν and G are *Young's modulus*, *Poisson's ratio* and *shear modulus*, respectively; they are related by Eq. (1.2.36).

Similarly, the generalized Hooke's law for isotropic materials reads

$$\begin{aligned} \sigma_r &= (\lambda + 2\mu)\varepsilon_r + \lambda\varepsilon_\alpha + \lambda\varepsilon_z, \\ \sigma_\alpha &= \lambda\varepsilon_r + (\lambda + 2\mu)\varepsilon_\alpha + \lambda\varepsilon_z, \\ \sigma_z &= \lambda\varepsilon_r + \lambda\varepsilon_\alpha + (\lambda + 2\mu)\varepsilon_z, \\ \tau_{\alpha z} &= \mu\gamma_{\alpha z}, \quad \tau_{zr} = \mu\gamma_{zr}, \quad \tau_{r\alpha} = \mu\gamma_{r\alpha}, \end{aligned} \tag{1.2.45}$$

in cylindrical coordinates, and

$$\begin{aligned} \sigma_\theta &= (\lambda + 2\mu)\varepsilon_\theta + \lambda\varepsilon_\alpha + \lambda\varepsilon_R, \\ \sigma_\alpha &= \lambda\varepsilon_\theta + (\lambda + 2\mu)\varepsilon_\alpha + \lambda\varepsilon_R, \\ \sigma_R &= \lambda\varepsilon_\theta + \lambda\varepsilon_\alpha + (\lambda + 2\mu)\varepsilon_R, \\ \tau_{\alpha R} &= \mu\gamma_{\alpha R}, \quad \tau_{R\theta} = \mu\gamma_{R\theta}, \quad \tau_{\theta\alpha} = \mu\gamma_{\theta\alpha}, \end{aligned} \tag{1.2.46}$$

in spherical coordinates.

In tensor form, the constitutive equations, Eq. (1.2.15), can also be written as

$$\varepsilon_{ij} = s_{ijkl}\sigma_{kl}, \tag{1.2.47}$$

where s_{ijkl} are the elastic compliances, which are components of a fourth-rank tensor; they have the following property of symmetry

$$s_{ijkl} = s_{jikl} = s_{ijlk} = s_{klij}. \tag{1.2.48}$$

Table 1.4 Material properties of transversely isotropic materials.

Material	c_{11}	c_{12}	c_{13}	c_{33}	c_{44}	ρ	Reference
	c_{ij}-GPa, ρ-kg/m^3						
Al_2O_3	460.2	174.7	127.4	509.5	126.9	4000*	Shi and Ramalingam (2001)
Barium-titanate	168	78	71	189	5.46	6020	Uyaner *et al.* (2000)
Beryllium (Be)	292.3	26.7	14	336.4	162.5	1848	Royer and Dieulesaint (2000)
Berea sandstone (measured under confining stress 20MPa and pore pressure 10 MPa)	21.17	4.34	3.84	17.34	13.97	2500	Hart (2000)
Cadmium	116	79	41	50.9	19.6	8642	Ahmad and Khan (2001)
Cadmium sulphide (CdS)	85.6	53.2	46.2	93.6	14.9	4824	Royer and Dieulesaint (2000)
Carbon-fiber	20	9.98	6.45	235	24	—	Kriz and Stinchcomb (1979)
Ceramic PZT-4	139	78	74	115	25.6	7500	Royer and Dieulesaint (2000)
Cobalt	307.0	165.0	103.0	358.1	75.3	8900	Kuo and Keer (1992)
Composite (60% Fiber)	13.6	7	5.47	144	6.01	—	Kriz and Stinchcomb (1979)
Concrete aged by chemical method, unsaturated	21.2	3.5	2.8	25.9	11.2	—	Panet *et al.* (2002)
Concrete aged by Freeze/thaw cycles, unsaturated	32.3	5.1	4.7	34.9	14.9	—	Panet *et al.* (2002)
E glass/epoxy	14.93	6.567	5.244	47.27	4.745	—	Behrens (1971)
GaS	157	33	15	36	8	3910*	Gardos (1990)
GaSe	103	29	12	34	9	5040*	Gardos (1990)
Gneiss rock (dry)	52	11	9	16	11	—	Bank (2002)
Gneiss rock (water-saturated)	99	46	45	78	15	—	Bank (2002)
Graphite	1060	180	15	37	0.35	2270*	Gardos (1990)
Graphite/epoxy	13.92	6.92	6.44	160.7	7.07	1700	Sáez and Domínguez (2000)
Human femur compact bone	46	39	39	53.3	3.3	—	Reilly and Burstein (1975)
InSe	73	27	30	36	12	5560*	Gardos (1990)
Magnesium	59.7	26.2	21.7	61.7	16.4	1740	Ahmad *et al.* (2002)
Mandible (cancellous bone)	0.299	0.070	0.124	0.954	0.115	—	Lin (2002)
Mandible (cortical bone)	17.45	6.79	10.18	27.38	5.9	—	Lin (2002)
MoS_2	238	-54	23	52	18.9	5000*	Gardos (1990)
$NbSe_2$	106	14	31	54	19.5	6480*	Gardos (1990)
Sapphire	496.8	163.6	110.9	498.1	147.4	3986	Wachtman *et al.* (1960)

SiC	479	97.8	55.3	521.4	148.4	3220*	Martin (1972)
T650/950-1	10.90	5.46	5.40	133.2	4.25	—	Melo and Radford (2002)
TiB_2	0.69	0.41	0.32	0.44	0.25	4500	Wang (2004)
Titanium (Ti)	162.4	92.0	69.0	180.7	46.7	4506	Royer and Dieulesaint (2000)
Wet bovine dentine	37.0	16.6	8.7	39.0	5.7	—	Lees and Rollins (1972)
Wet bovine enamel	115	42.4	30.0	125	22.8	—	Lees and Rollins (1972)
Wet bovine femur	17.0	10.2	9.72	29.6	3.6	—	Reilly and Burstein (1975)
Wood (Douglas Fir)	4.07	2.69	0.24	6.89	0.49	800	Ritter (1992)
Zinc	158.35	31.51	47.44	61.6	40.0	7140	Jain (1974)
Zinc oxide (ZnO)	209.7	121.1	105.1	210.9	42.5	5676	Royer and Dieulesaint (2000)

* Density quoted from NIMS materials Database (http://mits.nims.go.jp/db_top_eng.htm)

Table 1.4 lists the elastic constants of several transversely isotropic materials collected from various sources; the mass density ratios of some of them are also given. The reader is also referred to Payton (1983, Page 3) for elastic constants of some typical hexagonal crystals. It is noted that the handbook edited by Levy *et al.* (2001) presents more information about material properties of a wide range of solids.

1.3 BOUNDARY AND INITIAL CONDITIONS

The three sets of basic equations, *i.e.*, the geometric equations, the equations of motion and the constitutive equations, contain three sets of mechanical quantities: stresses, strains and displacements. The variation of these quantities with time is caused by the deformation of the solid under external stimuli, such as forces and temperature changes. Hence, we need to establish their relationships with the external stimuli, called *boundary conditions* and *initial conditions*, to obtain a meaningful solution to a specific engineering problem.

1.3.1 Boundary Conditions

For convenience, we divide the boundary conditions into three distinct types.

(1) Stress boundary conditions

If the whole surface, S, of a solid is subjected to known external stresses $(\overline{p}_x, \overline{p}_y, \overline{p}_z)$, where the subscripts x, y and z indicate the directions of the surface stresses applied, then Eq. (1.1.4) gives the stress boundary conditions

$$\begin{aligned} &\sigma_x \cos(n,x) + \tau_{xy} \cos(n,y) + \tau_{zx} \cos(n,z) = \bar{p}_x, \\ &\tau_{xy} \cos(n,x) + \sigma_y \cos(n,y) + \tau_{yz} \cos(n,z) = \bar{p}_y, \qquad \text{on} \quad S, \\ &\tau_{zx} \cos(n,x) + \tau_{yz} \cos(n,y) + \sigma_z \cos(n,z) = \bar{p}_z, \end{aligned} \tag{1.3.1}$$

where n stands for the direction of $\vec{n}$, the external normal to S. If we use n_i $(i = x, y, z)$ to denote the direction cosines of $\vec{n}$ with respect to the coordinate axes, the boundary conditions can be written as

$$\sigma_{ij} n_j = \bar{p}_i, \qquad \text{on} \quad S, \tag{1.3.2}$$

(2) Displacement boundary conditions

If the whole surface S of the solid has given displacements, $(\bar{u}, \bar{v}, \bar{w})$, the boundary conditions can be specified as

$$u = \bar{u}, \quad v = \bar{v}, \quad w = \bar{w}, \qquad \text{on} \quad S, \tag{1.3.3}$$

or in a more concise form,

$$u_i = \bar{u}_i, \qquad \text{on} \quad S. \tag{1.3.4}$$

(3) Mixed boundary conditions

When part of the surface, S_σ, of a solid is subjected to given stresses, while the rest of the surface, S_u, is under specified displacements, so that $S_\sigma \cup S_u = S$, we can use Eq. (1.3.1) to describe the boundary conditions on S_σ and Eq. (1.3.3) for those on S_u.

In some cases, we may know stresses in one or two directions at a surface point (*e.g.*, in x- and y-directions, $\bar{p}_x$ and $\bar{p}_y$ are given) but displacements in the remaining directions (*e.g.*, in z-direction, $\bar{w}$ is given). This is also a mixed type of boundary conditions. The simply supported condition imposed at the four straight edges of a rectangular plate, **Chapter 8**, is a typical example.

1.3.2 Initial Conditions

If a solid is subjected to a dynamic loading or a motion, we need to specify the initial conditions of the solid at the initial instant of investigation. Usually, we have two types of initial conditions at a specified point in a solid, *i.e.*, given initial displacements u_i^* and given initial velocity $\dot{u}_i^*$. Mathematically, they can be written as

$$t = t_0: \quad u_i = u_i^* \quad \text{and} \quad \dot{u}_i = \dot{u}_i^*. \tag{1.3.5}$$

If a solid is under static deformation, then there is no need to have initial conditions in solving the basic equations, as in this case none of the mechanical quantities is a function of time. Boundary conditions properly specified will be sufficient to determine the specific solution of a static problem.

Eringen and Sububi (1975) showed that the solution of an elastic problem based on the basic equations and boundary and initial conditions is unique[3].

1.4 THERMOELASTICITY

A temperature change will generate thermal deformation and thus thermal strains in a solid. When this happens, we must modify the stress-strain relations presented in **Section 1.2.3** to incorporate thermal effects. In this book we consider only the uncoupled theory of linear thermoelasticity (Nowacki, 1986). We can therefore use the *Duhamel-Neumann equation* to replace the previous generalized Hooke's law, Eq. (1.2.15) or Eq. (1.2.47), *i.e.*,

$$\sigma_{ij} = c_{ijkl}\varepsilon_{kl} + \beta_{ij}T, \tag{1.4.1}$$

or

$$\varepsilon_{ij} = s_{ijkl}\sigma_{kl} + \alpha_{ij}T, \tag{1.4.2}$$

where α_{ij} are the anisotropic *thermal expansion coefficients*, β_{ij} are the *thermal moduli* and T is the temperature change that can be solved independently from the *equation of heat conduction* as well as the thermal boundary conditions (see **Chapter 6** in this book). Both α_{ij} and β_{ij} are tensors of rank two and they have the following relationship:

[3] The uniqueness theorem lays the theoretical foundation of the *inverse* and *semi-inverse* methods in elasticity.

$$\alpha_{ij} = -s_{ijkl}\beta_{kl}, \quad \beta_{ij} = -c_{ijkl}\alpha_{kl}. \tag{1.4.3}$$

In matrix form, Eq. (1.4.1) can be written as

$$\{\sigma\} = [c]\{\varepsilon\} + T\{\beta\}, \tag{1.4.4}$$

where

$$\{\beta\} = [\beta_{11}, \beta_{22}, \beta_{33}, \beta_{23}, \beta_{31}, \beta_{12}]^{\mathrm{T}}. \tag{1.4.5}$$

As with the symmetry of elastic properties, there is also the symmetry of thermal properties. If we assume that the axes and planes of elastic symmetry coincide with the thermal ones[4], $\{\beta\}$ can be simplified. For instance, if there exists a plane of symmetry with z-axis as its principal axis, we get

$$\{\beta\} = [\beta_{11}, \beta_{22}, \beta_{33}, 0, 0, \beta_{12}]^{\mathrm{T}}. \tag{1.4.6}$$

Similarly, for an orthotropic solid, we have

$$\{\beta\} = [\beta_{11}, \beta_{22}, \beta_{33}, 0, 0, 0]^{\mathrm{T}} = [\beta_1, \beta_2, \beta_3, 0, 0, 0]^{\mathrm{T}}, \tag{1.4.7}$$

and

$$\begin{aligned} -\beta_1 &= c_{11}\alpha_{11} + c_{12}\alpha_{22} + c_{13}\alpha_{33}, \\ -\beta_2 &= c_{12}\alpha_{11} + c_{22}\alpha_{22} + c_{23}\alpha_{33}, \\ -\beta_3 &= c_{13}\alpha_{11} + c_{23}\alpha_{22} + c_{33}\alpha_{33}. \end{aligned} \tag{1.4.8}$$

In cylindrical and spherical coordinates, the above subscripts (1, 2, 3) represent (r, α, z) and (θ, α, R), respectively. Therefore, for orthotropic solids with respect to Cartesian, cylindrical and spherical coordinates, Eq. (1.4.4) can be respectively written as

$$\begin{aligned} &\sigma_x = c_{11}\varepsilon_x + c_{12}\varepsilon_y + c_{13}\varepsilon_z + \beta_1 T, \\ &\sigma_y = c_{12}\varepsilon_x + c_{22}\varepsilon_y + c_{23}\varepsilon_z + \beta_2 T, \\ &\sigma_z = c_{13}\varepsilon_x + c_{23}\varepsilon_y + c_{33}\varepsilon_z + \beta_3 T, \\ &\tau_{yz} = c_{44}\gamma_{yz}, \quad \tau_{zx} = c_{55}\gamma_{zx}, \quad \tau_{xy} = c_{66}\gamma_{xy}, \end{aligned} \tag{1.4.9}$$

---

[4] This assumption is reasonable but is not always exact. In some cases it may introduce certain inaccuracy of analysis.

$$\begin{aligned}
\sigma_r &= c_{11}\varepsilon_r + c_{12}\varepsilon_\alpha + c_{13}\varepsilon_z + \beta_1 T, \\
\sigma_\alpha &= c_{12}\varepsilon_r + c_{22}\varepsilon_\alpha + c_{23}\varepsilon_z + \beta_2 T, \\
\sigma_z &= c_{13}\varepsilon_r + c_{23}\varepsilon_\alpha + c_{33}\varepsilon_z + \beta_3 T, \\
\tau_{\alpha z} &= c_{44}\gamma_{\alpha z}, \quad \tau_{zr} = c_{55}\gamma_{zr}, \quad \tau_{r\alpha} = c_{66}\gamma_{r\alpha},
\end{aligned} \tag{1.4.10}$$

$$\begin{aligned}
\sigma_\theta &= c_{11}\varepsilon_\theta + c_{12}\varepsilon_\alpha + c_{13}\varepsilon_R + \beta_1 T, \\
\sigma_\alpha &= c_{12}\varepsilon_\theta + c_{22}\varepsilon_\alpha + c_{23}\varepsilon_R + \beta_2 T, \\
\sigma_R &= c_{13}\varepsilon_\theta + c_{23}\varepsilon_\alpha + c_{33}\varepsilon_R + \beta_3 T, \\
\tau_{\alpha R} &= c_{44}\gamma_{\alpha R}, \quad \tau_{R\theta} = c_{55}\gamma_{R\theta}, \quad \tau_{\theta\alpha} = c_{66}\gamma_{\theta\alpha}.
\end{aligned} \tag{1.4.11}$$

If the solid is a transversely isotropic material, we will get the same expressions of constitutive equation as listed from Eqs. (1.4.9) to (1.4.11), but we will have

$$c_{22} = c_{11}, \quad c_{23} = c_{13}, \quad c_{55} = c_{44}, \quad 2c_{66} = c_{11} - c_{12}, \quad \beta_2 = \beta_1, \tag{1.4.12}$$

in Cartesian, cylindrical, and spherical coordinate systems. If the material is isotropic, then for all the equations from Eqs. (1.4.9) to (1.4.11), we have

$$\begin{aligned}
&c_{12} = c_{13} = c_{23} = \lambda, \quad c_{11} = c_{22} = c_{33} = \lambda + 2\mu, \\
&c_{44} = c_{55} = c_{66} = \mu, \quad \beta_1 = \beta_2 = \beta_3 = \beta = -(3\lambda + 2\mu)\alpha,
\end{aligned} \tag{1.4.13}$$

where α is the coefficient of thermal expansion.

2

GENERAL SOLUTION FOR TRANSVERSELY ISOTROPIC PROBLEMS

This chapter discusses some basic methods for the general solution of transversely isotropic elasticity problems. Emphasis is placed on the *displacement method*, including that for solving the equations of motion of transversely and spherically isotropic materials. In the meantime, the *stress method* and *mixed method* using the state-space technique are developed.

2.1 GOVERNING EQUATIONS

2.1.1 Methods of Solution

The fifteen basic equations of elasticity, as summarized in Eqs. (1.2.4), (1.2.13) and (1.4.1), contain three groups of unknowns, *i.e.*, u_i, σ_{ij} and ε_{ij}, fifteen in total, where the constitutive equations, Eq. (1.4.1), are algebraic. When the strains in the constitutive equations are expressed in terms of displacements, we obtain

$$\sigma_{ij} = c_{ijkl} u_{k,l} + \beta_{ij} T , \tag{2.1.1}$$

or

$$u_{i,j} + u_{j,i} = 2 s_{ijkl} \sigma_{kl} + 2 \alpha_{ij} T . \tag{2.1.2}$$

In the derivation, we have applied the symmetry condition $c_{ijkl} = c_{ijlk}$. By combining Eq. (1.2.13) with Eq. (2.1.1), we get nine governing equations for general anisotropic elasticity problems, which are first order partial differential equations with nine unknowns, u_i and σ_{ij}.

In general, it is unrealistic to solve for the nine unknowns all together. We often take one group of the unknowns or some unknowns from different groups as the *basic variables*, find them first, and then derive the rest. In this way, we have four essential solution strategies: displacement method, stress method, strain method and mixed method.

In the displacement method, u_i (i = 1, 2, 3) are taken as the basic variables to be found. To do this, the other two groups of unknowns, *i.e.*, strains ε_{ij} and stresses σ_{ij}, must be eliminated from the basic equations. Similarly, in the stress method, we take stress components as the basic unknowns, find them first with corresponding stress boundary conditions and then calculate strains and displacements, which in turn must satisfy the displacement boundary conditions. The strain method is similar but in this case strains are first found. In many other cases, it may be easier to find some stresses and some displacements but not all the variables of a single group. This is the so-called mixed method. The solution procedures of the mixed method differ from case to case, but it is effective if incorporated with the semi-inverse method to be discussed later.

In principle, all these solution methods are equivalent. However, for different methods, the degrees of difficulty in solving the same problem may be very different. This mainly depends on the number of unknown variables, the complexity of the equations and the nature of the boundary conditions. For example, the displacement method is most common for anisotropic problems; the stress method is difficult to use due to the involvement of anisotropy, and is only suitable for some special cases, such as static problems, as to be discussed in **Section 2.3**.

In the following, we will discuss in detail the displacement and mixed methods, which will be used extensively in this book.

2.1.2 Governing Equations for the Displacement Method

To use the displacement method, it is necessary to eliminate stresses and strains from the basic equations. By substituting Eq. (2.1.1) into Eq. (1.2.13), we get

$$c_{ijkl}u_{k,lj} + F_i + \beta_{ij}T_{,j} = \rho\ddot{u}_i, \quad (i = 1, 2, 3). \tag{2.1.3}$$

The three displacement components, u_i, can be obtained by solving these three second-order partial differential equations. The stresses, σ_{ij}, can then be derived using Eq. (2.1.1).

For convenience, we prefer to list some explicit expressions of Eq. (2.1.3) below for orthotropic materials in various coordinate systems.

For orthotropic materials with the generalized Hooke's law in Cartesian coordinates expressed by Eq. (1.4.9), Eq. (2.1.3) can be expanded as

$$\begin{aligned}
&\left(c_{11}\frac{\partial^2}{\partial x^2}+c_{66}\frac{\partial^2}{\partial y^2}+c_{55}\frac{\partial^2}{\partial z^2}\right)u+(c_{12}+c_{66})\frac{\partial^2 v}{\partial x\partial y}\\
&\quad+(c_{13}+c_{55})\frac{\partial^2 w}{\partial x\partial z}+F_x+\beta_1\frac{\partial T}{\partial x}=\rho\ddot{u},\\
&(c_{12}+c_{66})\frac{\partial^2 u}{\partial x\partial y}+\left(c_{66}\frac{\partial^2}{\partial x^2}+c_{22}\frac{\partial^2}{\partial y^2}+c_{44}\frac{\partial^2}{\partial z^2}\right)v\\
&\quad+(c_{23}+c_{44})\frac{\partial^2 w}{\partial y\partial z}+F_y+\beta_2\frac{\partial T}{\partial y}=\rho\ddot{v},\\
&(c_{13}+c_{55})\frac{\partial^2 u}{\partial x\partial z}+(c_{23}+c_{44})\frac{\partial^2 v}{\partial y\partial z}+\left(c_{55}\frac{\partial^2}{\partial x^2}+c_{44}\frac{\partial^2}{\partial y^2}+c_{33}\frac{\partial^2}{\partial z^2}\right)w\\
&\quad+F_z+\beta_3\frac{\partial T}{\partial z}=\rho\ddot{w}.
\end{aligned}\tag{2.1.4}$$

For orthotropic materials with the generalized Hooke's law in cylindrical coordinates expressed by Eq. (1.4.10), Eq. (2.1.3) can be expanded as

$$\begin{aligned}
&\left[c_{11}\left(\frac{\partial^2}{\partial r^2}+\frac{1}{r}\frac{\partial}{\partial r}\right)-c_{22}\frac{1}{r^2}+c_{66}\frac{1}{r^2}\frac{\partial^2}{\partial\alpha^2}+c_{55}\frac{\partial^2}{\partial z^2}\right]u_r\\
&\quad+\left[(c_{12}+c_{66})\frac{1}{r}\frac{\partial^2}{\partial\alpha\partial r}-(c_{22}+c_{66})\frac{1}{r^2}\frac{\partial}{\partial\alpha}\right]u_\alpha\\
&\quad+\left[(c_{13}+c_{55})\frac{\partial^2}{\partial r\partial z}+(c_{13}-c_{23})\frac{1}{r}\frac{\partial}{\partial z}\right]w+F_r+\beta_1\frac{\partial T}{\partial r}+\frac{(\beta_1-\beta_2)T}{r}=\rho\ddot{u}_r,\\
&\left[(c_{12}+c_{66})\frac{1}{r}\frac{\partial^2}{\partial\alpha\partial r}+(c_{22}+c_{66})\frac{1}{r^2}\frac{\partial}{\partial\alpha}\right]u_r\\
&\quad+\left[c_{66}\left(\frac{\partial^2}{\partial r^2}+\frac{1}{r}\frac{\partial}{\partial r}-\frac{1}{r^2}\right)+c_{22}\frac{1}{r^2}\frac{\partial^2}{\partial\alpha^2}+c_{44}\frac{\partial^2}{\partial z^2}\right]u_\alpha\\
&\quad+(c_{23}+c_{44})\frac{1}{r}\frac{\partial^2 w}{\partial\alpha\partial z}+F_\alpha+\beta_2\frac{1}{r}\frac{\partial T}{\partial\alpha}=\rho\ddot{u}_\alpha,
\end{aligned}$$

$$
\begin{aligned}
&\left[(c_{13}+c_{55})\frac{\partial^2}{\partial r\partial z}+(c_{23}+c_{55})\frac{1}{r}\frac{\partial}{\partial z}\right]u_r+(c_{23}+c_{44})\frac{1}{r}\frac{\partial^2 u_\alpha}{\partial\alpha\partial z}\\
&\quad+\left[c_{55}\left(\frac{\partial^2}{\partial r^2}+\frac{1}{r}\frac{\partial}{\partial r}\right)+c_{44}\frac{1}{r^2}\frac{\partial^2}{\partial\alpha^2}+c_{33}\frac{\partial^2}{\partial z^2}\right]w+F_z+\beta_3\frac{\partial T}{\partial z}=\rho\ddot{w}.
\end{aligned}
\tag{2.1.5}
$$

For orthotropic materials with the generalized Hooke's law in spherical coordinates expressed by Eq. (1.4.11), Eq. (2.1.3) can be expanded as

$$
\begin{aligned}
&\left[c_{33}\nabla_3^2+c_{13}+c_{23}-c_{11}-c_{22}-2c_{12}+c_{55}\nabla_1^2+(c_{44}-c_{55})\frac{1}{\sin^2\theta}\frac{\partial^2}{\partial\alpha^2}\right]u_R\\
&\quad+\left[(c_{13}+c_{55})\nabla_2\left(\frac{\partial}{\partial\theta}+\cot\theta\right)+(c_{13}-c_{11}-c_{12}-c_{55})\frac{\partial}{\partial\theta}+(c_{23}-c_{22}-c_{12}-c_{55})\cot\theta\right]u_\theta\\
&\quad+\frac{1}{\sin\theta}[(c_{23}+c_{44})\nabla_2+c_{23}-c_{22}-c_{12}-c_{44}]\frac{\partial u_\alpha}{\partial\alpha}\\
&\quad+\beta_3R^2\frac{\partial T}{\partial R}+(2\beta_3-\beta_1-\beta_2)RT+R^2F_R=\rho R^2\ddot{u}_R,
\end{aligned}
$$

$$
\begin{aligned}
&[(c_{13}+c_{55})\nabla_2+2c_{55}+c_{11}+c_{12})]\frac{\partial u_R}{\partial\theta}+[(c_{13}-c_{23})\nabla_2+c_{11}-c_{22}]u_R\cot\theta\\
&\quad+\left[c_{55}\nabla_3^2-2c_{55}-c_{12}+c_{11}\nabla_1^2+(c_{66}-c_{11})\frac{1}{\sin^2\theta}\frac{\partial^2}{\partial\alpha^2}-c_{22}\cot^2\theta\right]u_\theta\\
&\quad+\frac{1}{\sin\theta}\left[(c_{12}+c_{66})\frac{\partial}{\partial\theta}-(c_{22}+c_{66})\cot\theta\right]\frac{\partial u_\alpha}{\partial\alpha}\\
&\quad+\beta_1R\frac{\partial T}{\partial\theta}+(\beta_1-\beta_2)RT\cot\theta+R^2F_\theta=\rho R^2\ddot{u}_\theta,\\
&\frac{1}{\sin\theta}[(c_{23}+c_{44})\nabla_2+2c_{44}+c_{22}+c_{12}]\frac{\partial u_R}{\partial\alpha}\\
&\quad+\frac{1}{\sin\theta}\left[(c_{66}+c_{12})\frac{\partial}{\partial\theta}+(c_{66}+c_{22})\cot\theta\right]\frac{\partial u_\theta}{\partial\alpha}\\
&\quad+\left[c_{44}\nabla_3^2-2c_{44}+c_{66}+c_{66}\nabla_1^2+(c_{22}-c_{66})\frac{1}{\sin^2\theta}\frac{\partial^2}{\partial\alpha^2}-c_{66}\cot^2\theta\right]u_\alpha\\
&\quad+\frac{1}{\sin\theta}\beta_2R\frac{\partial T}{\partial\alpha}+R^2F_\alpha=\rho R^2\ddot{u}_\alpha,
\end{aligned}
\tag{2.1.6}
$$

where

$$\nabla_2 = R\frac{\partial}{\partial R},\quad \nabla_2^2 = R\frac{\partial}{\partial R}R\frac{\partial}{\partial R},\quad \nabla_3^2 = \nabla_2^2 + \nabla_2,$$
$$\nabla_1^2 = \frac{\partial^2}{\partial \theta^2} + \cot\theta\frac{\partial}{\partial \theta} + \frac{1}{\sin^2\theta}\frac{\partial^2}{\partial \alpha^2}. \tag{2.1.7}$$

2.1.3 Equations for a Mixed Method – the State-Space Method

Mixed methods usually deal with nine equations with two groups of unknowns, *i.e.*, six stress components and three displacement components. There is a special mixed method, called the *state-space method*[1], which divides the nine unknowns into two sets: the set of *primary variables* (*state variables* or basic variables) consisting of three displacements and three stress components, and the set of *secondary variables* (*output variables* or *induced variables*) composed of the other three stress components. At the same time, the nine basic equations are also grouped into two sets: the first set containing six governing equations with the six primary variables and the second set containing the other three equations that include the primary variables and the linear combination of these variables and their derivatives. In a Cartesian coordinate system for instance, if we take u, v, w, τ_{zx}, τ_{yz}, σ_z as the primary variables, then the rest, *i.e.*, τ_{xy}, σ_x, σ_y, are the secondary variables. The corresponding primary and secondary sets of governing equations can be derived following the steps below.

The equations of motion and geometric equations can be written as

$$\frac{\partial}{\partial z}\{\sigma_2\} + [\partial_2]^{\mathrm{T}}\{\sigma_2\} + [\partial_1]^{\mathrm{T}}\{\sigma_1\} + \{F\} = \rho\frac{\partial^2}{\partial t^2}\{u\}, \tag{2.1.8}$$

$$\{\gamma_1\} = [\partial_1]\{u\},\quad \{\gamma_2\} = \frac{\partial}{\partial z}\{u\} + [\partial_2]\{u\}, \tag{2.1.9}$$

where

[1] The state-space method is also called the method of initial functions or the transfer matrix method. It has been used for many elasticity problems, *e.g.*, Vlasov and Leontev (1966), Bahar (1972, 1977), Kameswara Rao and Das (1977), Sundara Raja Iyengar and Raman (1977), Fan and Ye (1990*a*,*b*), Feng and Qin (1991), Spencer *et al.* (1992), Steele and Kim (1992), Zhong (1995), Ding, Chen and Xu (2001), and Tarn (2002).

$$\{\sigma_1\}=[\sigma_x,\sigma_y,\tau_{xy}]^{\mathrm{T}},\quad \{\sigma_2\}=[\sigma_z,\tau_{zx},\tau_{yz}]^{\mathrm{T}},$$
$$\{\gamma_1\}=[\varepsilon_x,\varepsilon_y,\gamma_{xy}]^{\mathrm{T}},\quad \{\gamma_2\}=[\varepsilon_z,\gamma_{zx},\gamma_{yz}]^{\mathrm{T}}, \tag{2.1.10}$$

$$\{u\}=[w,\ u,\ v]^{\mathrm{T}},\quad \{F\}=[F_z,\ F_x,\ F_y]^{\mathrm{T}}, \tag{2.1.11}$$

$$[\partial_1]=\begin{bmatrix} 0 & \dfrac{\partial}{\partial x} & 0 \\ 0 & 0 & \dfrac{\partial}{\partial y} \\ 0 & \dfrac{\partial}{\partial y} & \dfrac{\partial}{\partial x} \end{bmatrix},\quad [\partial_2]=\begin{bmatrix} 0 & 0 & 0 \\ \dfrac{\partial}{\partial x} & 0 & 0 \\ \dfrac{\partial}{\partial y} & 0 & 0 \end{bmatrix}. \tag{2.1.12}$$

Then, the constitutive relations, Eq. (1.4.4), can be rewritten as

$$\{\sigma_1\}=[c_1]\{\gamma_1\}+[c_2]\{\gamma_2\}+\{\beta_7\}T,$$
$$\{\sigma_2\}=[c_2]^{\mathrm{T}}\{\gamma_1\}+[c_4]\{\gamma_2\}+\{\beta_8\}T, \tag{2.1.13}$$

where

$$[c_1]=\begin{bmatrix} c_{11} & c_{12} & c_{16} \\ c_{12} & c_{22} & c_{26} \\ c_{16} & c_{26} & c_{66} \end{bmatrix},\quad [c_2]=\begin{bmatrix} c_{13} & c_{15} & c_{14} \\ c_{23} & c_{25} & c_{24} \\ c_{36} & c_{56} & c_{46} \end{bmatrix},\quad [c_4]=\begin{bmatrix} c_{33} & c_{35} & c_{34} \\ c_{35} & c_{55} & c_{45} \\ c_{34} & c_{45} & c_{44} \end{bmatrix},$$
$$\{\beta_7\}=[\beta_{11},\ \beta_{22},\ \beta_{12}]^{\mathrm{T}},\quad \{\beta_8\}=[\beta_{33},\ \beta_{31},\ \beta_{23}]^{\mathrm{T}}. \tag{2.1.14}$$

Solving for $\{\gamma_2\}$ using the second of Eq. (2.1.13) leads to

$$\{\gamma_2\}=[c_4]^{-1}(\{\sigma_2\}-[c_2]^{\mathrm{T}}\{\gamma_1\}-\{\beta_8\}T). \tag{2.1.15}$$

Substituting this into the first of Eq. (2.1.13) and making use of the first of Eq. (2.1.9), we get

$$\{\sigma_1\}=[B_1]\{\sigma_2\}+[B_2]\{u\}+\{\beta_9\}T, \tag{2.1.16}$$

where

$$[B_1]=[c_2][c_4]^{-1}, \qquad [B_2]=([c_1]-[B_1][c_2]^{\mathrm{T}})[\partial_1], \tag{2.1.17}$$

$$\{\beta_9\}=\{\beta_7\}-[B_1]\{\beta_8\}. \tag{2.1.18}$$

Now, by substituting Eq. (2.1.9) into Eq. (2.1.15), we obtain

$$\frac{\partial}{\partial z}\{u\}=[A_3]\{\sigma_2\}+[A_4]\{u\}-[c_4]^{-1}\{\beta_8\}T, \tag{2.1.19}$$

where

$$[A_3]=[c_4]^{-1}, \qquad [A_4]=-[\partial_2]-[B_1]^{\mathrm{T}}[\partial_1]. \tag{2.1.20}$$

On the other hand, if we substitute Eq. (2.1.16) into Eq. (2.1.8), we have

$$\frac{\partial}{\partial z}\{\sigma_2\}=[A_1]\{\sigma_2\}+[A_2]\{u\}-[\partial_1]^{\mathrm{T}}\{\beta_9\}T-\{F\}, \tag{2.1.21}$$

where

$$[A_1]=[A_4]^{\mathrm{T}}, \qquad [A_2]=[I]\rho\frac{\partial^2}{\partial t^2}-[\partial_1]^{\mathrm{T}}[B_2], \tag{2.1.22}$$

in which $[I]$ is the unit matrix of order three.

Combining Eqs. (2.1.19) with (2.1.21), we obtain

$$\frac{\partial}{\partial z}\{a\}=[A]\{a\}+\{\beta_4\}T+\{f\}, \tag{2.1.23}$$

where

$$\{a\}=\begin{Bmatrix}\sigma_2\\ u\end{Bmatrix}, \quad [A]=\begin{bmatrix}A_1 & A_2\\ A_3 & A_4\end{bmatrix}, \tag{2.1.24}$$

$$\{\beta_4\}=\begin{Bmatrix}-[\partial_1]^{\mathrm{T}}\{\beta_9\}\\ -[c_4]^{-1}\{\beta_8\}\end{Bmatrix}, \tag{2.1.25}$$

$$\{f\}=[-F_z,\,-F_x,\,-F_y,\,0,\,0,\,0]^{\mathrm{T}}. \tag{2.1.26}$$

Equation (2.1.23) is the governing equation for the primary variables, $\{a\}=[\sigma_z,\,\tau_{zx},\,\tau_{yz},\,w,\,u,\,v]^{\mathrm{T}}$, and Eq. (2.1.16) is that for the secondary variables, $\{\sigma_1\}$. Correspondingly, the boundary conditions described by Eq. (1.3.1) should be rewritten as

$$[N_3]\{\sigma_2\}+[N_4]\{u\}=\{P\}-[N_1]\{\beta_9\}T, \tag{2.1.27}$$

where

$$[N_3]=[N_2]+[N_1][B_1]\ ,\quad [N_4]=[N_1][B_2]\,, \tag{2.1.28}$$

$$[N_1]=\begin{bmatrix}0 & 0 & 0\\ \cos(n,x) & 0 & \cos(n,y)\\ 0 & \cos(n,y) & \cos(n,x)\end{bmatrix},\quad [N_2]=\begin{bmatrix}\cos(n,z) & \cos(n,x) & \cos(n,y)\\ 0 & \cos(n,z) & 0\\ 0 & 0 & \cos(n,z)\end{bmatrix}, \tag{2.1.29}$$

$$\{P\}=[\overline{p}_z,\,\overline{p}_x,\,\overline{p}_y]^{\mathrm{T}}. \tag{2.1.30}$$

When the material is orthotropic, we can obtain the following explicit matrices:

$$[B_1]=\frac{1}{c_{33}}[c_2]=\begin{bmatrix}c_{13}/c_{33} & 0 & 0\\ c_{23}/c_{33} & 0 & 0\\ 0 & 0 & 0\end{bmatrix}, \tag{2.1.31}$$

$$[B_2]=\begin{bmatrix}0 & \left(c_{11}-\dfrac{c_{13}^2}{c_{33}}\right)\dfrac{\partial}{\partial x} & \left(c_{12}-\dfrac{c_{13}c_{23}}{c_{33}}\right)\dfrac{\partial}{\partial y}\\ 0 & \left(c_{12}-\dfrac{c_{13}c_{23}}{c_{33}}\right)\dfrac{\partial}{\partial x} & \left(c_{22}-\dfrac{c_{23}^2}{c_{33}}\right)\dfrac{\partial}{\partial y}\\ 0 & c_{66}\dfrac{\partial}{\partial y} & c_{66}\dfrac{\partial}{\partial x}\end{bmatrix}, \tag{2.1.32}$$

$$[A_3]=\begin{bmatrix}1/c_{33} & 0 & 0\\ 0 & 1/c_{55} & 0\\ 0 & 0 & 1/c_{44}\end{bmatrix}, \tag{2.1.33}$$

$$[A_4]=\begin{bmatrix}0 & -\dfrac{c_{13}}{c_{33}}\dfrac{\partial}{\partial x} & -\dfrac{c_{23}}{c_{33}}\dfrac{\partial}{\partial y}\\ -\dfrac{\partial}{\partial x} & 0 & 0\\ -\dfrac{\partial}{\partial y} & 0 & 0\end{bmatrix}=[A_1]^{\mathrm{T}}, \tag{2.1.34}$$

$$[A_2]=\begin{bmatrix}\rho\dfrac{\partial^2}{\partial t^2} & 0 & 0\\ 0 & \rho\dfrac{\partial^2}{\partial t^2}-\left(c_{11}-\dfrac{c_{13}^2}{c_{33}}\right)\dfrac{\partial^2}{\partial x^2}-c_{66}\dfrac{\partial^2}{\partial y^2} & -\left(c_{12}+c_{66}-\dfrac{c_{13}c_{23}}{c_{33}}\right)\dfrac{\partial^2}{\partial x\partial y}\\ 0 & -\left(c_{12}+c_{66}-\dfrac{c_{13}c_{23}}{c_{33}}\right)\dfrac{\partial^2}{\partial x\partial y} & \rho\dfrac{\partial^2}{\partial t^2}-\left(c_{22}-\dfrac{c_{23}^2}{c_{33}}\right)\dfrac{\partial^2}{\partial y^2}-c_{66}\dfrac{\partial^2}{\partial x^2}\end{bmatrix}. \tag{2.1.35}$$

In this case, the operator matrix $[A]$ of Eq. (2.1.24) can be more concisely rewritten as $[A']$ if the components of state vector $\{a\}$ are properly rearranged. For example, if we introduce a new state vector as $\{a'\}=[\sigma_z, u, v, w, \tau_{zx}, \tau_{yz}]^{\mathrm{T}}$, we get

$$\{a'\}=\begin{bmatrix}k_1 & k_2\\ k_2 & k_1\end{bmatrix}\{a\}=[k]\{a\}, \tag{2.1.36}$$

where

$$[k_1]=\begin{bmatrix}1 & 0 & 0\\ 0 & 0 & 0\\ 0 & 0 & 0\end{bmatrix}, \quad [k_2]=\begin{bmatrix}0 & 0 & 0\\ 0 & 1 & 0\\ 0 & 0 & 1\end{bmatrix}. \tag{2.1.37}$$

With this rearrangement, Eq. (2.1.23) becomes

$$\frac{\partial}{\partial z}\{a'\}=[A']\{a'\}+\{\beta'\}T+\{f'\}, \tag{2.1.38}$$

where

$$[A'] = \begin{bmatrix} A_1' & A_2' \\ A_3' & A_4' \end{bmatrix} = [k][A][k], \tag{2.1.39}$$

$$\{\beta_4'\} = [k]\{\beta_4\}, \quad f' = [k]\{f\}. \tag{2.1.40}$$

Using Eqs. (2.1.33) to (2.1.37), we can show that

$$[A_1'] = [A_4'] = 0, \tag{2.1.41}$$

$$[A_2'] = \begin{bmatrix} \rho\dfrac{\partial^2}{\partial t^2} & -\dfrac{\partial}{\partial x} & -\dfrac{\partial}{\partial y} \\ -\dfrac{\partial}{\partial x} & \dfrac{1}{c_{55}} & 0 \\ -\dfrac{\partial}{\partial y} & 0 & \dfrac{1}{c_{44}} \end{bmatrix}, \tag{2.1.42}$$

$$[A_3'] = \begin{bmatrix} \dfrac{1}{c_{33}} & \dfrac{c_{13}}{c_{33}}\dfrac{\partial}{\partial x} & \dfrac{c_{23}}{c_{33}}\dfrac{\partial}{\partial y} \\ -\dfrac{c_{13}}{c_{33}}\dfrac{\partial}{\partial x} & \rho\dfrac{\partial^2}{\partial t^2} - \left(c_{11} - \dfrac{c_{13}^2}{c_{33}}\right)\dfrac{\partial^2}{\partial x^2} - c_{66}\dfrac{\partial^2}{\partial y^2} & -\left(c_{12} + c_{66} - \dfrac{c_{13}c_{23}}{c_{33}}\right)\dfrac{\partial^2}{\partial x \partial y} \\ -\dfrac{c_{23}}{c_{33}}\dfrac{\partial}{\partial y} & -\left(c_{12} + c_{66} - \dfrac{c_{13}c_{23}}{c_{33}}\right)\dfrac{\partial^2}{\partial x \partial y} & \rho\dfrac{\partial^2}{\partial t^2} - \left(c_{22} - \dfrac{c_{23}^2}{c_{33}}\right)\dfrac{\partial^2}{\partial y^2} - c_{66}\dfrac{\partial^2}{\partial x^2} \end{bmatrix}. \tag{2.1.43}$$

This result is consistent with that of Zheng and Zhang (1996).

2.2 DISPLACEMENT METHOD

We will find the solution for transversely isotropic problems using the governing equations developed above.

2.2.1 General Solution in Cartesian Coordinates

(1) General solution

For transversely isotropic materials, Eq. (2.1.4) reduces to

$$\begin{aligned}
&\left(c_{11}\frac{\partial^2}{\partial x^2}+c_{66}\frac{\partial^2}{\partial y^2}+c_{44}\frac{\partial^2}{\partial z^2}\right)u+(c_{12}+c_{66})\frac{\partial^2 v}{\partial x\partial y}\\
&\quad+(c_{13}+c_{44})\frac{\partial^2 w}{\partial x\partial z}+F_x+\beta_1\frac{\partial T}{\partial x}=\rho\ddot{u},\\
&(c_{12}+c_{66})\frac{\partial^2 u}{\partial x\partial y}+\left(c_{66}\frac{\partial^2}{\partial x^2}+c_{11}\frac{\partial^2}{\partial y^2}+c_{44}\frac{\partial^2}{\partial z^2}\right)v\\
&\quad+(c_{13}+c_{44})\frac{\partial^2 w}{\partial y\partial z}+F_y+\beta_1\frac{\partial T}{\partial y}=\rho\ddot{v},\\
&(c_{13}+c_{44})\frac{\partial}{\partial z}\left(\frac{\partial u}{\partial x}+\frac{\partial v}{\partial y}\right)+\left[c_{44}\left(\frac{\partial^2}{\partial x^2}+\frac{\partial^2}{\partial y^2}\right)+c_{33}\frac{\partial^2}{\partial z^2}\right]w\\
&\quad+F_z+\beta_3\frac{\partial T}{\partial z}=\rho\ddot{w}.
\end{aligned} \tag{2.2.1}$$

To obtain the solution more effectively, let us introduce two *displacement functions*, ψ and G, such that[2]

$$u=\frac{\partial\psi}{\partial y}-\frac{\partial G}{\partial x},\quad v=-\frac{\partial\psi}{\partial x}-\frac{\partial G}{\partial y},\quad w=w. \tag{2.2.2}$$

In cylindrical coordinates, they become

$$u_r=\frac{1}{r}\frac{\partial\psi}{\partial\alpha}-\frac{\partial G}{\partial r},\qquad u_\alpha=-\frac{\partial\psi}{\partial r}-\frac{1}{r}\frac{\partial G}{\partial\alpha},\qquad w=w. \tag{2.2.2a}$$

Similarly, we can express the body forces as

$$F_x=\frac{\partial V}{\partial y}-\frac{\partial U}{\partial x},\quad F_y=-\frac{\partial V}{\partial x}-\frac{\partial U}{\partial y},\quad F_z=F_z. \tag{2.2.3}$$

The substitution of Eqs. (2.2.2) and (2.2.3) into Eq. (2.2.1) leads to

$$\frac{\partial B}{\partial y}-\frac{\partial A}{\partial x}=0, \tag{2.2.4}$$

---

[2] More details about the uniqueness and possibility of other similar displacement presentations can be found in the works by Zou *et al.* (1994), Liang (1994), and Ding, Chen and Liang (1996).

$$\frac{\partial B}{\partial x}+\frac{\partial A}{\partial y}=0\,, \tag{2.2.5}$$

$$-(c_{13}+c_{44})\Lambda\frac{\partial G}{\partial z}+\left(c_{44}\Lambda+c_{33}\frac{\partial^2}{\partial z^2}-\rho\frac{\partial^2}{\partial t^2}\right)w+F_z+\beta_3\frac{\partial T}{\partial z}=0\,, \tag{2.2.6}$$

where

$$B=\left(c_{66}\Lambda+c_{44}\frac{\partial^2}{\partial z^2}-\rho\frac{\partial^2}{\partial t^2}\right)\psi+V\,, \tag{2.2.7}$$

$$A=\left(c_{11}\Lambda+c_{44}\frac{\partial^2}{\partial z^2}-\rho\frac{\partial^2}{\partial t^2}\right)G-(c_{13}+c_{44})\frac{\partial w}{\partial z}+U-\beta_1 T\,, \tag{2.2.8}$$

while

$$\Lambda=\frac{\partial^2}{\partial x^2}+\frac{\partial^2}{\partial y^2}\,. \tag{2.2.9}$$

On the other hand, Eq. (2.2.4) gives rise to

$$A=\frac{\partial H}{\partial y}\,,\quad B=\frac{\partial H}{\partial x}\,. \tag{2.2.10}$$

By substituting this into Eq. (2.2.5), we have

$$\Lambda H=0\,, \tag{2.2.11}$$

where $H(x, y, z, t)$ is an arbitrary function about z and t. Hence, the substitution of Eqs. (2.2.7) and (2.2.8) into Eq. (2.2.10) brings about

$$\left(c_{66}\Lambda+c_{44}\frac{\partial^2}{\partial z^2}-\rho\frac{\partial^2}{\partial t^2}\right)\psi+V=\frac{\partial H}{\partial x}\,, \tag{2.2.12}$$

$$\left(c_{11}\Lambda+c_{44}\frac{\partial^2}{\partial z^2}-\rho\frac{\partial^2}{\partial t^2}\right)G-(c_{13}+c_{44})\frac{\partial w}{\partial z}+U-\beta_1 T=\frac{\partial H}{\partial y}\,. \tag{2.2.13}$$

Zou *et al.* (1994), Liang (1994), and Ding, Chen and Liang (1996) have shown that without loss of generality, the right hand sides of these two equations can be set to zero. Thus Eq. (2.2.12) is simplified to

$$\left(c_{66}\Lambda + c_{44}\frac{\partial^2}{\partial z^2} - \rho\frac{\partial^2}{\partial t^2}\right)\psi = -V. \tag{2.2.14}$$

We resolve function ψ into

$$\psi = \psi_0 + \psi_1, \tag{2.2.15}$$

where ψ_0 is the general solution of the homogeneous equation of Eq. (2.2.14) and ψ_1 is the special solution of Eq. (2.2.14), *i.e.*,

$$\left(c_{66}\Lambda + c_{44}\frac{\partial^2}{\partial z^2} - \rho\frac{\partial^2}{\partial t^2}\right)\psi_0 = 0, \tag{2.2.16}$$

$$\left(c_{66}\Lambda + c_{44}\frac{\partial^2}{\partial z^2} - \rho\frac{\partial^2}{\partial t^2}\right)\psi_1 = -V. \tag{2.2.17}$$

On the other hand, Eq. (2.2.13) becomes

$$\left(c_{11}\Lambda + c_{44}\frac{\partial^2}{\partial z^2} - \rho\frac{\partial^2}{\partial t^2}\right)G - (c_{13} + c_{44})\frac{\partial w}{\partial z} = -U + \beta_1 T. \tag{2.2.18}$$

Solving this together with Eq. (2.2.6) for G and w, we get

$$G = G_0 + G_1 + G_2, \quad w = w_0 + w_1 + w_2, \tag{2.2.19}$$

where G_0 and w_0 are the general solution of the following equations:

$$\left(c_{11}\Lambda + c_{44}\frac{\partial^2}{\partial z^2} - \rho\frac{\partial^2}{\partial t^2}\right)G_0 - (c_{13} + c_{44})\frac{\partial w_0}{\partial z} = 0, \tag{2.2.20}$$

$$(c_{13} + c_{44})\Lambda\frac{\partial G_0}{\partial z} - \left(c_{44}\Lambda + c_{33}\frac{\partial^2}{\partial z^2} - \rho\frac{\partial^2}{\partial t^2}\right)w_0 = 0. \tag{2.2.21}$$

G_1 and w_1 are the special solution of the following non-homogeneous equations:

$$\begin{aligned}&\left(c_{11}\Lambda+c_{44}\frac{\partial^2}{\partial z^2}-\rho\frac{\partial^2}{\partial t^2}\right)G_1-(c_{13}+c_{44})\frac{\partial w_1}{\partial z}=-U+\beta_1 T,\\&(c_{13}+c_{44})\Lambda\frac{\partial G_1}{\partial z}-\left(c_{44}\Lambda+c_{33}\frac{\partial^2}{\partial z^2}-\rho\frac{\partial^2}{\partial t^2}\right)w_1=0,\end{aligned}\tag{2.2.22}$$

while G_2 and w_2 are the special solution of the following equations:

$$\begin{aligned}&\left(c_{11}\Lambda+c_{44}\frac{\partial^2}{\partial z^2}-\rho\frac{\partial^2}{\partial t^2}\right)G_2-(c_{13}+c_{44})\frac{\partial w_2}{\partial z}=0,\\&(c_{13}+c_{44})\Lambda\frac{\partial G_2}{\partial z}-\left(c_{44}\Lambda+c_{33}\frac{\partial^2}{\partial z^2}-\rho\frac{\partial^2}{\partial t^2}\right)w_2=F_z+\beta_3\frac{\partial T}{\partial z}.\end{aligned}\tag{2.2.23}$$

The general solution of Eqs. (2.2.20) and (2.2.21) can have various forms. One of them can be

$$G_0=(c_{13}+c_{44})\frac{\partial F_0}{\partial z},\quad w_0=\left(c_{11}\Lambda+c_{44}\frac{\partial^2}{\partial z^2}-\rho\frac{\partial^2}{\partial t^2}\right)F_0,\tag{2.2.24}$$

where F_0 is an arbitrary function. By substituting Eq. (2.2.24) into Eq. (2.2.21), we can easily obtain the condition that F_0 must follow, *i.e.*,

$$L_0F_0=0,\tag{2.2.25}$$

where

$$\begin{aligned}L_0=&\,c_{11}c_{44}\Lambda^2+[c_{11}c_{33}+c_{44}^2-(c_{13}+c_{44})^2]\Lambda\frac{\partial^2}{\partial z^2}+c_{33}c_{44}\frac{\partial^4}{\partial z^4}\\&-\rho\left[(c_{11}+c_{44})\Lambda+(c_{33}+c_{44})\frac{\partial^2}{\partial z^2}\right]\frac{\partial^2}{\partial t^2}+\rho^2\frac{\partial^4}{\partial t^4}.\end{aligned}\tag{2.2.26}$$

Another form of the general solution, G_0 and w_0, can be written as

$$G_0 = \left(c_{44}\Lambda + c_{33}\frac{\partial^2}{\partial z^2} - \rho\frac{\partial^2}{\partial t^2} \right) F_0, \quad w_0 = (c_{13} + c_{44})\Lambda \frac{\partial F_0}{\partial z}. \tag{2.2.27}$$

In this case, the condition that F_0 must satisfy is also given by Eq. (2.2.25).

Similarly, the special solutions of Eqs. (2.2.22) and (2.2.23) can be obtained using the following method. Let

$$G_1 = \left(c_{44}\Lambda + c_{33}\frac{\partial^2}{\partial z^2} - \rho\frac{\partial^2}{\partial t^2} \right) F_1, \quad w_1 = (c_{13} + c_{44})\Lambda \frac{\partial F_1}{\partial z}, \tag{2.2.28}$$

$$G_2 = (c_{13} + c_{44})\frac{\partial F_2}{\partial z}, \quad w_2 = \left(c_{11}\Lambda + c_{44}\frac{\partial^2}{\partial z^2} - \rho\frac{\partial^2}{\partial t^2} \right) F_2, \tag{2.2.29}$$

and

$$L_0 F_1 = -U + \beta_1 T, \tag{2.2.30}$$

$$L_0 F_2 = -F_z - \beta_3 \frac{\partial T}{\partial z}, \tag{2.2.31}$$

then G_1, w_1, G_2 and w_2 are determined when the special solutions of F_1 and F_2 of Eqs. (2.2.30) and (2.2.31) are substituted into Eqs. (2.2.28) and (2.2.29), respectively.

(2) Special case I: Absence of body force and temperature change

If there is no body force or temperature change, we need to consider only the corresponding homogeneous equations. Hence, from Eqs. (2.2.24) and (2.2.2), we have the following general solution:

$$\begin{aligned} u &= \frac{\partial \psi_0}{\partial y} - (c_{13} + c_{44})\frac{\partial^2 F_0}{\partial x \partial z}, \\ v &= -\frac{\partial \psi_0}{\partial x} - (c_{13} + c_{44})\frac{\partial^2 F_0}{\partial y \partial z}, \\ w &= \left(c_{11}\Lambda + c_{44}\frac{\partial^2}{\partial z^2} - \rho\frac{\partial^2}{\partial t^2} \right) F_0. \end{aligned} \tag{2.2.32}$$

If we let

$$\psi_0 = -\varphi, \quad (c_{13} + c_{44})F_0 = F, \tag{2.2.33}$$

Eq. (2.2.32) becomes

$$u = -\frac{\partial \varphi}{\partial y} - \frac{\partial^2 F}{\partial x \partial z}, \quad v = \frac{\partial \varphi}{\partial x} - \frac{\partial^2 F}{\partial y \partial z}, \quad w = \left(a\Lambda + b\frac{\partial^2}{\partial z^2} - c\frac{\partial^2}{\partial t^2} \right) F, \tag{2.2.34}$$

where

$$a = \frac{c_{11}}{c_{13} + c_{44}}, \quad b = \frac{c_{44}}{c_{13} + c_{44}}, \quad c = \frac{\rho}{c_{13} + c_{44}}, \tag{2.2.35}$$

and φ and F in Eq. (2.2.34) must satisfy the following conditions:

$$\left(c_{66}\Lambda + c_{44}\frac{\partial^2}{\partial z^2} - \rho\frac{\partial^2}{\partial t^2} \right) \varphi = 0, \tag{2.2.36}$$

$$L_0 F = 0. \tag{2.2.37}$$

Equations (2.2.34) to (2.2.37) are consistent with that obtained by Hu (1956) and reduce to those developed by Hu (1953) when the body is in static equilibrium.

(3) Special case II: Static problems

In this case, all physical quantities are independent of the time variable t, and hence we can rewrite Eqs. (2.2.34), (2.2.36) and (2.2.37) as

$$u = -\frac{\partial \varphi}{\partial y} - \frac{\partial^2 F}{\partial x \partial z}, \quad v = \frac{\partial \varphi}{\partial x} - \frac{\partial^2 F}{\partial y \partial z}, \quad w = \left(a\Lambda + b\frac{\partial^2}{\partial z^2} \right) F, \tag{2.2.38}$$

$$\left(c_{66}\Lambda + c_{44}\frac{\partial^2}{\partial z^2} \right) \varphi = 0, \tag{2.2.39}$$

$$L_s F = 0, \tag{2.2.40}$$

where

$$L_s = c_{11}c_{44}\Lambda^2 + [c_{11}c_{33} + c_{44}^2 - (c_{13} + c_{44})^2]\Lambda \frac{\partial^2}{\partial z^2} + c_{33}c_{44}\frac{\partial^4}{\partial z^4}. \tag{2.2.41}$$

This operator can be expressed as the product of two operators, *i.e*. Eq. (2.2.40) can be rewritten as

$$\left(\Lambda + \frac{\partial^2}{\partial z_1^2}\right)\left(\Lambda + \frac{\partial^2}{\partial z_2^2}\right)F = 0, \tag{2.2.42}$$

where $z_i = s_i z$ $(i = 1, 2)$ and s_i^2 are the roots of the following eigen-equation:

$$c_{33}c_{44}s^4 - [c_{11}c_{33} + c_{44}^2 - (c_{13} + c_{44})^2]s^2 + c_{11}c_{44} = 0. \tag{2.2.43}$$

Explicitly, s_1 and s_2 are

$$s_1 = \sqrt{\frac{(\overline{c}_{13} - c_{13})(\overline{c}_{13} + c_{13} + 2c_{44})}{4c_{33}c_{44}}} + \sqrt{\frac{(\overline{c}_{13} + c_{13})(\overline{c}_{13} - c_{13} - 2c_{44})}{4c_{33}c_{44}}}, \tag{2.2.44}$$

$$s_2 = \sqrt{\frac{(\overline{c}_{13} - c_{13})(\overline{c}_{13} + c_{13} + 2c_{44})}{4c_{33}c_{44}}} - \sqrt{\frac{(\overline{c}_{13} + c_{13})(\overline{c}_{13} - c_{13} - 2c_{44})}{4c_{33}c_{44}}}, \tag{2.2.45}$$

where

$$\overline{c}_{13} = \sqrt{c_{11}c_{33}}\,. \tag{2.2.46}$$

The elastic matrix is positive definite, so that for a transversely isotropic material, we have

$$\begin{gathered} c_{11} > 0,\ c_{33} > 0,\ c_{44} > 0,\ 2c_{66} = c_{11} - c_{12} > 0, \\ c_{11} + c_{12} > 0,\ (c_{11} + c_{12})c_{33} - 2c_{13}^2 > 0. \end{gathered} \tag{2.2.47}$$

Since $c_{11} > c_{12}$, we have $2c_{11}c_{33} > (c_{11} + c_{12})c_{33} > 2c_{13}^2$, *i.e.*, $\overline{c}_{13} > c_{13}$. According to Eqs. (2.2.44) and (2.2.45), hence, we have (i) $s_1 = s_2 > 0$ when $\overline{c}_{13} - c_{13} - 2c_{44} = 0$; (ii) s_1 and s_2 are conjugate complex numbers with $\mathrm{Re}[s_i] > 0$ $(i = 1, 2)$, when

$\bar{c}_{13} - c_{13} - 2c_{44} < 0$; and (iii) s_1 and s_2 are real and greater than zero when $\bar{c}_{13} - c_{13} - 2c_{44} > 0$.

The problem now is to find φ and F in Eq. (2.2.38) with the minimum effort and mathematical difficulties. The extended *Almansi's theorem* by Ding, Chen and Liang (1996)[3] presented below is particularly useful.

[*Theorem*] Let R be a domain in space (x, y, z) such that any line parallel to the z-axis intersects the boundary of R at no more than two points. If $F_n(x, y, z)$ satisfies

$$\prod_{i=1}^{n} \nabla_i^2 F_n = \nabla_1^2 \nabla_2^2 \cdots \nabla_{n-1}^2 \nabla_n^2 F_n = 0, \quad \text{in } R, \tag{2.2.48}$$

where

$$\nabla_i^2 = \Lambda + \frac{1}{s_i^2} \frac{\partial^2}{\partial z^2}, \tag{2.2.49}$$

then F_n has the form of

$$F_n(x,y,z) = F_{n-m-1}(x,y,z) + F^{(n)}(x,y,z) \sum_{i=0}^{m} z^m, \tag{2.2.50}$$

where F_{n-m-1} and $F^{(n)}$ satisfy the following conditions, respectively,

$$\prod_{i=1}^{n-m-1} \nabla_i^2 F_{n-m-1} = 0, \nabla_n^2 F^{(n)} = 0, \tag{2.2.51}$$

and m $(0 \le m \le n-1)$ is the total number of coefficients s_i^2 $(i = 1, 2, \cdots, n-1)$ that are equal to s_n^2.

---

[3] Almansi (1899) proved the theorem that bears his name: The solution of a differential equation with multiple three-dimensional Laplace operators can be represented by a linear combination of several three-dimensional harmonic functions. Eubanks and Sternberg (1953) extended Almansi's theorem to axisymmetric problems.

According to this theorem, the solution of Eq. (2.2.42) can be written as

$$F = F_1 + F_2, \qquad \text{when } s_1 \neq s_2, \tag{2.2.52}$$

$$F = F_1 + zF_2, \qquad \text{when } s_1 = s_2, \tag{2.2.53}$$

where F_1 and F_2 satisfy the following equations:

$$\left(\Lambda + \frac{\partial^2}{\partial z_i^2}\right) F_i = 0, \quad (i = 1,2). \tag{2.2.54}$$

Now the general solution of Eq. (2.2.38) can be simplified under different conditions.

(*a*) When $s_1 \neq s_2$, by substituting Eq. (2.2.52) into Eq. (2.2.38), using Eq. (2.2.54) and letting

$$\varphi_i = -\frac{\partial F_i}{\partial z}, \quad (i = 1, 2), \qquad \varphi = -\varphi_3, \tag{2.2.55}$$

we get

$$u = \sum_{i=1}^{2} \frac{\partial \varphi_i}{\partial x} + \frac{\partial \varphi_3}{\partial y}, \qquad v = \sum_{i=1}^{2} \frac{\partial \varphi_i}{\partial y} - \frac{\partial \varphi_3}{\partial x}, \qquad w = \sum_{i=1}^{2} \alpha_i \frac{\partial \varphi_i}{\partial z_i}, \tag{2.2.56}$$

where

$$\alpha_i = \frac{a - bs_i^2}{s_i} = \frac{c_{11} - c_{44} s_i^2}{(c_{13} + c_{44}) s_i}, \quad (i = 1, 2), \tag{2.2.57}$$

and according to Eqs. (2.2.39), (2.2.54) and (2.2.55), φ_i must satisfy

$$\left(\Lambda + \frac{\partial^2}{\partial z_i^2}\right) \varphi_i = 0, \quad (i = 1, 2, 3), \tag{2.2.58}$$

in which $z_i = s_i z$ $(i = 1, 2, 3)$, s_1 and s_2 are calculated by Eqs. (2.2.44) and (2.2.45), and

$$s_3 = \sqrt{c_{66} / c_{44}} \,. \tag{2.2.59}$$

The general solution (2.2.56) is consistent with that obtained by Hu (1954*a*) and Lodge (1955). If we further let $\varphi_3 = 0$, the solution reduces to *Elliott solution* (Elliott, 1948).

(*b*) When $s_1 = s_2$, using the F of Eq. (2.2.53) to replace that in Eq. (2.2.38) and then using Eq. (2.2.54) and letting

$$\varphi_1 = -\frac{\partial F_1}{\partial z_1} - F_2 \,, \quad \varphi_2 = -\frac{\partial F_2}{\partial z_2} \,, \quad \varphi = -\varphi_3 \,, \tag{2.2.60}$$

we can get the following general solution:

$$\begin{aligned} u &= \frac{\partial \varphi_1}{\partial x} + z_1 \frac{\partial \varphi_2}{\partial x} + \frac{\partial \varphi_3}{\partial y} \,, \\ v &= \frac{\partial \varphi_1}{\partial y} + z_1 \frac{\partial \varphi_2}{\partial y} - \frac{\partial \varphi_3}{\partial x} \,, \\ w &= \alpha_1 \left(\frac{\partial \varphi_1}{\partial z_1} + z_1 \frac{\partial \varphi_2}{\partial z_1} \right) - \alpha_3 \varphi_2 \,, \end{aligned} \tag{2.2.61}$$

where φ_i $(i = 1, 2, 3)$ are still the functions that satisfy Eq. (2.2.58) and

$$\alpha_3 = (a + b s_1^2) / s_1 \,. \tag{2.2.62}$$

With these displacement solutions and the constitutive equations (1.2.33), stress components in terms of φ_i can be obtained easily as follows.

(*a*) If $s_1 \neq s_2$, the displacements in Eq. (2.2.56) lead to

$$\begin{aligned} \sigma_x / c_{66} &= \sum_{i=1}^{2} k_{1i} \frac{\partial^2 \varphi_i}{\partial z_i^2} + 2 \left(\sum_{i=1}^{2} \frac{\partial^2 \varphi_i}{\partial x^2} + \frac{\partial^2 \varphi_3}{\partial x \partial y} \right), \\ \sigma_y / c_{66} &= \sum_{i=1}^{2} k_{1i} \frac{\partial^2 \varphi_i}{\partial z_i^2} + 2 \left(\sum_{i=1}^{2} \frac{\partial^2 \varphi_i}{\partial y^2} - \frac{\partial^2 \varphi_3}{\partial x \partial y} \right), \end{aligned}$$

$$\begin{aligned}
\sigma_z / c_{66} &= \sum_{i=1}^{2} k_{2i} \frac{\partial^2 \varphi_i}{\partial z_i^2}, \\
\tau_{yz} / c_{66} &= \sum_{i=1}^{2} k_{3i} \frac{\partial^2 \varphi_i}{\partial y \partial z_i} - \frac{1}{s_3} \frac{\partial^2 \varphi_3}{\partial x \partial z_3}, \\
\tau_{zx} / c_{66} &= \sum_{i=1}^{2} k_{3i} \frac{\partial^2 \varphi_i}{\partial x \partial z_i} + \frac{1}{s_3} \frac{\partial^2 \varphi_3}{\partial y \partial z_3}, \\
\tau_{xy} / c_{66} &= 2\sum_{i=1}^{2} \frac{\partial^2 \varphi_i}{\partial x \partial y} - \frac{\partial^2 \varphi_3}{\partial x^2} + \frac{\partial^2 \varphi_3}{\partial y^2},
\end{aligned} \tag{2.2.63}$$

where

$$\begin{aligned}
&k_{1i} = (c_{13}\alpha_i s_i - c_{12}) / c_{66}, \qquad k_{2i} = (c_{33}\alpha_i s_i - c_{13}) / c_{66}, \\
&k_{3i} = (\alpha_i + s_i) / s_3^2, \qquad (i = 1, 2).
\end{aligned} \tag{2.2.64}$$

The following identities can be verified

$$\begin{aligned}
&\alpha_1 \alpha_2 = s_1 s_2, \quad \alpha_1 k_{32} = s_2 k_{31}, \quad \alpha_2 k_{31} = s_1 k_{32}, \\
&k_{3i} = s_i k_{2i}, \quad k_{1i} + s_i k_{2i} = 2, \quad (i = 1, 2).
\end{aligned} \tag{2.2.65}$$

(*b*) If $s_1 = s_2$, the displacements in Eq. (2.2.61) give rise to

$$\begin{aligned}
\sigma_x / c_{66} &= k_{11}\left(\frac{\partial^2 \varphi_1}{\partial z_1^2} + z_1 \frac{\partial^2 \varphi_2}{\partial z_1^2}\right) + k_4 \frac{\partial \varphi_2}{\partial z_1} + 2\left(\frac{\partial^2 \varphi_1}{\partial x^2} + z_1 \frac{\partial^2 \varphi_2}{\partial x^2} + \frac{\partial^2 \varphi_3}{\partial x \partial y}\right), \\
\sigma_y / c_{66} &= k_{11}\left(\frac{\partial^2 \varphi_1}{\partial z_1^2} + z_1 \frac{\partial^2 \varphi_2}{\partial z_1^2}\right) + k_4 \frac{\partial \varphi_2}{\partial z_1} + 2\left(\frac{\partial^2 \varphi_1}{\partial y^2} + z_1 \frac{\partial^2 \varphi_2}{\partial y^2} - \frac{\partial^2 \varphi_3}{\partial x \partial y}\right), \\
\sigma_z / c_{66} &= k_{21}\left(\frac{\partial^2 \varphi_1}{\partial z_1^2} + z_1 \frac{\partial^2 \varphi_2}{\partial z_1^2}\right) + k_5 \frac{\partial \varphi_2}{\partial z_1}, \\
\tau_{yz} / c_{66} &= k_{31}\left(\frac{\partial^2 \varphi_1}{\partial y \partial z_1} + z_1 \frac{\partial^2 \varphi_2}{\partial y \partial z_1}\right) - \frac{1}{s_3} \frac{\partial^2 \varphi_3}{\partial x \partial z_3} + k_6 \frac{\partial \varphi_2}{\partial y}, \\
\tau_{zx} / c_{66} &= k_{31}\left(\frac{\partial^2 \varphi_1}{\partial x \partial z_1} + z_1 \frac{\partial^2 \varphi_2}{\partial x \partial z_1}\right) + \frac{1}{s_3} \frac{\partial^2 \varphi_3}{\partial y \partial z_3} + k_6 \frac{\partial \varphi_2}{\partial x}, \\
\tau_{xy} / c_{66} &= 2\left(\frac{\partial^2 \varphi_1}{\partial x \partial y} + z_1 \frac{\partial^2 \varphi_2}{\partial x \partial y}\right) - \frac{\partial^2 \varphi_3}{\partial x^2} + \frac{\partial^2 \varphi_3}{\partial y^2},
\end{aligned} \tag{2.2.66}$$

where

$$k_4 = c_{13}s_1(\alpha_1 - \alpha_3)/c_{66}\,,\ k_5 = c_{33}s_1(\alpha_1 - \alpha_3)/c_{66},\ k_6 = (s_1 - \alpha_3)/s_3^2. \tag{2.2.67}$$

These expressions can be written in a more compact form if we introduce a complex operator, *i.e.*,

$$\Delta = \frac{\partial}{\partial x} + \mathrm{i}\frac{\partial}{\partial y}, \tag{2.2.68}$$

where $\mathrm{i} = \sqrt{-1}$. The displacements and stresses in Eqs. (2.2.56), (2.2.61), (2.2.63) and (2.2.66) can then be rewritten as

(*a*) when $s_1 \neq s_2$,

$$u + \mathrm{i}v = \Delta(\varphi_1 + \varphi_2 - \mathrm{i}\varphi_3)\ ,\quad w = \alpha_1\frac{\partial \varphi_1}{\partial z_1} + \alpha_2\frac{\partial \varphi_2}{\partial z_2}, \tag{2.2.56a}$$

$$\begin{aligned}
&\sigma_x + \sigma_y = 2c_{66}\left[(k_{11} - 1)\frac{\partial^2 \varphi_1}{\partial z_1^2} + (k_{12} - 1)\frac{\partial^2 \varphi_2}{\partial z_2^2}\right],\\
&\sigma_x - \sigma_y + 2\mathrm{i}\tau_{xy} = 2c_{66}\Delta^2(\varphi_1 + \varphi_2 - \mathrm{i}\varphi_3),\\
&\tau_{zx} + \mathrm{i}\tau_{yz} = c_{66}\Delta\left(k_{31}\frac{\partial \varphi_1}{\partial z_1} + k_{32}\frac{\partial \varphi_2}{\partial z_2} - \frac{\mathrm{i}}{s_3}\frac{\partial \varphi_3}{\partial z_3}\right),\\
&\sigma_z = c_{66}\left(k_{21}\frac{\partial^2 \varphi_1}{\partial z_1^2} + k_{22}\frac{\partial^2 \varphi_2}{\partial z_2^2}\right),
\end{aligned} \tag{2.2.63a}$$

(*b*) when $s_1 = s_2$,

$$u + \mathrm{i}v = \Delta(\varphi_1 + z_1\varphi_2 - \mathrm{i}\varphi_3), w = \alpha_1\left(\frac{\partial \varphi_1}{\partial z_1} + z_1\frac{\partial \varphi_2}{\partial z_1}\right) - \alpha_3\varphi_2, \tag{2.2.61a}$$

$$\begin{aligned}
&\sigma_x + \sigma_y = 2c_{66}(k_{11} - 1)\left(\frac{\partial^2 \varphi_1}{\partial z_1^2} + z_1\frac{\partial^2 \varphi_2}{\partial z_1^2}\right) + 2k_4\frac{\partial \varphi_2}{\partial z_1},\\
&\sigma_x - \sigma_y + 2\mathrm{i}\tau_{xy} = 2c_{66}(\varphi_1 + z_1\varphi_2 - \mathrm{i}\varphi_3),\\
&\tau_{zx} + \mathrm{i}\tau_{yz} = c_{66}\Delta\left[k_{31}\left(\frac{\partial \varphi_1}{\partial z_1} + z_1\frac{\partial \varphi_2}{\partial z_1}\right) - \frac{\mathrm{i}}{s_3}\frac{\partial \varphi_3}{\partial z_3} + k_6\varphi_2\right],\\
&\sigma_z = c_{66}\left[k_{21}\left(\frac{\partial^2 \varphi_1}{\partial z_1^2} + z_1\frac{\partial^2 \varphi_2}{\partial z_1^2}\right) + k_5\frac{\partial \varphi_2}{\partial z_1}\right].
\end{aligned} \tag{2.2.66a}$$

2.2.2 General Solution in Cylindrical Coordinates

The solution in Cartesian coordinates can easily be transformed to other coordinate systems. In cylindrical coordinates (r, α, z), for instance, we have

(a) when $s_1 \neq s_2$,

$$
\begin{aligned}
u_r &= \sum_{i=1}^{2} \frac{\partial \varphi_i}{\partial r} + \frac{1}{r}\frac{\partial \varphi_3}{\partial \alpha}, \\
u_\alpha &= \sum_{i=1}^{2} \frac{1}{r}\frac{\partial \varphi_i}{\partial \alpha} - \frac{\partial \varphi_3}{\partial r}, \\
w &= \sum_{i=1}^{2} \alpha_i \frac{\partial \varphi_i}{\partial z_i},
\end{aligned}
\tag{2.2.69}
$$

$$
\begin{aligned}
\sigma_r / c_{66} &= \sum_{i=1}^{2} k_{1i} \frac{\partial^2 \varphi_i}{\partial z_i^2} + 2\left[\sum_{i=1}^{2} \frac{\partial^2 \varphi_i}{\partial r^2} + \frac{\partial}{\partial r}\left(\frac{1}{r}\frac{\partial \varphi_3}{\partial \alpha}\right)\right], \\
\sigma_\alpha / c_{66} &= \sum_{i=1}^{2} k_{1i} \frac{\partial^2 \varphi_i}{\partial z_i^2} + 2\left[\sum_{i=1}^{2}\left(\frac{1}{r}\frac{\partial \varphi_i}{\partial r} + \frac{1}{r^2}\frac{\partial^2 \varphi_i}{\partial \alpha^2}\right) - \frac{\partial}{\partial r}\left(\frac{1}{r}\frac{\partial \varphi_3}{\partial \alpha}\right)\right], \\
\sigma_z / c_{66} &= \sum_{i=1}^{2} k_{2i} \frac{\partial^2 \varphi_i}{\partial z_i^2}, \\
\tau_{\alpha z} / c_{66} &= \sum_{i=1}^{2} k_{3i} \frac{1}{r}\frac{\partial^2 \varphi_i}{\partial \alpha \partial z_i} - \frac{1}{s_3}\frac{\partial^2 \varphi_3}{\partial r \partial z_3}, \\
\tau_{zr} / c_{66} &= \sum_{i=1}^{2} k_{3i} \frac{\partial^2 \varphi_i}{\partial r \partial z_i} + \frac{1}{s_3}\frac{1}{r}\frac{\partial^2 \varphi_3}{\partial \alpha \partial z_3}, \\
\tau_{r\alpha} / c_{66} &= 2\sum_{i=1}^{2} \frac{\partial}{\partial r}\left(\frac{1}{r}\frac{\partial \varphi_i}{\partial \alpha}\right) + \frac{2}{r}\frac{\partial \varphi_3}{\partial r} + \frac{2}{r^2}\frac{\partial^2 \varphi_3}{\partial \alpha^2} + \frac{\partial^2 \varphi_3}{\partial z_3^2},
\end{aligned}
\tag{2.2.70}
$$

(b) when $s_1 = s_2$,

$$
\begin{aligned}
u_r &= \frac{\partial \varphi_1}{\partial r} + z_1 \frac{\partial \varphi_2}{\partial r} + \frac{1}{r}\frac{\partial \varphi_3}{\partial \alpha}, \\
u_\alpha &= \frac{1}{r}\frac{\partial \varphi_1}{\partial \alpha} + z_1 \frac{1}{r}\frac{\partial \varphi_2}{\partial \alpha} - \frac{\partial \varphi_3}{\partial r}, \\
w &= \alpha_1\left(\frac{\partial \varphi_1}{\partial z_1} + z_1 \frac{\partial \varphi_2}{\partial z_1}\right) - \alpha_3 \varphi_2,
\end{aligned}
\tag{2.2.71}
$$

$$\begin{aligned}
\sigma_r / c_{66} &= k_{11}\left(\frac{\partial^2\varphi_1}{\partial z_1^2}+z_1\frac{\partial^2\varphi_2}{\partial z_1^2}\right)+k_4\frac{\partial\varphi_2}{\partial z_1} \\
&\quad +2\left[\frac{\partial^2\varphi_1}{\partial r^2}+z_1\frac{\partial^2\varphi_2}{\partial r^2}+\frac{\partial}{\partial r}\frac{1}{r}\left(\frac{\partial\varphi_3}{\partial\alpha}\right)\right], \\
\sigma_\alpha / c_{66} &= k_{11}\left(\frac{\partial^2\varphi_1}{\partial z_1^2}+z_1\frac{\partial^2\varphi_2}{\partial z_1^2}\right)+k_4\frac{\partial\varphi_2}{\partial z_1} \\
&\quad +2\left[\frac{1}{r}\frac{\partial\varphi_1}{\partial r}+\frac{1}{r^2}\frac{\partial^2\varphi_1}{\partial\alpha^2}+z_1\left(\frac{1}{r}\frac{\partial\varphi_2}{\partial r}+\frac{1}{r^2}\frac{\partial^2\varphi_2}{\partial\alpha^2}\right)-\frac{\partial}{\partial r}\left(\frac{1}{r}\frac{\partial\varphi_3}{\partial\alpha}\right)\right], \\
\sigma_z / c_{66} &= k_{21}\left(\frac{\partial^2\varphi_1}{\partial z_1^2}+z_1\frac{\partial^2\varphi_2}{\partial z_1^2}\right)+k_5\frac{\partial\varphi_2}{\partial z_1}, \\
\tau_{\alpha z} / c_{66} &= k_{31}\left(\frac{1}{r}\frac{\partial^2\varphi_1}{\partial\alpha\partial z_1}+z_1\frac{1}{r}\frac{\partial^2\varphi_2}{\partial\alpha\partial z_1}\right)+k_6\frac{1}{r}\frac{\partial\varphi_2}{\partial\alpha}-\frac{1}{s_3}\frac{\partial^2\varphi_3}{\partial r\partial z_3}, \\
\tau_{zr} / c_{66} &= k_{31}\left(\frac{\partial^2\varphi_1}{\partial r\partial z_1}+z_1\frac{\partial^2\varphi_2}{\partial r\partial z_1}\right)+k_6\frac{\partial\varphi_2}{\partial r}+\frac{1}{s_3}\frac{1}{r}\frac{\partial^2\varphi_3}{\partial\alpha\partial z_3}, \\
\tau_{r\alpha} / c_{66} &= 2\frac{\partial}{\partial r}\left(\frac{1}{r}\frac{\partial\varphi_1}{\partial\alpha}+z_1\frac{1}{r}\frac{\partial\varphi_2}{\partial\alpha}\right)+2\left(\frac{1}{r}\frac{\partial\varphi_3}{\partial r}+\frac{1}{r^2}\frac{\partial^2\varphi_3}{\partial\alpha^2}\right)+\frac{\partial^2\varphi_3}{\partial z_3^2}.
\end{aligned} \tag{2.2.72}$$

It is worth mentioning that in cylindrical coordinates,

$$\Lambda = \frac{\partial^2}{\partial r^2}+\frac{1}{r}\frac{\partial}{\partial r}+\frac{1}{r^2}\frac{\partial^2}{\partial\alpha^2}. \tag{2.2.73}$$

This general solution can be further simplified in the following special deformation states.

(1) An axisymmetric body subjected to a pure torsion

In this case, let $\varphi_1 = \varphi_2 = 0$ and $\varphi_3 = \varphi_3(r, z_3)$. Equation (2.2.69) becomes the same as Eq. (2.2.71) and Eq. (2.2.70) becomes identical to Eq. (2.2.72), *i.e.*,

$$u_r = w = 0,\ u_\alpha = -\frac{\partial\varphi_3}{\partial r}, \qquad (2.2.69a), (2.2.71a)$$

$$\sigma_r = \sigma_\alpha = \sigma_z = \tau_{zr} = 0,$$
$$\tau_{\alpha z}/c_{66} = -\frac{1}{s_3}\frac{\partial^2 \varphi_3}{\partial r \partial z_3},$$
$$\tau_{r\alpha}/c_{66} = \frac{2}{r}\frac{\partial \varphi_3}{\partial r} + \frac{\partial^2 \varphi_3}{\partial z_3^2}. \qquad (2.2.70a), (2.2.72a)$$

(2) A body subjected to an axisymmetric deformation

Let $\varphi_3 = 0$, $\varphi_1 = \varphi_1(r, z_1)$, and $\varphi_2 = \varphi_2(r, z_2)$. When $s_1 \neq s_2$, Eqs. (2.2.69) and (2.2.70) reduce to

$$u_r = \sum_{i=1}^{2}\frac{\partial \varphi_i}{\partial r}, \quad w = \sum_{i=1}^{2}\alpha_i \frac{\partial \varphi_i}{\partial z_i}, \quad u_\alpha = 0, \qquad (2.2.69b)$$

$$\sigma_r / c_{66} = \sum_{i=1}^{2} k_{1i}\frac{\partial^2 \varphi_i}{\partial z_i^2} + 2\sum_{i=1}^{2}\frac{\partial^2 \varphi_i}{\partial r^2},$$
$$\sigma_\alpha / c_{66} = \sum_{i=1}^{2} k_{1i}\frac{\partial^2 \varphi_i}{\partial z_i^2} + 2\sum_{i=1}^{2}\frac{1}{r}\frac{\partial \varphi_i}{\partial r},$$
$$\sigma_z / c_{66} = \sum_{i=1}^{2} k_{2i}\frac{\partial^2 \varphi_i}{\partial z_i^2}, \qquad (2.2.70b)$$
$$\tau_{zr} / c_{66} = \sum_{i=1}^{2} k_{3i}\frac{\partial^2 \varphi_i}{\partial r \partial z_i}.$$

While when $s_1 = s_2$, Eqs. (2.2.71) and (2.2.72) give rise to

$$u_r = \frac{\partial \varphi_1}{\partial r} + z_1 \frac{\partial \varphi_2}{\partial r}, \quad w = \alpha_1\left(\frac{\partial \varphi_1}{\partial z_1} + z_1\frac{\partial \varphi_2}{\partial z_1}\right) - \alpha_3 \varphi_2, \quad u_\alpha = 0, \qquad (2.2.71b)$$

$$\sigma_r / c_{66} = k_{11}\left(\frac{\partial^2 \varphi_1}{\partial z_1^2} + z_1\frac{\partial^2 \varphi_2}{\partial z_1^2}\right) + k_4\frac{\partial \varphi_2}{\partial z_1} + 2\left(\frac{\partial^2 \varphi_1}{\partial r^2} + z_1\frac{\partial^2 \varphi_2}{\partial r^2}\right),$$
$$\sigma_\alpha / c_{66} = k_{11}\left(\frac{\partial^2 \varphi_1}{\partial z_1^2} + z_1\frac{\partial^2 \varphi_2}{\partial z_1^2}\right) + k_4\frac{\partial \varphi_2}{\partial z_1} + 2\left(\frac{1}{r}\frac{\partial \varphi_1}{\partial r} + z_1\frac{1}{r}\frac{\partial \varphi_2}{\partial r}\right), \qquad (2.2.72b)$$

$$\sigma_z / c_{66} = k_{21}\left(\frac{\partial^2 \varphi_1}{\partial z_1^2} + z_1 \frac{\partial^2 \varphi_2}{\partial z_1^2}\right) + k_5 \frac{\partial \varphi_2}{\partial z_1},$$

$$\tau_{zr} / c_{66} = k_{31}\left(\frac{\partial^2 \varphi_1}{\partial r \partial z_1} + z_1 \frac{\partial^2 \varphi_2}{\partial r \partial z_1}\right) + k_6 \frac{\partial \varphi_2}{\partial r}.$$

In these equations, φ_i still satisfy Eq. (2.2.58) except that

$$\Lambda = \frac{\partial^2}{\partial r^2} + \frac{1}{r}\frac{\partial}{\partial r}, \tag{2.2.74}$$

because the body is under axisymmetric deformation.

In anisotropic elasticity, stress method is usually adopted to solve some relatively simpler problems. A summarization of the earlier works in this respect can be found in the monograph of Lekhnitskii (1981). Based on the stress method, general solutions of axisymmetric problems of transversely isotropic media were derived in various forms by Ding (1987), Ding and Xu (1988), Wang and Wang (1989).

2.3 STRESS METHOD FOR AXISYMMETRIC PROBLEMS

When considering the deformation of a body of revolution, we can use a cylindrical coordinate system. The equilibrium equations for an axisymmetric problem in absence of body forces, according to Eq. (1.2.11), can be written as

$$\frac{\partial \sigma_r}{\partial r} + \frac{\partial \tau_{zr}}{\partial z} + \frac{\sigma_r - \sigma_\alpha}{r} = 0, \quad \frac{\partial \tau_{zr}}{\partial r} + \frac{\partial \sigma_z}{\partial z} + \frac{\tau_{zr}}{r} = 0. \tag{2.3.1}$$

If we introduce an arbitrary function G that obeys

$$\frac{\sigma_r - \sigma_\alpha}{r} = \frac{\partial G}{\partial r}, \tag{2.3.2}$$

we can rewrite these equilibrium equations as

$$\frac{\partial(\sigma_r + G)}{\partial r} + \frac{\partial \tau_{zr}}{\partial z} = 0, \quad \frac{1}{r}\frac{\partial}{\partial r}(r\tau_{zr}) + \frac{\partial \sigma_z}{\partial z} = 0. \tag{2.3.3}$$

Hence

$$\sigma_r + G = \frac{\partial P}{\partial z}, \quad \tau_{zr} = -\frac{\partial P}{\partial r},$$
$$r\tau_{zr} = -\frac{\partial Q}{\partial z}, \quad r\sigma_z = \frac{\partial Q}{\partial r}, \tag{2.3.4}$$

where P and Q are arbitrary functions. The second and third of Eq. (2.3.4) indicate that

$$\tau_{zr} = -\frac{\partial P}{\partial r} = -\frac{\partial}{\partial z}\left(\frac{Q}{r}\right).$$

Thus

$$P = \frac{\partial F}{\partial z}, \quad \frac{Q}{r} = \frac{\partial F}{\partial r}. \tag{2.3.5}$$

By substituting Eq. (2.3.5) into Eq. (2.3.4) and using Eq. (2.3.2), we get

$$\tau_{zr} = -\frac{\partial^2 F}{\partial r \partial z}, \quad \sigma_r = \frac{\partial^2 F}{\partial z^2} - G,$$
$$\sigma_z = \frac{\partial^2 F}{\partial r^2} + \frac{1}{r}\frac{\partial F}{\partial r}, \quad \sigma_\alpha = \frac{\partial^2 F}{\partial z^2} - \frac{\partial}{\partial r}(rG), \tag{2.3.6}$$

where F and G are arbitrary functions. Furthermore, if we introduce another arbitrary function, ψ, and let $rG = -\partial\psi/\partial r$, we can obtain a concise set of equations that are more convenient to use, *i.e.*,

$$\tau_{zr} = -\frac{\partial^2 F}{\partial r \partial z}, \quad \sigma_r = \frac{\partial^2 F}{\partial z^2} + \frac{1}{r}\frac{\partial \psi}{\partial r}, \quad \sigma_z = \frac{\partial^2 F}{\partial r^2} + \frac{1}{r}\frac{\partial F}{\partial r}, \quad \sigma_\alpha = \frac{\partial^2 F}{\partial z^2} + \frac{\partial^2 \psi}{\partial r^2}, \tag{2.3.7}$$

where F and ψ are called *stress functions*.

Using Eq. (1.2.2), we can obtain the strain-displacement relations as

$$\varepsilon_r = \frac{\partial u_r}{\partial r}, \quad \varepsilon_\alpha = \frac{u_r}{r}, \quad \varepsilon_z = \frac{\partial w}{\partial z}, \quad \gamma_{zr} = \frac{\partial w}{\partial r} + \frac{\partial u_r}{\partial z}. \tag{2.3.8}$$

Because of this, the compatibility equations of strains are obtained as already shown in Eq. (1.2.9), *i.e.*

$$\varepsilon_r - \varepsilon_\alpha = r\frac{\partial \varepsilon_\alpha}{\partial r}, \quad \frac{\partial \gamma_{zr}}{\partial z} - \frac{\partial \varepsilon_z}{\partial r} = r\frac{\partial^2 \varepsilon_\alpha}{\partial z^2}. \tag{2.3.9}$$

On the other hand, the generalized Hooke's law, Eq. (1.2.38), reduces to

$$\begin{aligned}
&\varepsilon_r = s_{11}\sigma_r + s_{12}\sigma_\alpha + s_{13}\sigma_z, \\
&\varepsilon_\alpha = s_{12}\sigma_r + s_{11}\sigma_\alpha + s_{13}\sigma_z, \\
&\varepsilon_z = s_{13}(\sigma_r + \sigma_\alpha) + s_{33}\sigma_z, \\
&\gamma_{zr} = s_{44}\tau_{zr}.
\end{aligned} \tag{2.3.10}$$

Therefore the compatibility equations in terms of stress components can be obtained by substituting Eq. (2.3.10) into Eq. (2.3.9), *i.e.*,

$$\begin{aligned}
&(s_{11} - s_{12})(\sigma_r - \sigma_\alpha) = r\frac{\partial}{\partial r}(s_{12}\sigma_r + s_{11}\sigma_\alpha + s_{13}\sigma_z), \\
&s_{44}\frac{\partial \tau_{zr}}{\partial z} - \frac{\partial}{\partial r}[s_{13}(\sigma_r + \sigma_\alpha) + s_{33}\sigma_z] = r\frac{\partial^2}{\partial z^2}(s_{12}\sigma_r + s_{11}\sigma_\alpha + s_{13}\sigma_z).
\end{aligned} \tag{2.3.11}$$

Equation (2.3.11) describes the conditions that the two stress functions, F and ψ, must obey. The substitution of Eq. (2.3.7) into Eq. (2.3.11) yields

$$\begin{aligned}
&(s_{12} - s_{11})r\frac{\partial}{\partial r}\left(\frac{1}{r}\frac{\partial \psi}{\partial r}\right) = \\
&\quad r\frac{\partial}{\partial r}\left[s_{12}\left(\frac{\partial^2 F}{\partial z^2} + \frac{1}{r}\frac{\partial \psi}{\partial r}\right) + s_{11}\left(\frac{\partial^2 F}{\partial z^2} + \frac{\partial^2 \psi}{\partial r^2}\right) + s_{13}\Lambda F\right],
\end{aligned} \tag{2.3.12}$$

$$\begin{aligned}
&-s_{44}\frac{\partial^3 F}{\partial r \partial z^2} - \frac{\partial}{\partial r}\left(s_{13}\Lambda\psi + 2s_{13}\frac{\partial^2 F}{\partial z^2} + s_{33}\Lambda F\right) = \\
&\quad r\frac{\partial^2}{\partial z^2}\left[s_{12}\left(\frac{\partial^2 F}{\partial z^2} + \frac{1}{r}\frac{\partial \psi}{\partial r}\right) + s_{11}\left(\frac{\partial^2 F}{\partial z^2} + \frac{\partial^2 \psi}{\partial r^2}\right) + s_{13}\Lambda F\right],
\end{aligned} \tag{2.3.13}$$

where

$$\Lambda = \frac{\partial^2}{\partial r^2} + \frac{1}{r}\frac{\partial}{\partial r}. \tag{2.3.14}$$

By integrating Eq. (2.3.12) about r, we get

$$s_{11}\Lambda\psi + (s_{11} + s_{12})\frac{\partial^2 F}{\partial z^2} + s_{13}\Lambda F = B(z), \tag{2.3.15}$$

where $B(z)$ is an arbitrary function of z. The terms containing F in the right hand side of Eq. (2.3.13) can therefore be eliminated by using Eq. (2.3.15). Hence, a further integration about r gives rise to

$$s_{13}\Lambda\psi - (s_{11} - s_{12})\frac{\partial^2\psi}{\partial z^2} + (2s_{13} + s_{44})\frac{\partial^2 F}{\partial z^2} + s_{33}\Lambda F = C(z) - \frac{1}{2}r^2 B''(z), \tag{2.3.16}$$

where a prime denotes differentiation with respect to z, and $C(z)$ is an arbitrary function of z. Making use of Eq. (2.3.15), we can cancel $\Lambda\psi$ in Eq. (2.3.16). We therefore obtain

$$\begin{aligned} s_{11}(s_{11} - s_{12})\frac{\partial^2\psi}{\partial z^2} - [s_{11}s_{44} + s_{13}(s_{11} - s_{12})]\frac{\partial^2 F}{\partial z^2} - (s_{11}s_{33} - s_{13}^2)\Lambda F \\ = s_{13}B(z) + s_{11}r^2 B''(z)/2 - s_{11}C(z). \end{aligned} \tag{2.3.17}$$

On the other hand, if we let

$$\psi = \frac{(s_{11} + s_{12})H}{s_{11}} = eH, \tag{2.3.18}$$

we can rewrite the expressions of stress components as

$$\tau_{zr} = -\frac{\partial^2 F}{\partial r \partial z}, \quad \sigma_r = \frac{\partial^2 F}{\partial z^2} + e\frac{\partial H}{r\partial r},$$
$$\sigma_z = \Lambda F, \qquad \sigma_\alpha = \frac{\partial^2 F}{\partial z^2} + e\frac{\partial^2 H}{\partial r^2}. \tag{2.3.19}$$

Similarly, Eqs. (2.3.15) and (2.3.17) can also be expressed as

$$d\Lambda H + a\Lambda F + d\frac{\partial^2 F}{\partial z^2} = dP(z), \tag{2.3.20}$$

$$d\frac{\partial^2 H}{\partial z^2} - \Lambda F - c\frac{\partial^2 F}{\partial z^2} = fP(z) + gr^2 P''(z) - dQ(z), \tag{2.3.21}$$

where

$$\begin{aligned}
a &= s_{13}(s_{11} - s_{12})/(s_{11}s_{33} - s_{13}^2),\\
c &= [s_{11}s_{44} + s_{13}(s_{11} - s_{12})]/(s_{11}s_{33} - s_{13}^2),\\
d &= (s_{11}^2 - s_{12}^2)/(s_{11}s_{33} - s_{13}^2),\\
f &= s_{13}(s_{11} + s_{12})/(s_{11}s_{33} - s_{13}^2),\\
g &= s_{11}(s_{11} + s_{12})/[2(s_{11}s_{33} - s_{13}^2)],
\end{aligned} \tag{2.3.22}$$

$$P(z) = B(z)/(s_{11} + s_{12}), \quad Q(z) = s_{11}C(z)/(s_{11}^2 - s_{12}^2). \tag{2.3.23}$$

Now the general solution of Eqs. (2.3.20) and (2.3.21) can be written as the sums of the general solutions of their homogeneous equations and the special solution of the non-homogeneous ones. Assume that H^* and F^* are the special solution, and have the following forms,

$$H^* = gr^2 H_2(z)/d + H_0(z), \quad F^* = F_0(z), \tag{2.3.24}$$

the governing equations for determining H_2, H_0 and F_0 can be obtained by substituting these into Eqs. (2.3.20) and (2.3.21), *i.e.*,

$$\begin{aligned}
&4gH_2(z) + dF_0''(z) = dP(z),\\
&H_2''(z) = P''(z),\\
&dH_0''(z) - cF_0''(z) = fP(z) - dQ(z).
\end{aligned} \tag{2.3.25}$$

This then leads to

$$
\begin{aligned}
&H_2(z)=P(z)+k_1z+k_0,\\
&dF_0''(z)=(d-4g)P(z)-4gk_1z-4gk_0,\\
&dH_0''(z)=fP(z)-dQ(z)+cF_0''(z),
\end{aligned}
\tag{2.3.26}
$$

where k_0 and k_1 are arbitrary constants. Hence, the stresses corresponding to H^* and F^*, according to Eqs. (2.3.24) and (2.3.19), are

$$\tau_{zr}^*=\sigma_z^*=0,\quad \sigma_r^*=\sigma_\alpha^*=m_1z+m_0, \tag{2.3.27}$$

where

$$m_1=2g(e-2)k_1/d,\quad m_0=2g(e-2)k_0/d. \tag{2.3.28}$$

Note that in the derivation, we have used the relationship $d-4g+2ge=0$ that is easy to confirm using Eq. (2.3.22).

Now consider the general solution of the homogeneous equations corresponding to Eqs. (2.3.20) and (2.3.21), *i.e.*, we are seeking the general solution of

$$d\Lambda H+a\Lambda F+d\frac{\partial^2 F}{\partial z^2}=0, \tag{2.3.29}$$

$$d\frac{\partial^2 H}{\partial z^2}-\Lambda F-c\frac{\partial^2 F}{\partial z^2}=0. \tag{2.3.30}$$

We can see that if we take

$$
\begin{aligned}
&H=\frac{r^2}{2(e-2)}(m_1z+m_0)-\frac{c}{d(e-2)}\left(\frac{1}{3}m_1z^3+m_0z^2\right),\\
&F=-\frac{1}{e-2}\left(\frac{1}{3}m_1z^3+m_0z^2\right),
\end{aligned}
\tag{2.3.31}
$$

they are indeed the solution of Eqs. (2.3.29) and (2.3.30) and bring about the following stress components, when using Eq. (2.3.19),

$$\tau_{zr}=\sigma_z=0,\quad \sigma_r=\sigma_\alpha=m_1z+m_0. \tag{2.3.32}$$

By comparing Eq. (2.3.32) with Eq. (2.3.27), we easily conclude that the stress state corresponding to the special solution of the non-homogeneous equations (2.3.20) and (2.3.21) can be obtained using the general solution of the homogeneous equations (2.3.29) and (2.3.30). In other words, from the point of view of stresses, we need only investigate the general solution of Eqs. (2.3.29) and (2.3.30) by ignoring the non-homogeneous equations (2.3.20) and (2.3.21).

We can also obtain displacements by integrating Eq. (2.3.8), with the aid of Eqs. (2.3.10), (2.3.19), (2.3.29) and (2.3.30). This yields

$$u_r = -e(s_{11} - s_{12})\frac{\partial H}{\partial r}, \quad w = \frac{\partial}{\partial z}[e(s_{11} - s_{12})H - s_{44}F]. \tag{2.3.33}$$

These results show that H and F can also be called displacement functions.

It is clear from this discussion that mathematically the solution of an axisymmetric problem means finding functions H and F satisfying Eqs. (2.3.29) and (2.3.30) and then finding displacements and stresses using Eqs. (2.3.33) and (2.3.19). Nevertheless, since we need to consider the solution of the partial differential equations, the solution process is not straightforward. The following is one of the ways to simplify the solution process further.

Multiplying Eq. (2.3.30) by a and then adding it to Eq. (2.3.29), we get

$$d\left(\Lambda + a\frac{\partial^2}{\partial z^2}\right)H + (d - ac)\frac{\partial^2 F}{\partial z^2} = 0. \tag{2.3.34}$$

Introduce the function Φ, which has the following relationships with H and F:

$$H = (d - ac)\frac{\partial^2 \Phi}{\partial z^2}, \quad F = -d\left(\Lambda + a\frac{\partial^2}{\partial z^2}\right)\Phi. \tag{2.3.35}$$

This satisfies Eq. (2.3.34) automatically. Substituting Eq. (2.3.35) into Eq. (2.3.29) or (2.3.30) gives rise to

$$\left[\Lambda\Lambda + (a + c)\Lambda\frac{\partial^2}{\partial z^2} + d\frac{\partial^4}{\partial z^4}\right]\Phi = 0, \tag{2.3.36}$$

where[4]

$$d=\frac{1}{s_1^2 s_2^2},\quad a+c=\frac{1}{s_1^2}+\frac{1}{s_2^2}. \tag{2.3.37}$$

Thus Eq. (2.3.36) can be rewritten as

$$\left(\Lambda+\frac{\partial^2}{\partial z_1^2}\right)\left(\Lambda+\frac{\partial^2}{\partial z_2^2}\right)\Phi=0. \tag{2.3.38}$$

If we substitute Eq. (2.3.35) into (2.3.19) and let

$$d\frac{\partial\Phi}{\partial z}=\varphi, \tag{2.3.39}$$

we obtain

$$\begin{aligned}\tau_{zr}&=\frac{\partial}{\partial r}\left(\Lambda+a\frac{\partial^2}{\partial z^2}\right)\varphi,\quad \sigma_r=-\frac{\partial}{\partial z}\left(\frac{\partial^2}{\partial r^2}+b\frac{\partial}{r\partial r}+a\frac{\partial^2}{\partial z^2}\right)\varphi,\\ \sigma_z&=\frac{\partial}{\partial z}\left(c\Lambda+d\frac{\partial^2}{\partial z^2}\right)\varphi,\quad \sigma_\alpha=-\frac{\partial}{\partial z}\left(b\frac{\partial^2}{\partial r^2}+\frac{\partial}{r\partial r}+a\frac{\partial^2}{\partial z^2}\right)\varphi,\end{aligned} \tag{2.3.40}$$

where φ satisfies

$$\left(\Lambda+\frac{\partial^2}{\partial z_1^2}\right)\left(\Lambda+\frac{\partial^2}{\partial z_2^2}\right)\varphi=0, \tag{2.3.41}$$

and

$$b=1-e(d-ac).$$

The completeness of this solution[5] was shown by Eubanks and Sternberg (1953) and many other solutions derived by various methods can be obtained directly. For example, Eqs. (2.3.40) and (2.3.41) are obviously equivalent to the *Lekhnitskii solution* (Lekhnitskii, 1940) and, when the material becomes isotropic, reduce to the Love solution (Love, 1927). With the extended Almansi's theorem for axisymmetric problems

[4] The expressions of d and $a+c$ could be derived from Eq. (78.14) in Lekhnitskii (1981).

[5] The general solution of axisymmetric problems and its completeness were discussed by Wang, Xu and Wang (1994).

(Eubanks and Sternberg, 1953), one can easily obtain Elliott solution described in Eqs. (2.2.69*b*) and (2.2.70*b*). Using this result, we can derive the extended *Michell solution* (Ding and Xu, 1988).

2.4 DISPLACEMENT METHOD FOR SPHERICALLY ISOTROPIC BODIES

In the following, we will introduce the general solution for spherically isotropic bodies using the displacement method.

2.4.1 General Solution

With Eq. (1.4.12), the governing equations for spherically isotropic bodies can be derived directly from Eq. (2.1.6), which is

$$\begin{aligned}&(c_{33}\nabla_3^2+2c_{13}-2c_{11}-2c_{12}+c_{44}\nabla_1^2)u_R\\&+[(c_{13}+c_{44})\nabla_2+(c_{13}-c_{11}-c_{12}-c_{44})]\left(\frac{\partial}{\partial\theta}+\cot\theta\right)u_\theta\\&+\frac{1}{\sin\theta}[(c_{13}+c_{44})\nabla_2+(c_{13}-c_{11}-c_{12}-c_{44})]\frac{\partial u_\alpha}{\partial\alpha}\\&+\beta_3R^2\frac{\partial T}{\partial R}+2(\beta_3-\beta_1)RT+R^2F_R=\rho R^2\ddot{u}_R,\end{aligned} \tag{2.4.1a}$$

$$\begin{aligned}&[(c_{13}+c_{44})\nabla_2+c_{11}+c_{12}+2c_{44}]\frac{\partial u_R}{\partial\theta}\\&+\left[(c_{44}\nabla_3^2-2c_{44}-c_{12}+c_{11}\nabla_1^2+(c_{66}-c_{11})\frac{1}{\sin^2\theta}\frac{\partial^2}{\partial\alpha^2}-c_{11}\cot^2\theta\right]u_\theta\\&+\frac{1}{\sin\theta}\frac{\partial}{\partial\alpha}\left[(c_{12}+c_{66})\frac{\partial}{\partial\theta}-(c_{11}+c_{66})\cot\theta\right]u_\alpha\\&+\beta_1R\frac{\partial T}{\partial\theta}+R^2F_\theta=\rho R^2\ddot{u}_\theta,\end{aligned} \tag{2.4.1b}$$

$$\begin{aligned}
&\frac{1}{\sin\theta}[(c_{13}+c_{44})\nabla_2+c_{11}+c_{12}+2c_{44})]\frac{\partial u_R}{\partial\alpha} \\
&+\frac{1}{\sin\theta}\left[(c_{12}+c_{66})\frac{\partial}{\partial\theta}+(c_{11}+c_{66})\cot\theta\right]\frac{\partial u_\theta}{\partial\alpha} \\
&+\left[c_{44}\nabla_3^2-2c_{44}+c_{66}+c_{66}\nabla_1^2+(c_{11}-c_{66})\frac{1}{\sin^2\theta}\frac{\partial^2}{\partial\alpha^2}-c_{66}\cot^2\theta\right]u_\alpha \\
&+\frac{\beta_1 R}{\sin\theta}\frac{\partial T}{\partial R}+R^2F_\alpha=\rho R^2\ddot{u}_\alpha,
\end{aligned} \tag{2.4.1c}$$

where

$$\begin{aligned}
&\nabla_2=R\frac{\partial}{\partial R},\quad \nabla_2^2=R\frac{\partial}{\partial R}R\frac{\partial}{\partial R},\quad \nabla_3^2=\nabla_2^2+\nabla_2, \\
&\nabla_1^2=\frac{\partial^2}{\partial\theta^2}+\cot\theta\frac{\partial}{\partial\theta}+\frac{1}{\sin^2\theta}\frac{\partial^2}{\partial\alpha^2}.
\end{aligned} \tag{2.4.2}$$

Let us introduce a set of displacement functions, w, G and ψ, which make

$$\begin{aligned}
u_R&=w, \\
u_\theta&=-\frac{1}{\sin\theta}\frac{\partial\psi}{\partial\alpha}-\frac{\partial G}{\partial\theta}, \\
u_\alpha&=\frac{\partial\psi}{\partial\theta}-\frac{1}{\sin\theta}\frac{\partial G}{\partial\alpha},
\end{aligned} \tag{2.4.3}$$

and let F_θ and F_α obey the following relationships:

$$\begin{aligned}
RF_\theta&=-\frac{1}{\sin\theta}\frac{\partial V}{\partial\alpha}-\frac{\partial U}{\partial\theta}, \\
RF_\alpha&=\frac{\partial V}{\partial\theta}-\frac{1}{\sin\theta}\frac{\partial U}{\partial\alpha}.
\end{aligned} \tag{2.4.4}$$

When the body forces are derived from potential, we have $V=0, F_R=-\partial U/\partial R$. Based on the above, *i.e.*, by substituting Eqs. (2.4.3) and (2.4.4) into Eq. (2.4.1), we can get

$$L_3w-L_4\nabla_1^2G+R^2F_R+2(\beta_3-\beta_1)RT+\beta_3R^2\frac{\partial T}{\partial R}-R^2\rho\frac{\partial^2 w}{\partial t^2}=0, \tag{2.4.5}$$

$$\frac{\partial}{\partial\theta}\left(A-RU+R\beta_1T+R^2\rho\frac{\partial^2G}{\partial t^2}\right)-\frac{1}{\sin\theta}\frac{\partial}{\partial\alpha}\left(B+RV-R^2\rho\frac{\partial^2\psi}{\partial t^2}\right)=0, \tag{2.4.6}$$

$$\frac{1}{\sin\theta}\frac{\partial}{\partial\alpha}\left(A-RU+R\beta_1T+R^2\rho\frac{\partial^2G}{\partial t^2}\right)+\frac{\partial}{\partial\theta}\left(B+RV-R^2\rho\frac{\partial^2\psi}{\partial t^2}\right)=0, \tag{2.4.7}$$

where

$$\begin{aligned}
&B=[c_{44}\nabla_3^2+2(c_{66}-c_{44})+c_{66}\nabla_1^2]\psi,\\
&A=L_1w-L_2G,\\
&L_1=(c_{13}+c_{44})\nabla_2+2(c_{11}+c_{44}-c_{66}),\\
&L_2=c_{44}\nabla_3^2+2(c_{66}-c_{44})+c_{11}\nabla_1^2,\\
&L_3=c_{33}\nabla_3^2+2(c_{13}-c_{11}-c_{12})+c_{44}\nabla_1^2,\\
&L_4=(c_{13}+c_{44})\nabla_2+c_{13}-c_{11}-c_{12}-c_{44}.
\end{aligned} \tag{2.4.8}$$

In deriving these equations, we have used the following relation:

$$\frac{\partial}{\partial\theta}\nabla_1^2=\frac{\partial^3}{\partial\theta^3}+\cot\theta\frac{\partial^2}{\partial\theta^2}-\frac{1}{\sin^2\theta}\frac{\partial}{\partial\theta}+\frac{1}{\sin^2\theta}\frac{\partial^3}{\partial\theta\partial\alpha^2}-\frac{2\cot\theta}{\sin^2\theta}\frac{\partial^2}{\partial\alpha^2}.$$

Hence, Eq. (2.4.6) gives rise to

$$\begin{aligned}
&B+RV-R^2\rho\frac{\partial^2\psi}{\partial t^2}=\sin\theta\frac{\partial H}{\partial\theta},\\
&A-RU+R\beta_1T+R^2\rho\frac{\partial^2G}{\partial t^2}=\frac{\partial H}{\partial\alpha}.
\end{aligned} \tag{2.4.9}$$

By substituting these into Eq. (2.4.7), we get the condition that function H must satisfy, *i.e.*,

$$\nabla_1^2H=0. \tag{2.4.10}$$

It can be shown (Hu, 1954*b*; Zou *et al.*, 1994; Chen, 1995) that without loss of generality, we can let the right hand side of Eq. (2.4.9) be zero.

Now, by making use of these displacement functions, w, G and ψ, and the body force functions, U and V, we can simplify the governing equations (2.4.1) to

$$[c_{44}\nabla_3^2 + 2(c_{66} - c_{44}) + c_{66}\nabla_1^2]\psi + RV - R^2\rho\frac{\partial^2\psi}{\partial t^2} = 0, \tag{2.4.11}$$

$$L_3 w - L_4\nabla_1^2 G + R^2 F_R + 2(\beta_3 - \beta_1)RT + \beta_3 R^2\frac{\partial T}{\partial R} - R^2\rho\frac{\partial^2 w}{\partial t^2} = 0, \tag{2.4.12}$$

$$L_1 w - L_2 G - RU + R\beta_1 T + R^2\rho\frac{\partial^2 G}{\partial t^2} = 0. \tag{2.4.13}$$

This set of partial differential equations is convenient for solution because Eq. (2.4.11) contains only the function ψ, while Eqs. (2.4.12) and (2.4.13) are independent of ψ.

The general solution of these equations is the summation of their special solution and the general solution of their homogeneous equations, *i.e.*,

$$\psi = \psi_0 + \psi_1, \quad w = w_0 + w_1 + w_2, \quad G = G_0 + G_1 + G_2, \tag{2.4.14}$$

where ψ_0 is the general solution of

$$\left[c_{44}\nabla_3^2 + 2(c_{66} - c_{44}) + c_{66}\nabla_1^2 - R^2\rho\frac{\partial^2}{\partial t^2}\right]\psi_0 = 0, \tag{2.4.15}$$

ψ_1 is the special solution of

$$\left[c_{44}\nabla_3^2 + 2(c_{66} - c_{44}) + c_{66}\nabla_1^2 - R^2\rho\frac{\partial^2}{\partial t^2}\right]\psi_1 = -RV, \tag{2.4.16}$$

w_0 and G_0 are the general solution of

$$\begin{aligned} &\left(L_3 - R^2\rho\frac{\partial^2}{\partial t^2}\right)w_0 - L_4\nabla_1^2 G_0 = 0, \\ &L_1 w_0 - \left(L_2 - R^2\rho\frac{\partial^2}{\partial t^2}\right)G_0 = 0, \end{aligned} \tag{2.4.17}$$

and w_1, G_1 and w_2, G_2 are the special solutions of the following two sets of equations, respectively

$$\left(L_3 - R^2\rho\frac{\partial^2}{\partial t^2}\right)w_1 - L_4\nabla_1^2 G_1 = 0,$$
$$L_1 w_1 - \left(L_2 - R^2\rho\frac{\partial^2}{\partial t^2}\right)G_1 = RU - R\beta_1 T, \tag{2.4.18}$$

$$\left(L_3 - R^2\rho\frac{\partial^2}{\partial t^2}\right)w_2 - L_4\nabla_1^2 G_2 = -\left[R^2 F_R + 2(\beta_3 - \beta_1)RT + \beta_3 R^2\frac{\partial T}{\partial R}\right],$$
$$L_1 w_2 - \left(L_2 - R^2\rho\frac{\partial^2}{\partial t^2}\right)G_2 = 0. \tag{2.4.19}$$

For static problems, the terms associated with time t in Eqs. (2.4.15)-(2.4.19) should be removed. In this case, a simpler general solution, which can also be used conveniently, can be derived. This will be shown immediately.

First, introduce an arbitrary function F_0 and let

$$w_0 = L_2 F_0, \quad G_0 = L_1 F_0. \tag{2.4.20}$$

The second of Eq. (2.4.17) is satisfied automatically. The satisfaction of the first of Eq. (2.4.17) requires that

$$L^* F_0 = 0, \tag{2.4.21}$$

where

$$\begin{aligned} L^* &= (L_2 L_3 - L_1 L_4 \nabla_1^2)/(c_{33}c_{44}) \\ &= \nabla_3^2\nabla_3^2 + 2D\nabla_3^2 + 4L + M\nabla_3^2\nabla_1^2 + 2(N+L)\nabla_1^2 + N\nabla_1^2\nabla_1^2, \end{aligned} \tag{2.4.22}$$

in which

$$\begin{aligned} D &= [c_{44}(c_{13} - c_{11} - c_{12}) + c_{33}(c_{66} - c_{44})]/(c_{33}c_{44}), \\ L &= (c_{66} - c_{44})(c_{13} - c_{11} - c_{12})/(c_{33}c_{44}), \\ M &= [c_{11}c_{33} - c_{13}(c_{13} + 2c_{44})]/(c_{33}c_{44}), \\ N &= c_{11}/c_{33}. \end{aligned} \tag{2.4.23}$$

Second, the special solutions, w_1 and G_1 of Eq. (2.4.18) as well as w_2 and G_2 of Eq. (2.4.19) in the static case can be written as

$$w_1 = L_4 \nabla_1^2 F_1 , \quad G_1 = L_3 F_1 , \tag{2.4.24}$$

$$w_2 = L_2 F_2 , \quad G_2 = L_1 F_2 , \tag{2.4.25}$$

where F_1 and F_2 are the special solutions of the following equations:

$$L^* F_i = P_i , \quad (i = 1, 2) , \tag{2.4.26}$$

in which

$$\begin{aligned} P_1 &= (R\beta_1 T - RU)/(c_{33}c_{44}), \\ P_2 &= -\left[R^2 F_R + 2(\beta_3 - \beta_1)RT + \beta_3 R^2 \frac{\partial T}{\partial R} \right] / (c_{33}c_{44}) . \end{aligned} \tag{2.4.27}$$

Equation (2.4.22) shows clearly that the operator L^* contains the sub-operators with respect to α, θ and R, respectively, and hence the displacement function F_0 should have a solution with separated variables, *i.e.*,

$$F_0 = X(R)Y(\theta, \alpha) . \tag{2.4.28}$$

A similar argument applies to the function ψ_0.

2.4.2 Relationship between Transversely Isotropic and Spherically Isotropic Solutions

When R is very large, a small surface element at a constant spherical radius can be approximated by a flat plane. Thus transverse isotropy can be viewed as the limiting case of spherical isotropy. Hence, by taking the following limits:

$$R \to \infty, \quad \theta \to 0, \tag{2.4.29}$$

we have

$$\cos\theta \to 1\ ,\quad \frac{1}{R}\frac{\partial}{\partial\theta} \to \frac{\partial}{\partial r},\qquad \frac{\partial}{\partial R} \to \frac{\partial}{\partial z} \tag{2.4.30}$$

and

$$u_\theta \to u_r\,,\quad u_R \to u_z = w\,. \tag{2.4.31}$$

We have used the fact that α is the common coordinate for cylindrical and spherical coordinate systems and $R\sin\theta = r$.

Now in Eq. (2.4.3), if we let

$$\psi = -\psi^* / R,\quad G = G^* / R\,, \tag{2.4.32}$$

we get the following limits:

$$u_z = w,\quad u_r = \frac{1}{r}\frac{\partial \psi^*}{\partial \alpha} - \frac{\partial G^*}{\partial r},\qquad u_\alpha = -\frac{\partial \psi^*}{\partial r} - \frac{1}{r}\frac{\partial G^*}{\partial \alpha}\,, \tag{2.4.33}$$

which are identical to Eq. (2.2.2*a*).

Letting the V in Eq. (2.4.4) be $-V^*$ and taking the limit as above, we obtain

$$F_r = \frac{1}{r}\frac{\partial V^*}{\partial \alpha} - \frac{\partial U}{\partial r}\,,\qquad F_\alpha = -\frac{\partial V^*}{\partial r} - \frac{1}{r}\frac{\partial U}{\partial \alpha}\,. \tag{2.4.34}$$

Further, notice the following limits

$$\frac{1}{R^2}\nabla_1^2 \to \frac{\partial^2}{\partial r^2} + \frac{1}{r}\frac{\partial}{\partial r} + \frac{1}{r^2}\frac{\partial^2}{\partial \alpha^2} = \frac{\partial^2}{\partial x^2} + \frac{\partial^2}{\partial y^2} = \Lambda\,, \tag{2.4.35}$$

$$\frac{1}{R^2}\nabla_3^2 \to \frac{\partial^2}{\partial z^2}\,. \tag{2.4.36}$$

Equations (2.4.11)-(2.4.13) then become

$$-\left(c_{44}\frac{\partial^2}{\partial z^2} + c_{66}\Lambda\right)\psi^* - V^* + \rho\frac{\partial^2 \psi^*}{\partial t^2} = 0\,, \tag{2.4.37}$$

$$\left(c_{33}\frac{\partial^2}{\partial z^2}+c_{44}\Lambda-\rho\frac{\partial^2}{\partial t^2}\right)w-(c_{13}+c_{44})\Lambda\frac{\partial}{\partial z}G^*+F_z+\beta_3\frac{\partial T}{\partial z}=0, \tag{2.4.38}$$

$$(c_{13}+c_{44})\frac{\partial w}{\partial z}-\left(c_{44}\frac{\partial^2}{\partial z^2}+c_{11}\Lambda-\rho\frac{\partial^2}{\partial t^2}\right)G^*-U+\beta_1 T=0. \tag{2.4.39}$$

It is seen that Eqs. (2.4.37)-(2.4.39) are identical to Eqs. (2.2.14), (2.2.6) and (2.2.18), respectively.

We can therefore conclude that the general solution of spherically isotropic problems is the extension of the corresponding transversely isotropic ones.

3

PROBLEMS FOR INFINITE SOLIDS

This chapter presents a unified point force solution for problems associated with an infinite solid. The trial-and-error method and the general solution obtained in **Chapter 2** will be used to develop point force solutions when the material properties vary (*i.e.*, $s_1 = s_2$ and $s_1 \neq s_2$). The discussion will then be extended to problems for inclusions and spherically isotropic materials. The reader will find that the method of deriving the displacements and stresses is straightforward and explicit and the solutions obtained are easy to apply.

3.1 THE UNIFIED POINT FORCE SOLUTION

Consider a transversely isotropic solid in Cartesian coordinates (*x*, *y*, *z*) whose *z*-axis is perpendicular to the isotropic plane of the material. Any point force (or concentrated force) applied in the body can be resolved into three components T, Q and P in x-, y- and z-directions, respectively. Since *xOy* is an isotropic plane, when the solution due to force T is obtained, that due to Q is also ready. Thus in the following, we will concentrate on the solutions due to P along the *z*-axis and T in the x-direction.

3.1.1 A Point Force Perpendicular to the Isotropic Plane

Suppose the force is applied at the origin *O*. The problem is clearly axisymmetric. Thus when $s_1 \neq s_2$, we can take φ_i $(i = 1, 2, 3)$ of Eqs. (2.2.56) and (2.2.63) as

$$\varphi_3 = 0, \quad \varphi_i = A_i \operatorname{sign}(z) \ln R_i^*, \quad (i = 1, 2), \tag{3.1.1}$$

where $R_i^* = R_i + s_i |z|$, $R_i = \sqrt{r^2 + z_i^2}$, $r^2 = x^2 + y^2$ and A_i $(i = 1, 2)$ are constants to be determined. We can see that φ_i $(i = 1,2)$ are odd functions of the coordinate z.

The displacements and stresses can be derived directly by substituting Eq. (3.1.1) into Eqs. (2.2.56) and (2.2.63), *i.e.*,

$$u=\operatorname{sign}(z)\sum_{i=1}^{2}\frac{A_i x}{R_i R_i^*}, \tag{3.1.2}$$

$$v=\operatorname{sign}(z)\sum_{i=1}^{2}\frac{A_i y}{R_i R_i^*}, \tag{3.1.3}$$

$$w=\sum_{i=1}^{2}\frac{\alpha_i A_i}{R_i}, \tag{3.1.4}$$

$$\sigma_x=(c_{11}-c_{12})\operatorname{sign}(z)\sum_{i=1}^{2}A_i\left(\frac{1}{R_i R_i^*}-\frac{x^2}{R_i^3 R_i^*}-\frac{x^2}{R_i^2 R_i^{*2}}\right)-\sum_{i=1}^{2}\xi_i A_i\frac{z_i}{R_i^{\ 3}}, \tag{3.1.5}$$

$$\sigma_y=(c_{11}-c_{12})\operatorname{sign}(z)\sum_{i=1}^{2}A_i\left(\frac{1}{R_i R_i^*}-\frac{y^2}{R_i^3 R_i^*}-\frac{y^2}{R_i^2 R_i^{*2}}\right)-\sum_{i=1}^{2}\xi_i A_i\frac{z_i}{R_i^3}, \tag{3.1.6}$$

$$\tau_{xy}=-2c_{66}\operatorname{sign}(z)xy\sum_{i=1}^{2}A_i\left(\frac{1}{R_i^3 R_i^*}+\frac{1}{R_i^2 R_i^{*2}}\right), \tag{3.1.7}$$

$$\tau_{zx}=-\sum_{i=1}^{2}\frac{\omega_i A_i x}{R_i^3}, \tag{3.1.8}$$

$$\tau_{yz}=-\sum_{i=1}^{2}\frac{\omega_i A_i y}{R_i^3}, \tag{3.1.9}$$

$$\sigma_z=-\sum_{i=1}^{2}\frac{\vartheta_i A_i z_i}{R_i^3}. \tag{3.1.10}$$

where

$$\begin{aligned}
\xi_i &= k_{1i}c_{66}=c_{13}\alpha_i s_i-c_{12}, \\
\vartheta_i &= k_{2i}c_{66}=c_{33}\alpha_i s_i-c_{13}, \qquad (i=1,2). \\
\omega_i &= k_{3i}c_{66}=c_{44}(s_i+\alpha_i),
\end{aligned} \tag{3.1.11}$$

This solution from Eq. (3.1.2) to Eq. (3.1.10) shows that the displacement w and the stresses σ_z, τ_{zx}, τ_{yz} are continuous except at the origin. However, the continuity of the other displacement and stress components in the isotropic plane, $z=0$, is unclear and needs a further examination.

We can see that u, v and σ_x, σ_y, τ_{xy} are odd functions of z and hence vanish at $z=0$. This means that

$$\sum_{i=1}^{2} A_i = 0. \tag{3.1.12}$$

On the other hand, if we consider the *equilibrium of an elastic layer* bounded by planes $z=\pm h$, we obtain

$$\int_{-\infty}^{+\infty}\int_{-\infty}^{+\infty}[\sigma_z(x,y,h)-\sigma_z(x,y,-h)]\,\mathrm{d}x\,\mathrm{d}y+P=0. \tag{3.1.13}$$

By substituting Eq. (3.1.10) into Eq. (3.1.13), we get

$$4\pi\sum_{i=1}^{2}\vartheta_i A_i - P = 0. \tag{3.1.14}$$

Equations (3.1.12) and (3.1.14) then give rise to

$$A_1 = -A_2 = \frac{P}{4\pi(\vartheta_1-\vartheta_2)} = \frac{(c_{13}+c_{44})P}{4\pi c_{33}c_{44}(s_2^2-s_1^2)}. \tag{3.1.15}$$

These expressions for the coefficients show that the solution, Eqs. (3.1.2) to (3.1.10), will become $0/0$ when $s_1=s_2$. However, we will see that they approach finite limits. To show this, substitute Eq. (3.1.15) into Eqs. (3.1.2) to Eq. (3.1.4) and assume $r\neq 0$; this yields

$$\begin{aligned}
u &= \frac{z}{|z|}xA_1\left(\frac{1}{R_1R_1^*}-\frac{1}{R_2R_2^*}\right) = \frac{z}{|z|}xA_1\left(\frac{R_1-s_1|z|}{R_1}-\frac{R_2-s_2|z|}{R_2}\right)\frac{1}{r^2} \\
&= \frac{A_1(s_2R_1-s_1R_2)xz}{R_1R_2r^2} = \frac{A_1(s_2^2-s_1^2)xz}{R_1R_2(R_1s_2+R_2s_1)} \\
&= \frac{(c_{13}+c_{44})xzP}{4\pi c_{33}c_{44}R_1R_2(s_2R_1+s_1R_2)},
\end{aligned} \tag{3.1.16a}$$

$$v = \frac{(c_{13}+c_{44})yzP}{4\pi c_{33}c_{44}R_1R_2(s_2R_1+s_1R_2)}, \tag{3.1.16b}$$

$$w = \frac{r^2P}{4\pi c_{33}R_1R_2(s_2R_1+s_1R_2)} + \frac{s_1s_2[r^2+(s_1^2+s_2^2)z^2]P}{4\pi c_{44}R_1R_2(s_1R_1+s_2R_2)}. \tag{3.1.16c}$$

When $r=0$, these equations result in $u=v=0,\ w=P/(4\pi c_{44}|z|)$, which are the same as those produced by Eqs. (3.1.2) to (3.1.4) directly. Solution (3.1.16) is also valid for $s_1=s_2$. If the material is isotropic, then $s_1=s_2=1$. This is certainly a special case. We therefore call solution (3.1.16) the *unified solution* to a problem for a point force perpendicular to the isotropic plane of the material.

For an isotropic material, we can obtain the well-known *Kelvin solution* easily by substituting Eq. (1.4.13) into the unified solution, Eq. (3.1.16), *i.e.*,

$$\begin{aligned} u &= \frac{\lambda+\mu}{\mu(\lambda+2\mu)}\frac{P}{8\pi}\frac{xz}{R^3}, \quad v = \frac{\lambda+\mu}{\mu(\lambda+2\mu)}\frac{P}{8\pi}\frac{yz}{R^3}, \\ w &= \frac{\lambda+\mu}{\mu(\lambda+2\mu)}\frac{P}{8\pi}\frac{z^2}{R^3} + \frac{\lambda+3\mu}{\mu(\lambda+2\mu)}\frac{P}{8\pi}\frac{1}{R}, \end{aligned} \tag{3.1.17}$$

which are identical to the Kelvin solution (Kelvin, 1848) listed by Sokolnikoff (1956).

When $s_1=s_2$, as discussed above, there are other ways to find the solution. For example, if we take

$$\varphi_1 = \text{sign}(z)C_1\ln(R_1+s_1|z|), \quad \varphi_2 = C_2\frac{1}{R_1}, \quad \varphi_3 = 0, \tag{3.1.18}$$

and make use of the general solution, Eqs. (2.2.61) and (2.2.66), we get

$$C_1 = 0, \quad C_2 = \frac{P}{4\pi\vartheta_4}, \tag{3.1.19}$$

where

$$\vartheta_4 = k_5c_{66} = c_{33}\,s_1\,(\alpha_1+\alpha_4), \tag{3.1.20}$$

$$\alpha_4 = -\alpha_3 = -\frac{c_{11}+c_{44}s_1^2}{(c_{13}+c_{44})s_1}. \tag{3.1.21}$$

These will lead to the same result as given in Eq. (3.1.16) for the case $s_1 = s_2$.

With this unified solution for displacements, the expressions of stresses can be obtained easily. Nowadays, software for computerized symbolic derivations, such as MATHEMATICA or MAPLE, simplifies the formulation process.

3.1.2 A Point Force within the Isotropic Plane

Now consider the solution of an infinite solid subjected to point force T at the origin and in the xOy-plane, the isotropic plane of the transversely isotropic material. In this case, we can take the displacement functions as

$$\varphi_3 = \frac{D_3 y}{R_3 + s_3|z|}, \quad \varphi_i = \frac{D_i x}{R_i + s_i|z|}, \quad (i = 1,2), \tag{3.1.22}$$

where D_i (i =1, 2, 3) are undetermined constants. Thus using Eqs. (2.2.56) and (2.2.63) we obtain the following displacements and stresses:

$$u = D_3\left(\frac{1}{R_3^*} - \frac{y^2}{R_3 R_3^{*2}}\right) + \sum_{i=1}^{2} D_i\left(\frac{1}{R_i^*} - \frac{x^2}{R_i R_i^{*2}}\right), \tag{3.1.23}$$

$$v = \frac{D_3 xy}{R_3 R_3^{*2}} - xy\sum_{i=1}^{2}\frac{D_i}{R_i R_i^{*2}}, \tag{3.1.24}$$

$$w = -\operatorname{sign}(z)\, x\sum_{i=1}^{2}\frac{\alpha_i D_i}{R_i R_i^*}, \tag{3.1.25}$$

$$\begin{aligned}\sigma_x &= -(c_{11} - c_{12})D_3 x\left(\frac{1}{R_3 R_3^{*2}} - \frac{2y^2}{R_3^2 R_3^{*3}} - \frac{y^2}{R_3^3 R_3^{*2}}\right)\\ &+ x\sum_{i=1}^{2} D_i\left[\frac{\xi_i}{R_i^3} - (c_{11} - c_{12})\left(\frac{3}{R_i R_i^{*2}} - \frac{2x^2}{R_i^2 R_i^{*3}} - \frac{x^2}{R_i^3 R_i^{*2}}\right)\right],\end{aligned} \tag{3.1.26}$$

$$\begin{aligned}\sigma_y &= (c_{11} - c_{12})D_3 x\left(\frac{1}{R_3 R_3^{*2}} - \frac{2y^2}{R_3^2 R_3^{*3}} - \frac{y^2}{R_3^3 R_3^{*2}}\right)\\ &+ x\sum_{i=1}^{2} D_i\left[\frac{\xi_i}{R_i^3} - (c_{11} - c_{12})\left(\frac{1}{R_i R_i^{*2}} - \frac{2y^2}{R_i^2 R_i^{*3}} - \frac{y^2}{R_i^3 R_i^{*2}}\right)\right],\end{aligned} \tag{3.1.27}$$

$$\tau_{xy} = -c_{66} D_3 y \left(\frac{1}{R_3^3} - \frac{2}{R_3 R_3^{*2}} + \frac{4x^2}{R_3^2 R_3^{*3}} + \frac{2x^2}{R_3^3 R_3^{*2}} \right)$$
$$- 2c_{66} y \sum_{i=1}^{2} D_i \left(\frac{1}{R_i R_i^{*2}} - \frac{2x^2}{R_i^2 R_i^{*3}} - \frac{x^2}{R_i^3 R_i^{*2}} \right), \tag{3.1.28}$$

$$\tau_{zx} = -\omega_3 \operatorname{sign}(z) D_3 \left(\frac{1}{R_3 R_3^*} - \frac{y^2}{R_3^3 R_3^*} - \frac{y^2}{R_3^2 R_3^{*2}} \right)$$
$$- \operatorname{sign}(z) \sum_{i=1}^{2} \omega_i D_i \left(\frac{1}{R_i R_i^*} - \frac{x^2}{R_i^3 R_i^*} - \frac{x^2}{R_i^2 R_i^{*2}} \right), \tag{3.1.29}$$

$$\tau_{yz} = -\omega_3 \operatorname{sign}(z) D_3 xy \left(\frac{1}{R_3^3 R_3^*} + \frac{1}{R_3^2 R_3^{*2}} \right)$$
$$+ \operatorname{sign}(z) xy \sum_{i=1}^{2} \omega_i D_i \left(\frac{1}{R_i^3 R_i^*} + \frac{1}{R_i^2 R_i^{*2}} \right), \tag{3.1.30}$$

$$\sigma_z = \sum_{i=1}^{2} \vartheta_i \frac{D_i x}{R_i^3}, \tag{3.1.31}$$

where

$$\omega_3 = c_{44} s_3 . \tag{3.1.32}$$

We know that w, τ_{zx} and τ_{yz} are continuous at $z = 0$, so that

$$\sum_{i=1}^{2} \alpha_i D_i = 0, \tag{3.1.33}$$

$$\omega_3 D_3 - \sum_{i=1}^{2} \omega_i D_i = 0. \tag{3.1.34}$$

Using Eqs. (3.1.32) and (3.1.33), we can simplify Eq. (3.1.34) further to

$$\sum_{i=1}^{3} s_i D_i = 0. \tag{3.1.35}$$

Similarly, by taking the static equilibrium of the elastic layer in the vicinity of *xOy*-plane, we get

$$\int_{-\infty}^{+\infty}\int_{-\infty}^{+\infty}[\tau_{zx}(x, y, h)-\tau_{zx}(x, y, -h)]\mathrm{d}x\,\mathrm{d}y+T=0. \tag{3.1.36}$$

The shear stresses in this equation are given by Eq. (3.1.29). Hence, we have

$$-\omega_3 D_3-\omega_1 D_1-\omega_2 D_2+\frac{T}{2\pi}=0. \tag{3.1.37}$$

With the expressions of ω_i in Eq. (3.1.32), this equation becomes

$$-s_3 D_3-s_1 D_1-s_2 D_2+\frac{T}{2\pi c_{44}}=0. \tag{3.1.38}$$

Then by solving Eqs. (3.1.33), (3.1.35) and (3.1.38) together, we finally get

$$\begin{aligned} D_1&=\frac{\alpha_2 T}{4\pi c_{44}(s_1\alpha_2-s_2\alpha_1)}=\frac{s_1(c_{11}-c_{44}s_2^2)T}{4\pi c_{11}c_{44}(s_1^2-s_2^2)},\\ D_2&=-\frac{\alpha_1 T}{4\pi c_{44}(s_1\alpha_2-s_2\alpha_1)}=-\frac{s_2(c_{11}-c_{44}s_1^2)T}{4\pi c_{11}c_{44}(s_1^2-s_2^2)},\\ D_3&=\frac{T}{4\pi c_{44}s_3}. \end{aligned} \tag{3.1.39}$$

It is clear that if we substitute these coefficients into Eqs. (3.1.23) to (3.1.31) for displacements and stresses, we will have the situation of $0/0$ when $s_1=s_2$. To overcome the problem, we proceed as in the last section. Rewrite Eqs. (3.1.23) to (3.1.25) as

$$\frac{u}{T}=\frac{1}{4\pi c_{44}s_3}\left(\frac{1}{R_3^*}-\frac{y^2}{R_3R_3^{*2}}\right)+\frac{1}{4\pi(s_1R_2+s_2R_1)}\left(\frac{1}{c_{33}s_1s_2}+\frac{r^2}{c_{44}R_1^*R_2^*}\right)+v_1x^2, \tag{3.1.40a}$$

$$\frac{v}{T}=\frac{xy}{4\pi c_{44}s_3R_3R_3^{*2}}+v_1xy, \tag{3.1.40b}$$

$$\frac{w}{T}=\frac{c_{13}+c_{44}}{4\pi c_{33}c_{44}}\frac{xz}{R_1R_2(s_1R_2+s_2R_1)}, \tag{3.1.40c}$$

where

$$
\begin{aligned}
v_1 = & -\frac{r^2 + (s_1^2 + s_2^2)z^2}{4\pi c_{33} R_1 R_2 s_1 s_2 (s_1 R_2 + s_2 R_1)(R_1 R_2 + s_1 s_2 z^2)} \\
& + \frac{1}{4\pi c_{44} R_1 R_1^{*2} R_2 R_2^{*2}} \left[\frac{s_1 s_2 z^2 (R_1^{*2} + R_2^{*2})}{(s_1 R_1 + s_2 R_2)} - \frac{r^4}{(s_1 R_2 + s_2 R_1)} \right].
\end{aligned}
\tag{3.1.41}
$$

Equation (3.1.40) now allows $s_1 = s_2$. Again, the Kelvin solution (Kelvin, 1848) can be obtained if we note the relationships among the constants when the material is isotropic. Hence, by substituting Eq. (1.4.13) into Eq. (3.1.40), we get

$$
\begin{aligned}
u &= \frac{\lambda + \mu}{\mu(\lambda + 2\mu)} \frac{T}{8\pi} \frac{x^2}{R^3} + \frac{\lambda + 3\mu}{\mu(\lambda + 2\mu)} \frac{T}{8\pi} \frac{1}{R}, \\
v &= \frac{\lambda + \mu}{\mu(\lambda + 2\mu)} \frac{T}{8\pi} \frac{xy}{R^3}, \\
w &= \frac{\lambda + \mu}{\mu(\lambda + 2\mu)} \frac{T}{8\pi} \frac{xz}{R^3}.
\end{aligned}
\tag{3.1.42}
$$

As with the discussion in the last section, there are other ways to find the solution when $s_1 = s_2$. For example, we may take

$$
\varphi_1 = \frac{G_1 x}{R_1 + s_1|z|}, \quad \varphi_2 = \text{sign}(z)\frac{G_2 x}{R_1(R_1 + s_1|z|)}, \quad \varphi_3 = \frac{G_3 y}{R_3 + s_3|z|}, \tag{3.1.43}
$$

and make use of the general solution, Eqs. (2.2.61) and (2.2.66), where G_i (i = 1, 2, 3) are constants. Then by following the same procedures as before, we will have

$$
G_3 = -\frac{T}{4\pi c_{44} s_3}, \quad G_1 = -\frac{\alpha_4 T}{4\pi c_{44} s_1 (\alpha_1 - \alpha_4)}, \quad G_2 = -\frac{\alpha_1 T}{4\pi c_{44} s_1 (\alpha_1 - \alpha_4)}. \tag{3.1.44}
$$

This will lead to the same solution as given in Eq. (3.1.40) for the case with $s_1 = s_2$.

The solutions described by Eqs. (3.1.16) and (3.1.40) are the unified displacement solutions for either isotropic or transversely isotropic material, and there is no need to worry about the eigenvalue condition of $s_1 = s_2$ or $s_1 \neq s_2$. This is particularly important for materials with very close values of s_1 and s_2. The unified solutions developed here will not introduce any numerical error for materials with any combinations of s_1 and s_2.

When $P = T = 1$, we call the above unified solutions the *unified fundamental solutions*; they form the basis of the solutions to many complex problems (Fabrikant, 2004). They are also useful in the boundary element solution (Ding, Liang and Chen, 1997).

3.2 THE POINT FORCE SOLUTION FOR AN INFINITE SOLID COMPOSED OF TWO HALF-SPACES

Consider an infinite solid composed of two transversely isotropic half-spaces with different material constants. The interface of the two half-spaces is parallel to their isotropic planes. Let the coordinate plane xOy coincide with the interface and the z-axis point downwards, Fig. 3.1. Then the half-space above the interface occupies $z \le 0$ and that below has $z \ge 0$. For convenience, we use a superscript $(\ )'$ to denote the variables in the upper half-space.

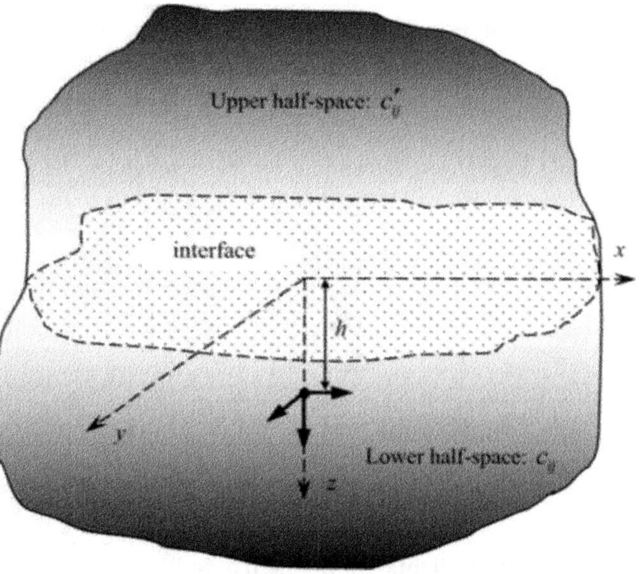

Fig. 3.1 An infinite solid composed of two half-spaces.

When the two half-spaces are ideally bonded at the interface $z = 0$ such that the material becomes continuous across the interface, we have

$$\begin{aligned} &u = u', \quad v = v', \quad w = w', \\ &\sigma_z = \sigma'_z, \quad \tau_{zx} = \tau'_{zx}, \quad \tau_{yz} = \tau'_{yz}. \end{aligned} \tag{3.2.1}$$

On the other hand, if the half-spaces are in *smooth contact*, *i.e.*, in a complete contact without friction, then

$$w = w', \quad \sigma_z = \sigma'_z,$$
$$\tau_{zx} = \tau'_{zx} = \tau_{yz} = \tau'_{yz} = 0. \tag{3.2.2}$$

In the following, we will develop the solution for displacements and stresses for the composite body when a point force is applied on the z-axis in the lower half-space ($z \geq 0$), *i.e.* the point where the force acts is $(0, 0, h)$ with $h > 0$.

We can always express the displacements in the lower half-space as

$$u = u_1 + u_2, \quad v = v_1 + v_2, \quad w = w_1 + w_2, \quad (z \geq 0), \tag{3.2.3}$$

Here u_1, v_1 and w_1 are displacements due to the point force applied at $(0, 0, h)$ in an infinite body, and u_2 , v_2 and w_2 are displacements which make Eq. (3.2.3) satisfy the boundary conditions on the interface. We will focus on the development of u_2 , v_2 and w_2 in this section.

We need to discuss four cases, depending on the eigenvalues of the two half-space materials. For convenience, we will use the following notations in the formulation:

$$\begin{aligned} z_i &= s_i z, \quad h_i = s_i h, \quad z'_i = s'_i z \ , \\ z_{ij} &= z_i + h_j, \quad R_{ij} = \sqrt{x^2 + y^2 + z_{ij}^2} \ , \\ \bar{z}_{ij} &= z_i - h_j, \quad \bar{R}_{ij} = \sqrt{x^2 + y^2 + \bar{z}_{ij}^2} \ , \\ z'_{ij} &= z'_i - h_j, \quad R'_{ij} = \sqrt{x^2 + y^2 + z'^2_{ij}} \ , \end{aligned} \tag{3.2.4}$$

where $i, j = 1, 2, 3$.

3.2.1 A Point Force Perpendicular to the Isotropic Plane

The problem is axisymmetric and hence $\varphi_3 = \varphi'_3 = 0$. The displacements and stresses are as follows:

(*a*) $s_1 \neq s_2$

Since the point force is acting at $z = h$, u_1, v_1 and w_1 can be obtained directly using the solution obtained in **Section 3.1**, provided that the variable z there is replaced by $z - h$. Hence, letting

$$\varphi_i = A_i \operatorname{sign}(z-h) \ln\left(\bar{R}_{ii} + s_i |z-h|\right) + \sum_{j=1}^{2} A_{ij} \ln(R_{ij} + z_{ij}), \quad (i = 1,2), \tag{3.2.5}$$

where A_1 and A_2 are given in Eq. (3.1.15), we then derive from Eq. (2.2.56),

$$
\begin{aligned}
u &= \sum_{i=1}^{2}\left[\operatorname{sign}(z-h)\frac{A_i x}{\bar{R}_{ii}(\bar{R}_{ii}+s_i|z-h|)}+\sum_{j=1}^{2}\frac{A_{ij}x}{R_{ij}(R_{ij}+z_{ij})}\right],\\
v &= \sum_{i=1}^{2}\left[\operatorname{sign}(z-h)\frac{A_i y}{\bar{R}_{ii}(\bar{R}_{ii}+s_i|z-h|)}+\sum_{j=1}^{2}\frac{A_{ij}y}{R_{ij}(R_{ij}+z_{ij})}\right],\\
w &= \sum_{i=1}^{2}\alpha_i\left(\frac{A_i}{\bar{R}_{ii}}+\sum_{j=1}^{2}\frac{A_{ij}}{R_{ij}}\right).
\end{aligned}
\tag{3.2.6}
$$

Using Eq. (2.2.63), we can obtain the corresponding stress components as follows:

$$
\begin{aligned}
\tau_{zx} &= -x\sum_{i=1}^{2}\omega_i\left(\frac{A_i}{\bar{R}_{ii}^3}+\sum_{j=1}^{2}\frac{A_{ij}}{R_{ij}^3}\right),\\
\tau_{yz} &= -y\sum_{i=1}^{2}\omega_i\left(\frac{A_i}{\bar{R}_{ii}^3}+\sum_{j=1}^{2}\frac{A_{ij}}{R_{ij}^3}\right),\\
\sigma_z &= -\sum_{i=1}^{2}\vartheta_i\left(\frac{A_i\bar{z}_{ii}}{\bar{R}_{ii}^3}+\sum_{j=1}^{2}\frac{A_{ij}z_{ij}}{R_{ij}^3}\right).
\end{aligned}
\tag{3.2.7}
$$

In the upper half-space, *i.e.*, $z \le 0$, we can have two cases, depending on the eigenvalues s_1' and s_2'. By taking

$$
\varphi_i' = \sum_{j=1}^{2} A_{ij}'\ln(R_{ij}'-z_{ij}'),\quad (i=1,2),\quad \text{when } s_1' \ne s_2',
\tag{3.2.8}
$$

or

$$
\varphi_1' = \sum_{j=1}^{2} C_{1j}'\ln(R_{1j}'-z_{1j}'),\quad \varphi_2' = \sum_{j=1}^{2}\frac{C_{2j}'}{R_{1j}'},\quad \text{when } s_1' = s_2',
\tag{3.2.9}
$$

we can obtain the displacements and stresses in the upper half-space by using Eqs. (2.2.56), (2.2.63), (2.2.61) and (2.2.66). Hence, when $s_1' \ne s_2'$,

$$
\begin{aligned}
u' &= \sum_{i=1}^{2}\sum_{j=1}^{2}\frac{A_{ij}'x}{R_{ij}'(R_{ij}'-z_{ij}')}, & \tau_{zx}' &= x\sum_{i=1}^{2}\omega_i'\sum_{j=1}^{2}\frac{A_{ij}'}{R_{ij}'^3},\\
v' &= \sum_{i=1}^{2}\sum_{j=1}^{2}\frac{A_{ij}'y}{R_{ij}'(R_{ij}'-z_{ij}')}, & \tau_{yz}' &= y\sum_{i=1}^{2}\omega_i'\sum_{j=1}^{2}\frac{A_{ij}'}{R_{ij}'^3},
\end{aligned}
$$

$$w' = -\sum_{i=1}^{2} \alpha'_i \sum_{j=1}^{2} \frac{A'_{ij}}{R'_{ij}}, \quad \sigma'_z = \sum_{i=1}^{2} \vartheta'_i \sum_{j=1}^{2} \frac{A'_{ij} z'_{ij}}{R'^3_{ij}}, \tag{3.2.10}$$

but when $s'_1 = s'_2$,

$$u' = x\sum_{j=1}^{2}\left[\frac{C'_{1j}}{R'_{1j}(R'_{1j} - z'_{1j})} - z'_1 \frac{C'_{2j}}{R'^3_{1j}}\right], \quad v' = y\sum_{j=1}^{2}\left[\frac{C'_{1j}}{R'_{1j}(R'_{1j} - z'_{1j})} - z'_1 \frac{C'_{2j}}{R'^3_{1j}}\right],$$
$$w' = \sum_{j=1}^{2}\left(-\alpha'_1 \frac{C'_{1j}}{R'_{1j}} + \alpha'_4 \frac{C'_{2j}}{R'_{1j}} - z'_1 z'_{1j} \alpha'_1 \frac{C'_{2j}}{R'^3_{1j}}\right), \tag{3.2.11a}$$

$$\tau'_{zx} = x\sum_{j=1}^{2}\left(\omega'_1 \frac{C'_{1j}}{R'^3_{1j}} - \omega'_4 \frac{C'_{2j}}{R'^3_{1j}} + 3z'_1 z'_{1j} \omega'_1 \frac{C'_{2j}}{R'^5_{1j}}\right),$$
$$\tau'_{yz} = y\sum_{j=1}^{2}\left(\omega'_1 \frac{C'_{1j}}{R'^3_{1j}} - \omega'_4 \frac{C'_{2j}}{R'^3_{1j}} + 3z'_1 z'_{1j} \omega'_1 \frac{C'_{2j}}{R'^5_{1j}}\right), \tag{3.2.11b}$$
$$\sigma'_z = \sum_{j=1}^{2}\left[\vartheta'_1 \frac{C'_{1j} z'_{1j}}{R'^3_{1j}} - \vartheta'_4 \frac{C'_{2j} z'_{1j}}{R'^3_{1j}} + \vartheta'_1 C'_{2j} z'_1 \sum_{j=1}^{3} \left(\frac{3z'^2_{1j}}{R'^5_{1j}} - \frac{1}{R'^3_{1j}}\right)\right].$$

In these solutions,

$$\omega'_4 = k'_6 c'_{66} = c'_{44}(s'_1 + \alpha'_4). \tag{3.2.12}$$

The undetermined constants can be obtained using the contact conditions at the interface. When the two half-spaces are ideally bonded, the conditions in Eq. (3.2.1) lead to

$$-A_i + \sum_{j=1}^{2} A_{ji} = \sum_{j=1}^{2} A'_{ji} \quad \text{or} \quad C'_{1i},$$
$$\alpha_i A_i + \sum_{j=1}^{2} \alpha_j A_{ji} = -\sum_{j=1}^{2} \alpha'_j A'_{ji} \quad \text{or} \quad -\alpha'_1 C'_{1i} + \alpha'_4 C'_{2i},$$
$$-\omega_i A_i - \sum_{j=1}^{2} \omega_j A_{ji} = \sum_{j=1}^{2} \omega'_j A'_{ji} \quad \text{or} \quad \omega'_1 C'_{1i} - \omega'_4 C'_{2i}, \tag{3.2.13}$$
$$\vartheta_i A_i - \sum_{j=1}^{2} \vartheta_j A_{ji} = \sum_{j=1}^{2} \vartheta'_j A'_{ji} \quad \text{or} \quad -\vartheta'_1 C'_{1i} + \vartheta'_4 C'_{2i},$$

where $i=1,2$. For each case, $s_1' \neq s_2'$ or $s_1' = s_2'$, we have eight equations to solve for the undetermined constants A_{ij}, A_{ij}' $(i, j=1, 2)$ or A_{ij}, C_{ij}' $(i, j=1, 2)$. However, the solutions will be lengthy and we do not list them here. When calculating the displacements and stresses numerically, it is easier to solve the equations listed in Eq. (3.2.13) directly.

If the two half-spaces are in smooth contact as defined by Eq. (3.2.2), we get

$$\begin{aligned}
&\alpha_i A_i + \sum_{j=1}^{2} \alpha_j A_{ji} = -\sum_{j=1}^{2} \alpha_j' A_{ji}' \quad \text{or} \quad -\alpha_1' C_{1i}' + \alpha_4' C_{2i}' , \\
&\vartheta_i A_i - \sum_{j=1}^{2} \vartheta_j A_{ji} = \sum_{j=1}^{2} \vartheta_j' A_{ji}' \quad \text{or} \quad -\vartheta_1' C_{1i}' + \vartheta_4' C_{2i}' , \\
&-\omega_i A_i - \sum_{j=1}^{2} \omega_j A_{ji} = 0 , \\
&\sum_{j=1}^{2} \omega_j' A_{ji}' = 0 \quad \text{or} \quad \omega_1' C_{1i}' - \omega_4' C_{2i}' = 0 ,
\end{aligned} \tag{3.2.14}$$

where $i=1, 2$. In a similar way, we can determine constants A_{ij}, A_{ij}' $(i, j=1, 2)$ or A_{ij}, C_{ij}' $(i, j=1, 2)$.

(b) $s_1 = s_2$

In this case, we may take

$$\begin{aligned}
&\varphi_1 = \text{sign}(z-h) C_1 \ln(\bar{R}_{11} + s_1 |z-h|) + C_{11} \ln(R_{11} + z_{11}), \\
&\varphi_2 = C_2 \frac{1}{\bar{R}_{11}} + \frac{C_{21}}{R_{11}},
\end{aligned} \tag{3.2.15}$$

where C_1 and C_2 are given in Eq. (3.1.19), corresponding to the solution of an infinite body subjected to a concentrated force. With these φ_1 and φ_2, the displacements are given by Eq. (2.2.61), *i.e.*,

$$u = \text{sign}(z-h) \frac{C_1 x}{\bar{R}_{11}(\bar{R}_{11} + s_1 |z-h|)} - \frac{C_2 z_1 x}{\bar{R}_{11}^3} + \frac{C_{11} x}{R_{11}(R_{11} + z_{11})} - z_1 x \frac{C_{21}}{R_{11}^3}, \tag{3.2.16a}$$

$$v = \text{sign}(z-h) \frac{C_1 y}{\bar{R}_{11}(\bar{R}_{11} + s_1 |z-h|)} - \frac{C_2 z_1 y}{\bar{R}_{11}^3} + \frac{C_{11} y}{R_{11}(R_{11} + z_{11})} - z_1 y \frac{C_{21}}{R_{11}^3}, \tag{3.2.16b}$$

$$w = \frac{C_1 \alpha_1}{\bar{R}_{11}} - C_2 \left(\frac{\alpha_1 z_1 \bar{z}_{11}}{\bar{R}_{11}^3} - \frac{\alpha_4}{\bar{R}_{11}} \right) + \alpha_1 \frac{C_{11}}{R_{11}} - \alpha_1 z_1 \frac{C_{21} z_{11}}{R_{11}^3} + \alpha_4 \frac{C_{21}}{R_{11}}, \tag{3.2.16c}$$

and similarly, the stresses can be obtained from Eq. (2.2.66), *i.e.*,

$$\tau_{zx} = -\frac{C_1\omega_1 x}{\bar{R}_{11}^3} + C_2 x\left(\frac{3\omega_1 z_1 \bar{z}_{11}}{\bar{R}_{11}^5} - \frac{\omega_4}{\bar{R}_{11}^3}\right) - \omega_1 x\frac{C_{11}}{R_{11}^3} - \omega_4 x\frac{C_{21}}{R_{11}^3} + \omega_1 z_1 x\frac{3C_{21}z_{11}}{R_{11}^5}, \tag{3.2.17a}$$

$$\tau_{yz} = -\frac{C_1\omega_1 y}{\bar{R}_{11}^3} + C_2 y\left(\frac{3\omega_1 z_1 \bar{z}_{11}}{\bar{R}_{11}^5} - \frac{\omega_4}{\bar{R}_{11}^3}\right) - \omega_1 y\frac{C_{11}}{R_{11}^3} - \omega_4 y\frac{C_{21}}{R_{11}^3} + \omega_1 z_1 y\frac{3C_{21}z_{11}}{R_{11}^5}, \tag{3.2.17b}$$

$$\sigma_z = -\frac{C_1\vartheta_1\bar{z}_{11}}{\bar{R}_{11}^3} - \vartheta_4\frac{C_2\bar{z}_{11}}{\bar{R}_{11}^3} - \vartheta_1\frac{C_{11}z_{11}}{R_{11}^3} - \vartheta_4\frac{C_{21}z_{11}}{R_{11}^3} - \vartheta_1 z_1\left(\frac{C_2}{\bar{R}_{11}^3} - \frac{3C_2\bar{z}_{11}^2}{\bar{R}_{11}^5} + \frac{C_{21}}{R_{11}^3} - \frac{3C_{21}z_{11}^2}{R_{11}^5}\right), \tag{3.2.17c}$$

where $\omega_4 = k_6 c_{66} = c_{44}(s_1 + \alpha_4)$.

For the upper half-space ($z \le 0$), depending on the eigenvalues, we can take

$$\varphi_i' = H_{i1}' \ln(R_{i1}' - z_{i1}'), \quad (i = 1, 2), \quad \text{when } s_1' \ne s_2', \tag{3.2.18}$$

or

$$\varphi_1' = K_{11}' \ln(R_{11}' - z_{11}'), \quad \varphi_2' = \frac{K_{21}'}{R_{11}'}, \quad \text{when } s_1' = s_2'. \tag{3.2.19}$$

The displacements and stresses are given directly by Eqs. (2.2.56), (2.2.63), (2.2.61) and (2.2.66). According to the conditions in Eq. (3.2.1), the coefficients in these equations are specified by the following relations:

$$-C_1 + C_{11} = \sum_{j=1}^{2} H_{j1}' \quad \text{or} \quad K_{11}', \tag{3.2.20}$$

$$\alpha_1 C_1 + \alpha_4 C_2 + \alpha_1 C_{11} + \alpha_4 C_{21} = -\sum_{j=1}^{2} \alpha_j' H_{j1}' \quad \text{or} \quad -\alpha_1' K_{11}' + \alpha_4' K_{21}', \tag{3.2.21}$$

$$-\omega_1 C_1 - \omega_4 C_2 - \omega_1 C_{11} - \omega_4 C_{21} = \sum_{j=1}^{2} \omega_j' H_{j1}' \quad \text{or} \quad \omega_1' K_{11}' - \omega_4' K_{21}', \tag{3.2.22}$$

$$\vartheta_1 C_1 + \vartheta_4 C_2 - \vartheta_1 C_{11} - \vartheta_4 C_{21} = -\sum_{j=1}^{2} \vartheta_j' H_{j1}' \quad \text{or} \quad -\vartheta_1' K_{11}' + \vartheta_4' K_{21}'. \tag{3.2.23}$$

When the two half-spaces are in smooth contact, according to the conditions specified in Eq. (3.2.2), we have

$$\alpha_1 C_1 + \alpha_4 C_2 + \alpha_1 C_{11} + \alpha_4 C_{21} = -\sum_{j=1}^{2} \alpha'_j H'_{j1} \quad \text{or} \quad -\alpha'_1 K'_{11} + \alpha'_4 K'_{21} \quad , \tag{3.2.24}$$

$$-\omega_1 C_1 - \omega_4 C_2 - \omega_1 C_{11} - \omega_4 C_{21} = 0\,, \tag{3.2.25}$$

$$\sum_{j=1}^{2} \omega'_j H'_{j1} = 0 \quad \text{or} \quad \omega'_1 K'_{11} - \omega'_4 K'_{21} = 0\,, \tag{3.2.26}$$

$$\vartheta_1 C_1 + \vartheta_4 C_2 - \vartheta_1 C_{11} - \vartheta_4 C_{21} = -\sum_{j=1}^{2} \vartheta'_j H'_{j1} \quad \text{or} \quad -\vartheta'_1 K'_{11} + \vartheta'_4 K'_{21}\,. \tag{3.2.27}$$

These equations determine the coefficients C_{j1} and H'_{j1} $(j = 1, 2)$ or C_{j1} and K'_{j1} $(j = 1, 2)$.

3.2.2 A Point Force within the Isotropic Plane

As with the solution under a point force perpendicular to the isotropic plane, we also need to discuss various cases:

(a) $s_1 \neq s_2$

As before, the point force is applied at $z = h$. Thus the solution for u_1, v_1 and w_1 are readily obtained when variable z in Eqs. (3.1.22) to (3.1.31) is replaced by $(z-h)$. Letting

$$\begin{aligned} \varphi_i &= \frac{D_i x}{\bar{R}_{ii} + s_i |z-h|} + \sum_{j=1}^{2} \frac{D_{ij} x}{R_{ij} + z_{ij}}, \qquad (i = 1,\ 2)\,, \\ \varphi_3 &= \frac{D_3 y}{\bar{R}_{33} + s_3 |z-h|} + \frac{D_{33} y}{R_{33} + z_{33}}, \end{aligned} \tag{3.2.28}$$

we obtain from Eq. (2.2.56),

$$\begin{aligned} u = {} & D_3 \left[\frac{1}{\bar{R}_{33} + s_3 |z-h|} - \frac{y^2}{\bar{R}_{33}(\bar{R}_{33} + s_3 |z-h|)^2} \right] + D_{33} \left[\frac{1}{R_{33} + z_{33}} - \frac{y^2}{R_{33}(R_{33} + z_{33})^2} \right] \\ & + \sum_{i=1}^{2} \left\{ D_i \left[\frac{1}{\bar{R}_{ii} + s_i |z-h|} - \frac{x^2}{\bar{R}_{ii}(\bar{R}_{ii} + s_i |z-h|)^2} \right] + \sum_{j=1}^{2} D_{ij} \left[\frac{1}{R_{ij} + z_{ij}} - \frac{x^2}{R_{ij}(R_{ij} + z_{ij})^2} \right] \right\}, \end{aligned} \tag{3.2.29}$$

$$v = \frac{D_3 xy}{\overline{R}_{33}(\overline{R}_{33} + s_3|z-h|)^2} + \frac{D_{33} xy}{R_{33}(R_{33} + z_{33})^2} - xy\sum_{i=1}^{2}\left[\frac{D_i}{\overline{R}_{ii}(\overline{R}_{ii} + s_i|z-h|)^2} + \sum_{j=1}^{2}\frac{D_{ij}}{R_{ij}(R_{ij} + z_{ij})^2}\right], \tag{3.2.30}$$

$$w = -x\sum_{i=1}^{2}\alpha_i\left[\text{sign}(z-h)\frac{D_i}{\overline{R}_{ii}(\overline{R}_{ii} + s_i|z-h|)} + \sum_{j=1}^{2}\frac{D_{ij}}{R_{ij}(R_{ij} + z_{ij})}\right]. \tag{3.2.31}$$

The stress components can be derived by using Eq. (2.2.63), which gives rise to

$$\begin{aligned}\tau_{zx} = &-\omega_3\left\{\text{sign}(z-h)\,D_3\left[\frac{1}{\overline{R}_{33}(\overline{R}_{33} + s_3|z-h|)} - \frac{y^2}{\overline{R}_{33}^3(\overline{R}_{33} + s_3|z-h|)} - \frac{y^2}{\overline{R}_{33}^2(\overline{R}_{33} + s_3|z-h|)^2}\right]\right.\\ &\left.+D_{33}\left[\frac{1}{R_{33}(R_{33}+z_{33})} - \frac{y^2}{R_{33}^3(R_{33}+z_{33})} - \frac{y^2}{R_{33}^2(R_{33}+z_{33})^2}\right]\right\}\\ &-\sum_{i=1}^{2}\omega_i\left\{\text{sign}(z-h)\,D_i\left[\frac{1}{\overline{R}_{ii}(\overline{R}_{ii}+s_i|z-h|)} - \frac{x^2}{\overline{R}_{ii}^3(\overline{R}_{ii}+s_i|z-h|)} - \frac{x^2}{\overline{R}_{ii}^2(\overline{R}_{ii}+s_i|z-h|)^2}\right]\right.\\ &\left.+\sum_{j=1}^{2}D_{ij}\left[\frac{1}{R_{ij}(R_{ij}+z_{ij})} - \frac{x^2}{R_{ij}^3(R_{ij}+z_{ij})} - \frac{x^2}{R_{ij}^2(R_{ij}+z_{ij})^2}\right]\right\},\end{aligned} \tag{3.2.32}$$

$$\begin{aligned}\tau_{yz} = &-\omega_3 xy\left\{\text{sign}(z-h)\,D_3\left[\frac{1}{\overline{R}_{33}^3(\overline{R}_{33} + s_3|z-h|)} + \frac{1}{\overline{R}_{33}^2(\overline{R}_{33} + s_3|z-h|)^2}\right]\right.\\ &\left.+D_{33}\left[\frac{1}{R_{33}^3(R_{33} + z_{33})} + \frac{1}{R_{33}^2(R_{33} + z_{33})^2}\right]\right\}\\ &+xy\sum_{i=1}^{2}\omega_i\left\{\text{sign}(z-h)\,D_i\left[\frac{1}{\overline{R}_{ii}^3(\overline{R}_{ii} + s_i|z-h|)} + \frac{1}{\overline{R}_{ii}^2(\overline{R}_{ii} + s_i|z-h|)^2}\right]\right.\\ &\left.+\sum_{j=1}^{2}D_{ij}\left[\frac{1}{R_{ij}^3(R_{ij} + z_{ij})} + \frac{1}{R_{ij}^2(R_{ij} + z_{ij})^2}\right]\right\},\end{aligned} \tag{3.2.33}$$

$$\sigma_z = x\sum_{i=1}^{2}\vartheta_i\left(\frac{D_i}{\overline{R}_{ii}^3} + \sum_{j=1}^{2}\frac{D_{ij}}{R_{ij}^3}\right), \tag{3.2.34}$$

where D_1, D_2 and D_3 are defined by Eq. (3.1.39) and D_{33} and D_{ij} $(i, j = 1, 2)$ are the five constants to be determined by the corresponding boundary conditions.

For the upper half-space ($z \leq 0$), we can take the displacement functions as

$$\varphi_i' = \sum_{j=1}^{2} \frac{L_{ij}' x}{R_{ij}' - z_{ij}'} \quad (i=1,2), \quad \varphi_3' = \frac{L_{33}' y}{R_{33}' - z_{33}'}, \quad \text{when } s_1' \neq s_2', \tag{3.2.35}$$

$$\varphi_1' = \sum_{j=1}^{2} \frac{N_{1j}' x}{R_{1j}' - z_{1j}'}, \quad \varphi_2' = \sum_{j=1}^{2} \frac{N_{2j}' x}{R_{1j}'(R_{1j}' - z_{1j}')}, \quad \varphi_3' = \frac{N_{33}' y}{R_{33}' - z_{33}'}, \quad \text{when } s_1' = s_2'. \tag{3.2.36}$$

The expressions of displacements and stresses can then be obtained using Eqs. (2.2.56), (2.2.63), (2.2.61) and (2.2.66). When $s_1' \neq s_2'$, we have

$$u' = L_{33}' \left[\frac{1}{R_{33}' - z_{33}'} - \frac{y^2}{R_{33}'(R_{33}' - z_{33}')^2} \right] + \sum_{i=1}^{2} \sum_{j=1}^{2} L_{ij}' \left[\frac{1}{R_{ij}' - z_{ij}'} - \frac{x^2}{R_{ij}'(R_{ij}' - z_{ij}')^2} \right], \tag{3.2.37}$$

$$v' = \frac{L_{33}' xy}{R_{33}'(R_{33}' - z_{33}')^2} - xy \sum_{i=1}^{2} \sum_{j=1}^{2} \frac{L_{ij}'}{R_{ij}'(R_{ij}' - z_{ij}')^2}, \tag{3.2.38}$$

$$w' = x \sum_{i=1}^{2} \alpha_i' \sum_{j=1}^{2} \frac{L_{ij}'}{R_{ij}'(R_{ij}' - z_{ij}')}, \tag{3.2.39}$$

$$\begin{aligned} \tau_{zx}' = {} & \omega_3' L_{33}' \left[\frac{1}{R_{33}'(R_{33}' - z_{33}')} - \frac{y^2}{R_{33}'^3 (R_{33}' - z_{33}')} - \frac{y^2}{R_{33}'^2 (R_{33}' - z_{33}')^2} \right] \\ & + \sum_{i=1}^{2} \omega_i' \sum_{j=1}^{2} L_{ij}' \left[\frac{1}{R_{ij}'(R_{ij}' - z_{ij}')} - \frac{x^2}{R_{ij}'^3 (R_{ij}' - z_{ij}')} - \frac{x^2}{R_{ij}'^2 (R_{ij}' - z_{ij}')^2} \right], \end{aligned} \tag{3.2.40}$$

$$\begin{aligned} \tau_{yz}' = {} & \omega_3' xy L_{33}' \left[\frac{1}{R_{33}'^3 (R_{33}' - z_{33}')} + \frac{1}{R_{33}'^2 (R_{33}' - z_{33}')^2} \right] \\ & - xy \sum_{i=1}^{2} \omega_i' \sum_{j=1}^{2} L_{ij}' \left[\frac{1}{R_{ij}'^3 (R_{ij}' - z_{ij}')} + \frac{1}{R_{ij}'^2 (R_{ij}' - z_{ij}')^2} \right], \end{aligned} \tag{3.2.41}$$

$$\sigma_z' = x \sum_{i=1}^{2} \vartheta_i' \sum_{j=1}^{2} \frac{L_{ij}'}{R_{ij}'^3}. \tag{3.2.42}$$

When $s_1' = s_2'$, we get

$$u' = N'_{33}\left[\frac{1}{R'_{33} - z'_{33}} - \frac{y^2}{R'_{33}(R'_{33} - z'_{33})^2}\right] + \sum_{j=1}^{2}\left\{N'_{1j}\left[\frac{1}{R'_{1j} - z'_{1j}} - \frac{x^2}{R'_{1j}(R'_{1j} - z'_{1j})^2}\right]\right.$$
$$\left. + z'_1 N'_{2j}\left[\frac{1}{R'_{1j}(R'_{1j} - z'_{1j})} - \frac{x^2}{R'^3_{1j}(R'_{1j} - z'_{1j})} - \frac{x^2}{R'^2_{1j}(R'_{1j} - z'_{1j})^2}\right]\right\}, \tag{3.2.43}$$

$$v' = -xy\sum_{j=1}^{2}\left\{\frac{N'_{1j}}{R'_{1j}(R'_{1j} - z'_{1j})^2} + z'_1 N'_{2j}\left[\frac{1}{R'^3_{1j}(R'_{1j} - z'_{1j})} + \frac{1}{R'^2_{1j}(R'_{1j} - z'_{1j})^2}\right]\right\}$$
$$+ \frac{N'_{33}xy}{R'_{33}(R'_{33} - z'_{33})^2}, \tag{3.2.44}$$

$$w' = x\sum_{j=1}^{2}\left[\alpha'_1\frac{N'_{1j}}{R'_{1j}(R'_{1j} - z'_{1j})} + \alpha'_4\frac{N'_{2j}}{R'_{1j}(R'_{1j} - z'_{1j})} + \frac{z'_1}{R'^3_{1j}}\alpha'_1 N'_{2j}\right], \tag{3.2.45}$$

$$\tau'_{zx} = \omega'_3 N'_{33}\left[\frac{1}{R'_{33}(R'_{33} - z'_{33})} - \frac{y^2}{R'^3_{33}(R'_{33} - z'_{33})} - \frac{y^2}{R'^2_{33}(R'_{33} - z'_{33})^2}\right]$$
$$+ \sum_{j=1}^{2}\left\{\omega'_1 N'_{1j}\left[\frac{1}{R'_{1j}(R'_{1j} - z'_{1j})} - \frac{x^2}{R'^3_{1j}(R'_{1j} - z'_{1j})} - \frac{x^2}{R'^2_{1j}(R'_{1j} - z'_{1j})^2}\right]\right.$$
$$+ \omega'_4 N'_{2j}\left[\frac{1}{R'_{1j}(R'_{1j} - z'_{1j})} - \frac{x^2}{R'^3_{1j}(R'_{1j} - z'_{1j})} - \frac{x^2}{R'^2_{1j}(R'_{1j} - z'_{1j})^2}\right]$$
$$\left. + z'_1\omega'_1 N'_{2j}\left(\frac{1}{R'^3_{1j}} - \frac{3x^2}{R'^5_{1j}}\right)\right\}, \tag{3.2.46}$$

$$\tau'_{yz} = \omega'_3 xy N'_{33}\left[\frac{1}{R'^3_{33}(R'_{33} - z'_{33})} + \frac{1}{R'^2_{33}(R'_{33} - z'_{33})^2}\right]$$
$$- xy\sum_{j=1}^{2}\left\{\omega'_1 N'_{1j}\left[\frac{1}{R'^3_{1j}(R'_{1j} - z'_{1j})} + \frac{1}{R'^2_{1j}(R'_{1j} - z'_{1j})^2}\right]\right.$$
$$\left. + \omega'_4 N'_{2j}\left[\frac{1}{R'^3_{1j}(R'_{1j} - z'_{1j})} + \frac{1}{R'^2_{1j}(R'_{1j} - z'_{1j})^2}\right] + \frac{3z'_1}{R'^5_{1j}}\omega'_1 N'_{2j}\right\}, \tag{3.2.47}$$

$$\sigma'_z = x\sum_{j=1}^{2}\left[\frac{1}{R'^3_{1j}}(\vartheta'_1 N'_{1j} + \vartheta'_4 N'_{2j}) - 3z'_1\frac{z'_{1j}}{R'^5_{1j}}\vartheta'_1 N'_{2j}\right]. \tag{3.2.48}$$

Again, when the two half-spaces are ideally bonded at the interface $z = 0$, the conditions in Eq. (3.2.1) lead to

$$D_3 + D_{33} = L'_{33} \quad \text{or} \quad N'_{33}, \tag{3.2.49}$$

$$D_i + \sum_{j=1}^{2} D_{ji} = \sum_{j=1}^{2} L'_{ji} \quad \text{or} \quad N'_{1i}, \tag{3.2.50}$$

$$\alpha_i D_i - \sum_{j=1}^{2} \alpha_j D_{ji} = \sum_{j=1}^{2} \alpha'_j L'_{ji} \quad \text{or} \quad \alpha'_1 N'_{1i} + \alpha'_4 N'_{2i}, \tag{3.2.51}$$

$$\omega_3 (D_{33} - D_3) = \omega'_3 L'_{33} \quad \text{or} \quad \omega'_3 N'_{33}, \tag{3.2.52}$$

$$-\omega_i D_i + \sum_{j=1}^{2} \omega_j D_{ji} = -\sum_{j=1}^{2} \omega'_j L'_{ji} \quad \text{or} \quad -\omega'_1 N_{1i} - \omega'_4 N_{2i}, \tag{3.2.53}$$

$$\vartheta_i D_i + \sum_{j=1}^{2} \vartheta_j D_{ji} = \sum_{j=1}^{2} \vartheta'_j L'_{ji} \quad \text{or} \quad \vartheta'_1 N'_{1i} + \vartheta'_4 N'_{2i}, \tag{3.2.54}$$

where $i, j = 1, 2$. These ten equations determine the coefficients D_{33}, L'_{33}, D_{ij}, L'_{ij} $(i, j = 1, 2)$ or D_{33}, N'_{33}, D_{ij}, N'_{ij} $(i, j = 1, 2)$.

Similarly, when the two half-spaces are in smooth contact at their interface, we can use the conditions listed in Eq. (3.2.2), which yield

$$\alpha_i D_i - \sum_{j=1}^{2} \alpha_j D_{ji} = \sum_{j=1}^{2} \alpha'_j L'_{ji} \quad \text{or} \quad \alpha'_1 N'_{1i} + \alpha'_4 N'_{2i}, \tag{3.2.55}$$

$$\vartheta_i D_i + \sum_{j=1}^{2} \vartheta_j D_{ji} = \sum_{j=1}^{2} \vartheta'_j L'_{ji} \quad \text{or} \quad \vartheta'_1 N'_{1i} + \vartheta'_4 N'_{2i}, \tag{3.2.56}$$

$$\omega_3 (D_{33} - D_3) = 0, \tag{3.2.57}$$

$$-\omega'_3 L'_{33} = 0 \quad \text{or} \quad -\omega'_3 N'_{33} = 0, \tag{3.2.58}$$

$$-\omega_i D_i + \sum_{j=1}^{2} \omega_j D_{ji} = 0, \tag{3.2.59}$$

$$-\sum_{j=1}^{2} \omega'_j L'_{ji} = 0 \quad \text{or} \quad -\omega'_1 N_{1i} - \omega'_4 N_{2i} = 0, \tag{3.2.60}$$

that determine the constants.

(b) $s_1 = s_2$

In this case, we may take the displacement functions as

$$\varphi_1 = \frac{G_1 x}{\bar{R}_{11} + s_1 |z-h|} + \frac{G_{11} x}{R_{11} + z_{11}},$$
$$\varphi_2 = \text{sign}(z-h) \frac{G_2 x}{\bar{R}_{11}(\bar{R}_{11} + s_1 |z-h|)} + \frac{G_{21} x}{R_{11}(R_{11} + z_{11})}, \tag{3.2.61}$$
$$\varphi_3 = \frac{G_3 y}{\bar{R}_{33} + s_3 |z-h|} + \frac{G_{33} y}{R_{33} + z_{33}},$$

where G_1, G_2 and G_3 are given in Eq. (3.1.44), indicating that the terms correspond to the solution of an infinite body under a point force. The substitution of Eq. (3.2.61) into Eq. (2.2.61) leads to the following displacement components:

$$\begin{aligned} u = {} & G_3 \left[\frac{1}{\bar{R}_{33} + s_3 |z-h|} - \frac{y^2}{\bar{R}_{33}(\bar{R}_{33} + s_3 |z-h|)^2} \right] + G_{33} \left[\frac{1}{R_{33} + z_{33}} - \frac{y^2}{R_{33}(R_{33} + z_{33})^2} \right] \\ & + G_1 \left[\frac{1}{\bar{R}_{11} + s_1 |z-h|} - \frac{x^2}{\bar{R}_{11}(\bar{R}_{11} + s_1 |z-h|)^2} \right] + z_1 \text{sign}(z-h) G_2 \left[\frac{1}{\bar{R}_{11}(\bar{R}_{11} + s_1 |z-h|)} \right. \\ & \left. - \frac{x^2}{\bar{R}_{11}^3(\bar{R}_{11} + s_1 |z-h|)} - \frac{x^2}{\bar{R}_{11}^2(\bar{R}_{11} + s_1 |z-h|)^2} \right] + \frac{G_{11}}{R_{11} + z_{11}} \\ & - \frac{G_{11} x^2}{R_{11}(R_{11} + z_{11})^2} + z_1 \left[\frac{G_{21}}{R_{11}(R_{11} + z_{11})} - \frac{G_{21} x^2}{R_{11}^3(R_{11} + z_{11})} - \frac{G_{21} x^2}{R_{11}^2(R_{11} + z_{11})^2} \right], \end{aligned} \tag{3.2.62}$$

$$\begin{aligned} v = {} & \frac{G_3 xy}{\bar{R}_{33}(\bar{R}_{33} + s_3 |z-h|)^2} + \frac{G_{33} xy}{R_{33}(R_{33} + z_{33})^2} - \frac{G_1 xy}{\bar{R}_{11}(\bar{R}_{11} + s_1 |z-h|)^2} \\ & - G_2 xyz_1 \text{sign}(z-h) \left[\frac{1}{\bar{R}_{11}^3(\bar{R}_{11} + s_1 |z-h|)} + \frac{1}{\bar{R}_{11}^2(\bar{R}_{11} + s_1 |z-h|)^2} \right] - \frac{G_{11} xy}{R_{11}(R_{11} + z_{11})^2} \\ & - xyz_1 \left[\frac{G_{21}}{R_{11}^3(R_{11} + z_{11})} + \frac{G_{21}}{R_{11}^2(R_{11} + z_{11})^2} \right], \end{aligned} \tag{3.2.63}$$

$$\begin{aligned} w = {} & -\text{sign}(z-h) \frac{x}{\bar{R}_{11}(\bar{R}_{11} + s_1 |z-h|)} (\alpha_1 G_1 - \alpha_4 G_2) - \frac{\alpha_1 G_2 x z_1}{\bar{R}_{11}^3} \\ & - \alpha_1 x \frac{G_{11}}{R_{11}(R_{11} + z_{11})} + \alpha_4 x \frac{G_{21}}{R_{11}(R_{11} + z_{11})} - \alpha_1 z_1 x \frac{G_{21}}{R_{11}^3}. \end{aligned} \tag{3.2.64}$$

The corresponding stress expressions can then be obtained as

$$
\begin{aligned}
\tau_{zx} =& -\omega_3\left\{\operatorname{sign}(z-h)\,G_3\left[\frac{1}{\bar{R}_{33}(\bar{R}_{33}+s_3|z-h|)}-\frac{y^2}{\bar{R}_{33}^3(\bar{R}_{33}+s_3|z-h|)}-\frac{y^2}{\bar{R}_{33}^2(\bar{R}_{33}+s_3|z-h|)^2}\right]\right.\\
&\left.+G_{33}\left[\frac{1}{R_{33}(R_{33}+z_{33})}-\frac{y^2}{R_{33}^3(R_{33}+z_{33})}-\frac{y^2}{R_{33}^2(R_{33}+z_{33})^2}\right]\right\}-\operatorname{sign}(z-h)\,(\omega_1 G_1-\omega_4 G_2)\\
&\times\left[\frac{1}{\bar{R}_{11}(\bar{R}_{11}+s_1|z-h|)}-\frac{x^2}{\bar{R}_{11}^3(\bar{R}_{11}+s_1|z-h|)}-\frac{x^2}{\bar{R}_{11}^2(\bar{R}_{11}+s_1|z-h|)^2}\right]+G_2\omega_1 z_1\left(\frac{3x^2}{\bar{R}_{11}^5}-\frac{1}{\bar{R}_{11}^3}\right)\\
&-\omega_1\left[\frac{G_{11}}{R_{11}(R_{11}+z_{11})}-\frac{G_{11}x^2}{R_{11}^3(R_{11}+z_{11})}-\frac{G_{11}x^2}{R_{11}^2(R_{11}+z_{11})^2}\right]\\
&+\omega_4\left[\frac{G_{21}}{R_{11}(R_{11}+z_{11})}-\frac{G_{21}x^2}{R_{11}^3(R_{11}+z_{11})}-\frac{G_{21}x^2}{R_{11}^2(R_{11}+z_{11})^2}\right]\\
&+\omega_1 z_1\left(\frac{3G_{21}x^2}{R_{11}^5}-\frac{G_{21}}{R_{11}^3}\right),
\end{aligned}
\tag{3.2.65}
$$

$$
\begin{aligned}
\tau_{yz} =& -\omega_3 xy\left\{\operatorname{sign}(z-h)G_3\left[\frac{1}{\bar{R}_{33}^3(\bar{R}_{33}+s_3|z-h|)}+\frac{1}{\bar{R}_{33}^2(\bar{R}_{33}+s_3|z-h|)^2}\right]+G_{33}\left[\frac{1}{R_{33}^3(R_{33}+z_{33})}\right.\right.\\
&\left.\left.+\frac{1}{R_{33}^2(R_{33}+z_{33})^2}\right]\right\}+\operatorname{sign}(z-h)\,xy(\omega_1 G_1-\omega_4 G_2)\left[\frac{1}{\bar{R}_{11}^3(\bar{R}_{11}+s_1|z-h|)}\right.\\
&\left.+\frac{1}{\bar{R}_{11}^2(\bar{R}_{11}+s_1|z-h|)^2}\right]+\frac{3G_2\omega_{1m}xyz_1}{\bar{R}_{11}^5}+\omega_1 xy\left[\frac{G_{11}}{R_{11}^3(R_{11}+z_{11})}+\frac{G_{11}}{R_{11}^2(R_{11}+z_{11})^2}\right]\\
&-\omega_4 xy\left[\frac{G_{21}}{R_{11}^3(R_{11}+z_{11})}+\frac{G_{21}}{R_{11}^2(R_{11}+z_{11})^2}\right]+\omega_1 z_1 xy\frac{3G_{21}}{R_{11}^5},
\end{aligned}
\tag{3.2.66}
$$

$$
\sigma_z=\frac{G_1\vartheta_1 x}{\bar{R}_{11}^3}+G_2 x\left(\frac{3\vartheta_1 z_1 z_{11}}{\bar{R}_{11}^5}-\frac{\vartheta_4}{\bar{R}_{11}^3}\right)+\vartheta_1 x\frac{G_{11}}{R_{11}^3}-\vartheta_4 x\frac{G_{21}}{R_{11}^3}+\vartheta_1 z_1 x\frac{3G_{21}z_{11}}{R_{11}^5}. \tag{3.2.67}
$$

For the upper half-space ($z \le 0$), we can take the following displacement functions according to the eigenvalues:

$$
\varphi_i'=\frac{S_{i1}'x}{R_{i1}'-z_{i1}'},\quad (i=1,2),\qquad \varphi_3'=\frac{S_{33}'y}{R_{33}'-z_{33}'},\quad \text{when } s_1'\neq s_2', \tag{3.2.68}
$$

and

$$\varphi_1' = \frac{\pi_{11}' x}{R_{11}' - z_{11}'}, \quad \varphi_2' = \frac{\pi_{21}' x}{R_{11}'(R_{11}' - z_{11}')}, \quad \varphi_3' = \frac{\pi_{33}' y}{R_{33}' - z_{33}'}, \quad \text{when } s_1' = s_2'. \tag{3.2.69}$$

The substitution of Eq. (3.2.68) or Eq. (3.2.69) into Eqs. (2.2.56) and (2.2.63) or Eqs. (2.2.61) and Eq. (2.2.66) gives rise to the stresses and displacements. As before, when the two half-spaces are bonded ideally at their interface, the conditions specified by Eq. (3.2.1) lead to

$$G_3 + G_{33} = S_{33}' \quad \text{or} \quad \pi_{33}', \tag{3.2.70}$$

$$G_1 + G_{11} = \sum_{j=1}^{2} S_{j1}' \quad \text{or} \quad \pi_{11}', \tag{3.2.71}$$

$$\alpha_1 G_1 - \alpha_4 G_2 - \alpha_1 G_{11} + \alpha_4 G_{21} = \sum_{j=1}^{2} \alpha_j' S_{j1}' \quad \text{or} \quad \alpha_1' \pi_{11}' + \alpha_4' \pi_{21}', \tag{3.2.72}$$

$$\omega_3 (G_3 - G_{33}) = \omega_3' S_{33}' \quad \text{or} \quad \omega_3' \pi_{33}', \tag{3.2.73}$$

$$-\omega_1 G_1 + \omega_4 G_2 + \omega_1 G_{11} - \omega_4 G_{21} = -\sum_{j=1}^{2} S_{j1}' \omega_j' \quad \text{or} \quad -\omega_1' \pi_{11}' - \omega_4' \pi_{21}', \tag{3.2.74}$$

$$\vartheta_1 G_1 - \vartheta_4 G_2 + \vartheta_1 G_{11} - \vartheta_4 G_{21} = \sum_{j=1}^{2} \vartheta_j' S_{j1}' \quad \text{or} \quad \pi_{11}' \vartheta_1' + \pi_{21}' \vartheta_4'. \tag{3.2.75}$$

These six equations determine the coefficients G_{33}, S_{33}', G_{1j}, and S_{1j}' $(j=1,\ 2)$ or G_{33}, π_{33}, G_{1j} and π_{1j}' $(j=1,\ 2)$.

Again, when the two half-spaces are in smooth contact at their interface, we use the conditions listed in Eq. (3.2.2), which bring about the following ten equations to determine the constants:

$$\alpha_1 G_1 - \alpha_4 G_2 - \alpha_1 G_{11} + \alpha_4 G_{21} = \sum_{j=1}^{2} \alpha_j' S_{j1}' \quad \text{or} \quad \alpha_1' \pi_{11}' + \alpha_4' \pi_{21}', \tag{3.2.76}$$

$$\omega_3 (G_3 - G_{33}) = 0, \tag{3.2.77}$$

$$\omega_3' S_{33}' = 0 \quad \text{or} \quad \omega_3' \pi_{33}' = 0, \tag{3.2.78}$$

$$-\omega_1 G_1 + \omega_4 G_2 + \omega_1 G_{11} - \omega_4 G_{21} = 0, \tag{3.2.79}$$

$$-\sum_{j=1}^{2} S_{j1}' \omega_j' = 0 \quad \text{or} \quad -\omega_1' \pi_{11}' - \omega_4' \pi_{21}' = 0, \tag{3.2.80}$$

$$\vartheta_1 G_1 - \vartheta_4 G_2 + \vartheta_1 G_{11} - \vartheta_4 G_{21} = \sum_{j=1}^{2} \vartheta'_j S'_{j1} \quad \text{or} \quad \pi'_{11}\vartheta'_1 + \pi'_{21}\vartheta'_4 . \tag{3.2.81}$$

3.2.3 Some Remarks

The method for deriving the displacements and stresses presented here has some merits. It is straightforward and explicit; the undetermined coefficients are easily found from the boundary conditions.

When the materials of the two half-spaces are the same, the results obtained above reduce directly to those in **Section 3.1** for an infinite transversely isotropic solid, provided that h is set to zero.

The solutions can be extended to a wide range of engineering problems, for instance, a laminated structure composed of different material layers.

3.3 AN INFINITE TRANSVERSELY ISOTROPIC SPACE WITH AN INCLUSION[1]

Assume that an infinite transversely isotropic solid has an ellipsoidal inclusion of revolution, as shown in Fig. 3.2, with its surface described by

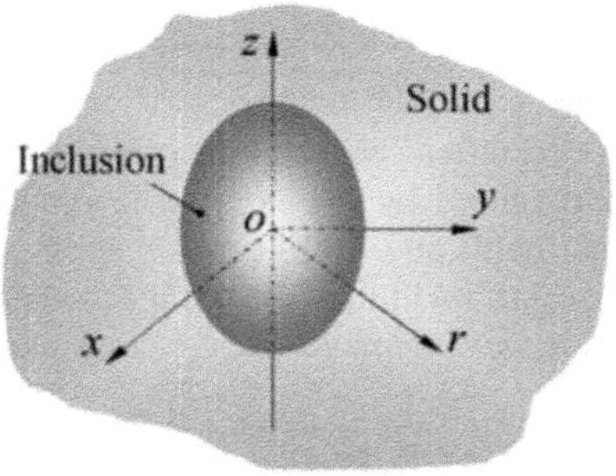

Figure 3.2 An ellipsoidal inclusion in an infinite solid.

$$\frac{z^2}{a^2} + \frac{r^2}{b^2} = 1 , \tag{3.3.1}$$

where $2a$ and $2b$ are the lengths of the major and minor axes of the ellipsoid. For convenience, we use superscript 'I' to indicate the mechanics quantities in the inclusion. Assume the problem is axisymmetric about the z-axis, the continuity of displacements and stresses across the solid-inclusion interface require that

[1] The solution procedure presented in this section mainly follows Chen's work (Chen, 1968).

$$u_r = u_r^I \ , \ w = w^I ,$$
$$(\sigma_r - \sigma_r^I)n_r + (\tau_{zr} - \tau_{zr}^I)n_z = 0, \tag{3.3.2}$$
$$(\tau_{zr} - \tau_{zr}^I)n_r + (\sigma_z - \sigma_z^I)n_z = 0,$$

where n_r and n_z are the direction cosines of the external normal of the interface in the *Orz* coordinate system.

The infinite space can be subjected to various loading conditions. For simplicity, we will discuss only the space under the following stresses at infinity:

$$\sigma_z = p_z, \ \sigma_r = \sigma_\alpha = p_r, \ \tau_{zr} = \tau_{r\alpha} = \tau_{\alpha z} = 0. \tag{3.3.3}$$

Clearly, such a loading condition defines an axisymmetric problem, and the general solutions of the displacements and stresses of the problem are already given in Eqs. (2.2.69*b*) and (2.2.70*b*).

Eshelby (1961) pointed out that the stress field inside the inclusion is uniform, *i.e.*,

$$u_r^I = B_1 r, \quad u_\alpha^I = 0, \quad w^I = B_2 z,$$
$$\sigma_r^I = \sigma_\alpha^I = (c_{11}^I + c_{12}^I)B_1 + c_{13}^I B_2, \tag{3.3.4}$$
$$\sigma_z^I = 2c_{13}^I B_1 + c_{33}^I B_2, \quad \tau_{zr}^I = 0,$$

where B_1 and B_2 are constants to be determined.

In the infinite body, the stress field can be considered as the superposition of two parts: the uniform stress field caused by the external loading described by Eq. (3.3.3) and the non-uniform stress field due to the interaction with the inclusion. For convenience, we use a superscript '*A*' to indicate the mechanics quantities related to the non-uniform stress field. In this way, we have

$$\sigma_r = p_r + \sigma_r^A, \quad \sigma_\alpha = p_r + \sigma_\alpha^A,$$
$$\sigma_z = p_z + \sigma_z^A, \quad \tau_{zr} = \tau_{zr}^A, \tag{3.3.5}$$
$$u_r = C_r r + u_r^A, \quad w = C_z z + u_z^A, \quad u_\alpha = 0,$$

where

$$C_r = (c_{33} p_r - c_{13} p_z)/[c_{33}(c_{11} + c_{12}) - 2c_{13}^2],$$
$$C_z = [(c_{11} + c_{12})p_z - 2c_{13} p_r]/[c_{33}(c_{11} + c_{12}) - 2c_{13}^2]. \tag{3.3.6}$$

To facilitate the use of the continuity conditions in Eq. (3.3.2), we can rewrite Eq. (3.3.1) as

$$\frac{z_i^2}{\left(\frac{a_i^2}{c_i^2}\right)}+\frac{r^2}{\left(\frac{b^2}{c_i^2}\right)}=c_i^2, \quad (i=1,2), \tag{3.3.7}$$

where

$$a_i = as_i, \quad c_i^2 = a_i^2 - b^2. \tag{3.3.8}$$

Introduce a new function $q_i(r, z)$ that satisfies

$$\frac{z_i^2}{q_i^2}+\frac{r^2}{q_i^2-1}=c_i^2, \quad (i=1,2); \tag{3.3.9}$$

Eq. (3.3.9) becomes identical to Eq. (3.3.1) when

$$q_i^2 = \rho_i^2 = \frac{a_i^2}{c_i^2} = \frac{a_i^2}{a_i^2-b^2}. \tag{3.3.10}$$

Thus at the space-inclusion interface, we have

$$\begin{aligned} &\frac{n_r}{n_z}=\frac{r}{z}\frac{a^2}{b^2}\equiv\frac{r}{s_i z_i}\frac{\rho_i^2}{(\rho_i^2-1)}, \\ &\left.\frac{\partial q_i}{\partial z_i}\right|_{q_i=\rho_i}=\frac{z_i}{\rho_i^3 D_i^2}, \quad \left.\frac{\partial q_i}{\partial r}\right|_{q_i=\rho_i}=\frac{r}{\rho_i(\rho_i^2-1)D_i^2}, \\ &D_i^2=\frac{r^2}{(\rho_i^2-1)^2}+\frac{z_i^2}{\rho_i^4}. \end{aligned} \tag{3.3.11}$$

To calculate the displacements and stresses using Eqs. (2.2.69*b*) and (2.2.70*b*), we need to determine the displacement function φ_i $(i=1,2)$ first. We can take

$$\varphi_i(r,z_i)=A_i H(r,z_i), \quad (i=1,2), \tag{3.3.12}$$

where A_i $(i=1,2)$ are undetermined constants and $H(r,z)$ is a harmonic function that is

$$H(r,z_i)=\frac{1}{2}[z_i^{\,2}\psi_1(q_i)+r^2\psi_2(q_i)-c_i^{\,2}\psi_0(q_i)], \tag{3.3.13}$$

where

$$\psi_0(q_i)=\frac{1}{2}\ln\left(\frac{q_i+1}{q_i-1}\right),\quad \psi_1(q_i)=\frac{1}{2}\ln\left(\frac{q_i+1}{q_i-1}\right)-\frac{1}{q_i},$$
$$\psi_2(q_i)=-\frac{1}{4}\ln\left(\frac{q_i+1}{q_i-1}\right)+\frac{1}{2}\frac{q_i}{q_i^2-1}. \tag{3.3.14}$$

Clearly,

$$\frac{\partial H(r,z_i)}{\partial z_i}=z_i\psi_1,\quad \frac{\partial H(r,z_i)}{\partial r}=r\psi_2. \tag{3.3.15}$$

Now the displacements and stresses can be obtained easily when the functions in Eqs. (3.3.12) to (3.3.14) are substituted into Eqs. (2.2.69*b*) and (2.2.70*b*). We therefore have

$$u_i^A=\sum_{i=1}^{2}A_i r\psi_2(q_i),\quad w^A=\sum_{i=1}^{2}A_i\alpha_i z_i\psi_1(q_i),$$
$$\frac{\sigma_r^A}{c_{66}}=\sum_{i=1}^{2}A_i s_i{}^2 k_{2i}\left[2\psi_2(q_i)+r\psi_2'(q_i)\frac{\partial q_i}{\partial r}\right]-2\sum_{i=1}^{2}A_i\psi_2(q_i),$$
$$\frac{\sigma_\alpha^A}{c_{66}}=-\sum_{i=1}^{2}A_i k_{1i}\left[2\psi_2(q_i)+r\psi_2'(q_i)\frac{\partial q_i}{\partial r}\right]+2\sum_{i=1}^{2}A_i\psi_2(q_i), \tag{3.3.16}$$
$$\frac{\sigma_z^A}{c_{66}}=-\sum_{i=1}^{2}A_i k_{2i}\left[2\psi_2(q_i)+r\psi_2'(q_i)\frac{\partial q_i}{\partial r}\right]=\sum_{i=1}^{2}A_i k_{2i}\left[\psi_1(q_i)+z_i\psi_1'(q_i)\frac{\partial q_i}{\partial z_i}\right],$$
$$\frac{\tau_{zr}^A}{c_{66}}=\sum_{i=1}^{2}A_i k_{3i}z_i\psi_1'(q_i)\frac{\partial q_i}{\partial r}=\sum_{i=1}^{2}A_i k_{3i}r\psi_2'(q_i)\frac{\partial q_i}{\partial z_i},$$

where $\psi_i'(q_j)$ stands for the derivative of $\psi_i(q_j)$ with respect to q_j.

To satisfy the continuity conditions of displacements and stresses across the space-inclusion interface, in view of Eqs. (3.3.4) and (3.3.5), we can substitute the above solution into the first, second and fourth equations in Eq. (3.3.2). This gives rise to

$$\sum_{i=1}^{2}A_i\psi_2(\rho_i)-B_1=-C_r,$$
$$\sum_{i=1}^{2}\alpha_i s_i A_i\psi_1(\rho_i)-B_2=-C_z, \tag{3.3.17}$$

$$\sum_{i=1}^{2}A_i\left[\psi_1(\rho_i)n_z+z_i\psi_1'(\rho_i)\left(\frac{\partial q_i}{\partial z_i}n_z+s_i\frac{\partial q_i}{\partial r}n_r\right)\right]_{q_i=\rho_i}=(2c_{13}^I B_1+c_{33}^I B_2-p_z)n_z, \tag{3.3.18}$$

where we have used the relationship among the material constants specified in Eq. (2.2.64). Using Eqs. (3.3.11) and (3.3.14), we can simplify Eq. (3.3.18) to

$$2\sum_{i=1}^{2}A_i k_{2i}\psi_2(\rho_i)+2c_{13}^I B_1+c_{33}^I B_2=p_z. \tag{3.3.19}$$

Furthermore, in view of Eqs. (3.3.4) and (3.3.5), when the third and sixth expressions in Eq. (3.3.16) are substituted into the third of Eq. (3.3.2), we get

$$\sum_{i=1}^{2}A_i[k_{2i}s_i^2\psi_1(\rho_i)+2\psi_2(\rho_i)]+(c_{11}^I+c_{12}^I)B_1+c_{12}^I B_2=p_r. \tag{3.3.20}$$

The constants A_i and B_i $(i=1, 2)$ can be found by using the four linear equations, Eqs. (3.3.17), (3.3.19) and (3.3.20).

Some stress components at certain locations are of special interest. For instance, the circumferential stress at the equator ($z=0$) can be obtained as

$$\sigma_z=c_{66}\sum_{i=1}^{2}A_i k_{2i}\psi_1(\rho_i)+p_z, \tag{3.3.21}$$

and that at the pole ($r=0$) is

$$\sigma_r=\sigma_\alpha=-2c_{66}\sum_{i=1}^{2}A_i k_{1i}\psi_2(\rho_i)+2c_{66}\sum_{i=1}^{2}A_i\psi_2(\rho_i)+p_r. \tag{3.3.22}$$

To discuss the stress distribution in detail, we need to know the specific conditions such as the property of the inclusion. In the following, we will take some extreme cases as the examples to show the applications of the solution obtained above.

(*a*) Rigid inclusion

In this case, the inclusion does not deform and all the displacement components in the inclusion vanish. Hence, Eq. (3.3.4) yields

$$B_1=B_2=0, \tag{3.3.23}$$

and therefore

$$\sum_{i=1}^{2} A_i \psi_2(\rho_i) = -C_r, \quad \sum_{i=1}^{2} \alpha_i s_i A_i \psi_1(\rho_i) = -C_z. \tag{3.3.24}$$

(*b*) Cavity

When the inclusion is a cavity, stresses vanish there. In this case, Eq. (3.3.4) also leads to $B_1 = B_2 = 0$. The other two constants, A_i $(i = 1, 2)$, are given by Eqs. (3.3.19) and (3.3.20), *i.e.*,

$$2\sum_{i=1}^{2} A_i k_{2i} \psi_2(\rho_i) = p_z, \quad \sum_{i=1}^{2} A_i [k_{2i} s_i^2 \psi_1(\rho_i) + 2\psi_2(\rho_i)] = p_r. \tag{3.3.25}$$

(*c*) Spherical inclusion

When the inclusion is a sphere, *i.e.*, $a = b$, we have

$$\frac{1}{\rho_i^2} = \frac{c_i^2}{a_i^2} = \frac{a_i^2 - a^2}{a_i^2} = 1 - \frac{1}{s_i^2}, \tag{3.3.26}$$

$$\psi_0(\rho_i) = \frac{1}{2}\ln\left(\frac{\rho_i + 1}{\rho_i - 1}\right) = \frac{1}{2}\ln\left[\frac{(1 + 1/\rho_i)^2}{(1 - 1/\rho_i^2)}\right] = \ln(s_i + \sqrt{s_i^2 - 1}),$$
$$\psi_1(\rho_i) = \psi_0(\rho_i) - \frac{1}{s_i}\sqrt{s_i^2 - 1}, \quad \psi_2(\rho_i) = -\frac{\psi_0(\rho_i) - s_i\sqrt{s_i^2 - 1}}{2}. \tag{3.3.27}$$

To show some quantitative figures, the stresses for a spherical inclusion were calculated with some selected materials whose properties are listed in Table 3.1 (Huntington, 1958). When a material is treated as an isotropic body, the Poisson's ratio is taken as 0.3. Tables 3.2 and 3.3 list the stresses at the equator and pole of the inclusion (Chen, 1968).

Table 3.1 Constants of various transversely isotropic materials (GPa).

Symbol	Material	c_{11}	c_{33}	c_{44}	c_{12}	c_{13}	Remarks
Zn	Zinc	161	61	38.3	34.2	50.1	
Mg	Magnesium	59.7	61.7	16.4	26.2	21.7	
SiO_2	β-quartz	116.6	110.4	36.06	16.7	32.8	Measured at 600°C
Cd	Cadmium	121	51.3	18.5	48.1	44.2	From different sources
		110	46.9	15.6	40.4	38.3	
Co	Cobalt	307	358.1	75.3	165	103	
$BaTiO_3$	Barium titanate ceramic	168	189	54.6	78.2	71	Electric field of $D = 0$
		166	162	42.9	76.6	77.5	Electric field of $E = 0$

It can be seen from Table 3.2 that the radial stress component σ_r at the pole for Zn, Cd and Co is much different from that for the isotropic material. Table 3.3 also shows that the axial stress component σ_z at the equator and the radial stress component σ_r at the pole for Zn and Cd are significantly different from the corresponding ones for the isotropic material. In particular, σ_z for Co is much greater than that for the isotropic case, indicating that the material anisotropy may lead to a state of stress with a lower safety factor.

Table 3.2 Circumferential stresses when $p_r = p_z = 1$ (GPa).

Inclusion material	σ_z at the equator ($r = a, z = 0$)	σ_r at the pole ($r = 0, z = a$)
Isotropic	15.00	15.00
Zn	14.05	6.542
Mg	15.59	16.95
SiO_2	16.03	13.23
Cd	15.24	8.556
Co	16.26	20.93
$BaTiO_3$	14.50	16.39

Table 3.3 Circumferential stresses when $p_r = 0$, $p_z = 1$ (GPa).

Inclusion material	σ_z at the equator ($r = a, z = 0$)	σ_r at the pole ($r = 0, z = a$)
Isotropic	20.46	-6.818
Zn	16.26	-10.19
Mg	21.29	-6.381
SiO_2	20.27	-4.903
Cd	17.91	-10.9
Co	23.06	-6.347
$BaTiO_3$	20.80	6.537

3.4 SPHERICALLY ISOTROPIC MATERIALS

In this section, we discuss two special problems when the space materials are spherically isotropic. One is an infinite space subjected to a point force and the other is the stress concentration caused by a spherical cavity (Chen, 1966*a*).

3.4.1 An Infinite Space Subjected to a Point Force

Assume that the point force, F_z, is at the origin and along the negative z-axis. We use the method given in **Section 2.4**. Consider a sphere of radius R with its centre at the origin. The static equilibrium of the sphere requires that the resultant force in the z-direction over the surface of the sphere must equal the point force F_z. That is,

$$2\pi \int_0^\pi (\sigma_R \cos\theta - \tau_{R\theta} \sin\theta) R^2 \sin\theta \, d\theta - F_z = 0 . \tag{3.4.1}$$

This equation states us that stresses σ_R and $\tau_{R\theta}$ are proportional to R^{-2}, which further indicates that the displacements should be proportional to R^{-1}. Now because the problem is axisymmetric, the function ψ_0 in Eq. (2.4.14) vanishes and the function F_0 in Eq. (2.4.20) can be taken as

$$F_0 = R^{-1} Y(\theta) . \tag{3.4.2}$$

By substituting this into Eq. (2.4.21), we get an equation involving $Y(\theta)$ whose characteristic roots are

$$l_1 = 1 , \quad l_2 = -2, \quad l_3 = (-1 + \sqrt{1 + 8L/N})/2 , \quad l_4 = -l_3 - 1 . \tag{3.4.3}$$

Hence, we can take

$$F_0 = c_1 R^{-1} P_1(\cos\theta) , \tag{3.4.4}$$

where c_1 is an unknown constant, and $P_1(\cos\theta)$ is a *Legendre polynomial*[2]. Using Eq. (2.4.20), we get

$$w_0 = x_1 R^{-1} c_1 P_1(\cos\theta), \qquad G_0 = y_1 R^{-1} c_1 P_1(\cos\theta) , \tag{3.4.5}$$

where

$$x_1 = -2(c_{11} + c_{44} - c_{66}) , \quad y_1 = 2c_{11} + c_{44} - c_{13} - 2c_{66} . \tag{3.4.6}$$

The displacements and stresses can be obtained easily when Eq. (3.4.5) is substituted into Eqs. (2.4.3), (1.2.3) and (1.2.40), *i.e.*,

$$u_R = x_1 R^{-1} c_1 \cos\theta, \quad u_\theta = y_1 R^{-1} c_1 \sin\theta, \quad u_\alpha = 0 , \tag{3.4.7}$$

---

[2] **Appendix B** presents a brief account of special functions that will be encountered in the book, *e.g.* *Legendre functions* and *Bessel functions*.

$$\begin{aligned}
\sigma_\theta &= R^{-2}[(c_{11}+c_{12}-c_{13})x_1+(c_{11}+c_{12})y_1]c_1\cos\theta,\\
\sigma_\alpha &= R^{-2}[(c_{11}+c_{12}-c_{13})x_1+(c_{11}+c_{12})y_1]c_1\cos\theta,\\
\sigma_R &= R^{-2}[(2c_{13}-c_{33})x_1+2c_{13}y_1]c_1\cos\theta,\\
\tau_{R\theta} &= -c_{44}R^{-2}(x_1+2y_1)c_1\sin\theta,\\
\tau_{\alpha R} &= \tau_{\theta\alpha}=0.
\end{aligned} \tag{3.4.8}$$

Now the integration in Eq. (3.4.1) can be completed with the σ_R and $\tau_{R\theta}$ given above, which gives rise to

$$2\pi\left\{\frac{2}{3}[(2c_{13}-c_{33})x_1+2c_{13}y_1]+\frac{4}{3}(x_1+2y_1)c_{44}\right\}c_1=F_z, \tag{3.4.9}$$

or,

$$c_1=\frac{3F_z}{4\pi[(2c_{13}-c_{33})x_1+2c_{13}y_1+2(x_1+2y_1)c_{44}]}, \tag{3.4.10}$$

and thus the displacements in Eq. (3.4.7) become

$$\begin{aligned}
u_R &= \frac{3F_zR^{-1}\dfrac{x_1}{y_1}}{4\pi\left[\dfrac{x_1}{y_1}(2c_{13}-c_{33}+2c_{44})+2c_{13}+4c_{44}\right]}\cos\theta,\\
u_\theta &= \frac{3F_zR^{-1}}{4\pi\left[\dfrac{x_1}{y_1}(2c_{13}-c_{33}+2c_{44})+2c_{13}+4c_{44}\right]}\sin\theta,\\
u_\alpha &= 0.
\end{aligned} \tag{3.4.11}$$

3.4.2 Stress Concentration near a Spherical Cavity

Assume that the centre of the spherical cavity of radius R_0 is at the coordinate origin. We consider the space subjected to a uniform tension, p, at infinity in the z-direction.

Again, the deformation is axisymmetric about the z-axis and thus $\psi_0=0$. The stress state in the space far away from the cavity is described by $\sigma_z=\sigma_z^\infty=p$. In the spherical coordinate system, the stress state can be written as

$$\sigma_R^\infty = \sigma_R^\infty \cos^2\theta = p\cos^2\theta, \quad \sigma_\theta^\infty = p\sin^2\theta,$$
$$\tau_{R\theta}^\infty = -p\sin\theta\cos\theta, \qquad \sigma_\alpha^\infty = 0. \tag{3.4.12}$$

The surface of the spherical cavity must be stress-free. Thus to satisfy this boundary condition at the cavity surface we must superpose the following stresses to those in Eq. (3.4.12):

$$\sigma_R = -\sigma_R^\infty = -p\cos^2\theta, \quad \tau_{R\theta} = -\tau_{R\theta}^\infty = p\sin\theta\cos\theta. \tag{3.4.13}$$

Using the Legendre function, we can rewrite these stresses as

$$\sigma_R = -\frac{1}{3}p - \frac{2}{3}pP_2(\cos\theta), \quad \tau_{R\theta} = \frac{1}{3}pP_2^1(\cos\theta). \tag{3.4.14}$$

The solution of the problem with these boundary conditions can be resolved into two parts, the constant boundary stress part and variable boundary stress part. The constant term in the stress field of Eq. (3.4.14) corresponds to a spherically symmetric problem whose cavity surface is subjected to

$$\sigma_R^0 = -p/3, \quad \tau_{R\theta}^0 = 0, \tag{3.4.14a}$$

where superscript 0' denotes that the stresses are the constant part of Eq. (3.4.14). With these special boundary conditions, the displacements can be obtained directly as

$$u_R^0 = c_0 R^n, \quad u_\theta^0 = u_\alpha^0 = 0, \tag{3.4.15}$$

where

$$n = -[1 + \sqrt{1 + 8(c_{11} + c_{12} - c_{13})/c_{33}}]/2, \tag{3.4.16}$$

Since $(c_{11} + c_{12})c_{33} - 2c_{13}^2 > 0$ [3], it can be shown that $8(c_{11} + c_{12})c_{33} - 8c_{13}c_{33} + c_{33}^2 = 8(c_{11} + c_{12})c_{33} - 16c_{13}^2 + (c_{33} - 4c_{13})^2 > 0$. Further noticing $c_{33} > 0$, we then conclude that

$$c_{33} + 8(c_{11} + c_{12} - c_{13}) > 0, \tag{3.4.17}$$

i.e. n is always a negative real number. The stresses are obtained as

[3] Equation (2.2.47).

$$\begin{aligned}\sigma_R^0 &= (2c_{13} + nc_{33})R^{n-1}c_0,\\ \sigma_\theta^0 &= \sigma_\alpha^0 = (c_{11} + c_{12} + nc_{13})R^{n-1}c_0,\\ \tau_{R\theta}^0 &= \tau_{\alpha R}^0 = \tau_{\theta\alpha}^0 = 0.\end{aligned} \tag{3.4.18}$$

Then the boundary conditions, Eq. (3.4.14*a*), give rise to

$$c_0 = -\frac{pR_0^{1-n}}{3(2c_{13} + nc_{33})}. \tag{3.4.19}$$

Thus the displacements, Eq. (3.4.15), and stresses, Eq. (3.4.18), are determined.

The variable terms in the boundary conditions specified by Eq. (3.4.14) contain (associated) Legendre functions $P_2(\cos\theta)$ and $P_2^1(\cos\theta)$. To solve this problem, let

$$F_0 = r(R)P_2(\cos\theta), \tag{3.4.20}$$

and substitute it into Eq. (2.4.21). This brings about a differential equation for $r(R)$. The characteristic roots of the equation are obtained as

$$\begin{aligned}n_1 &= \left[-1 + \sqrt{1 - 4P + 4\sqrt{P^2 - Q}}\right]/2, \quad n_2 = -n_1 - 1,\\ n_3 &= \left[-1 + \sqrt{1 - 4P - 4\sqrt{P^2 - Q}}\right]/2, \quad n_4 = -n_3 - 1,\end{aligned} \tag{3.4.21}$$

where

$$P = D - 3M, \quad Q = 8(3N - L), \tag{3.4.22}$$

in which, D, L, M and N were defined in Eq. (2.4.23). The solution for $r(R)$ is

$$r(R) = R^{n_i}. \tag{3.4.23}$$

The characteristic roots in Eq. (3.4.21) are functions of elastic constants and are quite possibly complex. First, consider the case of $P^2 - Q \geq 0$, which is satisfied by most materials listed in Table 1.4 such as Ceramic PZT-4, InSe, Magnesium, *etc.* In this case, we can verify that $1 - 4P > 4\sqrt{P^2 - Q}$ or, equivalently, $1 - 8P + 16Q > 0$ so that the four characteristic roots in Eq. (3.4.21) are all reals. Making use of Eqs. (3.4.22) and (2.4.23), we can show that this inequality is further equivalent to

$$\begin{aligned}&c_{44}(-120c_{12}+72c_{13}+9c_{33}+264c_{11})\\&\quad+(c_{11}-c_{12})(64c_{11}+64c_{12}-64c_{13}-4c_{33})-24c_{13}^2+24c_{11}c_{33}>0.\end{aligned} \tag{3.4.24}$$

The demonstration of this inequality is outlined below:

$$\begin{aligned}&c_{44}(-120c_{12}+72c_{13}+9c_{33}+264c_{11})\\&\quad+(c_{11}-c_{12})(64c_{11}+64c_{12}-64c_{13}-4c_{33})-24c_{13}^2+24c_{11}c_{33}\\&=c_{44}[120(c_{11}-c_{12})+72c_{13}+9c_{33}+144c_{11}]\\&\quad+(c_{11}-c_{12})[64(c_{11}+c_{12}-c_{13})-4c_{33}]-24c_{13}^2+24c_{11}c_{33}\\&>c_{44}[120(c_{11}-c_{12})+72c_{13}+9c_{33}+72c_{11}+72c_{12}]\\&\quad+(c_{11}-c_{12})(-8c_{33}-4c_{33})-24c_{13}^2+24c_{11}c_{33}\\&=c_{44}\{120(c_{11}-c_{12})+9[8(c_{11}+c_{12}+c_{13})+c_{33}]\}\\&\quad+12(c_{11}+c_{12})c_{33}-24c_{13}^2\\&>120c_{44}(c_{11}-c_{12})+12[(c_{11}+c_{12})c_{33}-2c_{13}^2]>0.\end{aligned} \tag{3.4.25}$$

In the derivation, we have employed Eq. (3.4.17), Eq. (2.2.47), as well as the inequality $8(c_{11}+c_{12}+c_{13})+c_{33}>0$, which can be verified in a way identical to that to obtain Eq. (3.4.17).

Thus, we can take

$$F_0=\sum_{i=2,4}c_iR^{n_i}P_2(\cos\theta). \tag{3.4.26}$$

Then the displacements and stresses can be derived as

$$\begin{aligned}u_R^p&=\sum_{i=2,4}c_ix_iR^{n_i}P_2(\cos\theta),\\u_\theta^p&=\sum_{i=2,4}c_iy_iR^{n_i}P_2^1(\cos\theta),\\u_\alpha^p&=0,\end{aligned} \tag{3.4.27}$$

$$\begin{aligned}\sigma_\theta^p&=\sum_{i=2,4}[(s_i+t_ic_{11})P_2(\cos\theta)+z_i\cot\theta P_2^1(\cos\theta)]R^{n_i-1}c_i,\\\sigma_\alpha^p&=\sum_{i=2,4}[(s_i+t_ic_{12})P_2(\cos\theta)-z_i\cot\theta P_2^1(\cos\theta)]R^{n_i-1}c_i,\\\sigma_R^p&=\sum_{i=2,4}(h_i+t_ic_{13})P_2(\cos\theta)R^{n_i-1}c_i,\\\tau_{R\theta}^p&=\sum_{i=2,4}c_{44}[y_i(n_i-1)-x_i]P_2^1(\cos\theta)R^{n_i-1}c_i,\\\tau_{\alpha R}^p&=\tau_{\theta\alpha}^p=0,\end{aligned} \tag{3.4.28}$$

where

$$\begin{aligned}
x_i &= n_i(n_i+1)c_{44} - 2(c_{44} - c_{66}) - 6c_{11}, \\
y_i &= n_i(c_{13} + c_{44}) + 2(c_{11} + c_{44} - c_{66}), \\
s_i &= x_i(c_{11} + c_{12} + n_i c_{13}), \ t_i = 6y_i, \\
z_i &= (c_{12} - c_{11})y_i, \ h_i = x_i(2c_{13} + n_i c_{33}),
\end{aligned} \tag{3.4.29}$$

where $i = 2, 4$. In these expressions, the constants c_2 and c_4 are determined by

$$\begin{aligned}
&(h_2 + t_2 c_{13})R_0^{n_2-1}c_2 + (h_4 + t_4 c_{13})R_0^{n_4-1}c_4 = -\frac{2p}{3}, \\
&c_{44}[y_2(n_2-1) - x_2]R_0^{n_2-1}c_2 + c_{44}[y_4(n_4-1) - x_4]R_0^{n_4-1}c_4 = \frac{p}{3}.
\end{aligned} \tag{3.4.30}$$

The complete stress solution of the original problem is the superposition of the following three parts:

$$\begin{aligned}
&\sigma_R = \sigma_R^0 + \sigma_R^p + \sigma_R^\infty, \ \sigma_\theta = \sigma_\theta^0 + \sigma_\theta^p + \sigma_\theta^\infty, \\
&\sigma_\alpha = \sigma_\alpha^0 + \sigma_\alpha^p, \ \tau_{R\theta} = \tau_{R\theta}^p + \tau_{R\theta}^\infty, \ \tau_{\theta\alpha} = \tau_{\alpha R} = 0.
\end{aligned} \tag{3.4.31}$$

When $P^2 - Q < 0$[4], n_1 and n_3 constitute a pair of complex conjugates, while n_2 and n_4 constitute another pair of complex conjugates. We take the function F_0 in the form as in Eq. (3.4.26), but with c_2 and c_4 being a pair of complex conjugates.

[4] For materials such as TiB_2 and Zinc in Table 1.4.

4

HALF-SPACE AND LAYERED MEDIA

The solution to the stresses and displacements in a half-space or a layered solid with transverse isotropy is fundamental to the development of the theory of elasticity and is of primary importance to many engineering applications such as the design of high strength foundations and underground structures. In this chapter, we will first present a unified solution for a point force applied on the surface of a half-space. Then we will discuss the solution for a transversely isotropic half-space subjected to a point force at the interior. The solution also will be extended to the problem of layered media using the state-space and Fourier transform methods.

4.1 UNIFIED SOLUTION FOR A HALF-SPACE SUJECTED TO A SURFACE POINT FORCE

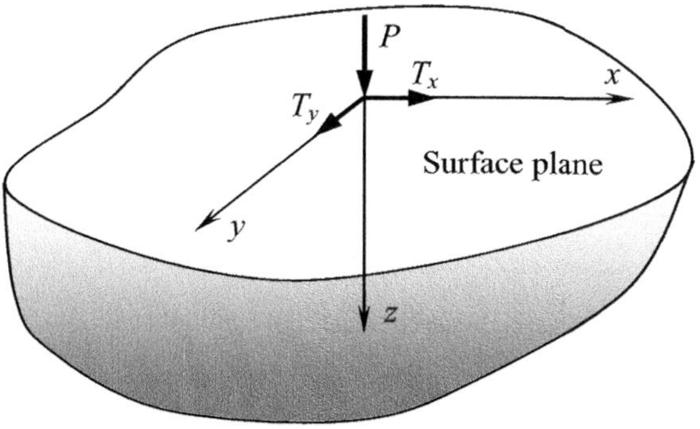

Fig. 4.1 A transversely isotropic half-space acted on by a surface point force.

Consider a transversely isotropic half-space with isotropic plane parallel to its boundary surface. We use the Cartesian coordinate system (x, y, z) and let the xOy-plane coincide with the half-space surface and the z-axis point towards the interior, Fig. 4.1. In this case, any point inside the half-space will have $z > 0$.

Assume that an arbitrary point force is applied at the origin. We can decompose the problem to three sub-problems by using the *principle of superposition*: the solution corresponding to a vertical force, P, in the positive z-direction, the solution to a tangential force $T_x = T$, in the x-direction, as well as the solution to a tangential force, $T_y = T$, in the y-direction. The last solution can be obtained from the second solution easily by a coordinate transform with x replaced by y, and y by $-x$, respectively.

4.1.1 A Point Force Normal to the Half-Space Surface

This is an axisymmetric problem. When $s_1 \neq s_2$, we take

$$\varphi_i = A_i \ln(R_i + z_i), \quad (i = 1, 2), \quad \varphi_3 = 0, \tag{4.1.1}$$

where A_i are constants to be determined, $R_i = \sqrt{r^2 + z_i^2}$, $r^2 = x^2 + y^2$, and $z_i = s_i z$. Substituting Eq. (4.1.1) into Eqs. (2.2.56) and (2.2.63) yields

$$u = x\sum_{i=1}^{2} \frac{A_i}{R_i R_i^*}, \quad v = y\sum_{i=1}^{2} \frac{A_i}{R_i R_i^*}, \quad w = \sum_{i=1}^{2} \frac{A_i \alpha_i}{R_i}, \tag{4.1.2}$$

$$\sigma_x = 2c_{66}\sum_{i=1}^{2} A_i \left(\frac{1}{R_i R_i^*} - \frac{x^2}{R_i^3 R_i^*} - \frac{x^2}{R_i^2 R_i^{*2}} \right) - c_{66}\sum_{i=1}^{2} k_{1i} \frac{A_i z_i}{R_i^3}, \tag{4.1.3}$$

$$\sigma_y = 2c_{66}\sum_{i=1}^{2} A_i \left(\frac{1}{R_i R_i^*} - \frac{y^2}{R_i^3 R_i^*} - \frac{y^2}{R_i^2 R_i^{*2}} \right) - c_{66}\sum_{i=1}^{2} k_{1i} \frac{A_i z_i}{R_i^3}, \tag{4.1.4}$$

$$\sigma_z = -c_{66}\sum_{i=1}^{2} k_{2i} \frac{A_i z_i}{R_i^3}, \tag{4.1.5}$$

$$\tau_{yz} = -c_{66} y \sum_{i=1}^{2} k_{3i} \frac{A_i}{R_i^3}, \quad \tau_{zx} = -c_{66} x \sum_{i=1}^{2} k_{3i} \frac{A_i}{R_i^3}, \tag{4.1.6}$$

$$\tau_{xy} = -2c_{66} xy \sum_{i=1}^{2} A_i \left(\frac{1}{R_i^3 R_i^*} + \frac{1}{R_i^2 R_i^{*2}} \right), \tag{4.1.7}$$

where $R_i^* = R_i + z_i$, α_i are given in Eq. (2.2.57), while k_{ij} are defined in Eq. (2.2.64).

Utilizing the boundary conditions $\tau_{yz} = \tau_{zx} = 0$ at $z = 0$, we obtain from Eq. (4.1.6)

$$k_{31}A_1 + k_{32}A_2 = 0. \tag{4.1.8}$$

It is seen from Eq. (4.1.5) that the boundary condition, $\sigma_z = 0$ at $z = 0$, is satisfied automatically except at the origin where the force is applied. Considering the equilibrium of an elastic layer bounded by the surface $z = 0$ and the plane $z = \varepsilon$ ($\varepsilon > 0$), we obtain

$$\int_{-\infty}^{+\infty}\int_{-\infty}^{+\infty} \sigma_z(x, y, \varepsilon)\,\mathrm{d}x\,\mathrm{d}y + P = 0. \tag{4.1.9}$$

Substituting Eq. (4.1.5) into this equation yields

$$k_{21}A_1 + k_{22}A_2 = \frac{P}{2\pi c_{66}}. \tag{4.1.10}$$

Equations (4.1.8) and (4.1.10) then give rise to

$$\begin{aligned} A_1 &= \frac{k_{32}P}{2\pi c_{66}(k_{32}k_{21} - k_{31}k_{22})} = -\frac{Ps_1s_2}{2\pi c_{66}(s_1 - s_2)k_{31}}, \\ A_2 &= \frac{-k_{31}P}{2\pi c_{66}(k_{32}k_{21} - k_{31}k_{22})} = \frac{Ps_1s_2}{2\pi c_{66}(s_1 - s_2)k_{32}}, \end{aligned} \tag{4.1.11}$$

in which the following identities have been used:

$$k_{3i} = s_i k_{2i}, \quad (i = 1, 2). \tag{4.1.12}$$

Substituting Eq. (4.1.11) into Eq. (4.1.2) yields

$$u = -\frac{s_1s_2Px}{2\pi(s_1 - s_2)c_{66}}\left(\frac{1}{k_{31}R_1R_1^*} - \frac{1}{k_{32}R_2R_2^*}\right),$$

$$v = \frac{y}{x}u,$$

$$w = -\frac{s_1 s_2 P}{2\pi(s_1 - s_2)c_{66}}\left(\frac{\alpha_1}{k_{31}R_1} - \frac{\alpha_2}{k_{32}R_2}\right). \tag{4.1.13}$$

Now we proceed as in **Section 3.1** to eliminate the factor $(s_1 - s_2)$ in the denominator in this equation:

$$\begin{aligned} s_1 s_2\left(\frac{1}{k_{31}R_1R_1^*} - \frac{1}{k_{32}R_2R_2^*}\right) &= \frac{(k_{32}R_2R_2^* - k_{31}R_1R_1^*)s_1s_2}{k_{31}k_{32}R_1R_1^*R_2R_2^*} \\ &= \frac{c_{44}[c_{11}(s_1R_2R_2^* - s_2R_1R_1^*) + c_{13}s_1s_2(s_2R_2R_2^* - s_1R_1R_1^*)]}{c_{66}(c_{13}+c_{44})k_{31}k_{32}R_1R_1^*R_2R_2^*}, \end{aligned} \tag{4.1.14}$$

$$\begin{aligned} s_1 s_2\left(\frac{\alpha_1}{k_{31}R_1} - \frac{\alpha_2}{k_{32}R_2}\right) &= \frac{s_1s_2(\alpha_1k_{32}R_2 - \alpha_2k_{31}R_1)}{k_{31}k_{32}R_1R_2} \\ &= \frac{s_2^2s_1k_{31}R_2 - s_1^2s_2k_{32}R_1}{k_{31}k_{32}R_1R_2} \\ &= \frac{c_{44}[c_{11}(s_2^2R_2 - s_1^2R_1) + c_{13}s_1^2s_2^2(R_2 - R_1)]}{c_{66}(c_{13}+c_{44})k_{31}k_{32}R_1R_2}, \end{aligned} \tag{4.1.15}$$

where the formulas in Eq. (2.2.65) as well as the following have been used:

$$s_i k_{3i} = \frac{c_{44}(c_{11} + c_{13}s_i^2)}{c_{66}(c_{13}+c_{44})}, \quad (i = 1, 2). \tag{4.1.16}$$

Since $R_i^* = R_i + z_i$, we obtain

$$s_1R_2R_2^* - s_2R_1R_1^* = (s_1 - s_2)\left[r^2 - s_1s_2z^2 - \frac{s_1s_2(s_1+s_2)}{R_1+R_2}z^3\right], \tag{4.1.17}$$

$$\begin{aligned} &s_2R_2R_2^* - s_1R_1R_1^* \\ &= (s_2 - s_1)\left\{r^2 + (s_1^2 + s_1s_2 + s_2^2)z^2 + \frac{(s_1+s_2)z}{(s_1^2R_1 + s_2^2R_2)}[(s_1^2+s_2^2)r^2 + (s_1^4 + s_1^2s_2^2 + s_2^4)z^2]\right\}, \end{aligned} \tag{4.1.18}$$

$$s_2^2R_2 - s_1^2R_1 = (s_2^2 - s_1^2)\frac{(s_1^2+s_2^2)r^2 + (s_1^4 + s_1^2s_2^2 + s_2^4)z^2}{s_1^2R_1 + s_2^2R_2}, \tag{4.1.19}$$

$$R_2 - R_1 = (s_2^2 - s_1^2)\frac{z^2}{R_1 + R_2}. \tag{4.1.20}$$

Substituting Eqs. (4.1.14) and (4.1.15) into Eq. (4.1.13) and making use of Eqs. (4.1.17) to (4.1.20), we obtain

$$\begin{aligned}
u &= \frac{Px}{2\pi(c_{11}c_{33} - c_{13}^2)R_1 R_1^* R_2 R_2^*}\left\{c_{11}\left(z^2 - \frac{r^2}{s_1 s_2} + \frac{s_1 + s_2}{R_1 + R_2}z^3\right)\right. \\
&\quad \left. + c_{13}\left[r^2 + z^2(s_1^2 + s_1 s_2 + s_2^2) + z(s_1 + s_2)\frac{(s_1^2 + s_2^2)r^2 + (s_1^4 + s_1^2 s_2^2 + s_2^4)z^2}{s_1^2 R_1 + s_2^2 R_2}\right]\right\}, \\
v &= \frac{y}{x}u, \\
w &= \frac{(s_1 + s_2)P}{2\pi s_1 s_2(c_{11}c_{33} - c_{13}^2)R_1 R_2}\left[c_{11}\frac{(s_1^2 + s_2^2)r^2 + (s_1^4 + s_1^2 s_2^2 + s_2^4)z^2}{s_1^2 R_1 + s_2^2 R_2} + c_{13}\frac{s_1^2 s_2^2 z^2}{R_1 + R_2}\right],
\end{aligned} \tag{4.1.21}$$

where we have also used the following identity to simplify the expression of $k_{31}k_{32}$

$$\frac{s_1^2 s_2^2}{(c_{11} + c_{13}s_1^2)(c_{11} + c_{13}s_2^2)} = \frac{c_{44}}{(c_{13} + c_{44})(c_{11}c_{33} - c_{13}^2)}. \tag{4.1.22}$$

It is clear that uncertainty will not appear when $s_1 = s_2$ in the solution given by Eq. (4.1.21). It is a unified solution for displacements that is suitable for both $s_1 \neq s_2$ and $s_1 = s_2$. The corresponding solution for isotropic materials, *i.e.* the famous *Boussinesq solution*, can be derived easily from it by setting $s_1 = s_2 = 1$ and using Eq. (1.4.13):

$$\begin{aligned}
u &= -\frac{xP}{4\pi(\lambda + \mu)RR^*} + \frac{xzP}{4\pi\mu R^3}, \\
v &= \frac{y}{x}u, \\
w &= \frac{(\lambda + 2\mu)P}{4\pi\mu(\lambda + \mu)R} + \frac{z^2 P}{4\pi\mu R^3},
\end{aligned} \tag{4.1.23}$$

where $R = \sqrt{r^2 + z^2}$ and $R^* = R + z$.

The surface settlement of the half-space caused by P is obtained from Eq. (4.1.21) readily

$$w\Big|_{z=0} = \frac{c_{11}(s_1 + s_2)}{2\pi s_1 s_2 (c_{11}c_{33} - c_{13}^2)} \frac{P}{r}. \tag{4.1.24}$$

Substituting the unified displacement solution, Eq. (4.1.21), into Eq. (1.2.33) leads to the unified solution for stresses.

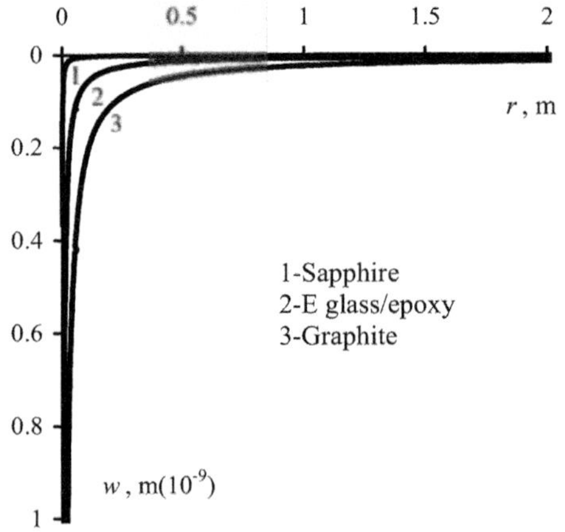

Fig. 4.2 Surface settlement due to a unit force normal to the half-space.

Figure 4.2 depicts the surface settlement of a transversely isotropic half-space due to a concentrated normal load (1 N) applied at the origin for three different materials, for which the elastic constants were given in Table 1.4. As shown in the figure, the magnitudes of settlement for the three materials are obviously different from each other.

4.1.2 A Point Force Tangential to the Half-Space Surface

Assume that a tangential point force, $T_x = T$, is applied at the origin along the x-direction. When $s_1 \neq s_2$, we take

$$\varphi_i = \frac{D_i x}{R_i + z_i}, \quad (i = 1,2), \qquad \varphi_3 = \frac{D_3 y}{R_3 + z_3}, \tag{4.1.25}$$

where D_i are constants to be determined. Substituting Eq. (4.1.25) into Eqs. (2.2.56) and (2.2.63) yields

$$u = \sum_{i=1}^{2} D_i \left(\frac{1}{R_i^*} - \frac{x^2}{R_i R_i^{*2}} \right) + D_3 \left(\frac{1}{R_3^*} - \frac{y^2}{R_3 R_3^{*2}} \right), \tag{4.1.26}$$

$$v = -xy \sum_{i=1}^{2} \frac{D_i}{R_i R_i^{*2}} + \frac{D_3 xy}{R_3 R_3^{*2}}, \tag{4.1.27}$$

$$w = -x \sum_{i=1}^{2} \frac{\alpha_i D_i}{R_i R_i^*}, \tag{4.1.28}$$

$$\begin{aligned} \sigma_x &= -2c_{66} D_3 x \left(\frac{1}{R_3 R_3^{*2}} - \frac{y^2}{R_3^3 R_3^{*2}} - \frac{2y^2}{R_3^2 R_3^{*3}} \right) \\ &\quad + c_{66} x \sum_{i=1}^{2} D_i \left(\frac{k_{1i}}{R_i^3} - \frac{6}{R_i R_i^{*2}} + \frac{2x^2}{R_i^3 R_i^{*2}} + \frac{4x^2}{R_i^2 R_i^{*3}} \right), \end{aligned} \tag{4.1.29}$$

$$\begin{aligned} \sigma_y &= 2c_{66} D_3 x \left(\frac{1}{R_3 R_3^{*2}} - \frac{y^2}{R_3^3 R_3^{*2}} - \frac{2y^2}{R_3^2 R_3^{*3}} \right) \\ &\quad + c_{66} x \sum_{i=1}^{2} D_i \left(\frac{k_{1i}}{R_i^3} - \frac{2}{R_i R_i^{*2}} + \frac{2y^2}{R_i^3 R_i^{*2}} + \frac{4y^2}{R_i^2 R_i^{*3}} \right), \end{aligned} \tag{4.1.30}$$

$$\sigma_z = c_{66} x \sum_{i=1}^{2} k_{2i} \frac{D_i}{R_i^3}, \tag{4.1.31}$$

$$\tau_{yz} = c_{66} xy \left[\sum_{i=1}^{2} k_{3i} D_i \left(\frac{1}{R_i^3 R_i^*} + \frac{1}{R_i^2 R_i^{*2}} \right) - \frac{D_3}{s_3} \left(\frac{1}{R_3^3 R_3^*} + \frac{1}{R_3^2 R_3^{*2}} \right) \right], \tag{4.1.32}$$

$$\begin{aligned} \tau_{zx} &= -c_{66} \sum_{i=1}^{2} k_{3i} D_i \left(\frac{1}{R_i R_i^*} - \frac{x^2}{R_i^3 R_i^*} - \frac{x^2}{R_i^2 R_i^{*2}} \right) \\ &\quad - \frac{c_{66} D_3}{s_3} \left(\frac{1}{R_3 R_3^*} - \frac{y^2}{R_3^3 R_3^*} - \frac{y^2}{R_3^2 R_3^{*2}} \right), \end{aligned} \tag{4.1.33}$$

$$\tau_{xy} = -2c_{66}y\sum_{i=1}^{2}D_i\left(\frac{1}{R_iR_i^{*2}} - \frac{2x^2}{R_i^2R_i^{*3}} - \frac{x^2}{R_i^3R_i^{*2}}\right)$$
$$+c_{66}D_3y\left(\frac{1}{R_3R_3^{*2}} - \frac{x^2}{R_3^3R_3^{*2}} - \frac{2x^2}{R_3^2R_3^{*3}}\right) - c_{66}D_3y\left(\frac{3}{R_3R_3^{*2}} - \frac{y^2}{R_3^3R_3^{*2}} - \frac{2y^2}{R_3^2R_3^{*3}}\right), \quad (4.1.34)$$

where $R_i^* = R_i + z_i$.

Using the boundary conditions, $\tau_{yz} = \tau_{zx} = \sigma_z = 0$ at $z = 0$, we obtain from Eqs. (4.1.31)-(4.1.33)

$$\sum_{i=1}^{2}k_{2i}D_i = 0, \quad (4.1.35)$$

$$\sum_{i=1}^{2}k_{3i}D_i - \frac{D_3}{s_3} = 0. \quad (4.1.36)$$

Further, consider an elastic layer bounded by the surface $z = 0$ and the plane $z = \varepsilon$ ($\varepsilon > 0$). The equilibrium of the layer demands

$$\int_{-\infty}^{+\infty}\int_{-\infty}^{+\infty}\tau_{zx}(x, y, \varepsilon)\,\mathrm{d}x\,\mathrm{d}y + T = 0. \quad (4.1.37)$$

Substituting Eq. (4.1.33) into this equation yields

$$\sum_{i=1}^{2}k_{3i}D_i + \frac{D_3}{s_3} = \frac{T}{\pi c_{66}}. \quad (4.1.38)$$

The three algebraic equations, Eqs. (4.1.35), (4.1.36) and (4.1.38), then give

$$D_1 = \frac{-k_{22}T}{2\pi c_{66}(k_{32}k_{21} - k_{31}k_{22})} = \frac{Ts_1}{2\pi c_{66}(s_1 - s_2)k_{31}},$$
$$D_2 = \frac{k_{21}P}{2\pi c_{66}(k_{32}k_{21} - k_{31}k_{22})} = -\frac{Ts_2}{2\pi c_{66}(s_1 - s_2)k_{32}},. \quad (4.1.39)$$
$$D_3 = \frac{Ts_3}{2\pi c_{66}}.$$

Substituting Eq. (4.1.39) into Eqs. (4.1.26)-(4.1.28), yields

$$
\begin{aligned}
u &= \frac{s_3 T}{2\pi c_{66}}\left(\frac{1}{R_3^*} - \frac{y^2}{R_3 R_3^{*2}}\right) \\
&\quad + \frac{T}{2\pi c_{66}(s_1 - s_2)}\left[\frac{s_1}{k_{31}R_1^*} - \frac{s_2}{k_{32}R_2^*} - x^2\left(\frac{s_1}{k_{31}R_1 R_1^{*2}} - \frac{s_2}{k_{32}R_2 R_2^{*2}}\right)\right], \\
v &= \frac{s_3 T}{2\pi c_{66}}\frac{xy}{R_3 R_3^{*2}} - \frac{Txy}{2\pi c_{66}(s_1 - s_2)}\left(\frac{s_1}{k_{31}R_1 R_1^{*2}} - \frac{s_2}{k_{32}R_2 R_2^{*2}}\right), \\
w &= -\frac{Tx}{2\pi c_{66}(s_1 - s_2)}\left(\frac{\alpha_1 s_1}{k_{31}R_1 R_1^*} - \frac{\alpha_2 s_2}{k_{32}R_2 R_2^*}\right).
\end{aligned}
\tag{4.1.40}
$$

Now we proceed as in the last section to eliminate the factor $(s_1 - s_2)$ in the denominator in Eq. (4.1.40). It can be shown that

$$
\begin{aligned}
&\frac{s_1}{k_{31}R_1^*} - \frac{s_2}{k_{32}R_2^*} = \frac{c_{44}[c_{11}(s_1^2 R_2^* - s_2^2 R_1^*) + c_{13}s_1^2 s_2^2 (s_2 - s_1)z]}{s_1 s_2 c_{66}(c_{13} + c_{44})k_{31}k_{32}R_1^* R_2^*}, \\
&\frac{s_1}{k_{31}R_1 R_1^{*2}} - \frac{s_2}{k_{32}R_2 R_2^{*2}} \\
&\qquad = \frac{c_{44}[c_{11}(s_1^2 R_2 R_2^{*2} - s_2^2 R_1 R_1^{*2}) + c_{13}s_1^2 s_2^2 (R_2 R_2^{*2} - R_1 R_1^{*2})]}{s_1 s_2 c_{66}(c_{13} + c_{44})k_{31}k_{32}R_1 R_2 R_1^{*2} R_2^{*2}}, \\
&\frac{\alpha_1 s_1}{k_{31}R_1 R_1^*} - \frac{\alpha_2 s_2}{k_{32}R_2 R_2^*} \\
&\qquad = \frac{c_{44}[c_{11}(s_2 R_2 R_2^* - s_1 R_1 R_1^*) + c_{13}s_1 s_2 (s_1 R_2 R_2^* - s_2 R_1 R_1^*)]}{c_{66}(c_{13} + c_{44})k_{31}k_{32}R_1 R_2 R_1^* R_2^*},
\end{aligned}
\tag{4.1.41}
$$

and

$$
s_1^2 R_2^* - s_2^2 R_1^* = (s_1 - s_2)\left[s_1 s_2 z + (s_1 + s_2)\frac{(s_1^2 + s_2^2)r^2 + s_1^2 s_2^2 z^2}{s_1^2 R_2 + s_2^2 R_1}\right],
$$

$$
\begin{aligned}
s_1^2 R_2 R_2^{*2} - s_2^2 R_1 R_1^{*2} &= (s_1 - s_2)\left\{ s_1 s_2 z\left[2(r^2 - s_1 s_2 z^2) - \frac{(s_1+s_2)s_1 s_2 z^3}{R_1 + R_2} \right] \right. \\
&\quad \left. + (s_1+s_2)\frac{(s_1^2+s_2^2)r^6 + 3s_1^2 s_2^2 r^4 z^2 - s_1^4 s_2^4 z^6}{s_1^2 R_2^3 + s_2^2 R_1^3} \right\}, \\
R_2 R_2^{*2} - R_1 R_1^{*2} &= -(s_1 - s_2)z\left\{ 2[r^2 + (s_1^2 + s_1 s_2 + s_2^2)z^2] \right. \\
&\quad + (s_1+s_2)z\frac{3r^4 + 3(s_1^2+s_2^2)r^2 z^2 + (s_1^4 + s_1^2 s_2^2 + s_2^4)z^4}{R_1^3 + R_2^3} \\
&\quad \left. + (s_1+s_2)z\frac{(s_1^2+s_2^2)r^2 + (s_1^4 + s_1^2 s_2^2 + s_2^4)z^2}{s_1^2 R_1 + s_2^2 R_2} \right\}, \\
s_2 R_2 R_2^{*} - s_1 R_1 R_1^{*} &= -(s_1 - s_2)\left\{ r^2 + (s_1^2 + s_1 s_2 + s_2^2)z^2 \right. \\
&\quad \left. + (s_1+s_2)z\frac{(s_1^2+s_2^2)r^2 + (s_1^4 + s_1^2 s_2^2 + s_2^4)z^2}{s_1^2 R_1 + s_2^2 R_2} \right\}, \\
s_1 R_2 R_2^{*} - s_2 R_1 R_1^{*} &= (s_1 - s_2)\left[r^2 - s_1 s_2 z^2 - (s_1+s_2)s_1 s_2 \frac{z^3}{R_1 + R_2} \right].
\end{aligned}
\tag{4.1.42}
$$

Substituting Eq. (4.1.41) into Eq. (4.1.40) and making use of Eqs. (4.1.42) and (4.1.22), we obtain

$$
\begin{aligned}
u &= \frac{s_3 T}{2\pi c_{66}}\left(\frac{1}{R_3^*} - \frac{y^2}{R_3 R_3^{*2}} \right) + \frac{c_{33} T}{2\pi c_{11}(c_{11}c_{33} - c_{13}^2)}(\tilde{u} - x^2 \tilde{v}), \\
v &= \frac{s_3 T}{2\pi c_{66}}\frac{xy}{R_3 R_3^{*2}} - \frac{c_{33} T xy}{2\pi c_{11}(c_{11}c_{33} - c_{13}^2)}\tilde{v}, \\
w &= -\frac{Tx}{2\pi s_1 s_2 (c_{11}c_{33} - c_{13}^2)}\tilde{w},
\end{aligned}
\tag{4.1.43}
$$

where

$$
\tilde{u} = \frac{1}{R_1^* R_2^*}\left\{ c_{11} s_1 s_2 z + \frac{c_{11}(s_1+s_2)[(s_1^2+s_2^2)r^2 + s_1^2 s_2^2 z^2]}{s_1^2 R_2 + s_2^2 R_1} - c_{13} s_1^2 s_2^2 z \right\}
\tag{4.1.44}
$$

$$\widetilde{v} = \frac{1}{R_1 R_2 R_1^{*2} R_2^{*2}} \left\{ 2c_{11} s_1 s_2 z (r^2 - s_1 s_2 z^2) \right.$$

$$- c_{11}(s_1 + s_2) \left[\frac{s_1^2 s_2^2 z^4}{R_1 + R_2} - \frac{(s_1^2 + s_2^2) r^6 + 3 s_1^2 s_2^2 r^4 z^2 - s_1^4 s_2^4 z^6}{s_1^2 R_2^3 + s_2^2 R_1^3} \right]$$

$$- 2c_{13} s_1^2 s_2^2 z [r^2 + (s_1^2 + s_1 s_2 + s_2^2) z^2] - c_{13}(s_1 + s_2) s_1^2 s_2^2 z$$

$$\left. \times \left[\frac{(s_1^2 + s_2^2) r^2 + (s_1^4 + s_1^2 s_2^2 + s_2^4) z^2}{s_1^2 R_1 + s_2^2 R_2} + \frac{3r^4 + 3(s_1^2 + s_2^2) r^2 z^2 - (s_1^4 + s_1^2 s_2^2 + s_2^4) z^4}{R_1^3 + R_2^3} \right] \right\}, \tag{4.1.45}$$

$$\widetilde{w} = \frac{1}{R_1 R_2 R_1^* R_2^*} \left\{ - c_{11} [r^2 + (s_1^2 + s_1 s_2 + s_2^2) z^2] \right.$$

$$- c_{11}(s_1 + s_2) z \frac{(s_1^2 + s_2^2) r^2 + (s_1^4 + s_1^2 s_2^2 + s_2^4) z^2}{s_1^2 R_1 + s_2^2 R_2} \tag{4.1.46}$$

$$\left. + 2 s_1 s_2 c_{13} \left[r^2 - s_1 s_2 z^2 - \frac{(s_1 + s_2) s_1 s_2 z^3}{R_1 + R_2} \right] \right\}.$$

Equation (4.1.43) gives the extended *Cerruti solution* for a transversely isotropic half-space, which is a unified solution suitable for both $s_1 \neq s_2$ and $s_1 = s_2$. For isotropic materials, Eq. (1.4.13) should be used, giving $s_1 = s_2 = s_3 = 1$. The Cerruti solution then is obtained from Eq. (4.1.43)

$$\begin{aligned} u &= \frac{T}{4\pi(\lambda + \mu)} \left(\frac{1}{R^*} - \frac{x^2}{R R^{*2}} \right) + \frac{T}{4\pi\mu R} + \frac{T x^2}{4\pi\mu R^3}, \\ v &= \frac{T}{4\pi\mu} \frac{xy}{R^3} - \frac{T}{4\pi(\lambda + \mu)} \frac{xy}{R R^{*2}}, \\ w &= \frac{T}{4\pi\mu} \frac{xz}{R^3} + \frac{T}{4\pi(\lambda + \mu)} \frac{x}{R R^*}. \end{aligned} \tag{4.1.47}$$

The surface settlement of the half-space caused by T is obtained from Eq. (4.1.43) readily

$$w\big|_{z=0} = \frac{(c_{11} - c_{13} s_1 s_2) T}{2\pi s_1 s_2 (c_{11} c_{33} - c_{13}^2)} \frac{x}{r^2} = \frac{T}{2\pi(\overline{c}_{13} + c_{13})} \frac{x}{r^2}. \tag{4.1.48}$$

This equation indicates that, when a transversely isotropic half-space is subjected to a tangential load, the surface settlement is related only to c_{13} and $\bar{c}_{13} = \sqrt{c_{11}c_{33}}$, while the surface settlement due to a normal load depends on c_{13}, c_{11}, c_{33} and c_{44}, as shown in Eq. (4.1.24).

Substituting the unified displacement solution (4.1.43) into Eq. (1.2.33) gives the unified solution for stresses.

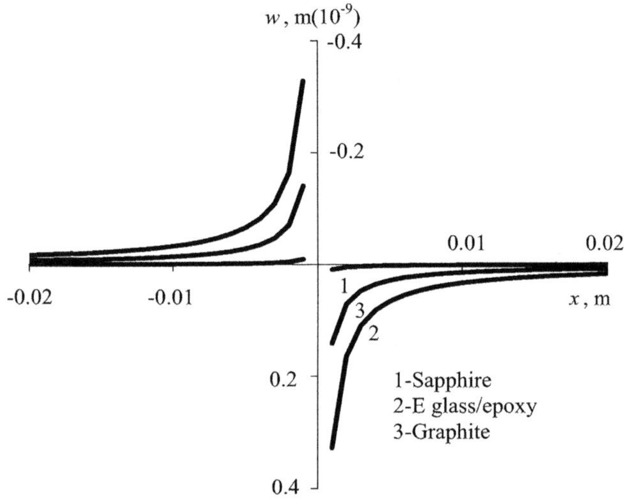

Fig. 4.3 Surface settlement due to a unit force tangential to the half-space.

The distributions of surface settlement along the x-axis of a half-space, subjected to a concentrated tangential force applied at the origin, are given in Fig. 4.3. The magnitudes of settlement for three transversely isotropic materials are again different from each other. An interesting phenomenon that is contrary to that in Fig. 4.2 is that, under the action of a tangential force, the settlement of the E glass/epoxy half-space is larger than that of the Graphite half-space. This just reflects the fact that, since an anisotropic material possesses different elastic constants along different directions, two anisotropic materials will generally have different relative rigidities in different directions.

4.2 A HALF-SPACE SUJECTED TO AN INTERIOR POINT FORCE

For the given displacements on the half-space surface, we have

$$u = v = w = 0, \quad \text{at } z = 0, \tag{4.2.1}$$

while for the given zero stresses on the half-space surface, we have

$$\sigma_z = \tau_{zx} = \tau_{yz} = 0, \quad \text{at } z = 0. \tag{4.2.2}$$

In the following, the solution of the problem of a concentrated load acts in the interior of a semi-infinite transversely isotropic half-space with boundary conditions shown in Eq. (4.2.1) is referred to as the extended *Lorentz solution*, while the solution corresponding to boundary conditions in Eq. (4.2.2) is referred to as the extended *Mindlin solution*.

In the derivations in **Section 3.1** and **Section 4.1**, to obtain unified solutions of an infinite space subjected to a point force and a semi-finite space subjected to a surface point force, respectively, we overcame the singularity by eliminating the term $(s_1 - s_2)$ in the denominators of the expressions. In the following, we can see that the denominators in the expressions for displacements, (u_2, v_2, w_2), of both the extended Lorentz and Mindlin solutions contain a term $(s_1 - s_2)^2$. In principle, the singularity thus caused can also be eliminated using the method in **Section 3.1** or **Section 4.1** though it is more difficult. Nevertheless, we prefer to use two different general solutions to derive the corresponding expressions of the point force solutions for a half-space for the two situations $s_1 \neq s_2$, and $s_1 = s_2$.

4.2.1 A Point Force Normal to the Half-Space Surface

Let the point force P act at the interior point $(0, 0, h)$. We need to discuss the two cases separately.

(1) $s_1 \neq s_2$

In this case, the displacements and stresses in the half space of $z \geq 0$ are given by Eqs. (3.2.6) and (3.2.7). Equations (3.2.6) and (4.2.1) yield

$$A_i - \sum_{j=1}^{2} A_{ji} = 0, \ (i = 1, 2), \tag{4.2.3}$$

$$\alpha_i A_i + \sum_{j=1}^{2} \alpha_j A_{ji} = 0, \ (i = 1, 2), \tag{4.2.4}$$

and hence

$$\begin{aligned}
A_{11} &= -A_1(\alpha_1+\alpha_2)/(\alpha_1-\alpha_2),\\
A_{12} &= -2A_2\alpha_2/(\alpha_1-\alpha_2) = 2A_1\alpha_2/(\alpha_1-\alpha_2),\\
A_{21} &= 2A_1\alpha_1/(\alpha_1-\alpha_2),\\
A_{22} &= A_2(\alpha_1+\alpha_2)/(\alpha_1-\alpha_2) = A_{11}.
\end{aligned} \tag{4.2.5}$$

Similarly, Eqs. (3.2.7) and (4.2.2) give rise to

$$\vartheta_i A_i - \sum_{j=1}^{2} \vartheta_j A_{ji} = 0\,, \quad (i=1,2)\,, \tag{4.2.6}$$

$$\omega_i A_i + \sum_{j=1}^{2} \omega_j A_{ji} = 0\,, \quad (i=1,2)\,, \tag{4.2.7}$$

which in turn leads to

$$\begin{aligned}
A_{11} &= -A_1(\vartheta_1\omega_2+\vartheta_2\omega_1)/D_P,\\
A_{12} &= -2A_2\vartheta_2\omega_2/D_P = 2A_1\vartheta_2\omega_2/D_P,\\
A_{21} &= 2A_1\vartheta_1\omega_1/D_P,\\
A_{22} &= A_2(\vartheta_1\omega_2+\vartheta_2\omega_1)/D_P = A_{11},
\end{aligned} \tag{4.2.8}$$

where

$$D_P = \vartheta_2\omega_1 - \vartheta_1\omega_2 = \vartheta_1\vartheta_2(s_1-s_2) = \frac{c_{44}(c_{11}c_{33}-c_{13}^2)}{c_{13}+c_{44}}(s_1-s_2)\,. \tag{4.2.9}$$

It is clear that the denominators of coefficients A_i contain the term (s_1-s_2) and thus the denominators of coefficients A_{ij} possess a singularity due to $(s_1-s_2)^2$. Therefore, we must pay attention to the numerical errors in calculation when the two eigenvalues are very close.

(2) $s_1 = s_2$

As the two eigenvalues are the same, we need to use the displacement and stress solutions given by Eqs. (3.2.16) and (3.2.17). Equations (3.2.16) and (4.2.1) lead to

$$\begin{aligned} &C_1 + C_{11} = 0, \\ &\alpha_1 C_1 + \alpha_4 C_2 + \alpha_1 C_{11} + \alpha_4 C_{21} = 0. \end{aligned} \tag{4.2.10}$$

Thus

$$C_{11} = -C_1, \ C_{21} = -C_2. \tag{4.2.11}$$

Similarly, Eqs. (3.2.17) and (4.2.2) yield

$$\vartheta_1 (C_1 - C_{11}) + \vartheta_4 (C_2 - C_{21}) = 0, \tag{4.2.12}$$

$$\omega_1 (C_1 + C_{11}) + \omega_4 (C_2 + C_{21}) = 0. \tag{4.2.13}$$

Thus

$$\begin{aligned} &C_{11} = -[C_1 (\vartheta_4 \omega_1 + \vartheta_1 \omega_4) + 2 C_2 \vartheta_4 \omega_4] / D_T, \\ &C_{21} = [2 C_1 \vartheta_1 \omega_1 + C_2 (\vartheta_4 \omega_1 + \vartheta_1 \omega_4)] / D_T, \end{aligned} \tag{4.2.14}$$

where

$$D_T = \vartheta_4 \omega_1 - \vartheta_1 \omega_4. \tag{4.2.15}$$

4.2.2 A Point Force Tangential to the Half-Space Surface

As usual, let the tangential point force act along the x-direction.

(1) $s_1 \neq s_2$

The displacements and stresses in this case are given by Eqs. (3.2.29) to (3.2.31) and (3.2.32) to (3.2.34), respectively. Equations (3.2.29) to (3.2.31) and Eq. (4.2.1) lead to

$$\begin{aligned} &D_{33} + D_3 = 0, \\ &D_i + \sum_{j=1}^{2} D_{ji} = 0, \quad (i = 1, 2), \\ &\alpha_i D_i - \sum_{j=1}^{2} \alpha_j D_{ji} = 0, \quad (i = 1, 2). \end{aligned} \tag{4.2.16}$$

Hence,

$$\begin{aligned}&D_{11}=D_1(\alpha_1+\alpha_2)/(\alpha_1-\alpha_2),\ D_{12}=2D_2\alpha_2/(\alpha_1-\alpha_2),\\&D_{21}=-2D_1\alpha_1/(\alpha_1-\alpha_2),\ D_{22}=-D_2(\alpha_1+\alpha_2)/(\alpha_1-\alpha_2),\ D_{33}=-D_3.\end{aligned} \tag{4.2.17}$$

Again in a similar way, Eqs. (3.2.32) to (3.2.34) and Eq. (4.2.2) lead to

$$D_{33}=D_3, \tag{4.2.18}$$

$$\vartheta_i D_i+\sum_{j=1}^{2}\vartheta_j D_{ji}=0,\quad (i=1,2), \tag{4.2.19}$$

$$-\omega_i D_i+\sum_{j=1}^{2}\omega_j D_{ji}=0,\quad (i=1,2). \tag{4.2.20}$$

Thus,

$$\begin{aligned}&D_{11}=D_1(\vartheta_1\omega_2+\vartheta_2\omega_1)/D_P,\ D_{12}=2D_2\vartheta_2\omega_2/D_P,\\&D_{21}=-2D_1\vartheta_1\omega_1/D_P,\ D_{22}=-D_2(\vartheta_1\omega_2+\vartheta_2\omega_1)/D_P,\ D_{33}=D_3,\end{aligned} \tag{4.2.21}$$

where D_P is defined by Eq. (4.2.9).

(2) $s_1=s_2$

The displacements and stresses are given by Eqs. (3.2.62) to (3.2.64) and (3.2.65) to (3.2.67), respectively. Equations (3.2.62) to (3.2.64), and the displacement boundary conditions, Eq. (4.2.1), lead to

$$\begin{aligned}&G_3+G_{33}=0,\\&G_1+G_{11}=0,\\&\alpha_1 G_1-\alpha_4 G_2-\alpha_1 G_{11}+\alpha_4 G_{21}=0.\end{aligned} \tag{4.2.22}$$

Hence,

$$\begin{aligned}&G_{33}=-G_3,\quad G_{11}=-G_1,\\&G_{21}=G_2-2G_1\alpha_1/\alpha_4.\end{aligned} \tag{4.2.23}$$

Equations (3.2.65) to (3.2.67), and the stress boundary conditions, Eq. (4.2.2), yield

$$G_{33} = G_3, \tag{4.2.24}$$

$$\omega_1(G_1 - G_{11}) + \omega_4(G_{21} - G_2) = 0, \tag{4.2.25}$$

$$\vartheta_1(G_1 + G_{11}) - \vartheta_4(G_2 + G_{21}) = 0, \tag{4.2.26}$$

and hence,

$$\begin{aligned} G_{11} &= [G_1(\vartheta_4\omega_1 + \vartheta_1\omega_4) - 2G_2\,\vartheta_4\omega_4]/D_T, \\ G_{21} &= [2G_1\,\vartheta_1\omega_1 - G_2(\vartheta_4\omega_1 + \vartheta_1\omega_4)]/D_T, \\ G_{33} &= G_3, \end{aligned} \tag{4.2.27}$$

where D_T is defined by Eq. (4.2.15).

4.3 GENERAL SOLUTION BY FOURIER TRANSFORM

The point force solutions in the last two sections were obtained by selecting suitable harmonic functions. Now let us try to use the Fourier transform method to develop the general solution again. This is another important method for solving transversely isotropic problems.

As with the formulation of Eq. (2.1.23) in **Section 2.1**, take

$$\mathbf{a} = [\sigma_z, \tau_{zx}, \tau_{yz}, w, u, v]^{\mathrm{T}}. \tag{4.3.1}$$

Then the state formulations can be written as

$$\frac{\partial \mathbf{a}}{\partial z} = \mathbf{A}\mathbf{a} + \mathbf{c}, \tag{4.3.2}$$

$$\mathbf{b} = \mathbf{B}\mathbf{a}, \tag{4.3.3}$$

where

$$\mathbf{c} = [-F_z, -F_x, -F_y, 0, 0, 0]^{\mathrm{T}}, \tag{4.3.4}$$

$$\mathbf{b}=[\sigma_x+\sigma_y,\ \sigma_x-\sigma_y,\ \tau_{xy}]^{\mathrm{T}}, \tag{4.3.5}$$

in which **A** and **B** are operator matrices defined by

$$\mathbf{A}=\begin{bmatrix} 0 & -\dfrac{\partial}{\partial x} & -\dfrac{\partial}{\partial y} & 0 & 0 & 0 \\ -\dfrac{c_{13}}{c_{33}}\dfrac{\partial}{\partial x} & 0 & 0 & 0 & \dfrac{c_{13}^2-c_{11}c_{33}}{c_{33}}\dfrac{\partial^2}{\partial x^2}-c_{66}\dfrac{\partial^2}{\partial y^2} & \dfrac{c_{13}^2+c_{33}c_{66}-c_{11}c_{33}}{c_{33}}\dfrac{\partial^2}{\partial x\partial y} \\ -\dfrac{c_{13}}{c_{33}}\dfrac{\partial}{\partial y} & 0 & 0 & 0 & \dfrac{c_{13}^2+c_{33}c_{66}-c_{11}c_{33}}{c_{33}}\dfrac{\partial^2}{\partial x\partial y} & \dfrac{c_{13}^2-c_{11}c_{33}}{c_{33}}\dfrac{\partial^2}{\partial y^2}-c_{66}\dfrac{\partial^2}{\partial x^2} \\ \dfrac{1}{c_{33}} & 0 & 0 & 0 & -\dfrac{c_{13}}{c_{33}}\dfrac{\partial}{\partial x} & -\dfrac{c_{13}}{c_{33}}\dfrac{\partial}{\partial y} \\ 0 & \dfrac{1}{c_{44}} & 0 & -\dfrac{\partial}{\partial x} & 0 & 0 \\ 0 & 0 & \dfrac{1}{c_{44}} & -\dfrac{\partial}{\partial y} & 0 & 0 \end{bmatrix}, \tag{4.3.6}$$

$$\mathbf{B}=\begin{bmatrix} \dfrac{2c_{13}}{c_{33}} & 0 & 0 & 0 & 2\left(c_{11}-c_{66}-\dfrac{c_{13}^2}{c_{33}}\right)\dfrac{\partial}{\partial x} & 2\left(c_{11}-c_{66}-\dfrac{c_{13}^2}{c_{33}}\right)\dfrac{\partial}{\partial y} \\ 0 & 0 & 0 & 0 & 2c_{66}\dfrac{\partial}{\partial x} & -2c_{66}\dfrac{\partial}{\partial y} \\ 0 & 0 & 0 & 0 & c_{66}\dfrac{\partial}{\partial y} & c_{66}\dfrac{\partial}{\partial x} \end{bmatrix}. \tag{4.3.7}$$

Equations (4.3.3) and (4.3.7) show that if we can find the individual elements of vector **a**, then the other three stress components in vector **b** can be derived directly by differentiation. Therefore **a** is called a basic variable (vector). By taking the *double Fourier transform*

$$\overline{f}(\alpha,\beta)=\frac{1}{2\pi}\int_{-\infty}^{+\infty}\int_{-\infty}^{+\infty}f(x,y)\,\mathrm{e}^{\mathrm{i}(\alpha x+\beta y)}\,\mathrm{d}x\,\mathrm{d}y, \tag{4.3.8}$$

we obtain from Eqs. (4.3.2) and Eq. (4.3.3)

$$\frac{\mathrm{d}\overline{\mathbf{a}}}{\mathrm{d}z}=\overline{\mathbf{A}}\overline{\mathbf{a}}+\overline{\mathbf{c}}, \tag{4.3.9}$$

$$\bar{\mathbf{b}} = \bar{\mathbf{B}}\bar{\mathbf{a}}, \tag{4.3.10}$$

where

$$\bar{\mathbf{a}} = [\bar{\sigma}_z, \bar{\tau}_{zx}, \bar{\tau}_{yz}, \bar{w}, \bar{u}, \bar{v}]^{\mathrm{T}}, \tag{4.3.11}$$

$$\bar{\mathbf{b}} = [\bar{\sigma}_x + \bar{\sigma}_y, \bar{\sigma}_x - \bar{\sigma}_y, \bar{\tau}_{xy}]^{\mathrm{T}}, \tag{4.3.12}$$

$$\bar{\mathbf{c}} = [-\bar{F}_z, -\bar{F}_x, -\bar{F}_y, 0, 0, 0]^{\mathrm{T}}, \tag{4.3.13}$$

$$\bar{\mathbf{A}} = \begin{bmatrix} 0 & \mathrm{i}\alpha & \mathrm{i}\beta & 0 & 0 & 0 \\ \mathrm{i}\alpha\dfrac{c_{13}}{c_{33}} & 0 & 0 & 0 & c_{66}\beta^2 - \dfrac{c_{13}^2 - c_{11}c_{33}}{c_{33}}\alpha^2 & -\dfrac{c_{13}^2 + c_{33}c_{66} - c_{11}c_{33}}{c_{33}}\alpha\beta \\ \mathrm{i}\beta\dfrac{c_{13}}{c_{33}} & 0 & 0 & 0 & -\dfrac{c_{13}^2 + c_{33}c_{66} - c_{11}c_{33}}{c_{33}}\alpha\beta & c_{66}\alpha^2 - \dfrac{c_{13}^2 - c_{11}c_{33}}{c_{33}}\beta^2 \\ \dfrac{1}{c_{33}} & 0 & 0 & 0 & \mathrm{i}\alpha\dfrac{c_{13}}{c_{33}} & \mathrm{i}\beta\dfrac{c_{13}}{c_{33}} \\ 0 & \dfrac{1}{c_{44}} & 0 & \mathrm{i}\alpha & 0 & 0 \\ 0 & 0 & \dfrac{1}{c_{44}} & \mathrm{i}\beta & 0 & 0 \end{bmatrix}, \tag{4.3.14}$$

$$\bar{\mathbf{B}} = \begin{bmatrix} \dfrac{2c_{13}}{c_{33}} & 0 & 0 & 0 & -2\mathrm{i}\alpha\left(c_{11} - c_{66} - \dfrac{c_{13}^2}{c_{33}}\right) & -2\mathrm{i}\beta\left(c_{11} - c_{66} - \dfrac{c_{13}^2}{c_{33}}\right) \\ 0 & 0 & 0 & 0 & -2\mathrm{i}\alpha c_{66} & 2\mathrm{i}\beta c_{66} \\ 0 & 0 & 0 & 0 & -\mathrm{i}\beta c_{66} & -\mathrm{i}\alpha c_{66} \end{bmatrix}. \tag{4.3.15}$$

In these expressions, a variable with a bar represents its Fourier transform. Clearly, Eq. (4.3.9) is an ordinary differential equation with constant coefficients and six variables. The general solution of the equation can be described as

$$\bar{\mathbf{a}} = \mathbf{X}(z)\mathbf{X}^{-1}(0)\bar{\mathbf{a}}_0 + \mathbf{X}(z)\int_0^z \mathbf{X}^{-1}(s)\bar{\mathbf{c}}(s)\,\mathrm{d}s, \tag{4.3.16}$$

where $\bar{\mathbf{a}}_0$ is the value of $\bar{\mathbf{a}}(z)$ at $z = 0$ and $\mathbf{X}(z)$ is the solution of the homogeneous part of Eq. (4.3.9) which, as shown below, can be obtained by using the general solution, Eq. (2.2.32), derived in **Chapter 2**.

Under static loading, in Eq. (2.2.32) we can simply take

$$\psi_0 = \varphi, \quad F_0 = F. \tag{4.3.17}$$

Thus Eq. (2.2.32) simplifies to

$$\begin{aligned} u &= \frac{\partial \varphi}{\partial y} - (c_{13} + c_{44}) \frac{\partial^2 F}{\partial x \partial z}, \\ v &= -\frac{\partial \varphi}{\partial x} - (c_{13} + c_{44}) \frac{\partial^2 F}{\partial y \partial z}, \\ w &= \left(c_{11} \Lambda + c_{44} \frac{\partial^2}{\partial z^2} \right) F. \end{aligned} \tag{4.3.18}$$

Similarly, Eqs. (2.2.39) and (2.2.42) become

$$\left(\Lambda + \frac{1}{s_3^2} \frac{\partial^2}{\partial z^2} \right) \varphi = 0, \tag{4.3.19}$$

and

$$\left(\Lambda + \frac{1}{s_1^2} \frac{\partial^2}{\partial z^2} \right) \left(\Lambda + \frac{1}{s_2^2} \frac{\partial^2}{\partial z^2} \right) F = 0. \tag{4.3.20}$$

By taking the Fourier transform of Eqs. (4.3.18) to (4.3.20) and assuming that φ and F and their partial differentiations approach zero when $r^2 = x^2 + y^2 \to \infty$, we find that Eqs. (4.3.19) and (4.3.20) yield

$$\left(-\rho^2 s_3^2 + \frac{\mathrm{d}^2}{\mathrm{d} z^2} \right) \overline{\varphi} = 0, \tag{4.3.21}$$

$$\left(-\rho^2 s_1^2 + \frac{\mathrm{d}^2}{\mathrm{d} z^2} \right) \left(-\rho^2 s_2^2 + \frac{\mathrm{d}^2}{\mathrm{d} z^2} \right) \overline{F} = 0, \tag{4.3.22}$$

where

$$\rho^2 = \alpha^2 + \beta^2. \tag{4.3.23}$$

The solution of Eq. (4.3.21) can be written as

$$\overline{\varphi} = k_1 \exp(-\rho s_3 z) + k_2 \exp(\rho s_3 z)\,, \tag{4.3.24}$$

and that of Eq. (4.3.22) can be expressed as

$$\begin{aligned}
\overline{F} &= k_3 \exp(-\rho s_1 z) + k_4 \exp(\rho s_1 z) + k_5 \exp(-\rho s_2 z) + k_6 \exp(\rho s_2 z), \\
&\qquad \text{when } s_1 \neq s_2, \\
\overline{F} &= k_3 \exp(-\rho s_1 z) + k_4 \exp(\rho s_1 z) + k_5 z \exp(-\rho s_1 z) + k_6 z \exp(\rho s_1 z), \\
&\qquad \text{when } s_1 = s_2,
\end{aligned} \tag{4.3.25}$$

where k_i $(i = 1, 2, \cdots, 6)$ are constants to be determined. Hence, $\overline{w}$, $\overline{u}$ and $\overline{v}$ can be obtained by substituting Eqs. (4.3.24) and (4.3.25) into the transformed equation of Eq. (4.3.18). Consequently, $\overline{\sigma}_z$, $\overline{\tau}_{zx}$ and $\overline{\tau}_{yz}$ can be derived using Eqs. (1.2.1) and (1.2.33). With all these, the *solution matrix* $\mathbf{X}(z)$ defined by Eq. (4.3.16) can be obtained as

(*a*) when $s_1 \neq s_2$,

$$\mathbf{X} = \begin{bmatrix}
0 & 0 & 1 & 1 & 1 & 1 \\
\beta & \beta & \dfrac{\mathrm{i}\alpha s_1}{\rho} & -\dfrac{\mathrm{i}\alpha s_1}{\rho} & \dfrac{\mathrm{i}\alpha s_2}{\rho} & -\dfrac{\mathrm{i}\alpha s_2}{\rho} \\
-\alpha & -\alpha & \dfrac{\mathrm{i}\beta s_1}{\rho} & -\dfrac{\mathrm{i}\beta s_1}{\rho} & \dfrac{\mathrm{i}\beta s_2}{\rho} & -\dfrac{\mathrm{i}\beta s_2}{\rho} \\
0 & 0 & -\dfrac{b_{21} s_1}{\rho} & \dfrac{b_{21} s_1}{\rho} & -\dfrac{b_{22} s_2}{\rho} & \dfrac{b_{22} s_2}{\rho} \\
-\dfrac{\beta}{\rho c_{44} s_3} & \dfrac{\beta}{\rho c_{44} s_3} & -\dfrac{\mathrm{i}\alpha b_{31}}{\rho^2} & -\dfrac{\mathrm{i}\alpha b_{31}}{\rho^2} & -\dfrac{\mathrm{i}\alpha b_{32}}{\rho^2} & -\dfrac{\mathrm{i}\alpha b_{32}}{\rho^2} \\
\dfrac{\alpha}{\rho c_{44} s_3} & -\dfrac{\alpha}{\rho c_{44} s_3} & -\dfrac{\mathrm{i}\beta b_{31}}{\rho^2} & -\dfrac{\mathrm{i}\beta b_{31}}{\rho^2} & -\dfrac{\mathrm{i}\beta b_{32}}{\rho^2} & -\dfrac{\mathrm{i}\beta b_{32}}{\rho^2}
\end{bmatrix}$$

$$\times\ \mathrm{diag}\{\exp(-\rho s_3 z), \exp(\rho s_3 z), \exp(-\rho s_1 z), \exp(\rho s_1 z), \exp(-\rho s_2 z), \exp(\rho s_2 z)\}, \tag{4.3.26}$$

where

$$b_{21}=\frac{c_{13}+c_{44}}{c_{44}(c_{13}+c_{33}s_1^2)},\quad b_{22}=\frac{c_{13}+c_{44}}{c_{44}(c_{13}+c_{33}s_2^2)},$$
$$b_{31}=\frac{c_{33}s_1^2-c_{44}}{c_{44}(c_{13}+c_{33}s_1^2)},\quad b_{32}=\frac{c_{33}s_2^2-c_{44}}{c_{44}(c_{13}+c_{33}s_2^2)}. \tag{4.3.27}$$

(*b*) when $s_1=s_2=s$,

$$\mathbf{X}=\begin{bmatrix} 0 & 0 & 1 & 1 & z & z \\ \beta & \beta & \frac{\mathrm{i}\alpha s}{\rho} & -\frac{\mathrm{i}\alpha s}{\rho} & -\frac{\mathrm{i}\alpha}{\rho^2}+\frac{\mathrm{i}\alpha sz}{\rho} & -\frac{\mathrm{i}\alpha}{\rho^2}-\frac{\mathrm{i}\alpha sz}{\rho} \\ -\alpha & -\alpha & \frac{\mathrm{i}\beta s}{\rho} & -\frac{\mathrm{i}\beta s}{\rho} & -\frac{\mathrm{i}\beta}{\rho^2}+\frac{\mathrm{i}\beta sz}{\rho} & -\frac{\mathrm{i}\beta}{\rho^2}-\frac{\mathrm{i}\beta sz}{\rho} \\ 0 & 0 & -\frac{s}{2\rho c_{44}} & \frac{s}{2\rho c_{44}} & -\frac{b_{12}}{\rho^2}-\frac{sz}{2\rho c_{44}} & -\frac{b_{12}}{\rho^2}+\frac{sz}{2\rho c_{44}} \\ -\frac{\beta}{\rho c_{44}s_3} & \frac{\beta}{\rho c_{44}s_3} & -\frac{\mathrm{i}\alpha}{2\rho^2 c_{44}} & -\frac{\mathrm{i}\alpha}{2\rho^2 c_{44}} & \frac{\mathrm{i}\alpha s b_{11}}{\rho^3}-\frac{\mathrm{i}\alpha z}{2\rho^2 c_{44}} & -\frac{\mathrm{i}\alpha s b_{11}}{\rho^3}-\frac{\mathrm{i}\alpha z}{2\rho^2 c_{44}} \\ \frac{\alpha}{\rho c_{44}s_3} & -\frac{\alpha}{\rho c_{44}s_3} & -\frac{\mathrm{i}\beta}{2\rho^2 c_{44}} & -\frac{\mathrm{i}\beta}{2\rho^2 c_{44}} & \frac{\mathrm{i}\beta s b_{11}}{\rho^3}-\frac{\mathrm{i}\beta z}{2\rho^2 c_{44}} & -\frac{\mathrm{i}\beta s b_{11}}{\rho^3}-\frac{\mathrm{i}\beta z}{2\rho^2 c_{44}} \end{bmatrix}$$
$$\times\ \mathrm{diag}\{\exp(-\rho s_3 z),\exp(\rho s_3 z),\exp(-\rho s_1 z),\exp(\rho s_1 z),\exp(-\rho s_1 z),\exp(\rho s_1 z)\}, \tag{4.3.28}$$

where

$$b_{11}=\frac{c_{33}}{2c_{44}(c_{13}+c_{44})},\quad b_{12}=\frac{1}{2(c_{13}+c_{44})}. \tag{4.3.29}$$

It is easy to prove that Eq. (4.3.26) or (4.3.28) is the fundamental solution matrix of the following homogeneous equation

$$\frac{\mathrm{d}\bar{\mathbf{a}}}{\mathrm{d}z}=\bar{\mathbf{A}}\bar{\mathbf{a}}. \tag{4.3.9a}$$

With this general solution, we can consider some special loading conditions. In the following, we shall discuss an elastic solid subjected to a unit point force in the z-direction and the x-direction.

When the unit point force at point (0, 0, h) is in the z-direction, *i.e.*, $F_x=F_y=0$ and $F_z=\delta(x)\delta(y)\delta(z-h)$, Eq. (4.3.13) yields

$$\overline{\mathbf{c}}(z) = [-1/2\pi, 0, 0, 0, 0, 0]^{\mathrm{T}} \delta(z-h). \tag{4.3.30}$$

On the other hand, when a unit point force at point $(0, 0, h)$ is along the x-direction, *i.e.*, $F_x = \delta(x)\delta(y)\delta(z-h)$, $F_y = F_z = 0$, then Eq. (4.3.13) gives rise to

$$\overline{\mathbf{c}}(z) = [0, -1/2\pi, 0, 0, 0, 0]^{T} \delta(z-h). \tag{4.3.31}$$

Let the integration in Eq. (4.3.16) be denoted by

$$\mathbf{X}(z)\int_0^z \mathbf{X}^{-1}(s)\overline{\mathbf{c}}(s)\,\mathrm{d}s = N(z-h)\mathbf{X}(z)\mathbf{g}, \tag{4.3.32}$$

where

$$N(z-h) = \begin{cases} 0, & \forall z < h, \\ 1, & \forall z \geq h, \end{cases} \tag{4.3.33}$$

and function $\mathbf{g}$ has the following forms corresponding to the direction of the unit point force applied. In the former case with the force along z-direction, we have

$$\mathbf{g} = \mathbf{X}^{-1}(h)[-1/2\pi, 0, 0, 0, 0, 0]^{\mathrm{T}}$$
$$= \begin{cases} [0, 0, b_{41}\,\mathrm{e}^{\rho h s_1}, b_{41}\,\mathrm{e}^{-\rho h s_1}, -b_{42}\,\mathrm{e}^{\rho h s_2}, -b_{42}\,\mathrm{e}^{-\rho h s_2}]^{\mathrm{T}} & \text{for } s_1 \neq s_2, \\ [0, 0, \dfrac{b_{33}\rho h - s}{4\pi s}\mathrm{e}^{\rho h s}, -\dfrac{b_{33}\rho h + s}{4\pi s}\mathrm{e}^{-\rho h s}, -\dfrac{b_{33}\rho}{4\pi s}\mathrm{e}^{\rho h s}, -\dfrac{b_{33}\rho}{4\pi s}\mathrm{e}^{-\rho h s}]^{\mathrm{T}} & \text{for } s_1 = s_2 = s, \end{cases} \tag{4.3.34}$$

where

$$b_{33} = \frac{c_{13} + c_{44}}{c_{33}}, \quad b_{41} = \frac{(c_{13} + c_{33}s_1^2)(c_{33}s_2^2 - c_{44})}{4\pi c_{33}(c_{13} + c_{44})(s_1^2 - s_2^2)}, \quad b_{42} = \frac{(c_{13} + c_{33}s_2^2)(c_{33}s_1^2 - c_{44})}{4\pi c_{33}(c_{13} + c_{44})(s_1^2 - s_2^2)}. \tag{4.3.35}$$

If the unit force is applied along x-direction, function $\mathbf{g}$ becomes

$$\mathbf{g} = \mathbf{X}^{-1}(h)[0, -1/2\pi, 0, 0, 0, 0]^{\mathrm{T}}$$

$$= \begin{cases} \left[-\dfrac{\beta}{4\pi\rho^2}\mathrm{e}^{\rho h s_3}, -\dfrac{\beta}{4\pi\rho^2}\mathrm{e}^{-\rho h s_3}, -\dfrac{\mathrm{i}\alpha b_{51}}{\rho s_1}\mathrm{e}^{\rho h s_1}, \dfrac{\mathrm{i}\alpha b_{51}}{\rho s_1}\mathrm{e}^{-\rho h s_1}, \dfrac{\mathrm{i}\alpha b_{52}}{\rho s_2}\mathrm{e}^{\rho h s_2}, -\dfrac{\mathrm{i}\alpha b_{52}}{\rho s_2}\mathrm{e}^{-\rho h s_2}\right]^{\mathrm{T}} & \text{for } s_1 \neq s_2, \\ \left[-\dfrac{\beta}{4\pi\rho^2}\mathrm{e}^{\rho h s_3}, -\dfrac{\beta}{4\pi\rho^2}\mathrm{e}^{-\rho h s_3}, \dfrac{\mathrm{i}\alpha s c_{44}}{4\pi\rho c_{11}}\mathrm{e}^{\rho h s}, -\dfrac{\mathrm{i}\alpha s c_{44}}{4\pi\rho c_{11}}\mathrm{e}^{-\rho h s}, -\dfrac{\mathrm{i}\alpha(c_{13}+c_{44})}{4\pi s^2 c_{33}}\mathrm{e}^{\rho h s}, -\dfrac{\mathrm{i}\alpha(c_{13}+c_{44})}{4\pi s^2 c_{33}}\mathrm{e}^{-\rho h s}\right]^{\mathrm{T}} & \\ \quad + h\left[0, 0, \dfrac{\mathrm{i}\alpha s^2(c_{13}+c_{44})}{4\pi c_{11}}\mathrm{e}^{\rho h s}, \dfrac{\mathrm{i}\alpha s^2(c_{13}+c_{44})}{4\pi c_{11}}\mathrm{e}^{-\rho h s}, 0, 0\right]^{\mathrm{T}} & \text{for } s_1 = s_2 = s, \end{cases} \tag{4.3.36}$$

where

$$b_{51} = -\frac{(c_{13} + c_{33}s_1^2)}{4\pi c_{33}(s_1^2 - s_2^2)}, \quad b_{52} = -\frac{(c_{13} + c_{33}s_2^2)}{4\pi c_{33}(s_1^2 - s_2^2)}. \tag{4.3.37}$$

If we introduce an undetermined constant vector $\mathbf{k}$ and let

$$\mathbf{k} = \mathbf{X}^{-1}(0)\bar{\mathbf{a}}_0, \tag{4.3.38}$$

we obtain the general solution after Fourier transformation, *i.e.*,

$$\bar{\mathbf{a}} = \mathbf{X}(z)[\mathbf{k} + N(z-h)\mathbf{g}], \tag{4.3.39}$$

where vectors $\mathbf{k}$ and $\mathbf{g}$ individually have six components, k_i and g_i $(i = 1, 2, \cdots, 6)$.

For a given problem, we first need to find vector $\mathbf{k}$ using the boundary conditions. Then the solution can be derived by the inverse Fourier transform of Eq. (4.3.39), which will be a closed form if the integral in the inverse Fourier transform can be found analytically. Listed below are the expressions of k_i for some special cases.

(*a*) The solution for an infinite body

In this case, since $z \to \pm\infty$ at infinity, and the solution of the problem must be finite, Eqs. (4.3.26), (4.3.28) and (4.3.39) lead to

$$k_i = -g_i, \quad (i = 2, 4, 6), \quad k_i = 0, \quad (i = 1, 3, 5). \tag{4.3.40}$$

Thus the solution is

$$\bar{\mathbf{a}}_{\infty}(z)=\begin{cases}\mathbf{X}(z)[0,-g_2,0,-g_4,0,-g_6]^{\mathrm{T}}, & \text{for } z\le h,\\ \mathbf{X}(z)[g_1,0,g_3,0,g_5,0]^{\mathrm{T}}, & \text{for } z\ge h,\end{cases} \tag{4.3.41}$$

where subscript ∞ indicates that this is a solution for an infinite body.

(*b*) The solution for a half-space.

Let the origin of the coordinates be at the boundary surface and the positive direction of the z-axis point inwards the half-space. Again, when $z\to+\infty$, $\bar{\mathbf{a}}$ becomes finite. Hence,

$$k_i=-g_i,\quad (i=2,4,6). \tag{4.3.42}$$

If the boundary surface is subjected to surface stresses, then we will get

$$\begin{aligned}\bar{\sigma}_z(0)&=\mathbf{X}_1(0)\mathbf{k}=\bar{\sigma}_{z0},\\ \bar{\tau}_{zx}(0)&=\mathbf{X}_2(0)\mathbf{k}=\bar{\tau}_{zx0},\\ \bar{\tau}_{yz}(0)&=\mathbf{X}_3(0)\mathbf{k}=\bar{\tau}_{yz0},\end{aligned} \tag{4.3.43}$$

where $\mathbf{X}_i$ is a vector composed of the elements of the ith row in matrix $\mathbf{X}$. We can get a linear set of equations for k_1, k_3 and k_5 when Eq. (4.3.42) is used in these boundary conditions. For example, for $z\le h$, Eqs. (4.3.39) and (4.4.42) give

$$\bar{\mathbf{a}}=\mathbf{X}(z)\mathbf{k}=\mathbf{X}(z)\begin{bmatrix}k_1\\0\\k_3\\0\\k_5\\0\end{bmatrix}+\mathbf{X}(z)\begin{bmatrix}0\\-g_2\\0\\-g_4\\0\\-g_6\end{bmatrix}, \tag{4.3.44}$$

but for $z\ge h$, Eqs. (4.3.39) and (4.3.42) lead to

$$\bar{\mathbf{a}}=\mathbf{X}(z)(\mathbf{k}+\mathbf{g})=\mathbf{X}(z)\begin{bmatrix}k_1\\0\\k_3\\0\\k_5\\0\end{bmatrix}+\mathbf{X}(z)\begin{bmatrix}g_1\\0\\g_3\\0\\g_5\\0\end{bmatrix}. \tag{4.3.45}$$

The second term in the right-hand side of Eq. (4.3.44) or (4.3.45) is identical to that given in Eq. (4.3.41), indicating that the point force solution of the half space consists of two parts[1], *i.e.*, the general solution corresponding to an infinite body subjected to a point force, $\bar{\mathbf{a}}_\infty$, and the general solution of the homogeneous equations, $\Delta\bar{\mathbf{a}}$. Hence, the solution can be written as

$$\bar{\mathbf{a}} = \Delta\bar{\mathbf{a}} + \bar{\mathbf{a}}_\infty , \tag{4.3.46}$$

where

$$\Delta\bar{\mathbf{a}} = \mathbf{X}(z)[k_1, 0, k_3, 0, k_5, 0]^{\mathrm{T}} . \tag{4.3.47}$$

If the boundary surface is subjected to surface displacements, we get

$$\begin{aligned} \bar{w}(0) &= \mathbf{X}_4(0)\mathbf{k} = \bar{w}_0 , \\ \bar{u}(0) &= \mathbf{X}_5(0)\mathbf{k} = \bar{u}_0 , \\ \bar{v}(0) &= \mathbf{X}_6(0)\mathbf{k} = \bar{v}_0 . \end{aligned} \tag{4.3.48}$$

With other boundary conditions, such as that subjected to normal stress and tangential displacements, or normal displacement and tangential stresses, we can always obtain three linear equations to solve for $\mathbf{k}$ in conjunction with Eq. (4.3.42). In addition, we can always resolve the solution into the form of Eq. (4.3.46), *i.e.*, the sum of a general solution corresponding to an infinite body subjected to a point force and the general solution of the homogeneous equations, to simplify the formulation process.

(*c*) An infinite body composed of two half-spaces

To get the solution for this case, we can apply the technique used in **Section 3.2**. Let the interface of the two half-spaces be the xOy coordinate plane and let z-axis point downwards. Then the upper part of the infinite body occupies $z \le 0$. For convenience, we use a prime to denote a quantity in the upper half-space.

When body forces are ignored, in the upper half-space, we can use Eq. (4.3.39) to obtain

$$\bar{\mathbf{a}}' = \mathbf{X}'(z)\mathbf{k}' . \tag{4.3.49}$$

[1] This is consistent with the conclusion of Mindlin (1936) for isotropic materials.

When $z \to -\infty$, $\bar{\mathbf{a}}'$ should be finite, which gives rise to

$$k_i' = 0, \quad (i = 1, 3, 5). \tag{4.3.50}$$

In the lower half of the space, we obtain ($\bar{\mathbf{a}}$ is finite when $z \to +\infty$)

$$k_i = -g_i, \quad (i = 2, 4, 6). \tag{4.3.51}$$

If the two half-spaces are welded at their interface, the boundary conditions specified by Eq. (3.2.1) require that

$$\mathbf{X}'(0)\mathbf{k}' = \mathbf{X}(0)\mathbf{k}. \tag{4.3.52}$$

Vectors $\mathbf{k}$ and $\mathbf{k}'$ can then be solved using the twelve equations from Eqs. (4.3.50) to (4.3.52).

If the two half-spaces are in smooth contact, then the boundary conditions specified by Eq. (3.2.2) must be satisfied. This requires that

$$\begin{aligned} &\mathbf{X}_1'(0)\mathbf{k}' = \mathbf{X}_1(0)\mathbf{k}, \\ &\mathbf{X}_4'(0)\mathbf{k}' = \mathbf{X}_4(0)\mathbf{k}, \\ &\mathbf{X}_2'(0)\mathbf{k}' = \mathbf{X}_3'(0)\mathbf{k}' = 0, \\ &\mathbf{X}_2(0)\mathbf{k} = \mathbf{X}_3(0)\mathbf{k} = 0. \end{aligned} \tag{4.3.53}$$

Again, we have twelve equations, Eqs. (4.3.50), (4.3.51) and (4.3.53), that can be used to solve for $\mathbf{k}$ and $\mathbf{k}'$.

In a solution with any given boundary conditions, we must consider four cases due to the possible combinations of the material eigenvalues in the upper and lower half-spaces.

4.4 POINT FORCE SOLUTION OF AN ELASTIC LAYER

Layered materials with transverse isotropy are commonly encountered in a wide range of engineering fields, and thus the solutions to such problems are of practical importance. In this section we will focus on a single elastic layer with free boundary conditions. The solution will be extended to the multi-layer case in **Section 4.5**.

Consider an elastic layer of thickness H. Let the xOy coordinate plane coincide with the upper surface of the layer and z-axis point downwards. A point force is applied at $(0, 0, h)$. Using Eq. (4.3.32), we can rewrite Eq. (4.3.16) as

$$\bar{\mathbf{a}} = \mathbf{T}(z)\bar{\mathbf{a}}_0 + N(z-h)\overline{\mathbf{R}}(z), \tag{4.4.1}$$

where

$$\mathbf{T}(z) = \mathbf{X}(z)\mathbf{X}^{-1}(0), \tag{4.4.2}$$

$$\overline{\mathbf{R}}(z) = \mathbf{X}(z)\mathbf{g}, \tag{4.4.3}$$

and $\bar{\mathbf{a}}_0$ is the value of $\bar{\mathbf{a}}$ at $z = 0$, *i.e.*, at the upper surface of the layer. Equation (4.4.1) can be rewritten as

$$\begin{Bmatrix} \overline{\mathbf{\Sigma}} \\ \overline{\mathbf{U}} \end{Bmatrix} = \begin{bmatrix} \mathbf{T}_{11}(z) & \mathbf{T}_{12}(z) \\ \mathbf{T}_{21}(z) & \mathbf{T}_{22}(z) \end{bmatrix} \begin{Bmatrix} \overline{\mathbf{\Sigma}}_0 \\ \overline{\mathbf{U}}_0 \end{Bmatrix} + \begin{Bmatrix} N(z-h)\overline{\mathbf{R}}_1(z) \\ N(z-h)\overline{\mathbf{R}}_2(z) \end{Bmatrix}, \tag{4.4.4}$$

where $\overline{\mathbf{\Sigma}} = [\bar{\sigma}_z, \bar{\tau}_{zx}, \bar{\tau}_{yz}]^{\mathrm{T}}$, $\overline{\mathbf{U}} = [\bar{w}, \bar{u}, \bar{v}]^{\mathrm{T}}$ and so on. Because $0 \le h \le H$, at the lower surface of the elastic layer, we have $z = H$, and hence

$$\begin{Bmatrix} \overline{\mathbf{\Sigma}}_H \\ \overline{\mathbf{U}}_H \end{Bmatrix} = \begin{bmatrix} \mathbf{T}_{11}(H) & \mathbf{T}_{12}(H) \\ \mathbf{T}_{21}(H) & \mathbf{T}_{22}(H) \end{bmatrix} \begin{Bmatrix} \overline{\mathbf{\Sigma}}_0 \\ \overline{\mathbf{U}}_0 \end{Bmatrix} + \begin{Bmatrix} \overline{\mathbf{R}}_1(H) \\ \overline{\mathbf{R}}_2(H) \end{Bmatrix}. \tag{4.4.5}$$

This gives rise to

$$\overline{\mathbf{U}}_0 = \mathbf{T}_{12}^{-1}(H)[\overline{\mathbf{\Sigma}}_H - \mathbf{T}_{11}(H)\overline{\mathbf{\Sigma}}_0 - \overline{\mathbf{R}}_1(H)]. \tag{4.4.6}$$

Substituting it into Eq. (4.4.4) and letting $\overline{\mathbf{\Sigma}}_0 = \overline{\mathbf{\Sigma}}_H = 0$, we get

$$\begin{Bmatrix} \overline{\mathbf{\Sigma}} \\ \overline{\mathbf{U}} \end{Bmatrix} = N(z-h)\overline{\mathbf{R}}(z) + \overline{\mathbf{R}}_a(z), \tag{4.4.7}$$

where

$$\overline{\mathbf{R}}_a(z)=\begin{Bmatrix}-\mathbf{T}_{12}(z)\mathbf{T}_{12}^{-1}(H)\overline{R}_1(H)\\ -\mathbf{T}_{22}(z)\mathbf{T}_{12}^{-1}(H)\overline{R}_1(H)\end{Bmatrix}. \tag{4.4.8}$$

All the detailed expressions of the quantities in Eq. (4.4.7) have been derived in **Section 4.3**. Thus the solution after Fourier transform for the elastic layer is ready, and $\mathbf{a}$ is then obtained by inverse transform. As with the discussion in the last section, by considering the structures of Eqs. (4.4.3) and (4.4.7), we can also resolve the elastic layer solution to two parts, *i.e.*,

$$\overline{\mathbf{a}}=\overline{\mathbf{a}}_\infty+\overline{\mathbf{a}}_h, \tag{4.4.9}$$

where $\overline{\mathbf{a}}_\infty$ is the general solution for an infinite body subjected to a point force and $\overline{\mathbf{a}}_h$ is the specific solution for the elastic layer. In the following, we will discuss the detailed expressions for displacements in $\mathbf{a}_h$ when the eigenvalues of the layer material, s_1 and s_2, change. For convenience, we will use superscripts z or x to indicate the direction of the unit point force.

(1) $s_1\neq s_2$

When the point force is in z-direction, we have

$$u^z=\frac{x}{r}\int_0^\infty f_u^z J_1(\rho r)\,\mathrm{d}\rho\quad,\ v^z=\frac{y}{x}u^z,\quad w^z=\int_0^\infty f_w^z J_0(\rho r)\,\mathrm{d}\rho, \tag{4.4.10}$$

where

$$f_u^z=\frac{1}{t}(m_{31}r_1^z+m_{32}r_2^z)-b_{61}\,\mathrm{e}^{\rho s_1(z-h)}+b_{62}\,\mathrm{e}^{\rho s_2(z-h)}, \tag{4.4.11}$$

$$t=2-2\cosh(H\rho s_1)\cosh(H\rho s_2)+\left(\frac{s_1}{s_2}+\frac{s_2}{s_1}\right)\sinh(H\rho s_1)\sinh(H\rho s_2), \tag{4.4.12}$$

$$r_1^z=2b_{41}\cosh[(H-h)\rho s_1]-2b_{42}\cosh[(H-h)\rho s_2], \tag{4.4.13}$$

$$r_2^z=-2b_{41}s_1\sinh[(H-h)\rho s_1]+2b_{42}s_2\sinh[(H-h)\rho s_2], \tag{4.4.14}$$

$$b_{61}=b_{62}=\frac{c_{13}+c_{44}}{4\pi c_{33}c_{44}(s_1^2-s_2^2)}, \tag{4.4.15}$$

$$m_{31}=-b_{31}\left\{\cosh(H\rho s_2)\cosh(\rho s_1 z)-\cosh[\rho s_1(H-z)]-\frac{s_2}{s_1}\sinh(H\rho s_2)\sinh(\rho s_1 z)\right\}$$
$$-b_{32}\left\{\cosh(H\rho s_1)\cosh(\rho s_2 z)-\cosh[\rho s_2(H-z)]-\frac{s_1}{s_2}\sinh(H\rho s_1)\sinh(\rho s_2 z)\right\},\tag{4.4.16}$$

$$m_{32}=\frac{b_{31}}{s_1}\{\sinh[\rho s_1(H-z)]+\cosh(H\rho s_2)\sinh(\rho s_1 z)\}-\frac{b_{31}}{s_2}\cosh(\rho s_1 z)\sinh(H\rho s_2)$$
$$+\frac{b_{32}}{s_2}\{\sinh[\rho s_2(H-z)]+\cosh(H\rho s_1)\sinh(\rho s_2 z)\}-\frac{b_{32}}{s_1}\cosh(\rho s_2 z)\sinh(H\rho s_1),\tag{4.4.17}$$

$$f_w^z=\frac{1}{t}(m_{41}\,r_1^z+m_{42}\,r_2^z)+b_{71}s_1\,\mathrm{e}^{\rho s_1(z-h)}-b_{72}s_2\,\mathrm{e}^{\rho s_2(z-h)},\tag{4.4.18}$$

$$b_{71}=\frac{c_{33}s_2^2-c_{44}}{4\pi c_{33}c_{44}(s_1^2-s_2^2)},\quad b_{72}=\frac{c_{33}s_1^2-c_{44}}{4\pi c_{33}c_{44}(s_1^2-s_2^2)},\tag{4.4.19}$$

$$m_{41}=b_{21}s_1\{\sinh[\rho s_1(H-z)]+\cosh(H\rho s_2)\sinh(\rho s_1 z)\}-b_{21}s_2\cosh(\rho s_1 z)\sinh(H\rho s_2)$$
$$+b_{22}s_2\{\sinh[\rho s_2(H-z)]+\cosh(H\rho s_1)\sinh(\rho s_2 z)\}-b_{22}s_1\cosh(\rho s_2 z)\sinh(H\rho s_1),\tag{4.4.20}$$

$$m_{42}=-b_{21}\left\{\cosh(H\rho s_2)\cosh(\rho s_1 z)-\cosh[\rho s_1(H-z)]-\frac{s_1}{s_2}\sinh(H\rho s_2)\sinh(\rho s_1 z)\right\}$$
$$-b_{22}\left\{\cosh(H\rho s_1)\cosh(\rho s_2 z)-\cosh[\rho s_2(H-z)]-\frac{s_2}{s_1}\sinh(H\rho s_1)\sinh(\rho s_2 z)\right\}.\tag{4.4.21}$$

When the point force is in x-direction, the displacements are

$$u^x=\frac{y^2}{r^2}\int_0^\infty f_{u0}^x J_0(\rho r)\,\mathrm{d}\rho+\frac{x^2}{r^2}\int_0^\infty f_u^x J_0(\rho r)\,\mathrm{d}\rho$$
$$-\frac{y^2-x^2}{r^3}\left[\int_0^\infty f_{u0}^x[J_1(\rho r)/\rho]\mathrm{d}\rho-\int_0^\infty f_u^x[J_1(\rho r)/\rho]\,\mathrm{d}\rho\right],$$
$$v^x=\frac{xy}{r^2}\left\{\int_0^\infty (f_u^x-f_{u0}^x)J_0(\rho r)\mathrm{d}\rho-\frac{2}{r}\int_0^\infty (f_u^x-f_{u0}^x)[J_1(\rho r)/\rho]\mathrm{d}\rho\right\},\tag{4.4.22}$$
$$w^x=\frac{x}{r}\int_0^\infty f_w^x J_1(\rho r)\mathrm{d}\rho,$$

where

$$f_{u0}^{x} = \frac{\cosh[(H-h)\rho s_0]\sinh(\rho s_0 z)}{2\pi c_{44} s_0 \sinh(H\rho s_0)} - \frac{e^{\rho s_0 (z-h)}}{4\pi c_{44} s_0}, \tag{4.4.23}$$

$$f_u^x = \frac{1}{t}(-m_{31} r_1^x + m_{32} r_2^x) - (b_{72}/s_1)e^{\rho s_1 (z-h)} + (b_{71}/s_2)e^{\rho s_2 (z-h)}, \tag{4.4.24}$$

$$r_1^x = \frac{2b_{51}}{s_1}\sinh[(H-h)\rho s_1] - \frac{2b_{52}}{s_2}\sinh[(H-h)\rho s_2], \tag{4.4.25}$$

$$r_2^x = 2b_{51}\cosh[(H-h)\rho s_1] - 2b_{52}\cosh[(H-h)\rho s_2], \tag{4.4.26}$$

$$f_w^x = \frac{1}{t}(m_{41} r_1^x - m_{42} r_2^x) - b_{61}\, e^{\rho s_1 (z-h)} + b_{62}\, e^{\rho s_2 (z-h)}. \tag{4.4.27}$$

(2) $s_1 = s_2 = s$

In this case, the expressions of the displacements are still the same as those in the section described by Eqs. (4.4.10) and (4.4.22), provided that the following functions are used instead:

$$f_u^z = \frac{1}{2c_{44} t}(m_{31} r_1^z + m_{32} r_2^z) - \frac{\rho(z-h)(c_{13}+c_{44})}{8\pi s c_{33} c_{44}} e^{\rho s (z-h)}, \tag{4.4.28}$$

$$t = H^2 \rho^2 s^2 - \sinh(H\rho s)^2, \tag{4.4.29}$$

$$r_1^z = -\frac{1}{2\pi\rho}\cosh[(H-h)\rho s] + \frac{c_{13}+c_{44}}{2\pi c_{33} s}(H-h)\sinh[(H-h)\rho s], \tag{4.4.30}$$

$$r_2^z = \frac{c_{44}}{2\pi\rho c_{33} s}\sinh[(H-h)\rho s] - \frac{c_{13}+c_{44}}{2\pi c_{33}}(H-h)\cosh[(H-h)\rho s], \tag{4.4.31}$$

$$m_{31} = H\rho^3 s^2 z\cosh[\rho s(H-z)] + \frac{\rho c_{44}}{c_{13}+c_{44}}\sinh(H\rho s)\sinh(\rho s z) + \rho^2 s z\sinh(H\rho s)\cosh(\rho s z)$$

$$-\frac{H\rho^2 s^3 c_{33}}{c_{13}+c_{44}}\sinh[\rho s(H-z)] - H^2\rho\, s\cosh(H\rho s)\sinh(\rho s z), \tag{4.4.32}$$

$$m_{32}=\frac{\rho c_{33}s}{c_{13}+c_{44}}\cosh(\rho sz)\sinh(H\rho s)+H\rho^3 sz\sinh[\rho s(H-z)]-(H-z)\rho^2\sinh(H\rho s)\sinh(\rho sz)$$
$$-\frac{H\rho^2 s^2 c_{33}}{c_{13}+c_{44}}\cosh[\rho s(H-z)],$$

(4.4.33)

$$f_w^z=\frac{1}{2c_{44}t}(m_{41}r_1^z+m_{42}r_2^z)-\left[\frac{c_{13}+3c_{44}}{8\pi s c_{33}c_{44}}-\frac{\rho(z-h)(c_{13}+c_{44})}{8\pi c_{33}c_{44}}\right]\mathrm{e}^{\rho s(z-h)}, \tag{4.4.34}$$

$$m_{41}=\frac{\rho c_{33}s^3}{c_{13}+c_{44}}\cosh(\rho sz)\sinh(H\rho s)+H\rho^3 s^3 z\sinh[\rho s(H-z)]+\frac{H\rho^2 s^2 c_{44}}{c_{13}+c_{44}}\cosh[\rho s(H-z)]$$
$$+\rho^2 s^2\left[H\cosh(H\rho s)\cosh(\rho sz)-z\sinh(H\rho s)\sinh(\rho sz)\right],$$

(4.4.35)

$$m_{42}=H\rho^3 s^2 z\cosh[\rho s(H-z)]+\frac{\rho c_{44}}{c_{13}+c_{44}}\sinh(H\rho s)\sinh(\rho sz)$$
$$+(H-z)\rho^2 s\sinh(H\rho s)\cosh(\rho sz)+\frac{H\rho^2 s c_{44}}{c_{13}+c_{44}}\sinh[\rho s(H-z)], \tag{4.4.36}$$

$$r_1^x=-\frac{sc_{44}}{2\pi\rho c_{11}}\sinh[(H-h)\rho s]+\frac{c_{44}s^2-c_{11}}{2\pi c_{11}}(H-h)\cosh[(H-h)\rho s], \tag{4.4.37}$$

$$r_2^x=-\frac{1}{2\pi\rho}\cosh[(H-h)\rho s]-\frac{c_{13}+c_{44}}{2\pi s c_{33}}(H-h)\sinh[(H-h)\rho s], \tag{4.4.38}$$

$$f_u^x=\frac{1}{2c_{44}t}(-m_{31}r_1^x+m_{32}r_2^x)-\left[\frac{\rho(z-h)(c_{13}+c_{44})}{8\pi s^2 c_{33}c_{44}}+\frac{(c_{11}+s^2 c_{44})}{8\pi s\ c_{11}c_{44}}\right]\mathrm{e}^{\rho s(z-h)}, \tag{4.4.39}$$

$$f_w^x=\frac{1}{2c_{44}t}(m_{41}\ r_1^x-m_{42}\ r_2^x)-\frac{\rho(z-h)(c_{13}+c_{44})}{8\pi s c_{33}c_{44}}\mathrm{e}^{\rho s(z-h)}. \tag{4.4.40}$$

However, f_{u0}^x is the same as that defined by Eq. (4.4.23).

Now if we take

$$s_1=s_2=s_3=1\ ,\ c_{11}=c_{22}=c_{33}=\frac{(1-\nu)E}{(1+\nu)(1-2\nu)},$$
$$c_{12}=c_{13}=\frac{\nu E}{(1+\nu)(1-2\nu)}\ ,\ c_{44}=c_{66}=\frac{E}{2(1+\nu)}, \tag{4.4.41}$$

then the solution derived above reduces to that obtained by Benitez and Rosakis (1987) when the elastic layer is an isotropic material.

Let us use this solution to study the distribution of stress σ_z when an elastic layer is subjected to a point force *P* normal to the isotropic plane. In the numerical analysis, we need to consider the following two cases: (a) s_1 and s_2 are conjugate complex numbers, and (b) they are equal reals. Let us also consider two typical situations of loading: (i) the point force acts at the mid-plane of the layer, *i.e.*, $h = H/2$, and (ii) it is in the upper half of the layer, say $h = H/4$. The material constants of the elastic layer are given in Table 4.1.

Table 4.1 Material constants of the elastic layer (GPa).

	c_{11}	c_{13}	c_{33}	c_{44}	c_{66}	s_3	s_1	s_2
Case (a)	16.1	5.01	6.1	3.83	6.34	1.28	1.088+0.664 i	1.088-0.664 i
Case (b)	16.9	5.0	10.0	4.00	3.00	0.86	1.14	1.14

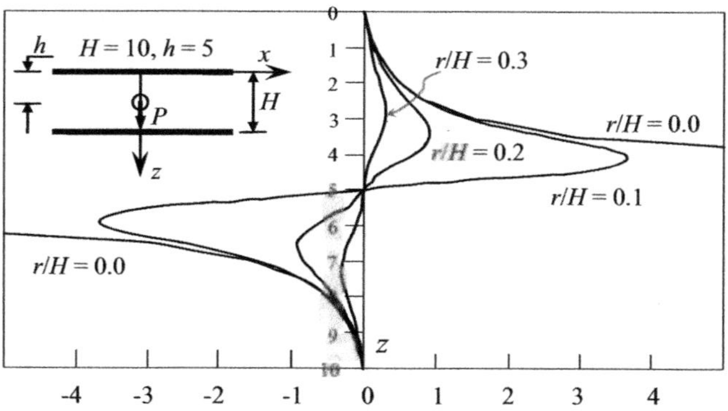

Fig. 4.4 Case (a) with $h/H = 1/2$.

Figures 4.4 to 4.7 show the distributions of the dimensionless stress $\sigma_z H^2 / P$ along the layer thickness (*z*-axis), and demonstrate the following interesting results:

(A) When $h = H/2$, *i.e.*, when the point force acts at the mid-plane of the elastic layer, stress σ_z has an anti-symmetric distribution about the mid-plane.

(B) There is a stress concentration near the point of application of force *P*, which means that the solution obtained reflects the mechanics well.

(C) When $r \to \infty$, $\sigma_z \to 0$. This verifies the validity of the solution in a way because stresses must approach zero at infinity. This also meets the condition for using the Fourier transform method.

(D) The stress boundary condition $\sigma_z = 0$ at $z = 0$ and $z = H$ is satisfied.

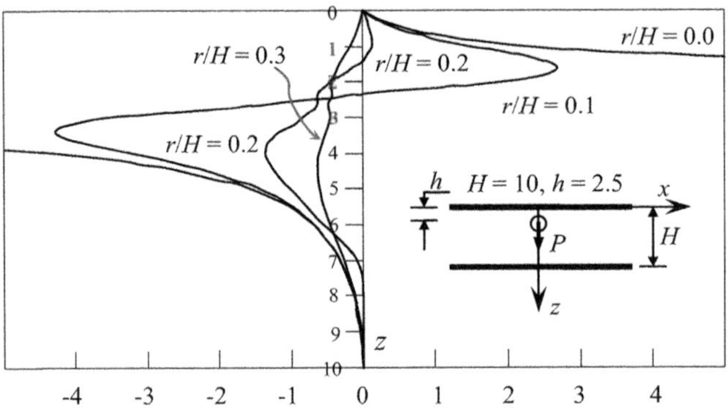

Fig. 4.5 Case (a) with $h/H = 1/4$.

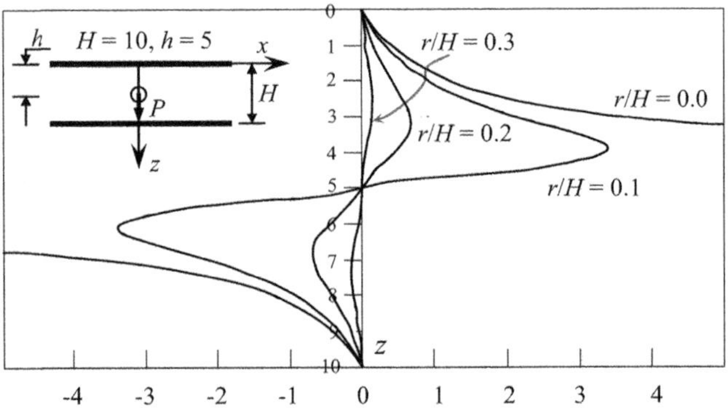

Fig. 4.6 Case (b) with $h/H = 1/2$.

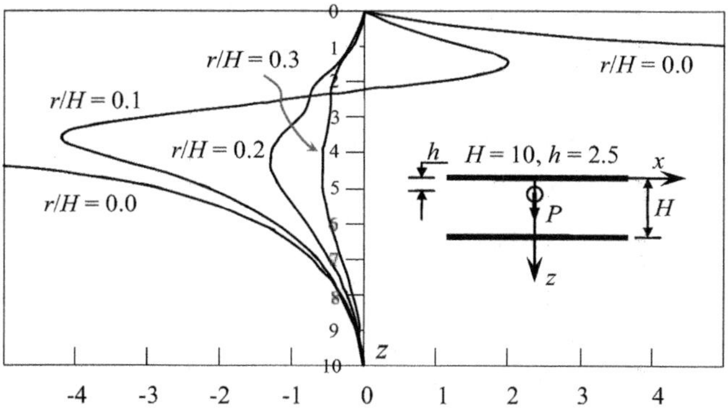

Fig. 4.7 Case (b) with $h/H = 1/4$.

4.5 LAYERED ELASTIC MEDIA

The mechanics of layered materials with transverse isotropy has been an important topic in civil engineering; it occurs in problems concerning the stress and deformation of soils, rocks and constructions foundations, roads and airports, and in aerospace engineering, such as the safety and failure analysis of multi-layered composites. It can also find its applications in the recent technological development of micro-electro-mechanical systems and devices with multi-layered coatings.

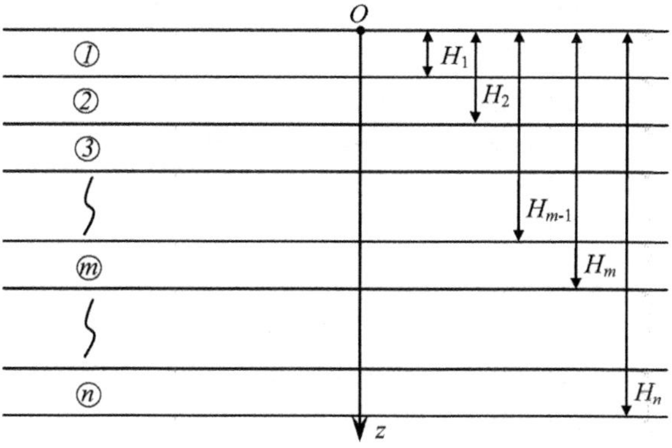

Fig. 4.8 An n-layered elastic medium.

Now let us extend the single layer solution obtained earlier to find the deformation of a solid with n layers, Fig. 4.8. Let the xOy plane of the coordinate system, $Oxyz$, coincide with the top surface of the solid and z-axis point downwards. The mth layer $(m = 1, 2, \cdots, n)$ is subjected to body forces (F_{xm}, F_{ym}, F_{zm}). For the n-layered medium, in Fig. 4.8, there are usually three boundary conditions either on the top surface $z = H_0 = 0$ or on the bottom surface $z = H_n$. For example, stresses σ_z, τ_{zx}, and τ_{yz} are usually prescribed at $z = 0$, and at $z = H_n$, if the layered medium contacts with a rigid base roughly, we have $u = v = w = 0$, while if it contacts with the base smoothly, we have $w = \tau_{zx} = \tau_{yz} = 0$. If the layered medium, at its bottom surface, is connected with an elastic half-space, we assume that body forces are absent in the half-space for simplicity.

Equation (4.3.16) in **Section 4.3** is valid for any layer in the solid, thus

$$\overline{\mathbf{a}}_m(z) = \mathbf{T}_m(z)[\overline{\mathbf{a}}_m(H_{m-1}) + \overline{\mathbf{R}}_m(z)], \quad (H_{m-1} \le z \le H_m; \ m = 1, 2, \cdots, n), \tag{4.5.1}$$

where

$$\mathbf{T}_m(z) = \mathbf{X}_m(z)\mathbf{X}_m^{-1}(H_{m-1}), \tag{4.5.2}$$

$$\overline{\mathbf{R}}_m(z) = \mathbf{X}_m(H_{m-1})\int_{H_{m-1}}^{z} \mathbf{X}_m^{-1}(s)\overline{\mathbf{c}}_m(s)\,\mathrm{d}s, \tag{4.5.3}$$

where $\mathbf{X}_m(z)$ was given in Eq. (4.3.26) or (4.3.28), and $\overline{\mathbf{a}}_m$ and $\overline{\mathbf{c}}_m$ were defined in Eqs. (4.3.11) and (4.3.13), respectively. Note that, we have affixed a subscript m here to indicate quantities associated with the mth layer.

Again for simplicity, we assume all the basic variables across any interface of two layers are continuous, *i.e.*,

$$\overline{\mathbf{a}}_{i+1}(H_i) = \overline{\mathbf{a}}_i(H_i), \quad (i = 1, 2, \cdots, n-1). \tag{4.5.4}$$

In the following, we will develop the solutions for the layered solid with its nth layer having finite thickness, or connecting with an elastic half-space.

(1) A finite n^{th} layer

Repeatedly making use of Eqs. (4.5.1) and (4.5.4), we get

$$\overline{\mathbf{a}}_m(z) = \mathbf{T}_m(z)\left[\mathbf{M}^1_{m-1}\overline{\mathbf{a}}_1(0) + \sum_{j=1}^{m-1}\mathbf{M}^j_{m-1}\overline{\mathbf{R}}_j(H_j) + \overline{\mathbf{R}}_m(z)\right], \quad (m = 1, 2, \cdots, n), \tag{4.5.5}$$

where

$$\mathbf{M}^j_m = \prod_{i=m}^{j}\mathbf{T}_i(H_i), \quad (j = 1, 2, \cdots, m). \tag{4.5.6}$$

Setting $m = n$ and $z = H_n$ in Eq. (4.5.5), yields

$$\overline{\mathbf{a}}_n(H_n) = \mathbf{M}^1_n\overline{\mathbf{a}}_1(0) + \sum_{j=1}^{n}\mathbf{M}^j_n\overline{\mathbf{R}}_j(H_j). \tag{4.5.7}$$

According to the six boundary conditions at $z = 0$ and $z = H_n$, we can solve for the six unknowns from Eq. (4.5.7).

(2) The nth layer connecting with an elastic half-space

A local coordinate system $O'xyz'$ with $z' = z - H_n$ is first established at the interface between the nth layer and the half-space. For the elastic half-space without body forces, the general solution in Eq. (4.3.16) can be simplified to

$$\overline{\mathbf{a}}'(z') = \mathbf{X}(z')\mathbf{k}, \tag{4.5.8}$$

where $\mathbf{X}(z')$ is the same as that in Eq. (4.3.26) or (4.3.28) except that z should be replaced by z', $\overline{\mathbf{a}}'$ is still given by Eq. (4.3.11) but here is for the half-space, and

$$\mathbf{k} = [k_1, k_2, k_3, k_4, k_5, k_6]^{\mathrm{T}} = \mathbf{X}^{-1}(0)\overline{\mathbf{a}}'(0). \tag{4.5.9}$$

Since $\overline{\mathbf{a}}'$ must be finite when $z' \to \infty$, it is required that $k_2 = k_4 = k_6 = 0$. Then Eq. (4.5.8) can be rewritten in a more compact form as

$$\overline{\mathbf{a}}'(z') = \mathbf{X}^*(z')\mathbf{k}^*, \tag{4.5.10}$$

where $\mathbf{X}^*(z')$ is a matrix of order 6×3, obtained from $\mathbf{X}(z')$ by deletion of the second, fourth, and sixth columns, and

$$\mathbf{k}^* = [k_1, k_3, k_5]^{\mathrm{T}}. \tag{4.5.11}$$

Setting $z'=0$ in Eq. (4.5.10) yields

$$\boldsymbol{\Sigma}'_0 = \mathbf{X}_1(0)\mathbf{k}^*, \tag{4.5.12}$$

$$\mathbf{U}'_0 = \mathbf{X}_2(0)\mathbf{k}^*, \tag{4.5.13}$$

where

$$\overline{\mathbf{a}}'(0) = \begin{Bmatrix} \boldsymbol{\Sigma}'_0 \\ \mathbf{U}'_0 \end{Bmatrix}, \quad \mathbf{X}^*(0) = \begin{Bmatrix} \mathbf{X}_1(0) \\ \mathbf{X}_2(0) \end{Bmatrix}, \tag{4.5.14}$$

$$\boldsymbol{\Sigma}'_0 = [\overline{\sigma}'_z(0), \overline{\tau}'_{zx}(0), \overline{\tau}'_{yz}(0)]^{\mathrm{T}}, \quad \mathbf{U}'_0 = [\overline{w}'(0), \overline{u}'(0), \overline{v}'(0)]^{\mathrm{T}}, \tag{4.5.15}$$

$$\mathbf{X}_1(0) = \begin{bmatrix} 0 & 1 & 1 \\ \beta & \mathrm{i}\alpha s_1/\rho & \mathrm{i}\alpha s_2/\rho \\ -\alpha & \mathrm{i}\beta s_1/\rho & \mathrm{i}\beta s_2/\rho \end{bmatrix}, \quad s_1 \neq s_2, \tag{4.5.16a}$$

$$\mathbf{X}_1(0) = \begin{bmatrix} 0 & 1 & 0 \\ \beta & \mathrm{i}\alpha s/\rho & -\mathrm{i}\alpha/\rho^2 \\ -\alpha & \mathrm{i}\beta s/\rho & -\mathrm{i}\beta/\rho^2 \end{bmatrix}, \quad s_1 = s_2 = s, \tag{4.5.16b}$$

$$\mathbf{X}_2(0) = \begin{bmatrix} 0 & -b_{21}s_1/\rho & -b_{22}s_2/\rho \\ -\beta/(\rho c_{44}s_3) & -\mathrm{i}\alpha b_{31}/\rho^2 & -\mathrm{i}\alpha b_{32}/\rho^2 \\ \alpha/(\rho c_{44}s_3) & -\mathrm{i}\beta b_{31}/\rho^2 & -\mathrm{i}\beta b_{32}/\rho^2 \end{bmatrix}, \quad s_1 \neq s_2, \tag{4.5.17a}$$

$$\mathbf{X}_2(0) = \begin{bmatrix} 0 & -s/(2\rho c_{44}) & -b_{12}/\rho^2 \\ -\beta/(\rho c_{44}s_3) & -\mathrm{i}\alpha/(2\rho^2 c_{44}) & \mathrm{i}\alpha s b_{11}/\rho^3 \\ \alpha/(\rho c_{44}s_3) & -\mathrm{i}\beta/(2\rho^2 c_{44}) & \mathrm{i}\beta s b_{11}/\rho^3 \end{bmatrix}, \quad s_1 = s_2 = s. \tag{4.5.17b}$$

If the nth layer contacts with the half-space at $z = H_n$ (or $z'=0$), where the continuous conditions as in Eq. (3.2.1) hold, we have

$$\overline{\mathbf{a}}(H_n) = \overline{\mathbf{a}}'(0). \tag{4.5.18}$$

Substituting Eqs. (4.5.7) and (4.5.10) into Eq. (4.5.18) yields

$$\mathbf{X}^*(0)\mathbf{k}^* = \mathbf{M}_n^1 \overline{\mathbf{a}}_1(0) + \sum_{j=1}^{n} \mathbf{M}_n^j \overline{\mathbf{R}}_j(H_j). \tag{4.5.19}$$

Then according to the three boundary conditions at $z = 0$, the six unknowns can be found from Eq. (4.5.19).

If the interface conditions as in Eq. (3.2.2) hold on the interface, we can obtain from Eq. (4.5.12)

$$\mathbf{k}^* = \mathbf{X}_1^{-1}(0)\mathbf{\Sigma}_0', \tag{4.5.20}$$

where

$$\mathbf{X}_1^{-1}(0) = \begin{bmatrix} 0 & \dfrac{\beta}{\rho^2} & \dfrac{\alpha}{\rho^2} \\ -\dfrac{s_2}{s_1 - s_2} & -\dfrac{\mathrm{i}\alpha}{\rho(s_1 - s_2)} & -\dfrac{\mathrm{i}\beta}{\rho(s_1 - s_2)} \\ \dfrac{s_1}{s_1 - s_2} & \dfrac{\mathrm{i}\alpha}{\rho(s_1 - s_2)} & \dfrac{\mathrm{i}\beta}{\rho(s_1 - s_2)} \end{bmatrix}, \quad s_1 \neq s_2, \tag{4.5.21a}$$

$$\mathbf{X}_1^{-1}(0) = \begin{bmatrix} 0 & \beta/\rho^2 & -\alpha/\rho^2 \\ 1 & 0 & 0 \\ s\rho & \mathrm{i}\alpha & \mathrm{i}\beta \end{bmatrix}, \quad s_1 = s_2 = s. \tag{4.5.21b}$$

Substituting Eq. (4.5.20) into Eq. (4.5.13), yields

$$\mathbf{U}_0' = \mathbf{N}(0)\mathbf{\Sigma}_0', \tag{4.5.22}$$

where

$$\mathbf{N}(0) = \mathbf{X}_2(0)\mathbf{X}_1^{-1}(0) = [N_{ij}], \quad (i, j = 1, 2, 3). \tag{4.5.23}$$

Noticing $\overline{\tau}_{zx}'(0) = \overline{\tau}_{yz}'(0) = 0$, we obtain from Eq. (4.5.22)

$$\overline{w}'(0)=N_{11}\overline{\sigma}_z'(0)\,,\quad \overline{u}'(0)=N_{21}\overline{\sigma}_z'(0)\,,\quad \overline{v}'(0)=N_{31}\overline{\sigma}_z'(0)\,. \tag{4.5.24}$$

Thus, the remaining four equations in Eq. (3.2.2) can be rewritten as

$$\begin{aligned}&\overline{\sigma}_{zn}(H_n)=\overline{\sigma}_z'(0)\,, &&\overline{\tau}_{zxn}(H_n)=0\,,\\ &\overline{w}_n(H_n)=N_{13}\overline{\sigma}_z'(0)\,, &&\overline{\tau}_{yzn}(H_n)=0\,.\end{aligned} \tag{4.5.25}$$

Now, by substituting the expressions for $\overline{\sigma}_{zn}(H_n)$, $\overline{\tau}_{zxn}(H_n)$, $\overline{\tau}_{yzn}(H_n)$ and $\overline{w}_n(H_n)$ in Eq. (4.5.7) into Eq. (4.5.25), and making use of the three boundary conditions at $z=0$, we can determine the four unknowns from Eq. (4.5.25).

5

EQUILIBRIUM OF BODIES OF REVOLUTION

Bodies of revolution, such as circular and annular plates, solid and hollow cylinders, solid and hollow spheres, solid and hollow cones, *etc.*, are frequently encountered in engineering. In this chapter, we will develop several analytical solutions for circular and annular plates and for solid and hollow cones, based on the general solution derived in **Chapter 2**. These include the solutions for a circular (annular) plate subjected to axial tension, radial compression, pure bending, uniform transverse load (with simple supports), or uniform rotation, and for a solid (hollow) cone subjected to concentrated loading at its apex. The problems of cylinders and spheres will be discussed in later chapters.

5.1 SOME HARMONIC FUNCTIONS

The *harmonic equation* in cylindrical coordinates reads as

$$\left(\frac{\partial^2}{\partial r^2}+\frac{1}{r}\frac{\partial}{\partial r}+\frac{1}{r^2}\frac{\partial^2}{\partial \alpha^2}+\frac{\partial^2}{\partial z^2}\right)\varphi=0\,. \qquad (5.1.1)$$

It can be shown that the functions to be given below all satisfy Eq. (5.1.1).

5.1.1 Harmonic Polynomials

$$\begin{aligned}&\varphi_0=1,\quad \varphi_1=z,\quad \varphi_2=z^2-\tfrac{1}{2}r^2,\quad \varphi_3=z^3-\tfrac{3}{2}r^2z,\\ &\varphi_4=z^4-3r^2z^2+\tfrac{3}{8}r^4,\quad \varphi_5=z^5-5r^2z^3+\tfrac{15}{8}r^4z.\end{aligned} \qquad (5.1.2)$$

5.1.2 Harmonic Functions Containing $\ln(r/r_1)$

$$\ln(r/r_1),\quad z\ln(r/r_1),\quad (z^2-\tfrac{1}{2}r^2)\ln(r/r_1)+\tfrac{1}{2}r^2,$$

$$
\begin{aligned}
&(z^3 - \tfrac{3}{2}r^2 z)\ln(r/r_1) + \tfrac{3}{2}r^2 z, \\
&(z^4 - 3r^2 z^2 + \tfrac{3}{8}r^4)\ln(r/r_1) + 3r^2 z^2 - \tfrac{9}{16}r^4, \\
&(z^5 - 5r^2 z^3 + \tfrac{15}{8}r^4 z)\ln(r/r_1) + 5r^2 z^3 - \tfrac{45}{16}r^4 z,
\end{aligned}
\tag{5.1.3}
$$

where r_1 is a constant.

5.1.3 Harmonic Functions Containing R

$$
\begin{aligned}
&1/R, \quad \ln(R+z), \quad \ln(R-z), \\
&\frac{r\sin\alpha}{R+z}, \quad \frac{r\sin\alpha}{R-z}, \quad \frac{r\cos\alpha}{R+z}, \quad \frac{r\cos\alpha}{R-z} \\
&\frac{r\sin\alpha}{R(R+z)}, \quad \frac{r\sin\alpha}{R(R-z)}, \quad \frac{r\cos\alpha}{R(R+z)}, \quad \frac{r\cos\alpha}{R(R-z)},
\end{aligned}
\tag{5.1.4}
$$

where $R = \sqrt{r^2 + z^2}$.

For the problems to be studied in this chapter, we will use these harmonic functions to construct the displacement functions φ_i in the general solution as presented in Eqs. (2.2.69) and (2.2.70). By comparing Eq. (5.1.1) with Eq. (2.2.58), we find that the harmonic functions given in Eqs. (5.1.2)-(5.1.4), on substitution of z_i for z, will satisfy Eq. (2.2.58). Note that, we will consider the case $s_1 \neq s_2$ only in the following discussion.

5.2 AN ANNULAR (CIRCULAR) PLATE SUBJECTED TO AXIAL TENSION AND RADIAL COMPRESSION

Consider an annular plate of inner radius r_0, outer radius r_1, and thickness h, with the middle surface parallel to the isotropic plane and consistent with the *xOy* coordinate plane. The geometry of the annular plate and the cylindrical coordinates are shown in Fig. 5.1.

In view of Eqs. (5.1.2) and (5.1.3), we take

$$
\varphi_i = A_{2i}\left(z_i^2 - \frac{1}{2}r^2\right) + B_0 \ln\left(\frac{r}{r_1}\right), \quad (i = 1, 2), \tag{5.2.1}
$$

where A_{21}, A_{22}, and B_0 are undetermined constants.

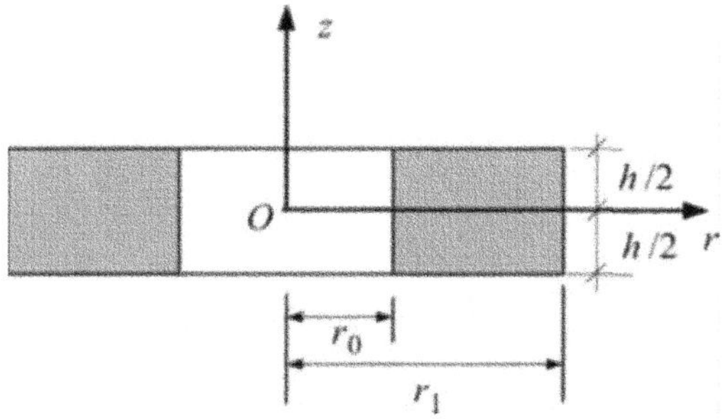

Fig. 5.1 The geometry of problem and the coordinates.

Substituting Eq. (5.2.1) into Eqs. (2.2.69*b*) and (2.2.70*b*) yields

$$u_r = -\sum_{i=1}^{2} A_{2i} r + B_0 \frac{1}{r}, \quad w = 2\sum_{i=1}^{2} \alpha_i A_{2i} z_i, \quad u_\alpha = 0, \tag{5.2.2}$$

$$\begin{aligned} \sigma_r &= 2c_{66}\left(\sum_{i=1}^{2} e_i A_{2i} - B_0 \frac{1}{r^2}\right), \quad \sigma_\alpha = 2c_{66}\left(\sum_{i=1}^{2} e_i A_{2i} + B_0 \frac{1}{r^2}\right), \\ \sigma_z &= 2c_{66} \sum_{i=1}^{2} k_{2i} A_{2i}, \quad \tau_{zr} = \tau_{r\alpha} = \tau_{\alpha z} = 0, \end{aligned} \tag{5.2.3}$$

where

$$e_i = k_{1i} - 1. \tag{5.2.4}$$

The boundary conditions are

$$\begin{aligned} &\sigma_r = -q_k, \quad \tau_{zr} = 0, \quad \text{at } r = r_k, \quad (k = 0, 1), \\ &\sigma_z = p, \quad \tau_{zr} = 0, \quad \text{at } z = \pm h/2. \end{aligned} \tag{5.2.5}$$

where q_0, q_1, and p are the uniform pressures prescribed on the inner and outer cylindrical surfaces, and the uniform tension stress applied on the upper and lower surfaces, respectively.

Substituting the expressions for σ_r and σ_z in Eq. (5.2.3) into Eq. (5.2.5) yields

$$2c_{66}r_k^2\sum_{i=1}^{2}e_iA_{2i} - 2c_{66}B_0 = -r_k^2 q_k, \quad (k=0,1)\ ,$$
$$2c_{66}\sum_{i=1}^{2}k_{2i}A_{2i} = p, \tag{5.2.6}$$

which lead to

$$A_{21} = \frac{pe_2 - k_{22}Q}{2c_{66}(k_{21}e_2 - k_{22}e_1)}, \quad A_{22} = \frac{k_{21}Q - pe_1}{2c_{66}(k_{21}e_2 - k_{22}e_1)},$$
$$B_0 = \frac{r_0^2 r_1^2(q_0 - q_1)}{2c_{66}(r_1^2 - r_0^2)}, \quad Q = \frac{r_0^2 q_0 - r_1^2 q_1}{r_1^2 - r_0^2}. \tag{5.2.7}$$

Substituting Eq. (5.2.7) into Eq. (5.2.3) gives rise to the non-vanishing stresses

$$\sigma_r = Q - r_0^2(Q + q_0)\frac{1}{r^2}, \quad \sigma_\alpha = Q + r_0^2(Q + q_0)\frac{1}{r^2}, \quad \sigma_z = p. \tag{5.2.8}$$

Setting $r_0 = 0$ in Eqs. (5.2.8), (5.2.2) and (5.2.7) gives the solution for a circular plate. When $q_0 = q_1 = 0$ and $p \neq 0$, the solution for an annular (circular) plate subjected to axial tension is obtained; otherwise, the solution of a radially compressed plate is obtained.

It is seen from Eq. (5.2.8) that the stress state in the plate is independent of the elastic constants, and is identical to that in an isotropic plate. Nevertheless, the displacements change with the elastic properties of the plate.

5.3 AN ANNULAR (CIRCULAR) PLATE SUBJECTED TO PURE BENDING

In view of Eqs. (5.1.2) and (5.1.3), we take

$$\varphi_i = A_{3i}\left(z_i^3 - \frac{3}{2}r^2 z_i\right) + B_{1i}z_i\ln\left(\frac{r}{r_1}\right), \quad (i=1,2), \tag{5.3.1}$$

where A_{3i} and B_{1i} are undetermined constants.

Substituting Eq. (5.3.1) into Eqs. (2.2.69*b*) and (2.2.70*b*) yields

$$u_r = \sum_{i=1}^{2}\left(-3A_{3i}rz_i + B_{1i}\frac{z_i}{r}\right), \quad w = \sum_{i=1}^{2}\alpha_i\left[3A_{3i}\left(z_i^2 - \frac{1}{2}r^2\right) + B_{1i}\ln\left(\frac{r}{r_1}\right)\right], \tag{5.3.2}$$

$$\begin{aligned} &\sigma_r = 2c_{66}\sum_{i=1}^{2}\left(3e_iA_{3i}z_i - B_{1i}\frac{z_i}{r^2}\right), \quad \sigma_\alpha = 2c_{66}\sum_{i=1}^{2}\left(3e_iA_{3i}z_i + B_{1i}\frac{z_i}{r^2}\right), \\ &\sigma_z = 6c_{66}\sum_{i=1}^{2}k_{2i}A_{3i}z_i, \quad \tau_{zr} = c_{66}\sum_{i=1}^{2}k_{3i}\left(-3A_{3i}r + B_{1i}\frac{1}{r}\right), \end{aligned} \tag{5.3.3}$$

where e_i were shown in Eq. (5.2.4).

The boundary conditions are

$$\begin{aligned} &\sigma_z = \tau_{zr} = 0, \quad \text{at } z = \pm h/2, \\ &\tau_{zr} = 0, \quad \int_{-h/2}^{h/2} z\sigma_r\,\mathrm{d}z = M_k, \quad \text{at } r = r_k, \quad (k = 0, 1), \end{aligned} \tag{5.3.4}$$

where M_0 and M_1 are the distributed moments prescribed on the inner and outer cylindrical surfaces of the annular plate.

Substituting the expressions for τ_{zr}, σ_z, and σ_r in Eq. (5.3.3) into Eq. (5.3.4), yields

$$\begin{aligned} &\sum_{i=1}^{2}k_{3i}A_{3i} = 0, \quad \sum_{i=1}^{2}k_{3i}B_{1i} = 0, \\ &\sum_{i=1}^{2}(3r_k^2s_ie_iA_{3i} - s_iB_{1i}) = \frac{6M_kr_k^2}{c_{66}h^3}, \quad (k = 0, 1), \end{aligned} \tag{5.3.5}$$

where use has been made of $s_ik_{2i} = k_{3i}$ in Eq. (2.2.65). We get immediately from Eq. (5.3.5)

$$\begin{aligned} &A_{31} = \frac{k_{32}P}{s_1e_1k_{32} - s_2e_{21}k_{31}}, \quad A_{32} = -\frac{k_{31}P}{s_1e_1k_{32} - s_2e_{21}k_{31}}, \\ &B_{11} = \frac{k_{32}Q}{s_1k_{32} - s_2k_{31}}, \quad B_{12} = -\frac{k_{31}Q}{s_1k_{32} - s_2k_{31}}, \end{aligned} \tag{5.3.6}$$

where

$$P = \frac{2(r_1^2 M_1 - r_0^2 M_0)}{c_{66}(r_1^2 - r_0^2)h^3}, \qquad Q = \frac{6r_0^2 r_1^2 (M_1 - M_0)}{c_{66}(r_1^2 - r_0^2)h^3}. \tag{5.3.7}$$

Substituting Eq. (5.3.6) into Eq. (5.3.3) gives

$$\begin{aligned}
\sigma_r &= \frac{12z}{(r_1^2 - r_0^2)h^3}\left[r_1^2 M_1 - r_0^2 M_0 - r_0^2 r_1^2 (M_1 - M_0)\frac{1}{r^2}\right], \\
\sigma_\alpha &= \frac{12z}{(r_1^2 - r_0^2)h^3}\left[r_1^2 M_1 - r_0^2 M_0 + r_0^2 r_1^2 (M_1 - M_0)\frac{1}{r^2}\right], \\
\sigma_z &= \tau_{zr} = 0.
\end{aligned} \tag{5.3.8}$$

Setting $r_0 = 0$ in Eqs. (5.3.8), (5.3.2) and (5.3.6) gives the solution for a circular plate in bure bending.

While the stresses in the plate is independent of the elastic constants, and hence is identical to that in an isotropic plate, as seen from Eq. (5.3.8), the displacements depend on the elastic properties of the plate.

5.4 A SIMPLY-SUPPORTED ANNULAR (CIRCULAR) PLATE UNDER UNIFORM TRANSVERSE LOADING

In view of Eqs. (5.1.2) and (5.1.3), we take

$$\begin{aligned}
\varphi_i &= F_{3i}\left(z_i^3 - \frac{3}{2}r^2 z_i\right) + F_{5i}\left(z_i^5 - 5r^2 z_i^3 + \frac{15}{8}r^4 z_i\right) \\
&\quad + G_{1i} z_i \ln\left(\frac{r}{r_1}\right) + G_{3i}\left[\left(z_i^3 - \frac{3}{2}r^2 z_i\right)\ln\left(\frac{r}{r_1}\right) + \frac{3}{2}r^2 z_i\right], \quad (i = 1, 2),
\end{aligned} \tag{5.4.1}$$

where F_{3i}, F_{5i}, G_{1i} and G_{3i} are undetermined constants.

Substituting Eq. (5.4.1) into Eqs. (2.2.69*b*) and (2.2.70*b*), and adding a rigid body axial displacement w_0, which does not affect the stresses, we obtain

$$
\begin{aligned}
u_r &= -\sum_{i=1}^{2}\left[3F_{3i}rz_i + 5F_{5i}\left(2rz_i^3 + \frac{3}{2}r^3 z_i\right)\right] \\
&\quad + \sum_{i=1}^{2}\left\{G_{1i}\frac{z_i}{r} + G_{3i}\left[-3rz_i\ln\left(\frac{r}{r_1}\right) + \frac{z_i^3}{r} + \frac{3}{2}rz_i\right]\right\}, \\
w &= \sum_{i=1}^{2}\alpha_i\left[3F_{3i}\left(z_i^2 - \frac{1}{2}r^2\right) + 5F_{5i}\left(z_i^4 - 3r^2z_i^2 + \frac{3}{8}r^4\right)\right] \\
&\quad + \sum_{i=1}^{2}\alpha_i\left\{G_{1i}\ln\left(\frac{r}{r_1}\right) + 3G_{3i}\left[\left(z_i^2 - \frac{1}{2}r^2\right)\ln\left(\frac{r}{r_1}\right) + \frac{1}{2}r^2\right]\right\} + w_0,
\end{aligned}
\tag{5.4.2}
$$

$$
\begin{aligned}
\sigma_r / c_{66} &= \sum_{i=1}^{2}\left[6e_iF_{3i}z_i + 5F_{5i}(4e_iz_i^3 - 3f_ir^2z_i)\right] \\
&\quad + \sum_{i=1}^{2}\left\{-2G_{1i}\frac{z_i}{r^2} + G_{3i}\left[6e_iz_i\ln\left(\frac{r}{r_1}\right) - \frac{2z_i^3}{r^2} - 3z_i\right]\right\}, \\
\sigma_\alpha / c_{66} &= \sum_{i=1}^{2}\left[6e_iF_{3i}z_i + 5F_{5i}(4e_iz_i^3 - 3g_ir^2z_i)\right] \\
&\quad + \sum_{i=1}^{2}\left\{2G_{1i}\frac{z_i}{r^2} + G_{3i}\left[6e_iz_i\ln\left(\frac{r}{r_1}\right) + \frac{2z_i^3}{r^2} + 3z_i\right]\right\}, \\
\sigma_z / c_{66} &= \sum_{i=1}^{2}2k_{2i}\left[3F_{3i}z_i + 5F_{5i}(2z_i^3 - 3r^2z_i) + 3G_{3i}z_i\ln\left(\frac{r}{r_1}\right)\right], \\
\tau_{zr} / c_{66} &= -\sum_{i=1}^{2}3k_{3i}\left[F_{3i}r + 5F_{5i}\left(2rz_i^2 - \frac{1}{2}r^3\right)\right] \\
&\quad + \sum_{i=1}^{2}k_{3i}\left\{G_{1i}\frac{1}{r} - 3G_{3i}\left[r\ln\left(\frac{r}{r_1}\right) - \frac{z_i^2}{r} - \frac{r}{2}\right]\right\},
\end{aligned}
\tag{5.4.3}
$$

where e_i were defined in Eq. (5.2.4), and

$$
f_i = 2e_i - 1, \quad g_i = 2e_i + 1. \tag{5.4.4}
$$

The boundary conditions are

$$
\begin{aligned}
&\tau_{zr} = 0, \quad \sigma_z = \pm p, \quad \text{at } z = \pm h/2, \\
&\int_{-h/2}^{h/2} z\sigma_r \,\mathrm{d}z = 0, \quad w\big|_{z=0} = 0, \quad \text{at } r = r_k, \quad (k = 0, 1),
\end{aligned}
\tag{5.4.5}
$$

where p is the uniform loading applied on the upper and lower surfaces of the annular plate.

Substituting the expressions for τ_{zr}, σ_z, and σ_r in Eq. (5.4.3) and the expression for w in Eq. (5.4.2) into Eq. (5.4.5), yields

$$\begin{aligned}
&\sum_{i=1}^{2} k_{3i} G_{3i} = 0, \quad \sum_{i=1}^{2} k_{3i} G_{1i} + \frac{3h^2}{4} \sum_{i=1}^{2} s_i^2 k_{3i} G_{3i} = 0, \\
&\sum_{i=1}^{2} k_{3i} F_{5i} = 0, \quad 2\sum_{i=1}^{2} k_{3i} F_{3i} + 5h^2 \sum_{i=1}^{2} s_i^2 k_{3i} F_{5i} = 0, \\
&3h \sum_{i=1}^{2} s_i k_{2i} F_{3i} + \frac{5h^3}{2} \sum_{i=1}^{2} s_i^2 k_{2i} F_{5i} = \frac{p}{c_{66}}, \\
&r_k^2 \sum_{i=1}^{2} 3s_i \left[2e_i F_{3i} + (s_i^2 e_i h^2 - 5 f_i r_k^2) F_{5i}\right] - 2\sum_{i=1}^{2} s_i G_{1i} \\
&\quad + 3\sum_{i=1}^{2} s_i \left[2e_i r_k^2 \ln\left(\frac{r_k}{r_1}\right) - \frac{1}{10} s_i^2 h^2 - r_k^2\right] G_{3i} = 0, \quad (k = 0, 1), \\
&2w_0 - 3r_k^2 \sum_{i=1}^{2} 3\alpha_i \left(F_{3i} - \frac{5}{4} r_k^2 F_{5i}\right) + 2\ln\left(\frac{r_k}{r_1}\right) \sum_{i=1}^{2} \alpha_i G_{1i} \\
&\quad + 3r_k^2 \left[1 - \ln\left(\frac{r_k}{r_1}\right)\right] \sum_{i=1}^{2} \alpha_i G_{3i} = 0, \quad (k = 0, 1).
\end{aligned} \tag{5.4.6}$$

In deriving this equation, we have employed $k_{3i} = s_i k_{2i}$. It is clear that we can solve for the nine unknowns, *i.e.* F_{3i}, F_{5i}, G_{1i}, G_{3i}, $(i = 1, 2)$ and w_0, from Eq. (5.4.6). For a circular plate, we shall take $G_{1i} = G_{3i} = 0$ in relevant formulations, and simultaneously discard the equations involving r_0 in Eq. (5.4.6). Then the five unknowns F_{3i}, F_{5i} and w_0 can be obtained from the remaining five algebraic equations.

By superposing this solution on the solution obtained in **Section 5.2** for axial tension, we obtain the solution for an annular (circular) plate with traction-free at $z = -h/2$, subjected to a uniform load $2p$ at $z = h/2$.

5.5 A UNIFORMLY ROTATING ANNULAR (CIRCULAR) PLATE

Consider an annular plate rotating about the symmetric axis with a constant angular velocity ω. This is an axisymmetric problem, for which all displacement and stress components are independent of the coordinate α. For transversely isotropic materials, the governing equations in terms of displacements can be obtained from Eqs. (1.4.12) and (2.1.5) as

$$\left[c_{11}\left(\frac{\partial^2}{\partial r^2}+\frac{1}{r}\frac{\partial}{\partial r}-\frac{1}{r^2}\right)+c_{44}\frac{\partial^2}{\partial z^2}\right]u_r+(c_{13}+c_{44})\frac{\partial^2 w}{\partial r\partial z}+F_r=0,$$
$$(c_{13}+c_{44})\left(\frac{\partial^2}{\partial r\partial z}+\frac{1}{r}\frac{\partial}{\partial z}\right)u_r+\left[c_{44}\left(\frac{\partial^2}{\partial r^2}+\frac{1}{r}\frac{\partial}{\partial r}\right)+c_{33}\frac{\partial^2}{\partial z^2}\right]w+F_z=0. \quad (5.5.1)$$

For the rotating annular plate, we only consider the centrifugal force as the body force, *i.e.*

$$F_r=\rho r\omega^2,\quad F_z=0, \quad (5.5.2)$$

where ρ is the density. It is easy to find the particular solution to Eq. (5.5.1):

$$u_r=-\frac{1}{8c_{11}}\rho\omega^2 r^3,\quad w=0. \quad (5.5.3)$$

Substituting Eq. (5.5.3) into Eq. (1.2.2), and noticing Eq. (1.2.37), we obtain the corresponding stress components as

$$\sigma_r=h_1 r^2,\quad \sigma_\alpha=h_2 r^2,\quad \sigma_z=h_3 r^2,\quad \tau_{zr}=0, \quad (5.5.4)$$

where

$$h_1=-\frac{(3c_{11}+c_{12})}{8c_{11}}\rho\omega^2,\quad h_2=-\frac{(c_{11}+3c_{12})}{8c_{11}}\rho\omega^2,\quad h_3=-\frac{c_{13}}{2c_{11}}\rho\omega^2. \quad (5.5.5)$$

The general solution to the homogenous equation of Eq. (5.5.1) takes the form as shown in Eq. (2.2.69*b*). In view of Eqs. (5.1.2) and (5.1.3), we take the displacement functions as

$$\varphi_i = F_{2i}\left(z_i^2 - \frac{1}{2}r^2\right) + F_{4i}\left(z_i^4 - 3r^2 z_i^2 + \frac{3}{8}r^4\right) + G_0 \ln\left(\frac{r}{r_1}\right), \quad (i=1,2), \tag{5.5.6}$$

where F_{2i}, F_{4i} and G_0 are undetermined constants.

Substituting Eq. (5.5.6) into Eqs. (2.2.69*b*) and (2.2.70*b*), and superposing the particular solution, Eqs. (5.5.3) and (5.5.4), on the results, we obtain

$$\begin{aligned} u_r &= -\sum_{i=1}^{2} F_{2i} r + 3\sum_{i=1}^{2} F_{4i}\left(-2rz_i^2 + \frac{1}{2}r^3\right) + G_0\frac{1}{r} - \frac{\rho\omega^2}{8c_{11}}r^3, \\ w &= 2\sum_{i=1}^{2}\alpha_i F_{2i} z_i + 2\sum_{i=1}^{2}\alpha_i F_{4i}(2z_i^3 - 3r^2 z_i), \end{aligned} \tag{5.5.7}$$

$$\begin{aligned} \sigma_r &= 2c_{66}\sum_{i=1}^{2} e_i F_{2i} + 3c_{66}\sum_{i=1}^{2} F_{4i}(4e_i z_i^2 - f_i r^2) - 2c_{66}G_0\frac{1}{r^2} + h_1 r^2, \\ \sigma_\alpha &= 2c_{66}\sum_{i=1}^{2} e_i F_{2i} + 3c_{66}\sum_{i=1}^{2} F_{4i}(4e_i z_i^2 - g_i r^2) + 2c_{66}G_0\frac{1}{r^2} + h_2 r^2, \\ \sigma_z &= 2c_{66}\sum_{i=1}^{2} k_{2i}F_{2i} + 6c_{66}\sum_{i=1}^{2} k_{2i}F_{4i}(2z_i^2 - r^2) + h_3 r^2, \\ \tau_{zr} &= -12c_{66}\sum_{i=1}^{2} k_{3i}F_{4i} r z_i, \end{aligned} \tag{5.5.8}$$

where f_i and g_i, and e_i were given by Eqs. (5.4.4) and (5.2.4), respectively.

The boundary conditions are

$$\begin{aligned} &\tau_{zr} = \sigma_z = 0, \quad \text{at } z = \pm h/2, \\ &\tau_{zr} = 0, \quad \int_{-h/2}^{h/2}\sigma_r \, \mathrm{d}z = 0, \quad \text{at } r = r_k, \quad (k=0,1). \end{aligned} \tag{5.5.9}$$

Substituting the expressions for τ_{zr}, σ_z, and σ_r in Eq. (5.5.8) into Eq. (5.5.9), yields

$$\begin{aligned} &\sum_{i=1}^{2} s_i k_{3i} F_{4i} = 0, \quad 2\sum_{i=1}^{2} k_{2i}F_{2i} + 3h^2\sum_{i=1}^{2} s_i^2 k_{2i} F_{4i} = 0, \\ &6c_{66}\sum_{i=1}^{2} k_{2i}F_{4i} - h_3 = 0, \end{aligned} \tag{5.5.10}$$

$$r_k^2\sum_{i=1}^{2}\left[2e_iF_{2i}-(3f_ir_k^2-s_i^2e_ih^2)F_{4i}\right]-2G_0+\frac{h_1r_k^4}{c_{66}}=0,\quad (k=0,1).$$

Now we can solve for the five unknowns, *i.e.* F_{2i}, F_{4i}, $(i=1,2)$ and G_0, from Eq. (5.5.10). For a circular plate, we shall take $G_0=0$ in relevant formulations and simultaneously discard the equations involving r_0 in Eq. (5.5.10). Then the remaining four algebraic equations give the four unknowns F_{2i} and F_{4i}.

If treating this problem as a plane-stress problem, *i.e.*

$$\sigma_z=\tau_{\alpha z}=\tau_{zr}=0, \tag{5.5.11}$$

then from the constitutive relations in Eq. (1.2.38), we derive

$$\varepsilon_r=s_{11}\sigma_r+s_{12}\sigma_\alpha,\quad \varepsilon_\alpha=s_{12}\sigma_r+s_{11}\sigma_\alpha,\quad \gamma_{r\alpha}=s_{66}\tau_{r\alpha}, \tag{5.5.12}$$

where $s_{66}=2(s_{11}-s_{12})$. This corresponds to a plane-stress problem of an isotropic material with the following Young's modulus E and Poisson's ratio ν

$$E=\frac{1}{s_{11}}=\frac{(c_{11}-c_{12})[(c_{11}+c_{12})c_{33}-2c_{13}^2]}{c_{11}c_{33}-c_{13}^2},\quad \nu=-\frac{s_{12}}{s_{11}}=\frac{c_{12}c_{33}-c_{13}^2}{c_{11}c_{33}-c_{13}^2}. \tag{5.5.13}$$

Here the converse relation between the stiffness matrix in Eq. (1.2.37) and the compliance matrix in Eq. (1.2.38) has been used.

The solution of an isotropic solid disk rotating at a constant angular velocity ω can be found in the textbook of Timoshenko and Goodier (1970). The following maximum stresses occur at the center of the disk,

$$\sigma_r=\sigma_\theta=\frac{3+\nu}{8}\rho\omega^2b^2 \tag{5.5.14}$$

where b is the radius of the disk, ρ the density, and ν the Poisson's ratio.

Table 5.1 gives the values of radial stress σ_r at the center of a circular plate rotating at $\omega=300$ turns per minute for several combinations of parameters, where $D=2r_1$ is the diameter of the disk. The three-dimensional solution calculated from Eq. (5.5.8) is compared with the plane-stress solution calculated from Eq. (5.5.14). It is seen that the radial stress varies significantly with material properties. For example, the radial stress at the center of a disk made of Sapphire is more than two times that in the Graphite/epoxy disk. When the thickness-to-diameter ratio h/D is small, the plane-stress solution agrees very well with the three-dimensional solution. With the increase of h/D, the stress predicted by the three-dimensional theory increases tardily, while the plane-stress solution remains unchanged.

Table 5.1 The radial stress σ_r (Pa) at the center of a rotating circular plate.*

Material / h/D	Graphite		Graphite/epoxy		Sapphire	
	Plane-stress solution	3D solution	Plane-stress solution	3D solution	Plane-stress solution	3D solution
0.01	2215.9	2216.0	1828.6	1828.6	4049.9	4049.9
0.05	2215.9	2217.0	1828.6	1828.7	4049.9	4051.1
0.10	2215.9	2220.3	1828.6	1829.1	4049.9	4054.6
0.20	2215.9	2233.5	1828.6	1830.3	4049.9	4068.8
0.30	2215.9	2255.6	1828.6	1832.4	4049.9	4092.4

* Angular velocity $\omega = 300$ turns/minute, diameter of plate $D = 2r_1 = 10$ cm.

5.6 TRANSVERSELY ISOTROPIC CONES

Consider a transversely isotropic elastic cone of conical angle $2\beta_0$ subjected to a concentrated force at its tip. The lateral surface of the cone is stress-free. Let the origin of the cylindrical coordinate system, (r, α, z), be at the cone tip and the z-axis be aligned with the cone axis pointing towards the interior of the cone, as illustrated in Fig. 5.2. For convenience, we define

$$R_i^* = R_i + s_i z , \qquad (i = 1, 2, 3) .$$

In this section, we will focus on the case of $s_1 \neq s_2$, but will consider the problems under either an axial or a transverse force.

5.6.1 Compression of a Cone under an Axial Force

The general solution of the problem in cylindrical coordinates was given in Eqs. (2.2.69) and (2.2.70). Now, because the deformation of the cone is axisymmetric, we can take

$$\varphi_3 = 0, \quad \varphi_i = A_i \ln(R_i + s_i z), \quad (i = 1,2), \tag{5.6.1}$$

where A_i $(i = 1, 2)$ are constants to be determined. The displacements and stresses can be obtained directly when these functions are substituted into Eqs. (2.2.69*b*) and (2.2.70*b*), which gives rise to

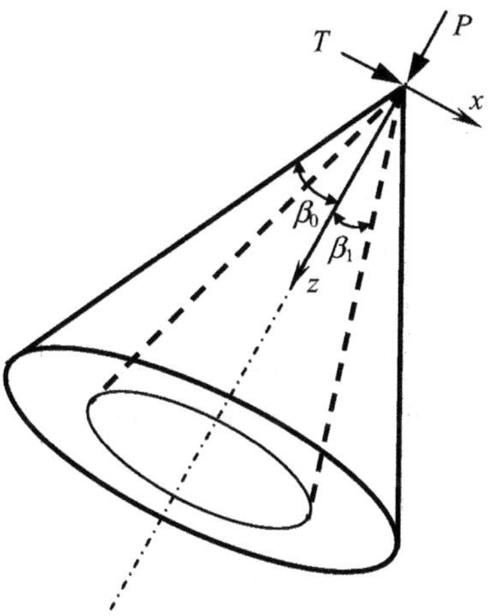

Fig. 5.2 A conical shell subjected to concentrated forces at its apex.

$$u_r = \sum_{i=1}^{2} \frac{A_i r}{R_i R_i^*}, \tag{5.6.2}$$

$$u_\alpha = 0, \tag{5.6.3}$$

$$w = \sum_{i=1}^{2} \frac{\alpha_i A_i}{R_i}, \tag{5.6.4}$$

$$\frac{\sigma_r}{c_{66}} = 2\sum_{i=1}^{2} A_i \left(\frac{1}{R_i R_i^*} - \frac{r^2}{R_i^3 R_i^*} - \frac{r^2}{R_i^2 R_i^{*2}} \right) - \sum_{i=1}^{2} k_{1i} A_i \frac{z_i}{R_i^3}, \tag{5.6.5}$$

$$\frac{\sigma_\alpha}{c_{66}} = 2\sum_{i=1}^{2} \frac{A_i}{R_i R_i^*} - \sum_{i=1}^{2} k_{1i} A_i \frac{z_i}{R_i^3}, \tag{5.6.6}$$

$$\frac{\sigma_z}{c_{66}} = -\sum_{i=1}^{2} k_{2i} A_i \frac{z_i}{R_i^3}, \tag{5.6.7}$$

$$\frac{\tau_{zr}}{c_{66}} = -\sum_{i=1}^{2} \frac{k_{3i} A_i}{R_i^3} r, \tag{5.6.8}$$

$$\tau_{\alpha z} = \tau_{r\alpha} = 0. \tag{5.6.9}$$

Since the lateral surface is traction-free,

$$\begin{aligned} &\tau_{zr} - \sigma_r \cot\beta_0 = 0, \\ &\tau_{\alpha z} - \tau_{r\alpha} \cot\beta_0 = 0, \qquad \text{when } r : z = 1 : \cot\beta_0. \\ &\sigma_z - \tau_{zr} \cot\beta_0 = 0, \end{aligned} \tag{5.6.10}$$

Equation (5.6.9) indicates that the second condition in Eq. (5.6.10) is satisfied automatically and similarly the third condition in Eq. (5.6.10) is also satisfied because of Eqs. (5.6.7) and (5.6.8) and the expressions of k_{2i} and k_{3i} given in Eq. (2.2.64). The only one left to examine in Eq. (5.6.10) is the first condition.

The equilibrium condition for the cone requires that the resultant force on an arbitrary section at $z = \text{const.}$ should be equal to P, the external force, *i.e.*

$$\int_0^{z\tan\beta_0} \int_0^{2\pi} \sigma_z r \mathrm{d}r \mathrm{d}\alpha = -P, \tag{5.6.11}$$

and

$$\begin{aligned} &\int_0^{z\tan\beta_0} \int_0^{2\pi} (\tau_{zr} \cos\alpha - \tau_{\alpha z} \sin\alpha) r \mathrm{d}r \mathrm{d}\alpha = 0, \\ &\int_0^{z\tan\beta_0} \int_0^{2\pi} (\tau_{zr} \sin\alpha + \tau_{\alpha z} \cos\alpha) r \mathrm{d}r \mathrm{d}\alpha = 0. \end{aligned} \tag{5.6.12}$$

Since the stress components are independent of α, Eq. (5.6.12) is automatically satisfied. By substituting Eq. (5.6.7) into Eq. (5.6.11) and noticing the following integral

$$\int_0^{z\tan\beta_0} \frac{z_i}{R_i^3} r \mathrm{d}r = 1 - \frac{s_i}{\sqrt{s_i^2 + \tan^2\beta_0}}, \tag{5.6.13}$$

we obtain

$$2\pi\sum_{i=1}^{2}k_{2i}A_i\left(\frac{s_i}{\sqrt{\tan^2\beta_0+s_i^2}}-1\right)=-2\pi\sum_{i=1}^{2}\frac{k_{2i}A_i}{\overline{R}_i\overline{R}_i^*}\tan^2\beta_0=-\frac{P}{c_{66}},\tag{5.6.14}$$

where

$$\overline{R}_i=\sqrt{\tan^2\beta_0+s_i^2},\ \ \overline{R}_i^*=\overline{R}_i+s_i.\tag{5.6.15}$$

Then by substituting Eqs. (5.6.5) and (5.6.8) into the first in Eq. (5.6.10), we get

$$2\sum_{i=1}^{2}A_i\left(\frac{1}{\overline{R}_i\overline{R}_i^*}-\frac{\tan^2\beta_0}{\overline{R}_i^3\overline{R}_i^*}-\frac{\tan^2\beta_0}{\overline{R}_i^2\overline{R}_i^{*2}}\right)-\sum_{i=1}^{2}\frac{A_i}{\overline{R}_i^3}(s_ik_{1i}-k_{3i}\tan^2\beta_0)=0,\tag{5.6.16}$$

where A_1 and A_2 can be obtained from Eqs. (5.6.14) and (5.6.16). For convenience, let us introduce the following notations

$$\begin{aligned}\overline{A}_{1i}&=\frac{k_{2i}\tan^2\beta_0}{\overline{R}_i\overline{R}_i^*},\\ \overline{A}_{2i}&=2\left(\frac{1}{\overline{R}_i\overline{R}_i^*}-\frac{\tan^2\beta_0}{\overline{R}_i^3\overline{R}_i^*}-\frac{\tan^2\beta_0}{\overline{R}_i^3\overline{R}_i^{*2}}\right)-\frac{(s_ik_{1i}-k_{3i}\tan^2\beta_0)}{\overline{R}_i^3},\end{aligned}\tag{5.6.17}$$

where $i=1,2$. Hence A_1 and A_2 can be written as

$$\begin{aligned}A_1&=\frac{\overline{A}_{22}P}{2\pi(\overline{A}_{11}\overline{A}_{22}-\overline{A}_{12}\overline{A}_{21})c_{66}}=\frac{P}{2\pi(\overline{A}_{11}-\overline{A}_{12}\overline{A}_{21}/\overline{A}_{22})c_{66}},\\ A_2&=-\frac{\overline{A}_{21}P}{2\pi(\overline{A}_{11}\overline{A}_{22}-\overline{A}_{12}\overline{A}_{21})c_{66}}=-\frac{P}{2\pi(\overline{A}_{11}\overline{A}_{22}/\overline{A}_{21}-\overline{A}_{12})c_{66}}.\end{aligned}\tag{5.6.18}$$

When $\beta_0=\pi/2$, the cone becomes a half-space with the isotropic plane parallel to the surface where the vertical concentrated force is applied. Since $\tan^2\beta_0\to\infty$,

$$\overline{A}_{1i}=k_{2i},\ \ (i=1,2),\quad \frac{\overline{A}_{21}}{\overline{A}_{22}}=\frac{k_{31}}{k_{32}}=\frac{s_1k_{21}}{s_2k_{22}}.\tag{5.6.19}$$

By substituting Eq. (5.6.19) into Eq. (5.6.18), we obtain

$$A_1 = -\frac{s_2 P}{2\pi c_{66}(s_1 - s_2)k_{21}}, \quad A_2 = \frac{s_1 P}{2\pi c_{66}(s_1 - s_2)k_{22}}. \tag{5.6.20}$$

which agree well with Eq. (4.1.11).

As in **Section 4.1**, a unified solution of a transversely isotropic half-space subjected to a concentrated force vertical to its surface can be developed.

Now consider a conical shell with $0 < 2\beta_1 < 2\beta_0 < 2\pi$, where $2\beta_1$ is the inner angle as shown in Fig. 5.2. In this case, we can take

$$\varphi_3 = 0, \quad \varphi_i = A_i \ln(R_i + s_i z) + B_i \ln(R_i - s_i z), \quad (i = 1,2),$$

where A_i and B_i are unknown constants to be determined by Eqs. (5.6.10) to (5.6.12) and the traction-free condition of the inner lateral surface. In this case, however, the lower limit of the integral with respect to r in Eqs. (5.6.11) and (5.6.12) should be $z \tan\beta_1$.

5.6.2 Bending of a Cone under a Transverse Force

We can take

$$\varphi_3 = \frac{D_3 r \sin\alpha}{R_3 + s_3 z}, \quad \varphi_i = \frac{D_i r \cos\alpha}{R_i + s_i z}, \quad (i = 1, 2), \tag{5.6.21}$$

where, D_1, D_2 and D_3 are arbitrary constants to be determined. By substituting Eq. (5.6.21) into Eqs. (2.2.69) and (2.2.70), we obtain

$$u_r = \cos\alpha\left(\sum_{i=1}^{2} \frac{D_i z}{R_i R_i^*} + \frac{1}{R_3^*}\right), \tag{5.6.22}$$

$$u_\alpha = -\sin\alpha\left(\sum_{i=1}^{2} \frac{D_i}{R_i^*} + \frac{z}{R_3 R_3^*}\right), \tag{5.6.23}$$

$$w = -r\cos\alpha \sum_{i=1}^{2} \frac{\alpha_i D_i}{R_i}, \tag{5.6.24}$$

$$\frac{\sigma_r}{c_{66}} = r\cos\alpha \left[\sum_{i=1}^{2} k_{1i} D_i \frac{1}{R_i^3} - \sum_{i=1}^{2} 2D_i \frac{z_i(z_i + 2R_i)}{R_i^3 R_i^{*2}} - 2D_3 \frac{1}{R_3 R_3^{*2}} \right], \tag{5.6.25}$$

$$\frac{\sigma_\alpha}{c_{66}} = r\cos\alpha \left(\sum_{i=1}^{2} \frac{k_{1i} D_i}{R_i^3} - \sum_{i=1}^{2} \frac{2D_i}{R_i R_i^{*2}} + \frac{2D_3}{R_3 R_3^{*2}} \right), \tag{5.6.26}$$

$$\frac{\sigma_z}{c_{66}} = r\cos\alpha \sum_{i=1}^{2} \frac{k_{2i} D_i}{R_i^3}, \tag{5.6.27}$$

$$\frac{\tau_{\alpha z}}{c_{66}} = \sin\alpha \left(\sum_{i=1}^{2} k_{3i} D_i \frac{1}{R_i R_i^*} - \frac{1}{s_3} D_3 \frac{r^2 - z_3 R_3}{R_3^3 R_3^*} \right), \tag{5.6.28}$$

$$\frac{\tau_{zr}}{c_{66}} = \cos\alpha \left(\sum_{i=1}^{2} k_{3i} D_i \frac{r^2 - z_i R_i}{R_i^3 R_i^*} - \frac{1}{s_3} D_3 \frac{1}{R_3 R_3^*} \right), \tag{5.6.29}$$

$$\frac{\tau_{r\alpha}}{c_{66}} = \sin\alpha \left(2\sum_{i=1}^{2} D_i \frac{r}{R_i R_i^{*2}} + 2D_3 \frac{z_3}{r R_3 R_3^*} - 2D_3 \frac{1}{r R_3^*} + D_3 \frac{r}{R_3^3} \right). \tag{5.6.30}$$

The stresses must satisfy the boundary conditions at the lateral surface, Eq. (5.6.10), and the equilibrium with the external force T, *i.e.*,

$$\int_0^{z\tan\beta_0} \int_0^{2\pi} (\tau_{zr} \cos\alpha - \tau_{\alpha z} \sin\alpha) r \mathrm{d}r \mathrm{d}\alpha = -T, \tag{5.6.31}$$

and

$$\begin{gathered} \int_0^{z\tan\beta_0} \int_0^{2\pi} (\tau_{zr} \sin\alpha + \tau_{\alpha z} \cos\alpha) r \mathrm{d}r \mathrm{d}\alpha = 0, \\ \int_0^{z\tan\beta_0} \int_0^{2\pi} \sigma_z r \mathrm{d}r \mathrm{d}\alpha = 0. \end{gathered} \tag{5.6.32}$$

The conditions specified by Eq. (5.6.32) are clearly satisfied if we notice Eqs. (5.6.28), (5.6.29) and (5.6.27) as well as the property of trigonometric functions. The integration of Eq. (5.6.31) with respect to α leads to

$$\int_0^{z\tan\beta_0} \left(\sum_{i=1}^{2} k_{3i} D_i \frac{z_i}{R_i^3} + \frac{1}{s_3} D_3 \frac{z_3}{R_3^3} \right) r \mathrm{d}r = \frac{T}{\pi c_{66}}. \tag{5.6.33}$$

Then using Eq. (5.6.13), we get

$$\sum_{i=1}^{2} k_{3i} D_i \left(1 - \frac{s_i}{\sqrt{s_i^2 + \tan^2 \beta_0}}\right) + \frac{1}{s_3} D_3 \left(1 - \frac{s_3}{\sqrt{s_3^2 + \tan^2 \beta_0}}\right) = \frac{T}{\pi c_{66}}. \tag{5.6.34}$$

On the other hand, by substituting Eqs. (5.6.25) and (5.6.27) to (5.6.30) into Eq. (5.6.10), we obtain three equations for D_i, of which only two are independent. Thus we can solve for D_i $(i = 1, 2, 3)$ from these two independent ones and Eq. (5.6.34).

Again, when $\beta_0 = \pi/2$, the cone becomes a half-space subjected to a concentrated force applied on the surface tangentially. In this case, Eq. (5.6.10) becomes

$$\sigma_z = 0, \ \tau_{\alpha z} = 0, \ \tau_{zr} = 0, \tag{5.6.35}$$

at $z = 0$ and $r \neq 0$, which leads to

$$\sum_{i=1}^{2} k_{2i} D_i = 0, \quad \sum_{i=1}^{2} k_{3i} D_i - \frac{1}{s_3} D_3 = 0. \tag{5.6.36}$$

Thus Eq. (5.6.34) becomes

$$\sum_{i=1}^{2} k_{3i} D_i + \frac{1}{s_3} D_3 = \frac{T}{\pi c_{66}}. \tag{5.6.37}$$

By making use of Eq. (2.2.64), the above two equations give rise to

$$D_1 = \frac{T}{2\pi c_{66} k_{21}(s_1 - s_2)}, \quad D_2 = -\frac{T}{2\pi c_{66} k_{22}(s_1 - s_2)}, \quad D_3 = \frac{T s_3}{2\pi c_{66}}. \tag{5.6.38}$$

which are identical to Eq. (4.1.39).

Similarly, for a conical shell we can take

$$\varphi_i = \frac{D_i r \cos\alpha}{R_i + s_i z} + \frac{E_i r \cos\alpha}{R_i - s_i z}, \quad (i = 1, 2),$$
$$\varphi_3 = \frac{D_3 r \sin\alpha}{R_3 + s_3 z} + \frac{E_3 r \sin\alpha}{R_3 - s_3 z}, \tag{5.6.39}$$

and the solution can be obtained correspondingly.

5.7 SPHERICALLY ISOTROPIC CONES

5.7.1 Equilibrium and Boundary Conditions

According to Eqs. (2.4.20) and (2.4.3), we have

$$u_R = L_2 F,$$
$$u_\theta = -\frac{1}{\sin\theta}\frac{\partial\psi}{\partial\alpha} - \frac{\partial(L_1 F)}{\partial\theta},$$
$$u_\alpha = \frac{\partial\psi}{\partial\theta} - \frac{1}{\sin\theta}\frac{\partial(L_1 F)}{\partial\alpha}, \tag{5.7.1}$$

where L_1 and L_2 were defined in Eq. (2.4.8). For the static problem ignoring the body forces, the two functions, ψ and F, satisfy

$$[c_{44}\nabla_3^2 - 2(c_{44} - c_{66}) + c_{66}\nabla_1^2]\psi = 0, \tag{5.7.2}$$

$$[\nabla_3^2\nabla_3^2 + 2D\nabla_3^2 + 4L + M\nabla_3^2\nabla_1^2 + 2(N+L)\nabla_1^2 + N\nabla_1^2\nabla_1^2]F = 0. \tag{5.7.3}$$

Equations (5.7.1), (1.2.3) and (1.2.40) yield the stresses

$$\sigma_\theta = c_{11}\varepsilon_\theta + c_{12}\varepsilon_\alpha + c_{13}\varepsilon_R,$$
$$\sigma_\alpha = c_{12}\varepsilon_\theta + c_{11}\varepsilon_\alpha + c_{13}\varepsilon_R,$$
$$\sigma_R = c_{13}(\varepsilon_\theta + \varepsilon_\alpha) + c_{33}\varepsilon_R, \tag{5.7.4}$$
$$\tau_{R\theta} = \frac{c_{44}}{R}\left\{\frac{1}{\sin\theta}(1-\nabla_2)\frac{\partial\psi}{\partial\alpha} + \frac{\partial}{\partial\theta}[(1-\nabla_2)L_1 + L_2]F\right\},$$

$$\tau_{\alpha R} = \frac{c_{44}}{R}\left\{-(1-\nabla_2)\frac{\partial \psi}{\partial \theta} + \frac{\partial}{\sin\theta\partial\alpha}[(1-\nabla_2)L_1 + L_2]F\right\},$$

$$\tau_{\theta\alpha} = \frac{c_{66}}{R}\left[\left(2\frac{\partial^2}{\partial\theta^2} - \nabla_1^2\right)\psi - \frac{2}{\sin\theta}\left(\frac{\partial}{\partial\theta} - \cot\theta\right)\frac{\partial}{\partial\alpha}(L_1 F)\right],$$

where

$$\begin{aligned}
\varepsilon_R &= \frac{\partial}{\partial R}(L_2 F), \\
\varepsilon_\theta &= -\frac{\partial}{R\partial\theta}\left(\frac{1}{\sin\theta}\frac{\partial\psi}{\partial\alpha}\right) - \frac{1}{R}\left(\frac{\partial^2}{\partial\theta^2}L_1 - L_2\right)F, \\
\varepsilon_\alpha &= \frac{1}{R\sin\theta}\left\{\left(\frac{\partial}{\partial\theta} - \cot\theta\right)\frac{\partial\psi}{\partial\alpha} + \left[L_2\sin\theta - \left(\frac{1}{\sin\theta}\frac{\partial^2}{\partial\alpha^2} + \cos\theta\frac{\partial}{\partial\theta}\right)L_1\right]F\right\}.
\end{aligned} \tag{5.7.5}$$

In the present case, we can take

$$F = R^n Y(\theta,\alpha). \tag{5.7.6}$$

Hence Eq. (5.7.3) brings about

$$\nabla_1^2\nabla_1^2 Y + 2S(n)\nabla_1^2 Y + K(n)Y = 0, \tag{5.7.7}$$

where

$$\begin{aligned}
2S(n) &= [Mn(n+1) + 2(N+L)]/N, \\
K(n) &= [n^2(n+1)^2 + 2Dn(n+1) + 4L]/N.
\end{aligned} \tag{5.7.8}$$

Assuming $Y(\theta,\alpha) = S_l^m(\theta,\alpha)$, here $S_l^m(\theta,\alpha)$ are *spherical harmonics* that satisfy[1]

$$\nabla_1^2 S_l^m(\theta,\alpha) + l(l+1)S_l^m(\theta,\alpha) = 0, \tag{5.7.9}$$

we obtain four eigenvalues from Eq. (5.7.7)

---

[1] cf. **Appendix B**

$$l_1 = [-1+\sqrt{1+4S+4\sqrt{S^2-K}}\,]/2\ ,\ l_2 = -l_1 - 1\ ,$$
$$l_3 = [-1+\sqrt{1+4S-4\sqrt{S^2-K}}\,]/2,\ \ l_4 = -l_3 - 1\,. \tag{5.7.10}$$

Similarly, we can take

$$\psi = R^n S_l^m(\theta,\alpha)\,. \tag{5.7.11}$$

Then Eq. (5.7.2) leads to

$$c_{66}\nabla_1^2 S_l^m(\theta,\alpha) + [2c_{66} + (n-1)(n+2)c_{44}]S_l^m(\theta,\alpha) = 0\,. \tag{5.7.12}$$

Hence two corresponding eigenvalues are obtained

$$l_5 = [-1+\sqrt{9+4[n(n+1)-2]c_{44}/c_{66}}\,]/2\,,\ l_6 = -l_5 - 1\,. \tag{5.7.13}$$

When a conical shell of $0 < 2\beta_1 < 2\beta_0 < 2\pi$ is subjected to a concentrated force $\mathbf{P} = p_x\mathbf{i} + p_y\mathbf{j} + p_z\mathbf{k}$ and a concentrated moment $\mathbf{M} = M_x\mathbf{i} + M_y\mathbf{j} + M_z\mathbf{k}$ at the apex O, where $\mathbf{i}$, $\mathbf{j}$ and $\mathbf{k}$ are the three unit coordinate vectors in the Cartesian coordinate system, the boundary conditions at the lateral surfaces of the shell are

$$\theta = \beta_0: \quad \sigma_\theta = \bar{\sigma}_\theta\,, \quad \tau_{R\theta} = \bar{\tau}_{R\theta}\,, \quad \tau_{\theta\alpha} = \bar{\tau}_{\theta\alpha}\,, \tag{5.7.14}$$
$$\theta = \beta_1: \quad \sigma_\theta = \sigma_\theta^*\,, \quad \tau_{R\theta} = \tau_{R\theta}^*\,, \quad \tau_{\theta\alpha} = \tau_{\theta\alpha}^*\,, \tag{5.7.15}$$

where the quantities with an overbar or a star are known.

Imagine that a conical segment with radius R and centre O is isolated. The equilibrium of the segment requires that

$$\begin{aligned}
&\mathbf{P} + \int_0^{2\pi}\int_{\beta_1}^{\beta_0} (\sigma_R\mathbf{e}_R + \tau_{R\theta}\mathbf{e}_\theta + \tau_{\alpha R}\mathbf{e}_a)R^2\sin\theta\,\mathrm{d}\theta\,\mathrm{d}\alpha \\
&+ \int_0^{2\pi}\int_0^{R} (\bar{\tau}_{R\theta}\mathbf{e}_R + \bar{\sigma}_\theta\mathbf{e}_\theta + \bar{\tau}_{\theta\alpha}\mathbf{e}_\alpha)R\,\mathrm{d}R\,\mathrm{d}\alpha\sin\beta_0 \\
&- \int_0^{2\pi}\int_0^{R} (\tau_{R\theta}^*\mathbf{e}_R + \sigma_\theta^*\mathbf{e}_\theta + \tau_{\theta\alpha}^*\mathbf{e}_\alpha)R\,\mathrm{d}R\,\mathrm{d}\alpha\sin\beta_1 = 0,
\end{aligned} \tag{5.7.16}$$

$$\begin{aligned}\mathbf{M}+\int_0^{2\pi}\int_{\beta_1}^{\beta_0}(\tau_{R\theta}\mathbf{e}_\alpha-\tau_{\alpha R}\mathbf{e}_\theta)R^3\sin\theta\,\mathrm{d}\alpha\,\mathrm{d}\theta\\+\int_0^{2\pi}\int_0^{R}(\bar{\sigma}_\theta\mathbf{e}_\alpha-\bar{\tau}_{\theta\alpha}\mathbf{e}_\theta)R^2\,\mathrm{d}R\,\mathrm{d}\alpha\sin\beta_0\\-\int_0^{2\pi}\int_0^{R}(\sigma^*_\theta\mathbf{e}_\alpha-\tau^*_{\theta\alpha}\mathbf{e}_\theta)R^2\,\mathrm{d}R\,\mathrm{d}\alpha\sin\beta_1=0,\end{aligned}\tag{5.7.17}$$

where

$$\begin{aligned}\mathbf{e}_R&=\sin\theta\cos\alpha\,\mathbf{i}+\sin\theta\sin\alpha\,\mathbf{j}+\cos\theta\,\mathbf{k}\,,\\\mathbf{e}_\theta&=\cos\theta\cos\alpha\,\mathbf{i}+\cos\theta\sin\alpha\,\mathbf{j}-\sin\theta\,\mathbf{k}\,,\\\mathbf{e}_\alpha&=-\sin\alpha\,\mathbf{i}+\cos\alpha\,\mathbf{j}\,.\end{aligned}\tag{5.7.18}$$

For a cone ($\beta_1=0$), only Eq. (5.7.14) should be considered and $\sigma^*_\theta=\tau^*_{\theta\alpha}=\tau^*_{R\theta}=0$ in Eqs. (5.7.16) and (5.7.17).

5.7.2 A Cone under Tip Forces

(1) Compression

If the cone is subjected to a concentrated force, $\mathbf{P}=p_z\,\mathbf{k}$, the problem is axisymmetric so that $\tau_{\alpha R}=\tau_{\theta\alpha}=0$ and σ_θ, σ_α, σ_R and $\tau_{R\theta}$ are independent of α. Since

$$\int_0^{2\pi}\cos\alpha\,\mathrm{d}\alpha=\int_0^{2\pi}\sin\alpha\,\mathrm{d}\alpha=0\,,$$

it can be seen that five equations in Eqs. (5.7.16) and (5.7.17) are satisfied automatically and the remaining one becomes

$$\begin{aligned}p_z/2\pi+R^2\int_0^{\beta_0}(\sigma_R\cos\theta-\tau_{R\theta}\sin\theta)\sin\theta\,\mathrm{d}\theta\\+\int_0^{R}(\bar{\tau}_{R\theta}\cos\beta_0-\bar{\sigma}_\theta\sin\beta_0)R\,\mathrm{d}R\sin\beta_0=0\,.\end{aligned}\tag{5.7.19}$$

Since $\bar{\tau}_{R\theta}=\bar{\sigma}_\theta=0$,

$$p_z/2\pi+R^2\int_0^{\beta_0}(\sigma_R\cos\theta-\tau_{R\theta}\sin\theta)\sin\theta\,\mathrm{d}\theta=0\,.\tag{5.7.20}$$

It can be shown from Eq. (5.7.20) that σ_R and $\tau_{R\theta}$ are proportional to R^{-2}. Hence, we can take

$$\psi = 0\,,\ F = R^{-1}[c_1 P_{l_1}(\cos\theta) + c_3 P_1(\cos\theta)]\,, \tag{5.7.21}$$

$$l_1 = [-1 + \sqrt{1 + 8L/N}]/2\,, \tag{5.7.22}$$

where c_1 and c_3 are undetermined constants, and $P_n(\cos\theta)$ is a Legendre function of order n.

By substituting Eq. (5.7.21) into Eqs. (5.7.1) and (5.7.4) and noticing that

$$P_n^1(\cos\theta) = \sin\theta \frac{\mathrm{d}P_n(\cos\theta)}{\mathrm{d}\cos\theta} = -\frac{\mathrm{d}P_n(\cos\theta)}{\mathrm{d}\theta}\,, \tag{5.7.23}$$

$$\frac{\mathrm{d}P_n^1(\cos\theta)}{\mathrm{d}\theta} = -\frac{\mathrm{d}^2 P_n(\cos\theta)}{\mathrm{d}\theta^2} = n(n+1)P_n(\cos\theta) - \cot\theta\, P_n^1(\cos\theta)\,, \tag{5.7.24}$$

we obtain

$$\begin{aligned} u_R &= R^{-1}[x_1 c_1 P_{l_1}(t) + x_2 c_3 P_1(t)], \\ u_\theta &= R^{-1} x_3 [c_1 P_{l_1}^1(t) + c_3 P_1^1(t)], \\ u_\alpha &= 0, \end{aligned} \tag{5.7.25}$$

$$\begin{aligned} \sigma_\theta &= R^{-2}\{[(x_1 E_1 + c_{11} E_2) P_{l_1}(t) + E_3 \cot\theta\, P_{l_1}^1(t)]c_1 \\ &\quad + [x_2 E_1 + x_3 E_4] P_1(t) c_3\}, \\ \sigma_\alpha &= R^{-2}\{[(x_1 E_1 + c_{12} E_2) P_{l_1}(t) - E_3 \cot\theta\, P_{l_1}^1(t)]c_1 \\ &\quad + [x_2 E_1 + x_3 E_4] P_1(t) c_3\}, \\ \sigma_R &= R^{-2}\{[x_1(2c_{13} - c_{33}) + c_{13} E_2] P_{l_1}(t) c_1 \\ &\quad + [x_2(2c_{13} - c_{33}) + 2x_3 c_{13}] P_1(t) c_3\}, \\ \tau_{R\theta} &= -c_{44} R^{-2}\{(2x_3 + x_1) P_{l_1}^1(t) c_1 + (2x_3 + x_2) P_1^1(t) c_3\}, \\ \tau_{\alpha R} &= \tau_{\theta\alpha} = 0\,, \end{aligned} \tag{5.7.26}$$

where $t = \cos\theta$, $P_n^1(t)$ is the associated Legendre function of the first kind and

$$
\begin{aligned}
x_1 &= -l_1(l_1+1)c_{11} - 2c_{44} + 2c_{66}, \\
x_2 &= -2(c_{11} + c_{44} - c_{66}), \\
x_3 &= 2c_{11} + c_{44} - 2c_{66} - c_{13},
\end{aligned} \tag{5.7.27}
$$

$$
\begin{aligned}
&E_1 = c_{11} + c_{12} - c_{13}, \; E_2 = x_3 l_1(l_1+1), \\
&E_3 = (c_{12} - c_{11})x_3 \;\;, E_4 = c_{11} + c_{12}.
\end{aligned} \tag{5.7.28}
$$

In these expressions, constants c_1 and c_3 can be determined by Eq. (5.7.20) and the appropriate boundary conditions. As shown in Eq. (5.7.19), the satisfaction of either $\overline{\sigma}_\theta = 0$ or $\overline{\tau}_{R\theta} = 0$ immediately leads to the satisfaction of the other one because of Eq. (5.7.20). Thus we can just consider one boundary condition only, *i.e.* $\overline{\tau}_{R\theta} = 0$, from which and Eq. (5.7.20), we obtain

$$
\begin{aligned}
&p_z / 2\pi + g_1 c_1 + g_2 c_3 = 0, \\
&g_3 c_1 + g_4 c_3 = 0.
\end{aligned} \tag{5.7.29}
$$

If $g_1 g_4 - g_2 g_3 \neq 0$, we have

$$
c_1 = p_z g_4 / [2\pi(g_2 g_3 - g_1 g_4)], \;\; c_3 = -p_z g_3 / [2\pi(g_2 g_3 - g_1 g_4)], \tag{5.7.30}
$$

where g_i $(i = 1,2,3,4)$ are defined at the end of this section. As expected, the solution satisfies the other boundary condition $\overline{\sigma}_\theta = 0$ automatically.

(2) Bending

If the tip force is $\mathbf{P} = p_x\,\mathbf{i}$, the cone is subjected to bending symmetrical about the xOz coordinate plane. Thus σ_θ, σ_α, σ_R and $\tau_{R\theta}$ are proportional to $\cos\alpha$ but $\tau_{\alpha R}$ and $\tau_{\theta\alpha}$ are proportional to $\sin\alpha$. Because of $\int_0^{2\pi} \sin\alpha\cos\alpha\,\mathrm{d}\alpha = 0$ and $\int_0^{2\pi} \sin^2\alpha\,\mathrm{d}\alpha = \int_0^{2\pi} \cos^2\alpha\,\mathrm{d}\alpha = \pi$, four equations in Eqs. (5.7.16) and (5.7.17) are satisfied automatically. The remaining two to examine are

$$p_x/\pi + R^2\int_0^{\beta_0}(\sigma_R\sin\theta/\cos\alpha + \tau_{R\theta}\cos\theta/\cos\alpha - \tau_{\alpha R}/\sin\alpha)\sin\theta\,\mathrm{d}\theta$$
$$+\int_0^R(\overline{\tau}_{R\theta}\sin\beta_0/\cos\alpha + \overline{\sigma}_\theta\cos\beta_0/\cos\alpha - \overline{\tau}_{\theta\alpha}/\sin\alpha)R\,\mathrm{d}R\sin\beta_0 = 0, \tag{5.7.31}$$

and

$$R^3\int_0^{\beta_0}(\tau_{R\theta}/\cos\alpha - \tau_{\alpha R}\cos\theta/\sin\alpha)\sin\theta\,\mathrm{d}\theta$$
$$+\int_0^R(\overline{\sigma}_\theta/\cos\alpha - \overline{\tau}_{\theta\alpha}\cos\beta_0/\sin\alpha)R^2\,\mathrm{d}R\sin\beta_0 = 0. \tag{5.7.32}$$

Since $\overline{\tau}_{R\theta} = \overline{\sigma}_\theta = \overline{\tau}_{\theta\alpha} = 0$,

$$p_x/\pi + R^2\int_0^{\beta_0}(\sigma_R\sin\theta/\cos\alpha + \tau_{R\theta}\cos\theta/\cos\alpha - \tau_{\alpha R}/\sin\alpha)\sin\theta\,\mathrm{d}\theta = 0, \tag{5.7.33}$$

and

$$\int_0^{\beta_0}(\tau_{R\theta}/\cos\alpha - \tau_{\alpha R}\cos\theta/\sin\alpha)\sin\theta\,\mathrm{d}\theta = 0. \tag{5.7.34}$$

It is known from Eq. (5.7.33) that σ_R, $\tau_{R\theta}$ and $\tau_{\alpha R}$ are proportional to R^{-2}. We can therefore take

$$\psi = c_5R^{-1}P_{l_5}^1(t)\sin\alpha, \tag{5.7.35}$$

$$F = R^{-1}[c_1P_{l_1}^1(t) + c_3P_1^1(t)]\cos\alpha, \tag{5.7.36}$$

where c_1, c_3 and c_5 are undetermined constants, l_1 is given by Eq. (5.7.22) and

$$l_5 = [-1+\sqrt{9-8c_{44}/c_{66}}]/2. \tag{5.7.37}$$

By substituting Eqs. (5.7.35) and (5.7.36) into Eqs. (5.7.1) and (5.7.4), we obtain

$$
\begin{aligned}
u_R &= R^{-1}[x_1c_1P_{l_1}^1(t)+x_2c_3P_1^1(t)]\cos\alpha,\\
u_\theta &= -R^{-1}\{x_3[l_1(l_1+1)P_{l_1}(t)-\cot\theta\, P_{l_1}^1(t)]c_1+x_3c_3P_1(t)\\
&\quad +P_{l_5}^1(t)c_5/\sin\theta]\cos\alpha,\\
u_\alpha &= R^{-1}\{\mathrm{x}_3[P_{l_1}(t)c_1+P_1^1(t)c_3]/\sin\theta+[l_5(l_5+1)P_{l_5}(t)\\
&\quad -\cot\theta\, P_{l_5}^1(t)]c_5\}\sin\alpha,
\end{aligned}
\tag{5.7.38}
$$

$$
\begin{aligned}
\sigma_\theta &= R^{-2}\{[(x_1E_1+c_{11}E_2+(1+\cos^2\theta)E_3/\sin^2\theta)P_{l_1}^1(t)\\
&\quad -E_2E_3\cot\theta\, P_{l_1}(t)/x_3]c_1+[(x_2+x_3)E_4-x_2c_{13}]P_1^1(t)c_3\\
&\quad +(c_{12}-c_{11})E_8c_5\}\cos\alpha,\\
\sigma_\alpha &= R^{-2}\{[(x_1E_1+c_{12}E_2+(1+\cos^2\theta)E_3/\sin^2\theta)P_{l_1}^1(t)\\
&\quad +E_2E_3\cot\theta\, P_{l_1}(t)/x_3]c_1+[(x_2+x_3)E_4-x_2c_{13}]P_1^1(t)c_3\\
&\quad +(c_{11}-c_{12})E_8c_5\}\cos\alpha,\\
\sigma_R &= R^{-2}\{[x_1(2c_{13}-c_{33})+c_{13}E_2]P_{l_1}^1(t)c_1+[2c_{13}(x_2+x_3)\\
&\quad -x_2c_{33}]P_1^1(t)c_3\}\cos\alpha,\\
\tau_{\alpha R} &= -c_{44}R^{-2}\{(x_1+2x_3)P_{l_1}^1(t)c_1/\sin\theta+(x_2+2x_3)c_3+2E_6c_5\}\sin\alpha,\\
\tau_{R\theta} &= c_{44}R^{-2}\{(x_1+2x_3)E_5c_1+(x_2+2x_3)P_1(t)c_3+2P_{l_5}^1(t)c_5/\sin\theta\}\cos\alpha,\\
\tau_{\theta\alpha} &= c_{66}R^{-2}\{2x_3E_7c_1+[2(1+\cos^2\theta)P_{l_5}^1(t)/\sin^2\theta-l_5(l_5+1)P_{l_5}^1(t)\\
&\quad -2l_5(l_5+1)\cot\theta\, P_{l_5}^1(t)]c_5\}\sin\alpha,
\end{aligned}
\tag{5.7.39}
$$

where x_i $(i=1, 2, 3)$ were defined by Eq. (5.7.27), E_i $(i=1,2,3,4)$ by Eq. (5.7.28) and

$$
\begin{aligned}
&E_5=l_1(l_1+1)P_{l_1}(t)-\cot\theta\, P_{l_1}^1(t)\,,\quad E_6=l_5(l_5+1)P_{l_5}(t)-\cot\theta\, P_{l_5}^1(t),\\
&E_7=[E_5-\cot\theta\, P_{l_1}^1(t)]/\sin\theta,\qquad E_8=[E_6-\cot\theta\, P_{l_5}^1(t)]/\sin\theta.
\end{aligned}
\tag{5.7.40}
$$

We can see that if Eq. (5.7.33) is satisfied, only two are independent among the three conditions, $\bar{\tau}_{R\theta}=0, \bar{\tau}_{\theta\alpha}=0$ and $\bar{\sigma}_\theta=0$, as shown in Eq. (5.7.31). Thus the two independent boundary conditions $\bar{\tau}_{R\theta}=0$ and $\bar{\tau}_{\theta\alpha}=0$ should be considered. The substitution of Eq. (5.7.39) into Eq. (5.7.33) and these two boundary conditions, leads to

$$
\begin{aligned}
&p_x/\pi+f_1c_1+f_2c_3+f_3c_5=0,\\
&f_4c_1+f_5c_3+f_6c_5=0,\\
&f_7c_1+f_8c_5=0.
\end{aligned}
\tag{5.7.41}
$$

If the coefficient determinant of Eq. (5.7.41) does not vanish, we get

$$
\begin{aligned}
&c_1 = -(p_x/\pi)[f_1 - f_3 f_7 / f_8 - f_2(f_4 f_8 - f_6 f_7)/ f_5 f_6], \\
&c_3 = -[f_4 - f_6 f_7 / f_8] c_1 / f_5, \quad c_5 = -f_7 c_1 / f_8,
\end{aligned} \tag{5.7.42}
$$

where f_i $(i = 1, 2, \cdots, 8)$ are given at the end of this section. It can be verified that Eq. (5.7.34) and $\bar{\sigma}_\theta = 0$ have been satisfied automatically.

5.7.3 A Cone under Concentrated Moments at Its Apex

(1) Bending

When the moment is $\mathbf{M} = M_y\ \mathbf{j}$, the cone is subjected to bending symmetrical about the xOz coordinate plane. The relation between the stresses and the angle α is the same as that for the bending problem considered in **Section 5.7.2**. Thus four equations in Eqs. (5.7.16) and (5.7.17) are satisfied automatically and the remaining two become

$$
\begin{aligned}
&R^2 \int_0^{\beta_0} (\sigma_R \sin\theta / \cos\alpha + \tau_{R\theta} \cos\theta / \cos\alpha - \tau_{\alpha R} / \sin\alpha) \sin\theta \, \mathrm{d}\theta \\
&\quad + \int_0^R (\bar{\tau}_{R\theta} \sin\beta_0 / \cos\alpha + \bar{\sigma}_\theta \cos\beta_0 / \cos\alpha - \bar{\tau}_{\theta\alpha} / \sin\alpha) R \, \mathrm{d}R \sin\beta_0 = 0,
\end{aligned} \tag{5.7.43}
$$

and

$$
\begin{aligned}
&M_y / \pi + R^3 \int_0^{\beta_0} (\tau_{R\theta} / \cos\alpha - \tau_{\alpha R} \cos\theta / \sin\alpha) \sin\theta \, \mathrm{d}\theta \\
&\quad + \int_0^R (\bar{\sigma}_\theta / \cos\alpha - \bar{\tau}_{\theta\alpha} \cos\beta_0 / \sin\alpha) R^2 \, \mathrm{d}R \sin\beta_0 = 0.
\end{aligned} \tag{5.7.44}
$$

Since $\bar{\tau}_{R\theta} = \bar{\sigma}_\theta = \bar{\tau}_{\theta\alpha} = 0$,

$$
M_y / \pi + R^3 \int_0^{\beta_0} (\tau_{R\theta} / \cos\alpha - \tau_{\alpha R} \cos\theta / \sin\alpha) \sin\theta \, \mathrm{d}\theta = 0, \tag{5.7.45}
$$

and

$$\int_0^{\beta_0} (\sigma_R \sin\theta / \cos\alpha + \tau_{R\theta} \cos\theta / \cos\alpha - \tau_{\alpha R} / \sin\alpha) \sin\theta \, \mathrm{d}\theta = 0 \ . \tag{5.7.46}$$

It can be shown from Eq. (5.7.45) that $\tau_{R\theta}$ and $\tau_{\alpha R}$ are proportional to R^{-3}, thus we can take

$$\psi = c_5 R^{-2} P_1^1(t) \sin\alpha , \tag{5.7.47}$$

$$F = R^{-2}[c_1 P_{l_1}^1(t) + c_3 P_{l_3}^1(t)] \cos\alpha , \tag{5.7.48}$$

where c_1, c_3 and c_5 are undetermined constants and l_1 and l_3 can be obtained *via* Eq. (5.7.10) when $n = -2$.

By substituting Eqs. (5.7.47) and (5.7.48) into Eqs. (5.7.1) and (5.7.4), we obtain

$$\begin{aligned}
u_R &= R^{-2}[y_1 c_1 P_{l_1}^1(t) + y_2 c_3 P_{l_3}^1(t)] \cos\alpha, \\
u_\theta &= -R^{-2}[y_3 D_5 c_1 + y_3 D_6 c_3 + c_5] \cos\alpha, \\
u_\alpha &= R^{-1}\{[y_3 c_1 P_{l_1}^1(t) + y_3 c_3 P_{l_3}^1(t)] / \sin\theta + c_5 P_1(t)\} \sin\alpha,
\end{aligned} \tag{5.7.49}$$

$$\begin{aligned}
\sigma_\theta &= R^{-3}\{[(y_1 D_1 + c_{11} D_3 + y_3 D_2 (1+\cos^2\theta)/\sin^2\theta) P_{l_1}^1(t) \\
&\quad - D_2 D_3 \cot\theta \, P_{l_1}(t)] c_1 + [(y_2 D_1 + c_{11} D_4 \\
&\quad + y_3 D_2 (1+\cos^2\theta)/\sin^2\theta \, P_{l_3}^1(t) - D_2 D_4 \cot\theta \, P_{l_3}(t)] c_3\} \cos\alpha, \\
\sigma_\alpha &= R^{-3}\{[(y_1 D_1 + c_{12} D_3 - y_3 D_2 (1+\cos^2\theta)/\sin^2\theta) P_{l_1}^1(t) \\
&\quad + D_2 D_3 \cot\theta \, P_{l_1}(t)] c_1 + [(y_2 D_1 + c_{12} D_4 \\
&\quad - y_3 D_2 (1+\cos^2\theta)/\sin^2\theta P_{l_3}^1(t) + D_2 D_4 \cot\theta \, P_{l_3}(t)] c_3\} \cos\alpha, \\
\sigma_R &= R^{-3}\{[2y_1(c_{13} - c_{33}) + c_{13} D_3] P_{l_1}^1(t) c_1 + [2y_2(c_{13} - c_{33}) \\
&\quad + c_{13} D_4] P_{l_3}^1(t) c_3\} \cos\alpha, \\
\tau_{\alpha R} &= -c_{44} R^{-3}\{(y_1 + 3y_3) P_{l_1}^1(t) c_1 / \sin\theta \\
&\quad + (y_2 + 3y_3) P_{l_3}^1(t) c_3 / \sin\theta + 3P_1(t) c_5\} \sin\alpha, \\
\tau_{R\theta} &= c_{44} R^{-3}\{(y_1 + 3y_3) D_5 c_1 + (y_2 + 2y_3) D_6 c_3 + 3c_5\} \cos\alpha, \\
\tau_{\theta\alpha} &= 2c_{66} R^{-3} y_3 \{D_7 c_1 + D_8 c_3\} \sin\alpha,
\end{aligned} \tag{5.7.50}$$

where

$$\begin{aligned} y_1 &= 2c_{66} - l_1(l_1+1)c_{11}, \\ y_2 &= 2c_{66} - l_3(l_3+1)c_{11}, \\ y_3 &= 2(c_{11} - c_{66} - c_{13}), \end{aligned} \tag{5.7.51}$$

$$\begin{aligned} &D_1 = c_{11} + c_{12} - 2c_{13},\ D_2 = c_{12} - c_{11},\ D_3 = y_3 l_1(l_1+1),\ D_4 = y_4 l_3(l_3+1), \\ &D_5 = l_1(l_1+1)P_{l_1}(t) - \cot\theta\, P_{l_1}^1(t),\quad D_6 = l_3(l_3+1)P_{l_3}(t) - \cot\theta\, P_{l_3}^1(t), \\ &D_7 = [D_5 - \cot\theta\, P_{l_1}^1(t)]/\sin\theta,\quad D_8 = [D_6 - \cot\theta\, P_{l_3}^1(t)]/\sin\theta. \end{aligned} \tag{5.7.52}$$

As shown in Eq. (5.7.44), the satisfaction of either $\bar{\sigma}_\theta = 0$ or $\bar{\tau}_{R\theta} = 0$ immediately leads to the satisfaction of the other one because of Eq. (5.7.45). Thus we can just consider two boundary conditions $\bar{\tau}_{R\theta} = \bar{\tau}_{\theta\alpha} = 0$, from which and Eq. (5.7.45), we obtain

$$\begin{aligned} &M_y/\pi + h_5 c_1 + h_6 c_3 + h_7 c_5 = 0, \\ &h_1 c_1 + h_2 c_3 + 3c_5 = 0, \\ &h_3 c_1 + h_4 c_3 = 0. \end{aligned} \tag{5.7.53}$$

Then, we have

$$\begin{aligned} &c_1 = -(M_y/\pi)[h_5 - h_3 h_6/f_4 - h_7(h_1 - h_2 h_3)/h_4)/3], \\ &c_3 = -h_3 c_1/h_4,\quad c_5 = -(h_1 - h_2 h_3/h_4)c_1/3, \end{aligned} \tag{5.7.54}$$

where h_i $(i = 1, 2, \cdots, 7)$ are defined at the end of this section. It can be verified that Eq. (5.7.46) and $\bar{\sigma}_\theta = 0$ have been satisfied automatically.

(2) Torsion

When the moment applied is $\mathbf{M} = M_z\,\mathbf{k}$, the cone is under torsion such that $\sigma_R = \sigma_\alpha = \sigma_\theta = \tau_{R\theta} = 0$ and $\tau_{\alpha R}$ and $\tau_{\theta\alpha}$ are independent of α. It can be shown in this case that five equations in Eqs. (5.7.16) and (5.7.17) are satisfied automatically. The remaining one to examine is

$$M_z/2\pi + R^3 \int_0^{\beta_0} \tau_{\alpha R} \sin^2\theta\,\mathrm{d}\theta + \int_0^R \bar{\tau}_{\theta\alpha} R^2\,\mathrm{d}R \sin^2\beta_0 = 0. \tag{5.7.55}$$

Since $\bar{\tau}_{\theta\alpha} = 0$,

$$M_z / 2\pi + R^3 \int_0^{\beta_0} \tau_{\alpha R} \sin^2 \theta \,\mathrm{d}\theta = 0 . \tag{5.7.56}$$

Using Eq. (5.7.56), we can show that $\tau_{\alpha R}$ is proportional to R^{-3}. It is therefore reasonable to take

$$\psi = c_1 R^{-2} P_1(t) , \quad F = 0 . \tag{5.7.57}$$

The substitution of Eq. (5.7.57) into Eqs. (5.7.1) and (5.7.4) leads to

$$u_R = u_\theta = 0, \quad u_\alpha = -c_1 \sin\theta / R^2 , \tag{5.7.58}$$

$$\begin{aligned} &\sigma_R = \sigma_\alpha = \sigma_\theta = \tau_{R\theta} = 0, \\ &\tau_{R\alpha} = 3c_{44} c_1 \sin\theta / R^3 , \end{aligned} \tag{5.7.59}$$

where constant c_1 can be obtained from Eq. (5.7.56), *i.e.*

$$c_1 = -M_z / [6\pi c_{44} (2/3 - \cos\beta_0 + \cos^3 \beta_0 / 3)] . \tag{5.7.60}$$

It can be verified that the boundary condition $\bar{\tau}_{\theta\alpha} = 0$ is satisfied automatically.

5.7.4 Conical Shells

Now the boundary conditions to consider should include the inner and outer surfaces. If a shell is under compression due to a tip force, $\mathbf{P} = p_z \mathbf{k}$, Eq. (5.7.19) becomes

$$\begin{aligned} p_z / 2\pi + R^2 &\int_{\beta_1}^{\beta_0} (\sigma_R \cos\theta - \tau_{R\theta} \sin\theta) \sin\theta \,\mathrm{d}\theta \\ &+ \int_0^R (\bar{\tau}_{R\theta} \cos\beta_0 - \bar{\sigma}_\theta \sin\beta_0) R \,\mathrm{d}R \sin\beta_0 \\ &- \int_0^R (\tau^*_{R\theta} \cos\beta_1 - \sigma^*_\theta \sin\beta_1) R \,\mathrm{d}R \sin\beta_1 = 0. \end{aligned} \tag{5.7.61}$$

Since $\bar{\tau}_{R\theta} = \bar{\sigma}_\theta = \tau^*_{R\theta} = \sigma^*_\theta = 0$,

$$p_z / 2\pi + R^2 \int_{\beta_1}^{\beta_0} (\sigma_R \cos\theta - \tau_{R\theta} \sin\theta) \sin\theta \,\mathrm{d}\theta = 0 \,. \tag{5.7.62}$$

As in the case of the cone discussed in **Section 5.7.2**, we can take

$$F = R^{-1}[c_1 P_{l_1}(\cos\theta) + c_3 P_1(\cos\theta) + c_2 Q_{l_1}(t) + c_4 Q_1(t)], \tag{5.7.63}$$

where $Q_n(\cos\theta)$ is the Legendre functions of the second kind of order n, c_i $(i = 1,2,3,4)$ are constants to be determined *via* Eq. (5.7.62) and the appropriate boundary conditions. Note that only three of the four conditions, $\overline{\sigma}_\theta = 0$, $\overline{\tau}_{R\theta} = 0$, $\sigma_\theta^* = 0$ and $\tau_{R\theta}^* = 0$, are independent because of Eqs. (5.7.61) and (5.7.62). Thus we can just consider the three independent ones, *i.e.* $\overline{\sigma}_\theta = \overline{\tau}_{R\theta} = 0$ and $\tau_{R\theta}^* = 0$. It can be verified that the remainder $\sigma_\theta^* = 0$ will be satisfied automatically.

If the shell is under bending caused by the tip force $\mathbf{P} = p_x \,\mathbf{i}$, Eqs. (5.7.31) and (5.7.32) become

$$\begin{aligned} p_x / \pi &+ R^2 \int_{\beta_1}^{\beta_0} (\sigma_R \sin\theta / \cos\alpha + \tau_{R\theta} \cos\theta / \cos\alpha - \tau_{\alpha R} / \sin\alpha) \sin\theta \,\mathrm{d}\theta \\ &+ \int_0^R (\overline{\tau}_{R\theta} \sin\beta_0 / \cos\alpha + \overline{\sigma}_\theta \cos\beta_0 / \cos\alpha - \overline{\tau}_{\theta\alpha} / \sin\alpha) R \,\mathrm{d}R \sin\beta_0 \\ &- \int_0^R (\tau_{R\theta}^* \sin\beta_1 / \cos\alpha + \sigma_\theta^* \cos\beta_1 / \cos\alpha - \tau_{\theta\alpha}^* / \sin\alpha) R \,\mathrm{d}R \sin\beta_1 = 0, \end{aligned} \tag{5.7.64}$$

and

$$\begin{aligned} R^3 &\int_{\beta_1}^{\beta_0} (\tau_{R\theta} / \cos\alpha - \tau_{\alpha R} \cos\theta / \sin\alpha) \sin\theta \,\mathrm{d}\theta \\ &+ \int_0^R (\overline{\sigma}_\theta / \cos\alpha - \overline{\tau}_{\theta\alpha} \cos\beta_0 / \sin\alpha) R^2 \,\mathrm{d}R \sin\beta_0 \\ &- \int_0^R (\sigma_\theta^* / \cos\alpha - \tau_{\theta\alpha}^* \cos\beta_1 / \sin\alpha) R^2 \,\mathrm{d}R \sin\beta_1 = 0. \end{aligned} \tag{5.7.65}$$

Since $\overline{\tau}_{R\theta} = \overline{\sigma}_\theta = \overline{\tau}_{\theta\alpha} = \tau_{R\theta}^* = \sigma_\theta^* = \tau_{\theta\alpha}^* = 0$,

$$p_x / \pi + R^2 \int_{\beta_1}^{\beta_0} (\sigma_R \sin\theta / \cos\alpha + \tau_{R\theta} \cos\theta / \cos\alpha - \tau_{R\alpha} / \sin\alpha) \sin\theta \,\mathrm{d}\theta = 0 \,, \tag{5.7.66}$$

and

$$\int_{\beta_1}^{\beta_0} (\tau_{R\theta} / \cos\alpha - \tau_{\alpha R} \cos\theta / \sin\alpha) \sin\theta \, \mathrm{d}\theta = 0. \tag{5.7.67}$$

Instead of Eqs. (5.7.35) and (5.7.36), we can take

$$\psi = R^{-1}[c_5 P_{l_5}^1(t) + c_6 Q_{l_5}^1(t)] \sin\alpha, \tag{5.7.68}$$

$$F = R^{-1}[c_1 P_{l_1}^1(t) + c_3 P_1^1(t) + c_2 Q_{l_1}^1(t) + c_4 Q_1^1(t)] \cos\alpha, \tag{5.7.69}$$

where $Q_n^l(t)$ is the associated Legendre function of the second kind and c_i $(i = 1, 2, \cdots, 6)$ are constants to be determined by Eqs. (5.7.66) and the appropriate boundary conditions. Note that only five are independent among the six conditions, $\overline{\tau}_{R\theta} = 0$, $\overline{\tau}_{\theta\alpha} = 0$, $\overline{\sigma}_\theta = 0$, $\tau_{R\theta}^* = 0$, $\tau_{\theta\alpha}^* = 0$ and $\sigma_\theta^* = 0$ because of Eqs. (5.7.64) and (5.7.66). Thus we can just consider the five independent ones, *i.e.* $\overline{\sigma}_\theta = \overline{\tau}_{R\theta} = \overline{\tau}_{\theta\alpha} = 0$ and $\tau_{R\theta}^* = \tau_{\theta\alpha}^* = 0$. It can be verified that the remaining, $\sigma_\theta^* = 0$, and Eq. (5.7.67) will be satisfied automatically.

When the shell bending is due to a moment, $\mathbf{M} = M_y \, \mathbf{j}$, Eqs. (5.7.43) and (5.7.44) become

$$\begin{aligned} & R^2 \int_{\beta_1}^{\beta_0} (\sigma_R \sin\theta / \cos\alpha + \tau_{R\theta} \cos\theta / \cos\alpha - \tau_{\alpha R} / \sin\alpha) \sin\theta \, \mathrm{d}\theta \\ & + \int_0^R (\overline{\tau}_{R\theta} \sin\beta_0 / \cos\alpha + \overline{\sigma}_\theta \cos\beta_0 / \cos\alpha - \overline{\tau}_{\theta\alpha} / \sin\alpha) R \, \mathrm{d}R \sin\beta_0 \\ & - \int_0^R (\tau_{R\theta}^* \sin\beta_1 / \cos\alpha + \sigma_\theta^* \cos\beta_1 / \cos\alpha - \tau_{\theta\alpha}^* / \sin\alpha) R \, \mathrm{d}R \sin\beta_1 = 0, \end{aligned} \tag{5.7.70}$$

and

$$\begin{aligned} & M_y / \pi + R^3 \int_{\beta_1}^{\beta_0} (\tau_{R\theta} / \cos\alpha - \tau_{\alpha R} \cos\theta / \sin\alpha) \sin\theta \, \mathrm{d}\theta \\ & + \int_0^R (\overline{\sigma}_\theta / \cos\alpha - \overline{\tau}_{\theta\alpha} \cos\beta_0 / \sin\alpha) R^2 \, \mathrm{d}R \sin\beta_0 \\ & - \int_0^R (\sigma_\theta^* / \cos\alpha - \tau_{\theta\alpha}^* \cos\beta_1 / \sin\alpha) R^2 \, \mathrm{d}R \sin\beta_1 = 0. \end{aligned} \tag{5.7.71}$$

Considering the boundary conditions, we have

$$M_y/\pi + R^3\int_{\beta_1}^{\beta_0}(\tau_{R\theta}/\cos\alpha - \tau_{\alpha R}\cos\theta/\sin\alpha)\sin\theta\,\mathrm{d}\theta = 0\,. \tag{5.7.72}$$

Instead of Eqs. (5.7.47) and (5.7.48), we should now take

$$\psi = R^{-2}[c_5 P_1^1(t) + c_6 Q_1^1(t)]\sin\alpha\,, \tag{5.7.73}$$

$$F = R^{-2}[c_1 P_{l_1}^1(t) + c_3 P_{l_3}^1(t) + c_2 Q_{l_1}^1(t) + c_4 Q_{l_3}^1(t)]\cos\alpha\,, \tag{5.7.74}$$

where c_i $(i = 1, 2, \cdots, 6)$ are constants to be determined by Eq. (5.7.72) and the appropriate boundary conditions, $\overline{\sigma}_\theta = \overline{\tau}_{R\theta} = \overline{\tau}_{\theta\alpha} = 0$ and $\tau_{R\theta}^* = \tau_{\theta\alpha}^* = 0$. The remainder $\sigma_\theta^* = 0$ and Eq. (5.7.70) will be satisfied automatically.

Finally, when the shell is under torsion, *i.e.*, when it is subjected to moment $\mathbf{M} = M_z\,\mathbf{k}$, we have

$$\begin{aligned} &M_z/2\pi + R^3\int_{\beta_1}^{\beta_0}\tau_{\alpha R}\sin^2\theta\,\mathrm{d}\theta + \int_0^R \overline{\tau}_{\theta\alpha}R^2\,\mathrm{d}R\sin^2\beta_0 \\ &\quad - \int_0^R \tau_{\theta\alpha}^* R^2\,\mathrm{d}R\sin^2\beta_1 = 0. \end{aligned} \tag{5.7.75}$$

Considering the boundary conditions, we have

$$M_z/2\pi + R^3\int_0^{\beta_0}\tau_{\alpha R}\sin^2\theta\,\mathrm{d}\theta = 0\,, \tag{5.7.76}$$

Now we can assume

$$\psi = R^{-2}[c_1 P_1(t) + c_2 Q_1(t)]\,, \tag{5.7.77}$$

where c_1 and c_2 are constants to be determined by Eq. (5.7.76) and the appropriate boundary condition, $\overline{\tau}_{\theta\alpha} = 0$. The remainder $\tau_{\theta\alpha}^* = 0$ will be satisfied automatically.

For convenience, in all these formulations, we have used the following notations:

$$D_{nm} = \int_0^{\beta_0} P_n(\cos\theta) P_m(\cos\theta)\sin\theta\,\mathrm{d}\theta$$
$$= \frac{\sin^2\beta_0[P_m(\cos\beta_0)\overline{P}_n(\cos\beta_0) - P_n(\cos\beta_0)\overline{P}_m(\cos\beta_0)]}{n(n+1)-m(m+1)},$$

$$C_{nm}^1 = \int_0^{\beta_0} P_n^1(\cos\theta) P_m^1(\cos\theta)\sin\theta\,\mathrm{d}\theta$$
$$= \frac{\sin^2\beta_0[P_m^1(\cos\beta_0)P_n^{-1}(\cos\beta_0) - P_n^1(\cos\beta_0)P_m^{-1}(\cos\beta_0)]}{n(n+1)-m(m+1)},$$

$$C_n^1 = \int_0^{\beta_0} P_n^1(\cos\theta)d\theta = -P_n(\cos\beta_0) + P_n(1),$$

$$C_{zn}^1 = \int_0^{\beta_0} P_n^1(\cos\theta)\cos\theta\,\mathrm{d}\theta = -P_n(\cos\beta_0)\cos\beta_0 + P_n(1) - D_{n0},$$

$$C_{xn}^1 = \int_0^{\beta_0} P_n^1(\cos\theta)\cos^2\theta\,\mathrm{d}\theta = -P_n(\cos\beta_0)\cos^2\beta_0 + P_n(1) - 2D_{n1},$$

$$\overline{P}_n(\cos\theta) = \frac{\mathrm{d}\,P_n(\cos\theta)}{\mathrm{d}(\cos\theta)},$$

$$\overline{P}_n^1(\cos\theta) = \frac{\mathrm{d}\,P_n^1(\cos\theta)}{\mathrm{d}(\cos\theta)},$$

$$g_1 = [x_1(2c_{13}-c_{33}) + c_{13}x_3 l_1(l_1+1)]D_{l_1 1} + c_{44}(2x_3+x_1)C_{l_1 1}^1,$$
$$g_2 = [x_2(2c_{13}-c_{33}) + 2c_{13}x_3)](1-\cos^3\beta_0)/3 + c_{44}(2x_3+x_2)(1/3-\cos\beta_0+\cos^3\beta_0/3),$$
$$g_3 = (2x_3+x_1)P_{l_1}^1(\cos\beta_0),$$
$$g_4 = (2x_3+x_2)\sin\beta_0,$$
$$f_1 = [x_1(2c_{13}-c_{33}) + c_{13}x_3 l_1(l_1+1)]C_{l_1 1}^1 + c_{44}(x_1+2x_3)[l_1(l_1+1)D_{l_1 1} - C_{xl_1}^1]$$
$$-c_{44}(x_1+2x_3)C_{l_1}^1,$$
$$f_2 = [2c_{13}(x_2+x_3) - x_2c_{33}](2/3-\cos\beta_0-\cos^3\beta_0/3) + c_{44}(x_2+2x_3)(1-\cos^3\beta_0)$$
$$-c_{44}(x_2+2x_3)(1-\cos\beta_0),$$
$$f_3 = 2c_{44}C_{zl_1}^1 + 2c_{44}[l_5(l_5+1)D_{l_5 0} - C_{zl_5}^1],$$
$$f_4 = (x_1+2x_3)[l_1(l_1+1)P_{l_1}(\cos\beta_0) - \cot\beta_0 P_{l_1}(\cos\beta_0)],$$
$$f_5 = (x_2+2x_3)\cos\beta_0,$$
$$f_6 = 2P_{l_5}^1(\cos\beta_0)/\sin\beta_0,$$
$$f_7 = 2x_3[l_1(l_1+1)P_{l_1}(\cos\beta_0) - 2\cot\beta_0 P_{l_1}(\cos\beta_0)],$$
$$f_8 = [2(1+\cos^2\beta_0)/\sin^2\beta_0 - l_5(l_5+1)]P_{l_5}^1(\cos\beta_0) - 2l_5(l_5+1)\cot\beta_0 P_{l_5}(\cos\beta_0),$$

$$h_1 = (y_1+3y_3)[l_1(l_1+1)P_{l_1}(\cos\beta_0) - \cot\beta_0 P_{l_1}^1(\cos\beta_0)],$$
$$h_2 = (y_2+3y_3)[l_3(l_3+1)P_{l_3}(\cos\beta_0) - \cot\beta_0 P_{l_3}^1(\cos\beta_0)],$$

$$
\begin{aligned}
h_3 &= 2y_3[l_1(l_1+1)P_{l_1}(\cos\beta_0) - 2\cot\beta_0 P_{l_1}^1(\cos\beta_0)],\\
h_4 &= 2y_3[l_3(l_3+1)P_{l_3}(\cos\beta_0) - 2\cot\beta_0 P_{l_3}^1(\cos\beta_0)],\\
h_5 &= c_{44}(y_1+3y_3)l_1(l_1+1)D_{l_1 0},\\
h_6 &= c_{44}(y_2+3y_3)l_3(l_3+1)D_{l_3 0},\\
h_7 &= c_{44}(4-3\cos\beta_0-\cos^3\beta_0).
\end{aligned}
\tag{5.7.78}
$$

6

THERMAL STRESSES

Thermal stresses appear in an internally or externally constrained solid when temperature changes. That is because the change of the temperature field generates non-uniform thermal deformation. When the deformation is in the elastic regime of the solid material, the associated thermal stress field is governed by the Duhamel-Neumann relations as discussed in **Section 1.4**. Strictly, deformation also causes temperature rise and thus deformation and temperature rise are coupled. In an elastic regime, however, such deformation-induced temperature change can be reasonably ignored in most materials and consequently the governing equations can be simplified mathematically. In this chapter, we will discuss the thermoelastic deformation of transversely isotropic and spherically isotropic solids. We will introduce the general solution first and then use some typical examples to demonstrate the procedures for solving engineering problems.

6.1 TRANSVERSELY ISOTROPIC MATERIALS

The displacement method outlined in **Section 2.2** can be applied here directly. According to Eqs. (2.2.2), (2.2.15) and (2.2.19), neglecting the body forces, we get

$$u = \frac{\partial \psi}{\partial y} - \frac{\partial G}{\partial x}, \quad v = -\frac{\partial \psi}{\partial x} - \frac{\partial G}{\partial y}, \quad w = w, \tag{6.1.1}$$

$$\psi = \psi_0, \quad G = G_0 + G_1 + G_2, \quad w = w_0 + w_1 + w_2, \tag{6.1.2}$$

where, in view of Eq. (2.2.16), ψ_0 satisfies

$$\left(c_{66}\Lambda + c_{44}\frac{\partial^2}{\partial z^2} - \rho\frac{\partial^2}{\partial t^2} \right)\psi_0 = 0, \tag{6.1.3}$$

and w_i and G_i are determined by

$$G_1=\left(c_{44}\Lambda+c_{33}\frac{\partial^2}{\partial z^2}-\rho\frac{\partial^2}{\partial t^2}\right)F_1\,,\quad w_1=(c_{13}+c_{44})\Lambda\frac{\partial F_1}{\partial z}\,, \tag{6.1.4}$$

$$G_i=(c_{13}+c_{44})\frac{\partial F_i}{\partial z}\,,\quad w_i=\left(c_{11}\Lambda+c_{44}\frac{\partial^2}{\partial z^2}-\rho\frac{\partial^2}{\partial t^2}\right)F_i\,,\quad (i=0,\,2)\,, \tag{6.1.5}$$

according to Eqs. (2.2.24), (2.2.28) and (2.2.29). In these equations, F_i should satisfy

$$L_0F_0=0\,, \tag{6.1.6}$$

$$L_0F_1=\beta_1 T\,, \tag{6.1.7}$$

$$L_0F_2=-\beta_3\frac{\partial T}{\partial z}\,, \tag{6.1.8}$$

because of Eqs. (2.2.25), (2.2.30) and (2.2.31). In Eqs. (6.1.6) to (6.1.8), L_0 is an operator defined by Eq. (2.2.26).

Functions ψ_0 and F_0 are the general solutions of Eqs. (6.1.3) and (6.1.6), respectively, while F_1 and F_2 are the particular solutions of Eqs. (6.1.7) and (6.1.8), respectively. Once F_1 is obtained, F_2 can be determined by

$$F_2=-\frac{\beta_3}{\beta_1}\frac{\partial F_1}{\partial z}\,. \tag{6.1.9}$$

With F_1 and F_2 obtained, w_1, w_2, G_1 and G_2 can be determined from Eqs. (6.1.4) and (6.1.5). Then with w and G calculated from Eq. (6.1.2), we readily obtain a *general thermoelasticity solution* using Eq. (6.1.1),

$$\begin{aligned}
u&=\frac{\partial\psi_0}{\partial y}-(c_{13}+c_{44})\frac{\partial^2F_0}{\partial x\partial z}-\frac{\partial}{\partial x}(a_1H_1+a_2H_2+a_3H_3),\\
v&=-\frac{\partial\psi_0}{\partial x}-(c_{13}+c_{44})\frac{\partial^2F_0}{\partial y\partial z}-\frac{\partial}{\partial y}(a_1H_1+a_2H_2+a_3H_3),\\
w&=\left(c_{11}\Lambda+c_{44}\frac{\partial^2}{\partial z^2}-\rho\frac{\partial^2}{\partial t^2}\right)F_0-\frac{\partial}{\partial z}(b_1H_1+b_2H_2+b_3H_3),
\end{aligned} \tag{6.1.10}$$

where ψ_0 and F_0 satisfy Eqs. (6.1.3) and (6.1.6), respectively, and

$$\begin{aligned} &a_1 = c_{44}, \quad a_2 = c_{33} - (c_{13} + c_{44})\beta_3 / \beta_1, \quad a_3 = \rho, \\ &b_1 = c_{11}\beta_3 / \beta_1 - c_{13} - c_{44}, \quad b_2 = c_{44}\beta_3 / \beta_1, \quad b_3 = \rho\beta_3 / \beta_1, \end{aligned} \tag{6.1.11}$$

$$H_1 = \Lambda F_1, \quad H_2 = \frac{\partial^2 F_1}{\partial z^2}, \quad H_3 = -\frac{\partial^2 F_1}{\partial t^2}, \tag{6.1.12}$$

where F_1 is the particular solution of Eq. (6.1.7).

In cylindrical coordinates (r, α, z), the third equation of Eq. (6.1.10) stays the same, while the first two become

$$\begin{aligned} u_r &= \frac{1}{r}\frac{\partial \psi_0}{\partial \alpha} - (c_{13} + c_{44})\frac{\partial^2 F_0}{\partial r \partial z} - \frac{\partial}{\partial r}(a_1 H_1 + a_2 H_2 + a_3 H_3), \\ u_\alpha &= -\frac{\partial \psi_0}{\partial r} - (c_{13} + c_{44})\frac{1}{r}\frac{\partial^2 F_0}{\partial \alpha \partial z} - \frac{1}{r}\frac{\partial}{\partial \alpha}(a_1 H_1 + a_2 H_2 + a_3 H_3). \end{aligned} \tag{6.1.13}$$

For a quasi-static problem, the terms associated with the partial differentiation with respect to time will disappear in all these equations. Using the results in **Section 2.2** to rewrite the terms containing F_0 in Eq. (6.1.10), we find that the general thermoelasticity solution become

$$u = \frac{\partial \psi_0}{\partial y} + \sum_{i=1}^{4}\frac{\partial \psi_i}{\partial x}, \quad v = -\frac{\partial \psi_0}{\partial x} + \sum_{i=1}^{4}\frac{\partial \psi_i}{\partial y}, \quad w = \sum_{i=1}^{4}\alpha_i \frac{\partial \psi_i}{\partial z}. \tag{6.1.14}$$

In cylindrical coordinates, the first two equations of Eq. (6.1.14) will have different forms. In this case, the general solution becomes

$$u_r = \frac{1}{r}\frac{\partial \psi_0}{\partial \alpha} + \sum_{i=1}^{4}\frac{\partial \psi_i}{\partial r}, \quad u_\alpha = -\frac{\partial \psi_0}{\partial r} + \frac{1}{r}\sum_{i=1}^{4}\frac{\partial \psi_i}{\partial \alpha}, \quad w = \sum_{i=1}^{4}\alpha_i \frac{\partial \psi_i}{\partial z}, \tag{6.1.15}$$

where ψ_i satisfy

$$\left(\Lambda + \frac{\partial^2}{\partial z_i^2}\right)\psi_i = 0, \quad (i = 0, 1, 2), \tag{6.1.16}$$

$$\psi_3 = -a_1 \Lambda F , \quad \psi_4 = -a_2 \frac{\partial^2 F}{\partial z^2} , \tag{6.1.17}$$

in which $z_i = s_i z$, $s_0 = \sqrt{c_{66} / c_{44}}$, s_1 and s_2 are given by Eqs. (2.2.44) and (2.2.45), respectively, and F is the particular solution of the following equation

$$\left(\Lambda + \frac{\partial^2}{\partial z_1^2} \right) \left(\Lambda + \frac{\partial^2}{\partial z_2^2} \right) F = \frac{\beta_1}{c_{11} c_{44}} T . \tag{6.1.18}$$

According to Eqs. (2.2.57) and (6.1.11), constants α_i in Eqs. (6.1.14) and (6.1.15) are defined by

$$\alpha_i = (a - b s_i^2) / s_i , \quad \alpha_{i+2} = b_i / a_i , \quad (i = 1, 2) . \tag{6.1.19}$$

Example 6.1: Steady-state thermal stresses in a half-space

Consider a transversely isotropic elastic half-space with its plane of isotropy being parallel with the *xOy* coordinate plane. Let the z-axis point to the half-space. The equation of heat conduction in steady-state is

$$\Lambda T + k^2 \frac{\partial^2 T}{\partial z^2} = 0 , \tag{6.1.20}$$

where

$$k^2 = k_z / k_x , \tag{6.1.21}$$

in which k_x and k_z are the *coefficients of thermal conduction* in the x- and z-directions, respectively.

The double Fourier transform was defined in Eq. (4.3.8), *i.e.*

$$\bar{f}(p, q, z) = \frac{1}{2\pi} \int_{-\infty}^{\infty} \int_{-\infty}^{\infty} f(x, y, z) \mathrm{e}^{\mathrm{i}(px+qy)} \mathrm{d}x \mathrm{d}y , \tag{6.1.22}$$

and the inverse transform is

$$f(x,y,z)=\frac{1}{2\pi}\int_{-\infty}^{\infty}\int_{-\infty}^{\infty}\bar{f}(p,q,z)\mathrm{e}^{-\mathrm{i}(px+qy)}\mathrm{d}p\,\mathrm{d}q\,. \tag{6.1.23}$$

The transformed equation of Eq. (6.1.20) is

$$k^2\frac{\mathrm{d}^2\overline{T}}{\mathrm{d}z^2}-\rho^2\overline{T}=0\,, \tag{6.1.24}$$

where $\rho^2=p^2+q^2$. Since the temperature should be finite at infinity, we obtain the solution to Eq. (6.1.24) as

$$\overline{T}=C_3\,\mathrm{e}^{-mz}\,, \tag{6.1.25}$$

where C_3 is an integral constant, and

$$m=\rho\,/\,k\,. \tag{6.1.26}$$

The boundary conditions of the temperature field at $z=0$ are usually of three main kinds: (1) the temperature, T, is given; (2) the *heat flux*, $k_z\partial T/\partial z$, is given; or (3) there is radiation from the surface, *i.e.* $k_z\partial T/\partial z+hT$ is known, here h is the *coefficient of heat radiation*. By performing Fourier transform on the proper boundary conditions and making use of Eq. (6.1.25), we can determine the constant C_3. Thus, $\overline{T}$ becomes a determinate function of z, p and q.

Taking Fourier transforms of Eqs. (6.1.16) and (6.1.18), we obtain

$$\left(\frac{\mathrm{d}^2}{\mathrm{d}z_i^2}-\rho^2\right)\overline{\psi}_i=0\;,\quad (i=0,1,2)\,, \tag{6.1.27}$$

$$\left(\frac{\mathrm{d}^2}{\mathrm{d}z_1^2}-\rho^2\right)\left(\frac{\mathrm{d}^2}{\mathrm{d}z_2^2}-\rho^2\right)\overline{F}=\frac{\beta_1}{c_{11}c_{44}}\overline{T}\,. \tag{6.1.28}$$

It is easy to obtain the general solution of Eq. (6.1.27) and the particular solution of Eq. (6.1.28). Since each physical quantity should be finite at infinity, we obtain

$$\overline{\psi}_i = C_i \,\mathrm{e}^{-\rho z}, \quad (i = 0, 1, 2), \tag{6.1.29}$$

$$\overline{F} = AC_3 \rho^{-4} \,\mathrm{e}^{-\rho z/k}, \tag{6.1.30}$$

where C_i $(i = 0, 1, 2)$ are integral constants, and

$$A = \frac{\beta_1 k^4 s_1^2 s_2^2}{c_{11} c_{44} (1 - k^2 s_1^2)(1 - k^2 s_2^2)}. \tag{6.1.31}$$

Generally, there are three boundary conditions at $z = 0$, which can be used to determine the three unknown constants C_i $(i = 0, 1, 2)$. For example, when tractions σ_z^*, τ_{yz}^* and τ_{zx}^* are prescribed, their Fourier transforms, $\overline{\sigma}_z^*$, $\overline{\tau}_{yz}^*$ and $\overline{\tau}_{zx}^*$, are known. Substituting Eqs. (6.1.29) and (6.1.30) into the transformed forms of Eqs. (6.1.14), (6.1.17), (1.2.1), (1.4.9) and (1.4.12), we obtain the expressions for stresses in the transformed domain, *i.e.* $\overline{\sigma}_x$, $\overline{\sigma}_y$, $\overline{\sigma}_z$, $\overline{\tau}_{yz}$, $\overline{\tau}_{zx}$, and $\overline{\tau}_{xy}$. Setting $\overline{\sigma}_z = \overline{\sigma}_z^*$, $\overline{\tau}_{yz} = \overline{\tau}_{yz}^*$, and $\overline{\tau}_{zx} = \overline{\tau}_{zx}^*$ at $z = 0$ gives three algebraic equations, from which the elastic constants $C_0(p, q)$, $C_1(p, q)$ and $C_2(p, q)$ can be obtained. Then ψ_i $(i = 0, 1, 2)$ and F can be derived by the inverse transform of Eqs. (6.1.29) and (6.1.30), which gives rise to

$$\psi_i = \frac{1}{2\pi} \int_{-\infty}^{\infty} \int_{-\infty}^{\infty} C_i(p, q) \mathrm{e}^{-\mathrm{i}(px+qy) - \rho z} \mathrm{d}p \,\mathrm{d}q, \quad (i = 0, 1, 2), \tag{6.1.32}$$

$$F = \frac{A}{2\pi} \int_{-\infty}^{\infty} \int_{-\infty}^{\infty} C_3(p, q) \rho^{-4} \,\mathrm{e}^{-\mathrm{i}(px+qy) - \rho z/k} \mathrm{d}p \,\mathrm{d}q. \tag{6.1.33}$$

Finally, the displacements are determined by Eq. (6.1.14), where the two functions, ψ_3 and ψ_4, are given by Eq. (6.1.17).

When the deformation is axisymmetric, these expressions can be simplified. In this case, because $x = r\cos\alpha$, $y = r\sin\alpha$, $p = \rho\cos\phi$, $q = \rho\sin\phi$, $px + qy = r\rho\cos(\phi - \alpha)$, $\psi_0 = 0$, $\psi_i = \psi_i(r, z), (i = 1, 2)$ and $F = F(r, z)$, we have $\overline{\psi}_i = \overline{\psi}_i(\rho, z)$, $\overline{F} = \overline{F}(\rho, z)$ and $C_i = C_i(\rho), (i = 1, 2, 3)$. The transform formula, Eq. (6.1.22), can then be rewritten as

$$\bar{f}(\rho,z)=\frac{1}{2\pi}\int_0^{2\pi}\int_0^{\infty} f(r,z)\mathrm{e}^{\mathrm{i}\rho r\cos(\phi-\alpha)}\, r\,\mathrm{d}r\,\mathrm{d}\alpha \\ =\int_0^{\infty} f(r,z)J_0(\rho r)r\,\mathrm{d}r, \tag{6.1.34}$$

when the following identity is used,

$$\int_0^{2\pi} \mathrm{e}^{\mathrm{i}\rho r\cos(\phi-\alpha)}\mathrm{d}\alpha = 2\pi J_0(\rho r), \tag{6.1.35}$$

where $J_0(x)$ is the Bessel function of the first kind of order zero. Similarly, the inverse formula, Eq. (6.1.23), becomes

$$f(r,z)=\int_0^{\infty}\bar{f}(\rho,z)J_0(\rho r)\rho\,\mathrm{d}\rho. \tag{6.1.36}$$

Equation (6.1.34) is usually known as the zeroth *Hankel transform* of function $f(r,z)$, and Eq. (6.1.36) is the corresponding inverse transform.

The derivation of Eqs. (6.1.24) to (6.1.31) is still valid for axisymmetric problem, and Eqs. (6.1.32) and (6.1.33) become

$$\psi_i(r,z)=\int_0^{\infty} C_i(\rho)\mathrm{e}^{-\rho z} J_0(\rho r)\rho\,\mathrm{d}\rho, \quad (i=1,2), \tag{6.1.37}$$

$$F(r,z)=A\int_0^{\infty} C_3(\rho)\rho^{-4}\mathrm{e}^{-\rho z/k} J_0(\rho r)\rho\,\mathrm{d}\rho. \tag{6.1.38}$$

Having obtained ψ_i as shown in Eq. (6.1.32) or Eq. (6.1.37) and F in Eq. (6.1.33) or Eq. (6.1.38), we can derive the displacements from Eqs. (6.1.17) and (6.1.14) or Eq. (6.1.15). Then the stresses can be obtained from the constitutive relations.

6.2 A DIFFERENT GENERAL SOLUTION FOR TRANSVERSELY ISOTROPIC THERMOELASTICITY[1]

6.2.1 General Solution for Dynamic Problems

The non-steady temperature field is governed by the following equation of heat conduction

[1] The derivation presented in this section was due to Chen and Ding (2003).

$$\left(k_x\Lambda + k_z\frac{\partial^2}{\partial z^2}\right)T = c\rho\frac{\partial T}{\partial t}, \tag{6.2.1}$$

where Λ is the planar Laplace operator given in Eq. (2.2.9), c is the *specific heat*, and ρ is the mass density. The equations of motion expressed by displacements were given in Eq. (2.2.1), where, in absence of body forces, $F_x = F_y = F_z = 0$.

To simplify the governing equations, we use the separation formulae given by Eq. (2.2.2). As with the derivation presented in **Section 2.2**, we get

$$\left(c_{66}\Lambda + c_{44}\frac{\partial^2}{\partial z^2} - \rho\frac{\partial^2}{\partial t^2}\right)\psi = 0, \tag{6.2.2}$$

$$\mathbf{D}\begin{Bmatrix} G \\ w \\ T \end{Bmatrix} = \begin{Bmatrix} 0 \\ 0 \\ 0 \end{Bmatrix}, \tag{6.2.3}$$

where

$$\mathbf{D} = \begin{bmatrix} c_{11}\Lambda + c_{44}\frac{\partial^2}{\partial z^2} - \rho\frac{\partial^2}{\partial t^2} & -(c_{13}+c_{44})\frac{\partial}{\partial z} & -\beta_1 \\ -(c_{13}+c_{44})\Delta\frac{\partial}{\partial z} & c_{44}\Lambda + c_{33}\frac{\partial^2}{\partial z^2} - \rho\frac{\partial^2}{\partial t^2} & \beta_3\frac{\partial}{\partial z} \\ 0 & 0 & k_x\Lambda + k_z\frac{\partial^2}{\partial z^2} - c\rho\frac{\partial}{\partial t} \end{bmatrix}. \tag{6.2.4}$$

It is easy to compute the determinant of this matrix:

$$|\mathbf{D}| = L_0\left(k_x\Lambda + k_z\frac{\partial^2}{\partial z^2} - c\rho\frac{\partial}{\partial t}\right). \tag{6.2.5}$$

where L_0 was given in Eq. (2.2.26). The following general solution of Eq. (6.2.3) can be derived

$$G = A_{31}F, \quad w = A_{32}F, \quad T = A_{33}F, \tag{6.2.6}$$

where F satisfies

$$|\mathbf{D}|F = 0, \tag{6.2.7}$$

and A_{3i} $(i = 1, 2, 3)$ are the algebraic co-minors of $\mathbf{D}$ corresponding to the third row, *i.e.*

$$\begin{aligned} A_{31} &= a_1\Lambda + b_1\frac{\partial^2}{\partial z^2} - \beta_1\rho\frac{\partial^2}{\partial t^2}, \\ A_{32} &= \left(a_2\Lambda + b_2\frac{\partial^2}{\partial z^2} + \beta_3\rho\frac{\partial^2}{\partial t^2}\right)\frac{\partial}{\partial z}, \\ A_{33} &= L_0, \end{aligned} \tag{6.2.8}$$

in which

$$\begin{aligned} &a_1 = \beta_1 c_{44}, \quad b_1 = -\beta_3(c_{13} + c_{44}) + \beta_1 c_{33}, \\ &a_2 = -\beta_3 c_{11} + \beta_1(c_{13} + c_{44}), \quad b_2 = -\beta_3 c_{44}. \end{aligned} \tag{6.2.9}$$

Then, from Eqs. (2.2.2), (6.2.6) and (6.2.8), we obtain a new general solution for displacements for transversely isotropic thermoelasticity:

$$\begin{aligned} u &= \frac{\partial\psi}{\partial y} - \left(a_1\Lambda + b_1\frac{\partial^2}{\partial z^2} - \beta_1\rho\frac{\partial^2}{\partial t^2}\right)\frac{\partial F}{\partial x}, \\ v &= -\frac{\partial\psi}{\partial x} - \left(a_1\Lambda + b_1\frac{\partial^2}{\partial z^2} - \beta_1\rho\frac{\partial^2}{\partial t^2}\right)\frac{\partial F}{\partial y}, \\ w &= \left(a_2\Lambda + b_2\frac{\partial^2}{\partial z^2} + \beta_3\rho\frac{\partial^2}{\partial t^2}\right)\frac{\partial F}{\partial z}, \\ T &= L_0 F, \end{aligned} \tag{6.2.10a}$$

where ψ and F satisfy Eqs. (6.2.2) and (6.2.7), respectively.

In cylindrical coordinates (r, α, z), this general solution takes the form

$$\begin{aligned}
u_r &= \frac{1}{r}\frac{\partial \psi}{\partial \alpha} - \left(a_1\Lambda + b_1 \frac{\partial^2}{\partial z^2} - \beta_1 \rho \frac{\partial^2}{\partial t^2} \right) \frac{\partial F}{\partial r}, \\
u_\alpha &= -\frac{\partial \psi}{\partial r} - \left(a_1\Lambda + b_1 \frac{\partial^2}{\partial z^2} - \beta_1 \rho \frac{\partial^2}{\partial t^2} \right) \frac{1}{r}\frac{\partial F}{\partial \alpha}, \\
w &= \left(a_2\Lambda + b_2 \frac{\partial^2}{\partial z^2} + \beta_3 \rho \frac{\partial^2}{\partial t^2} \right) \frac{\partial F}{\partial z}, \\
T &= L_0 F,
\end{aligned} \tag{6.2.10b}$$

where ψ and F still satisfy Eqs. (6.2.2) and (6.2.7), respectively, but with the planar Laplace operator expressed in cylindrical coordinates

$$\Lambda = \frac{\partial^2}{\partial r^2} + \frac{1}{r}\frac{\partial}{\partial r} + \frac{1}{r^2}\frac{\partial^2}{\partial \alpha^2}. \tag{6.2.11}$$

6.2.2 General Solution for Static Problems

For static problems, the general solution can be easily obtained from Eqs. (6.2.10*a*), (6.2.2) and (6.2.7) as

$$\begin{aligned}
u &= \frac{\partial \psi}{\partial y} - \left(a_1\Lambda + b_1 \frac{\partial^2}{\partial z^2} \right) \frac{\partial F}{\partial x}, \\
v &= -\frac{\partial \psi}{\partial x} - \left(a_1\Lambda + b_1 \frac{\partial^2}{\partial z^2} \right) \frac{\partial F}{\partial y}, \\
w &= \left(a_2\Lambda + b_2 \frac{\partial^2}{\partial z^2} \right) \frac{\partial F}{\partial z}, \\
T &= L_s F,
\end{aligned} \tag{6.2.12}$$

where L_s was given in Eq. (2.2.41), and ψ and F satisfy, respectively,

$$\left(c_{66}\Lambda + c_{44} \frac{\partial^2}{\partial z^2} \right) \psi = 0, \tag{6.2.13}$$

$$L_s \left(k_x\Lambda + k_z \frac{\partial^2}{\partial z^2} \right) F = 0. \tag{6.2.14}$$

As with the derivation in **Section 2.2**, these two equations can be rewritten as

$$\left(\Lambda+\frac{\partial^2}{\partial z_3^2}\right)\psi=0, \tag{6.2.15}$$

$$\left(\Lambda+\frac{\partial^2}{\partial z_1^2}\right)\left(\Lambda+\frac{\partial^2}{\partial z_2^2}\right)\left(\Lambda+\frac{\partial^2}{\partial z_0^2}\right)F=0, \tag{6.2.16}$$

where $z_i = s_i z$, $s_0=\sqrt{k_x/k_z}$, $s_3=\sqrt{c_{66}/c_{44}}$, and s_1 and s_2 were given by Eqs. (2.2.44) and (2.2.45), respectively.

According to Almansi's theorem, which was generalized in **Section 2.2**, the function F in Eq. (6.2.16) can be expressed as

$$F=\begin{cases} F_0+F_1+F_2\,, & \text{for } s_0\neq s_1\neq s_2\neq s_0, \\ F_0+F_1+zF_2\,, & \text{for } s_0\neq s_1=s_2, \\ F_0+zF_1+z^2F_2\,, & \text{for } s_0=s_1=s_2, \end{cases} \tag{6.2.17}$$

where F_i satisfy

$$\left(\Lambda+\frac{\partial^2}{\partial z_i^2}\right)F_i=0\,, \quad (i=0,1,2). \tag{6.2.18}$$

When $s_0\neq s_1\neq s_2\neq s_0$, substituting $F=\sum_{i=0}^{2}F_i$ into Eq. (6.2.12), and assuming

$$(b_1 s_i^2-a_1)\frac{\partial^2 F_i}{\partial z_i^2}=\varphi_i\,, \quad (i=0,1,2), \quad \psi=\varphi_3, \tag{6.2.19}$$

we obtain a general solution as

$$\begin{gathered} u=\frac{\partial\varphi_3}{\partial y}-\sum_{i=0}^{2}\frac{\partial\varphi_i}{\partial x}, \quad v=-\frac{\partial\varphi_3}{\partial x}-\sum_{i=0}^{2}\frac{\partial\varphi_i}{\partial y}, \\ w=\sum_{i=0}^{2}\alpha_{i1}\frac{\partial\varphi_i}{\partial z_i}, \quad T=\alpha_{02}\frac{\partial^2\varphi_0}{\partial z_0^2}=\sum_{i=0}^{2}\alpha_{i2}\frac{\partial^2\varphi_i}{\partial z_i^2}, \end{gathered} \tag{6.2.20}$$

where

$$\alpha_{i1} = \frac{(b_2 s_i^2 - a_2)s_i}{b_1 s_i^2 - a_1}, \quad (i = 0, 1, 2),$$
$$\alpha_{02} = \frac{c_{11}c_{44} - (c_{11}c_{33} - c_{13}^2 - 2c_{13}c_{44})s_0^2 + c_{33}c_{44}s_0^4}{b_1 s_0^2 - a_1}, \tag{6.2.21}$$
$$\alpha_{12} = \alpha_{22} = 0.$$

According to Eqs. (6.2.19), (6.2.18), and (6.2.13), functions φ_i in Eq. (6.2.20) satisfy

$$\left(\Lambda + \frac{\partial^2}{\partial z_i^2}\right)\varphi_i = 0, \quad (i = 0, 1, 2, 3). \tag{6.2.22}$$

The corresponding general solution in cylindrical coordinates is

$$u_r = \frac{1}{r}\frac{\partial \varphi_3}{\partial \alpha} - \sum_{i=0}^{2}\frac{\partial \varphi_i}{\partial r}, \quad u_\alpha = -\frac{\partial \varphi_3}{\partial r} - \frac{1}{r}\sum_{i=0}^{2}\frac{\partial \varphi_i}{\partial \alpha},$$
$$w = \sum_{i=0}^{2}\alpha_{i1}\frac{\partial \varphi_i}{\partial z_i}, \quad T = \sum_{i=0}^{2}\alpha_{i2}\frac{\partial^2 \varphi_i}{\partial z_i^2}. \tag{6.2.23}$$

Substituting the general solution for displacements, Eq. (6.2.20) into the constitutive relations, Eq. (1.4.9), and making use of Eq. (1.4.12), we get the general solution for stresses. For compactness, we adopt the notations introduced in Eqs. (2.2.56*a*) and (2.2.63*a*) to present the general solution for displacements and stresses as follows

$$u + \mathrm{i}v = -\Delta\left(\sum_{i=0}^{2}\varphi_i + \mathrm{i}\varphi_3\right), \quad w = \sum_{i=0}^{2}\alpha_{i1}\frac{\partial \varphi_i}{\partial z_i},$$
$$T = \sum_{i=0}^{2}\alpha_{i2}\frac{\partial^2 \varphi_i}{\partial z_i^2}, \quad \sigma_x + \sigma_y = \sum_{i=0}^{2}\gamma_{i1}\frac{\partial^2 \varphi_i}{\partial z_i^2},$$
$$\sigma_x - \sigma_y + 2\mathrm{i}\tau_{xy} = -2c_{66}\Delta^2\left(\sum_{i=0}^{2}\varphi_i + \mathrm{i}\varphi_3\right), \tag{6.2.24}$$
$$\sigma_z = \sum_{i=0}^{2}\gamma_{i2}\frac{\partial^2 \varphi_i}{\partial z_i^2}, \quad \tau_{zx} + \mathrm{i}\tau_{yz} = \Delta\left(\sum_{i=0}^{2}\gamma_{i3}\frac{\partial \varphi_i}{\partial z_i} - \mathrm{i}s_3 c_{44}\frac{\partial \varphi_3}{\partial z_3}\right),$$

where Δ was given in Eq. (2.2.68), and

$$\gamma_{i1} = 2(c_{11} - c_{66} + c_{13}s_i\alpha_{i1} + \beta_1\alpha_{i2}),$$
$$\gamma_{i2} = c_{13} + c_{33}s_i\alpha_{i1} + \beta_3\alpha_{i2}, \quad \gamma_{i3} = c_{44}(\alpha_{i1} - s_i). \tag{6.2.25}$$

For the other two cases $s_0 \neq s_1 = s_2$ and $s_0 = s_1 = s_2$, the general solutions expressed in terms of four quasi-harmonic functions can be obtained similarly.

Example 6.2: An infinite thermoelastic space containing a penny-shaped crack

Consider a flat crack S located on the coordinate plane $z = 0$, which is parallel to the isotropic plane, Fig. 6.1. Assume that the upper and lower surfaces of the crack have the same temperature distribution $\Theta(x, y)$. Thus, we can turn the problem to a mixed boundary value problem of the half-space $z \geq 0$, subjected to

$$\begin{aligned} &\sigma_z = 0, \quad T = \Theta(x,y), && \text{for } (x,y) \in S, \\ &w = \frac{\partial T}{\partial z} = 0, && \text{for } (x,y) \notin S, \\ &\tau_{zx} = \tau_{yz} = 0, && \text{for } -\infty < (x,y) < \infty. \end{aligned} \tag{6.2.26}$$

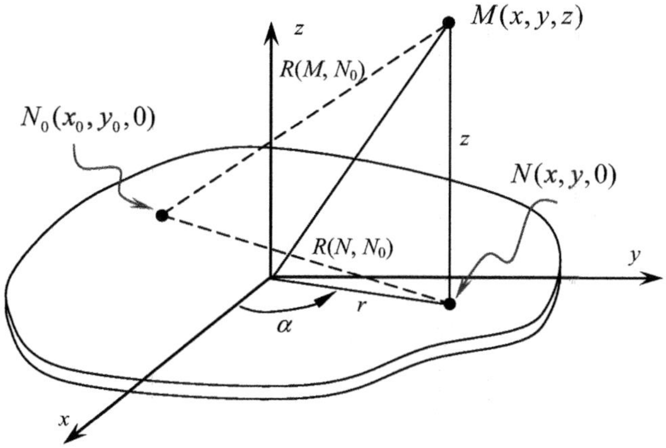

Fig. 6.1 A flat crack in an infinite solid.

Applying the *potential theory* proposed by Fabrikant (1989), we take

$$\varphi_3 = 0\,, \quad \varphi_i(z) = h_{i1}H_1(z_i) + h_{i2}H_2(z_i)\,, \quad (i = 0,1,2)\,, \tag{6.2.27}$$

where h_{i1} and h_{i2} are constants to be determined, and

$$\begin{aligned} H_1(x,y,z) &= \iint_S \frac{\omega(N_0)}{R(M,N_0)}\,\mathrm{d}S\,, \\ H_2(x,y,z) &= \iint_S \theta(N_0)\{z\ln[R(M,N_0)+z] - R(M,N_0)\}\,\mathrm{d}S\,, \end{aligned} \tag{6.2.28}$$

where $\omega(N_0)$ and $\theta(N_0)$ are respectively the crack surface displacement $w(x,y,0)$ and temperature gradient $\partial T(x,y,z)/\partial z|_{z=0}$, and $R(M,N_0)$ is the distance between point $M(x,y,z)$ and point $N_0(x_0,y_0,0)$.

By virtue of the property of the potential of a simple layer, we obtain

$$\begin{aligned} &\left.\frac{\partial H_1}{\partial z}\right|_{z=0} = 0\,, \quad \left.\frac{\partial^3 H_2}{\partial z^3}\right|_{z=0} = 0\,, \quad \text{for } (x,y) \notin S\,, \\ &\left.\frac{\partial H_1}{\partial z}\right|_{z=0} = -2\pi\omega = -2\pi w(x,y,0)\,, \quad \text{for } (x,y) \in S\,, \\ &\left.\frac{\partial^3 H_2}{\partial z^3}\right|_{z=0} = -2\pi\theta = -2\pi \left.\frac{\partial T(x,y,z)}{\partial z}\right|_{z=0}\,, \quad \text{for } (x,y) \in S\,. \end{aligned} \tag{6.2.29}$$

Substituting Eq. (6.2.27) into Eq. (6.2.24), and making use of Eq. (6.2.29), we get from the last four boundary conditions in Eq. (6.2.26),

$$\begin{aligned} &\sum_{i=0}^{2} \alpha_{i1} h_{i1} = -\frac{1}{2\pi}\,, \quad \sum_{i=0}^{2} \alpha_{i1} h_{i2} = 0\,, \\ &\sum_{i=0}^{2} \alpha_{i2} h_{i1} s_i = 0\,, \quad \sum_{i=0}^{2} \alpha_{i2} h_{i2} s_i = -\frac{1}{2\pi}\,, \\ &\sum_{i=0}^{2} \gamma_{i3} h_{ij} = 0\,, \quad (j = 1,2)\,. \end{aligned} \tag{6.2.30}$$

This gives

$$\begin{Bmatrix} h_{0j} \\ h_{1j} \\ h_{2j} \end{Bmatrix} = -\frac{1}{2\pi}\begin{bmatrix} \gamma_{03} & \gamma_{13} & \gamma_{23} \\ \alpha_{01} & \alpha_{11} & \alpha_{21} \\ \alpha_{02}s_1 & \alpha_{12}s_2 & \alpha_{22}s_3 \end{bmatrix}^{-1}\begin{Bmatrix} 0 \\ \delta_{1j} \\ \delta_{2j} \end{Bmatrix}, \quad (j=1,2), \tag{6.2.31}$$

where δ_{ij} is the *Kronecker delta.*

To satisfy the first two boundary conditions in Eq. (6.2.26), we derive

$$g_{11}\Lambda\iint_S \frac{\omega(N_0)}{R(N,N_0)}\mathrm{d}S - g_{12}\iint_S \frac{\theta(N_0)}{R(N,N_0)}\mathrm{d}S = 0, \tag{6.2.32}$$

$$\iint_S \frac{\theta(N_0)}{R(N,N_0)}\mathrm{d}S = -2\pi s_0\Theta(N), \tag{6.2.33}$$

where $R(N,N_0)$ is the distance between two points, $N(x,y,0)$ and $N_0(x_0,y_0,0)$, both on the crack surface, and

$$g_{1j} = \sum_{i=0}^{2}\gamma_{i1}h_{ij}, \quad (j=1,2). \tag{6.2.34}$$

Substituting Eq. (6.2.33) into Eq. (6.2.32) yields

$$\Lambda\iint_S \frac{\omega(N_0)}{R(N,N_0)}\mathrm{d}S = -2\pi s_0 g_{12}\Theta(N)/g_{11}. \tag{6.2.35}$$

It is seen that the integral equation, Eq. (6.2.33), governs the unknown crack surface temperature gradient $\theta(N_0)$, while the integro-differential equation, Eq. (6.2.35), governs the unknown crack surface displacement $\omega(N_0)$. They can be treated separately; in some cases, available results in potential theory (Fabrikant, 1989) can be used to obtain the analytical solutions.

Consider a penny-shaped crack with radius a, subjected to a uniformly distributed temperature $T(x,y,0)=\Theta_0$ at the crack surface. In cylindrical coordinates, the exact solutions of Eqs. (6.2.33) and (6.2.35) can be obtained as

$$\theta(r,\alpha) = -\frac{2s_0\Theta_0}{\pi\sqrt{a^2-r^2}}, \quad \omega(r,\alpha) = \frac{2s_0 g_{12}\Theta_0}{\pi g_{11}}\sqrt{a^2-r^2}. \tag{6.2.36}$$

Substituting Eq. (6.2.36) into Eq. (6.2.28) yields

$$
\begin{aligned}
H_1(r,\alpha,z) &= \frac{s_0 g_{12}\Theta_0}{g_{11}}\left[(2a^2+2z^2-r^2)\sin^{-1}\left(\frac{a}{l_2}\right)-\frac{2a^2-3l_1^2}{a}\sqrt{l_2^2-a^2}\right],\\
H_2(r,\alpha,z) &= -2s_0\Theta_0\left[\left(z^2-a^2-\frac{r^2}{2}\right)\sin^{-1}\left(\frac{a}{l_2}\right)-\frac{6a^2-3l_1^2}{2a}\sqrt{l_2^2-a^2}\right.\\
&\qquad\left.+2az\ln(l_2+\sqrt{l_2^2-r^2})\right],
\end{aligned}
\tag{6.2.37}
$$

where

$$
\begin{aligned}
l_1 &= \frac{1}{2}\sqrt{(r+a)^2+z^2}-\sqrt{(r-a)^2+z^2}],\\
l_2 &= \frac{1}{2}[\sqrt{(r+a)^2+z^2}+\sqrt{(r-a)^2+z^2}].
\end{aligned}
\tag{6.2.38}
$$

Substituting Eqs. (6.2.37) and (6.2.31) into Eq. (6.2.27) gives the expressions for displacement functions φ_i $(i=0,1,2)$. The displacements, stresses and temperature then can be obtained from Eq. (6.2.24) through simple differentiation of the displacement functions; they are all expressed in terms of elementary functions. For example, we can obtain

$$
\sigma_z = \frac{4s_0 g_{12}\Theta_0}{g_{11}}\sum_{i=0}^{2}\gamma_{i1}h_{i1}\left[\sin^{-1}\left(\frac{a}{l_{2i}}\right)-\frac{a\sqrt{l_{2i}^2-a^2}}{l_{2i}^2-l_{1i}^2}\right]-4s_0\Theta_0\sum_{i=0}^{2}\gamma_{i1}h_{i2}\sin^{-1}\left(\frac{a}{l_{2i}}\right),
\tag{6.2.39}
$$

where

$$
\begin{aligned}
l_{1i} &= \frac{1}{2}\sqrt{(r+a)^2+z_i^2}-\sqrt{(r-a)^2+z_i^2}],\\
l_{2i} &= \frac{1}{2}[\sqrt{(r+a)^2+z_i^2}+\sqrt{(r-a)^2+z_i^2}].
\end{aligned}
\tag{6.2.40}
$$

On the plane $z=0$, we have

$$\sigma_z\big|_{z=0} = -4s_0 g_{12}\Theta_0 \frac{a}{\sqrt{r^2 - a^2}}, \quad \text{for } r > a. \tag{6.2.41}$$

This is in agreement with that obtained by Tsai (1983*a*), who employed the Hankel transform technique. According to the definition of *stress intensity factor*, we have

$$K_{\mathrm{I}} = \lim_{r\to a}\left\{\sqrt{2\pi(r-a)}\sigma_z\big|_{z=0}\right\} = -4s_0\sqrt{\pi a}g_{21}\Theta_0. \tag{6.2.42}$$

6.3 SPHERICALLY ISOTROPIC MATERIALS

According to Eqs. (2.4.3) and (2.4.14), we have

$$\begin{aligned} u_R &= w, \\ u_\theta &= -\frac{1}{\sin\theta}\frac{\partial\psi}{\partial\alpha} - \frac{\partial G}{\partial\theta}, \\ u_\alpha &= \frac{\partial\psi}{\partial\theta} - \frac{1}{\sin\theta}\frac{\partial G}{\partial\alpha}, \end{aligned} \tag{6.3.1}$$

$$w = w_0 + w_1 + w_2, \quad G = G_0 + G_1 + G_2, \tag{6.3.2}$$

where ψ satisfies

$$[\nabla_3^2 + 2(c_{66}/c_{44} - 1) + (c_{66}/c_{44})\nabla_1^2]\psi = 0, \tag{6.3.3}$$

when considering Eq. (2.4.15).

With the aid of Eqs. (2.4.20), (2.4.21) and (2.4.24) to (2.4.27), we get the equations to determine w_i and G_i $(i = 0, 1, 2)$, *i.e.*,

$$w_1 = L_4\nabla_1^2 F_1, \quad G_1 = L_3 F_1, \tag{6.3.4}$$

$$w_i = L_2 F_i, \quad G_i = L_1 F_i, \quad (i = 0, 2), \tag{6.3.5}$$

where F_i satisfies

$$L^* F_0 = 0\,, \tag{6.3.6}$$

$$L^* F_1 = \frac{\beta_1}{c_{33}c_{44}} RT\,, \tag{6.3.7}$$

$$L^* F_2 = \frac{\beta_1}{c_{33}c_{44}} L_5(RT)\,, \tag{6.3.8}$$

where

$$L_5 = 2 - \frac{\beta_3}{\beta_1} - \frac{\beta_3}{\beta_1}\nabla_2\,. \tag{6.3.9}$$

In Eqs. (6.3.3) to (6.3.9), the operators ∇_3^2, ∇_1^2, ∇_2, L_1, L_2, L_3, L_4, and L^* were defined by Eqs. (2.4.2), (2.4.8) and (2.4.22).

The functions ψ and F_0 are the general solutions to Eqs. (6.3.3) and (6.3.6), respectively, while F_1 and F_2 are the particular solutions to Eqs. (6.3.7) and (6.3.8), respectively. Once F_1 is obtained, F_2 can be determined by

$$F_2 = L_5 F_1\,. \tag{6.3.10}$$

Therefore, the quasi-static general solution of spherically isotropic thermoelasticity is

$$\begin{aligned} u_R &= L_2 F_0 + L_6 F_1\,. \\ u_\theta &= -\frac{1}{\sin\theta}\frac{\partial \psi}{\partial \alpha} - \frac{\partial}{\partial \theta}(L_1 F_0 + L_7 F_1), \\ u_\alpha &= \frac{\partial \psi}{\partial \theta} - \frac{1}{\sin\theta}\frac{\partial}{\partial \alpha}(L_1 F_0 + L_7 F_1), \end{aligned} \tag{6.3.11}$$

where

$$L_6 = L_4 \nabla_1^2 + L_5 L_2\,, \quad L_7 = L_3 + L_5 L_1\,. \tag{6.3.12}$$

Example 6.3: Thermal deformation of a spherically isotropic hollow sphere

Assume that the inner and outer radii of the sphere are a and b, respectively. The temperature field applied is $T = T(R)$. Both the inner and outer surfaces of the sphere are traction-free.

The problem is spherically symmetric and hence we have $\psi=\psi(R)$, $F_0=F_0(R)$, and $F_1=F_1(R)$. From Eq. (6.3.11), $u_\theta=u_\alpha=0$ and

$$w=u_R=L_2F_0+L_5L_2F_1. \tag{6.3.13}$$

Meanwhile, Eqs. (6.3.6) and (6.3.7) are simplified to

$$L_2L_3F_0=0,\quad L_2L_3F_1=\beta_1RT. \tag{6.3.14}$$

Assume

$$L_2F_0=\Phi_0(R),\quad L_2F_1=\Phi_1(R). \tag{6.3.15}$$

Then Eqs. (6.3.13) and (6.3.14) become

$$w=\Phi_0+L_5\Phi_1, \tag{6.3.16}$$

$$L_3\Phi_0=0, \tag{6.3.17}$$

$$L_3\Phi_1=\beta_1RT, \tag{6.3.18}$$

where

$$L_3=c_{33}\nabla_3^2+2(c_{13}-c_{11}-c_{12}),\quad L_5=2-\frac{\beta_3}{\beta_1}-\frac{\beta_3}{\beta_1}\nabla_2. \tag{6.3.19}$$

The general solution to Eq. (6.3.17) and the particular solution to Eq. (6.3.18) are respectively

$$\Phi_0=C_1R^{m_1}+C_2R^{m_2}, \tag{6.3.20}$$

$$\Phi_1=\frac{\beta_1}{(m_1-m_2)c_{33}}\left[R^{m_1}\int_0^R\xi^{-m_1}T(\xi)\,\mathrm{d}\xi-R^{m_2}\int_0^R\xi^{-m_2}T(\xi)\,\mathrm{d}\xi\right], \tag{6.3.21}$$

where C_1 and C_2 are integral constants, and

$$m_{1,2} = -\frac{1}{2}[1 \pm \sqrt{1+8k}], \tag{6.3.22}$$

and

$$k = (c_{11} + c_{12} - c_{13}) / c_{33}. \tag{6.3.23}$$

According to Eq. (3.4.17), m_1 and m_2 are always real. By substituting Φ_0 and Φ_1 into Eq. (6.3.16), we get the expression of w. Thus the stresses can finally be obtained as

$$\begin{aligned} \sigma_\theta = \sigma_\alpha &= (c_{11} + c_{12})\frac{w}{R} + c_{13}\frac{\mathrm{d}w}{\mathrm{d}R} + \beta_1 T, \\ \sigma_R &= 2c_{13}\frac{w}{R} + c_{33}\frac{\mathrm{d}w}{\mathrm{d}R} + \beta_3 T. \end{aligned} \tag{6.3.24}$$

The integral constants C_1 and C_2 in Eq. (6.3.20) can be determined by the boundary conditions, $\sigma_R(a) = \sigma_R(b) = 0$.

Example 6.4: Effect of a harmonic temperature field

Let us still consider the hollow sphere discussed above but with a harmonic temperature field that can be expanded as

$$T = \sum_{n=0}^{\infty} T_n R^n S_n^m(\theta, \alpha), \tag{6.3.25}$$

where $S_n^m(\theta,\alpha)$ are spherical harmonics, satisfying $\nabla_1^2 S_n^m(\theta,\alpha) = -n(n+1)S_n^m(\theta,\alpha)$ (see **Appendix B**). For convenience, we can consider a single term of the series in Eq. (6.3.25) only. After obtaining the corresponding solution associated with the single term, the complete solution can be obtained easily in a series form by a simple superposition. Thus, Eq. (6.3.7) can be written as

$$L^* F_1 = A_n R^{n+} S_n^m(\theta, \alpha), \tag{6.3.26}$$

where

$$A_n = \beta_1 T_n / (c_{33} c_{44}). \tag{6.3.27}$$

The particular solution to Eq. (6.3.26) can be easily obtained as

$$F_1 = \frac{A_n}{\Delta_n} R^{n+1} S_n^m(\theta,\alpha), \tag{6.3.28}$$

where

$$\Delta_n = (n+2)[(n+2)(n+1)^2 + 2D(n+1) - M(n+1)^2 n - 2L(n-1) + Nn(n+1)(n-1)]. \tag{6.3.29}$$

The general solutions to Eqs. (6.3.3) and (6.3.6) are respectively

$$\begin{aligned} \psi &= (C_1 R^{m_1} + C_2 R^{m_2}) S_n^m(\theta,\alpha), \\ F_0 &= (C_3 R^{l_1} + C_4 R^{l_2} + C_5 R^{l_3} + C_6 R^{l_4}) S_n^m(\theta,\alpha), \end{aligned} \tag{6.3.30}$$

where C_i are integral constants and m_i and l_i are respectively the roots of the following two equations:

$$c_{44} m(m+1) + 2(c_{66} - c_{44}) - c_{66} n(n+1) = 0, \tag{6.3.31}$$

$$l^2 (l+1)^2 + 2P(n) l(l+1) + Q(n) = 0, \tag{6.3.32}$$

where

$$P(n) = [2D - Mn(n+1)]/2, \quad Q(n) = (n+2)(n-1)[Nn(n+1) - 2L]. \tag{6.3.33}$$

The substitution of ψ, F_0 and F_1 into Eq. (6.3.11) gives rise to the displacements and consequently the stresses. The integration constants, C_i, can be determined by the boundary conditions $\sigma_R(r,\theta,\alpha)\big|_{r=a,b} = 0$, $\tau_{R\theta}(r,\theta,\alpha)\big|_{r=a,b} = 0$ and $\tau_{\alpha R}(r,\theta,\alpha)\big|_{r=a,b} = 0$.

7

FRICTIONAL CONTACT

This chapter focuses on the applications of the established theories in solving frictional contact problems involving transversely isotropic materials. Following a summary of the fundamental theories of elastic contact, the solutions of four fundamental contact problems associated with spherical, cylindrical and conical punches are developed and discussed.

7.1 TWO ELASTIC BODIES IN CONTACT[1]

7.1.1 Mathematical Description of a Contact System

Assume that the elastic bodies ① and ② before loading are in contact at point O. Take O as the origin of two Cartesian coordinate systems, $Ox_1y_1z_1$ and $Ox_2y_2z_2$, and let axes z_1 and z_2 be along the common normal of the surfaces at point O pointing into bodies ① and ②, respectively. Thus axes x_1 and y_1 as well as x_2 and y_2 must be within the common tangential plane of the two surfaces at point O, as illustrated in Fig. 7.1. In addition, these axes are selected to be along the respective principal normal sections of the surfaces of bodies ① and ②.

If the surfaces of the elastic bodies can be described by

$$z_1 = F_1(x_1, y_1), \quad z_2 = F_2(x_2, y_2), \tag{7.1.1}$$

respectively, we can expand them in the neighbourhood of point O as *Taylor's series*. By ignoring the terms of orders higher than 2 and noting that at $x = y = 0$,

[1] This theory for isotropic materials is due to Hertz, see Timoshenko and Goodier (1970).

$$F_1 = F_2 = \frac{\partial F_1}{\partial x_1} = \frac{\partial F_1}{\partial y_1} = \frac{\partial F_2}{\partial x_2} = \frac{\partial F_2}{\partial y_2} = \frac{\partial^2 F_1}{\partial x_1 \partial y_1} = \frac{\partial^2 F_2}{\partial x_2 \partial y_2} = 0,$$

$$\frac{\partial^2 F_1}{\partial x_1^2} = K_{11}, \quad \frac{\partial^2 F_1}{\partial y_1^2} = K_{12}, \quad \frac{\partial^2 F_2}{\partial x_2^2} = K_{21}, \quad \frac{\partial^2 F_2}{\partial y_2^2} = K_{22}, \tag{7.1.2}$$

in the vicinity of point O, we can use the following approximate expressions,

$$z_1 = (K_{11}x_1^2 + K_{12}y_1^2)/2, \quad z_2 = (K_{21}x_2^2 + K_{22}y_2^2)/2, \tag{7.1.3}$$

to replace Eq. (7.1.1), where (K_{11}, K_{12}) and (K_{21}, K_{22}) are the principal curvatures of the surfaces of bodies ① and ② at point O.

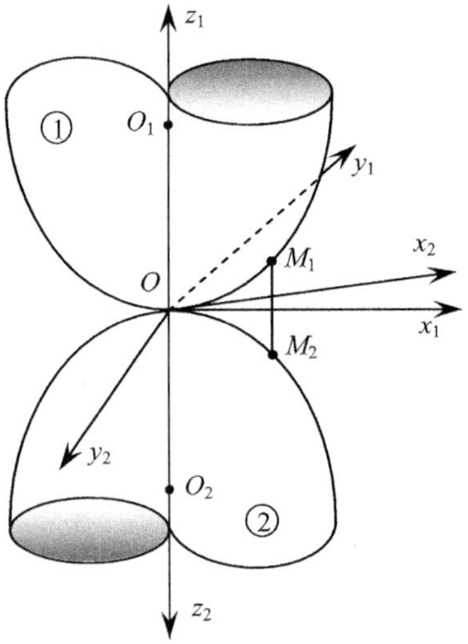

Fig. 7.1 Contact of two elastic bodies and the coordinate systems before loading.

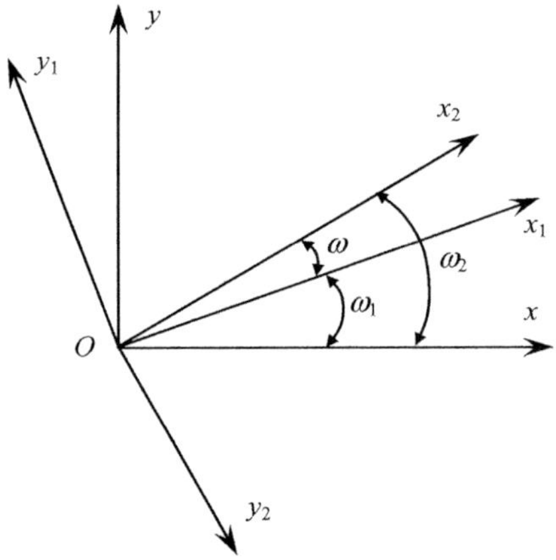

Fig. 7.2 Coordinate axes in the common tangential plane at origin O.

Generally, axes x_1 and x_2 do not coincide with each other. Let ω be the included angle of the two axes, introduce a new coordinate system, Oxy, within the tangential plane and assume that the included angle between x_1 and x is ω_1 while that between x_2 and x is ω_2, see Fig. 7.2. In so doing, by making use of the coordinate transformation, *i.e.*,

$$x_1 = x\cos\omega_1 + y\sin\omega_1\,, \quad y_1 = -x\sin\omega_1 + y\cos\omega_1\,,$$
$$x_2 = x\cos\omega_2 + y\sin\omega_2\,, \quad y_2 = x\sin\omega_2 - y\cos\omega_2\,, \tag{7.1.4}$$

the surface equation, Eq. (7.1.3), can be written in terms of the common coordinates x and y. Hence, the separation between point M_1, which is on the surface of body ① with coordinate (x, y), and point M_2, which is on the surface of body ② but with the same coordinate (x, y) as M_1, can be found to be $\overline{M_1 M_2} = z_1 + z_2$, or in terms of (x, y),

$$z_1 + z_2 = Ax^2 + By^2 + Cxy\,, \tag{7.1.5}$$

where coefficients A, B and C are calculated by

$$2A = K_{11}\cos^2\omega_1 + K_{12}\sin^2\omega_1 + K_{21}\cos^2\omega_2 + K_{22}\sin^2\omega_2\,,$$
$$2B = K_{11}\sin^2\omega_1 + K_{12}\cos^2\omega_1 + K_{21}\sin^2\omega_2 + K_{22}\cos^2\omega_2\,, \tag{7.1.6}$$
$$2C = (K_{11} - K_{12})\sin 2\omega_1 + (K_{21} - K_{22})\sin 2\omega_2\,.$$

We can make C zero by choosing the (x, y) system appropriately. In this case, Eq. (7.1.5) takes its simplest form, *i.e.*,

$$z_1 + z_2 = Ax^2 + By^2\,. \tag{7.1.7}$$

Since the separation between M_1 and M_2, *i.e.*, $z_1 + z_2$, cannot be negative, coefficients A and B in Eq. (7.1.7) cannot be negative. Considering any pair of two points with coordinate (x, y) and with $z_1 + z_2 = D = \text{const.}$, we can describe the projection of the trajectory of the two points on the common tangential plane by

$$Ax^2 + By^2 = D\,,$$

i.e.,

$$\frac{x^2}{D/A} + \frac{y^2}{D/B} = 1\,, \tag{7.1.8}$$

which is a set of similar ellipses whose major and minor axis lengths vary with D, the separation between points M_1 and M_2.

In a contact analysis, the x-axis and coefficients A and B can be determined using the following steps:

(a)Establish coordinate systems, $Ox_1'y_1'z_1$ and $Ox_2'y_2'z_2$, in bodies ① and ②, respectively, describe the surface equations of the bodies, find the principal planes of the surfaces and calculate K_{11}, K_{12}, K_{21}, and K_{22}. In the individual principal planes, establish coordinate axes x_1 and y_1 as well as axes x_2 and y_2 in such a way that the included angle between x_1-axis and x_2-axis follows $\omega \le 90^\circ$.

(b) Using the relationship shown in Fig. 7.2, *i.e.*,

$$\omega_2 = \omega_1 + \omega, \tag{7.1.9}$$

and bearing in mind that $C = 0$, we have

$$\tan 2\omega_1 = \frac{-(K_{21} - K_{22})\sin 2\omega}{K_{11} - K_{12} + (K_{21} - K_{22})\cos 2\omega}. \tag{7.1.10}$$

Thus the x-axis is determined when ω_1 is obtained from this equation. Equation (7.1.9) then specifies ω_2 and Eq. (7.1.6) provides coefficients A and B.

An alternative but more convenient way to calculate A and B is as follows. From Eq. (7.1.6), we can get

$$\begin{aligned} 2(A+B) &= K_{11} + K_{12} + K_{21} + K_{22} = \Sigma, \\ 2(A-B) &= (K_{11} - K_{12})\cos 2\omega_1 + (K_{21} - K_{22})\cos 2\omega_2. \end{aligned} \tag{7.1.11}$$

Thus the calculation of A and B involves $\cos 2\omega_1$ and $\cos 2\omega_2$ only. In this case, after the calculation of $\tan 2\omega_1$ from Eq. (7.1.10), there is no need to calculate ω_1 but go for $\cos 2\omega_1$ and $\sin 2\omega_1$ directly. The value of $\cos 2\omega_2$ can then be obtained from Eq. (7.1.9).

7.1.2 Deformation of Transversely Isotropic Bodies under Frictionless Contact

With this description of a contact system, we can now consider its deformation. Assume that bodies ① and ② are subjected to a pair of compressive forces, P, along the common normal at O. Because of the deformation, a *contact area* appears around point O; we need to find the shape and size of this contact area, and the pressure distribution[2] over it. The size of the contact area and the pressure distribution depend on the magnitude of the external force P. Note that the contact deformation is localized to the neighbourhood of the contact area. Far from the contact region, at points O_1 and O_2 in Fig. 7.1, the deformation is negligible.

Under the compression P, the deformation of bodies ① and ② near the contact area makes the separation between points O_1 and O_2 reduced by an amount δ. Similarly, the separation between points M_1 and M_2 also has a reduction of δ because of the translatory displacements of the two bodies. Assume that the contact pressure is $p(x,y)$, that the coordinate system established before loading, *Oxy*, is fixed in space, and that the displacement of point M_1 along the positive z_1-axis is w_1 and that of M_2 along the positive z_2-axis is w_2. Hence, the net movement of M_1 towards M_2 is $\delta - w_1 - w_2$. Since the original separation between M_1 and M_2 was $z_1 + z_2$, we find that if

$$z_1 + z_2 = \delta - (w_1 + w_2), \tag{7.1.12}$$

then two points will become in contact after the deformation. Otherwise, if

$$z_1 + z_2 > \delta - (w_1 + w_2), \tag{7.1.13}$$

the points will still be outside the contact area after the deformation.

When the dimension of the contact area, S, is much smaller than the surface dimensions of bodies ① and ②, which are transversely isotropic, two bodies may be treated as half-spaces, and the displacements, w_1 and w_2, can be calculated using Eq. (4.1.24), so that

$$w_1 = K_1 \iint_S \frac{p}{r} \mathrm{d}\xi \mathrm{d}\eta, \quad w_2 = K_2 \iint_S \frac{p}{r} \mathrm{d}\xi \mathrm{d}\eta, \tag{7.1.14}$$

where $r = [(x-\xi)^2 + (y-\eta)^2]^{1/2}$ and

---

[2] While introducing the basics of contact mechanics, we initially ignore friction.

$$K_i = \frac{1}{2\pi}\left[\frac{(s_1+s_2)c_{11}}{s_1 s_2 (c_{11}c_{33}-c_{13}^2)}\right]_i, \quad (i=1,2), \tag{7.1.15}$$

in which $i = 1, 2$ indicate bodies ① and ②, respectively. Now, by substituting Eqs. (7.1.7) and (7.1.14) into Eq. (7.1.12), we get

$$\delta - (Ax^2 + By^2) = \frac{c}{\pi}\iint_S \frac{p}{r}\,\mathrm{d}\xi\,\mathrm{d}\eta, \tag{7.1.16}$$

where

$$c = \frac{1}{2}\left[\frac{(s_1+s_2)c_{11}}{s_1 s_2 (c_{11}c_{33}-c_{13}^2)}\right]_1 + \frac{1}{2}\left[\frac{(s_1+s_2)c_{11}}{s_1 s_2 (c_{11}c_{33}-c_{13}^2)}\right]_2. \tag{7.1.17}$$

Equation (7.1.16) has the same form as that for isotropic materials (Hertz, 1895), except that coefficient c is defined differently. Thus using the contact mechanics theory for isotropic materials, we can write S and p in Eq. (7.1.16) as

$$p = p_0[1-(x/a)^2-(y/b)^2]^{1/2}, \quad \text{in } S: x^2/a^2 + y^2/b^2 \le 1, \tag{7.1.18}$$

where p_0 is the maximum pressure at $x = y = 0$. Substituting this into Eq. (7.1.16) leads to

$$\delta - (Ax^2 + By^2) = \frac{cp_0}{a}\left\{abK(e) - \frac{b}{a}D(e)x^2 - \frac{a}{b}[K(e)-D(e)]y^2\right\}, \tag{7.1.19}$$

where[3]

$$\begin{aligned} e &= \sqrt{1-(b/a)^2}, \\ K(e) &= \int_0^{\pi/2} \frac{\mathrm{d}\varphi}{\sqrt{1-e^2\sin^2\varphi}}, \\ E(e) &= \int_0^{\pi/2} \sqrt{1-e^2\sin^2\varphi}\,\mathrm{d}\varphi, \\ D(e) &= [K(e)-E(e)]/e^2. \end{aligned} \tag{7.1.20}$$

---

[3] The detailed derivation of Eq. (7.1.19) can be found in Section 11.9 of the book by Xie *et al.* (1988).

By comparing the two sides of Eq. (7.1.19), we find that

$$\delta = cp_0 bK(e)\,, \tag{7.1.21}$$

$$A = cp_0 \frac{b}{a^2} D(e)\,, \tag{7.1.22}$$

$$B = cp_0 \frac{1}{b}[K(e) - D(e)]\,. \tag{7.1.23}$$

These three equations contain four unknowns, *i.e.*, a, b, δ and p_0; the equilibrium equation provides the necessary fourth equation:

$$P = \iint_S p\,\mathrm{d}x\,\mathrm{d}y = \frac{2}{3}\pi abp_0\,. \tag{7.1.24}$$

Using Eqs. (7.1.22) and (7.1.23) to eliminate p_0, we get

$$\frac{A}{B} = (1 - e^2)\frac{D(e)}{K(e) - D(e)}\,. \tag{7.1.25}$$

Thus e is determined. On the other hand, Eq. (7.1.24) gives $p_0 = 3P/(2\pi ab)$ and if this is substituted into Eqs. (7.1.21) to (7.1.23), a, b and δ can be derived as

$$a = n_a \sqrt[3]{\frac{cP}{\Sigma}}\,, \quad n_a = \sqrt[3]{\frac{3}{\pi}\left(1 + \frac{B}{A}\right) D(e)}\,, \tag{7.1.26}$$

$$b = n_b \sqrt[3]{\frac{cP}{\Sigma}}\,, \quad n_b = \sqrt[3]{\frac{3}{\pi}\left(1 + \frac{B}{A}\right)[K(e) - D(e)]\sqrt{1 - e^2}}\,, \tag{7.1.27}$$

$$p_0 = n_p \sqrt[3]{\frac{\Sigma^2 P}{c^2}}, \quad n_p = \frac{3}{2\pi n_a n_b}, \tag{7.1.28}$$

$$\delta = n_\delta \sqrt[3]{P^2 c^2 \Sigma}, \quad n_\delta = \frac{3}{2\pi n_a} K(e). \tag{7.1.29}$$

This result leads to the following conclusions:

(*a*) The relationship between the external load P and the contact variables, *i.e.*, the maximum pressure p_0, the contact area parameters a and b and the approaching displacement δ, is not linear. Thus superposition cannot be used in calculating these variables.

(*b*) Parameters n_a, n_b, n_p and n_δ are determined when the ratio A/B is given.

(*c*) When $A/B = 1$, the contact area is circular.

(*d*) If e varies between 0 and 1, *i.e.* $a \geq b$, we can easily see from Eqs. (7.1.26) and (7.1.27) that $n_a \geq n_b$. Meanwhile, A/B varies between 1 and 0 according to Eq. (7.1.25).

For numerical calculation, when Eq. (7.1.11) gives rise to $A > B$, we interchange A and B. In this case, the x-axis will be in the direction of the minor axis of the elliptic contact area. Table 7.1 lists some values of A/B, n_a, n_b and n_δ, where θ is defined by

$$\cos\theta = \frac{A-B}{A+B}. \tag{7.1.30}$$

Table 7.1 The numerical relations among A/B, n_a, n_b and n_δ.

θ (°)	0	10	20	30	35	40	45	50
n_a	∞	7.5689	4.3247	3.1262	2.7439	2.4451	2.2047	2.0078
n_b	0	0.3652	0.4670	0.5643	0.6067	0.6491	0.6914	0.7338
n_δ		0.2788	0.3997	0.4760	0.5078	0.5363	0.5599	0.5805
θ (°)	55	60	65	70	75	80	85	90
n_a	1.8441	1.7010	1.5774	1.4698	1.3759	1.2912	1.2145	1.1447
n_b	0.7761	0.8208	0.8688	0.9181	0.9684	1.0222	1.1378	1.1447
n_δ	0.5988	0.6124	0.6264	0.6368	0.6444	0.6503	0.6539	0.6552

Let us now consider a few special cases of contact using these formulae.

(1) Two spheres in contact

In this case, both bodies ① and ② are spheres. When their radii are R_1 and R_2, respectively, we have

$$K_{11} = K_{12} = 1/R_1, \quad K_{21} = K_{22} = 1/R_2,$$

and ω can be arbitrary. Equation (7.1.11) then gives

$$A = B = \frac{1}{2}\left(\frac{1}{R_1} + \frac{1}{R_2}\right), \tag{7.1.31}$$

$$\Sigma = 2\left(\frac{1}{R_1} + \frac{1}{R_2}\right) = 2\frac{R_1 + R_2}{R_1 R_2}. \tag{7.1.32}$$

Equations (7.1.25) and (7.1.31) show that $e = 0$ which means the contact area is circular. Substituting Eq. (7.1.32) into Eqs. (7.1.26)-(7.1.29) yields

$$\begin{aligned} a &= 0.9086\sqrt[3]{cP\frac{R_1 R_2}{R_1 + R_2}}, \\ p_0 &= 0.5784\sqrt[3]{\frac{P}{c^2}\left(\frac{R_1 + R_2}{R_1 R_2}\right)^2}, \\ \delta &= 0.8255\sqrt[3]{(cP)^2\frac{R_1 + R_2}{R_1 R_2}}. \end{aligned} \tag{7.1.33}$$

When one of the spheres is much larger, say for example $R_2 \to \infty$, the problem becomes a contact between a sphere of radius $R_1 = R$ and a half-space. Thus Eq. (7.1.33) reduces to

$$\begin{aligned} a &= 0.9086\sqrt[3]{cPR}, \\ p_0 &= 0.5784\sqrt[3]{\frac{P}{c^2 R^2}}, \\ \delta &= 0.8255\sqrt[3]{\frac{c^2 P^2}{R}}. \end{aligned} \tag{7.1.34}$$

(2) Two circular cylinders in contact

Assume that the cylinders have the same radius, R, and their longitudinal axes are perpendicular to each other. In this case, the contact area is also circular. Thus

$$K_{11} = K_{21} = 1/R, \quad K_{12} = K_{22} = 0,$$

while $\omega = \pi/2$. Hence, Eqs. (7.1.9) and (7.1.10) give $\omega_1 = 0$ and $\omega_2 = \pi/2$, Eq. (7.1.11) yields $A = B = 1/(2R)$ and Eqs. (7.1.26) to (7.1.29) give rise to the results identical to Eq. (7.1.34).

(3) A sphere in contact with a spherically concave half-space

Let the radii of the sphere and concave surface be R_1 and R_2, respectively, where $R_1 < R_2$. The formulae of Eq. (7.1.33) are applicable provided that R_2 is replaced by $-R_2$ because the present surface of body ② is concave.

7.1.3 A Half-Space under Point Forces

The formulae in **Section 7.1.2** describe the contact area and pressure at the contact interface. In the mechanics analysis of a contact system, however, it is often necessary to know the interior stress and displacement fields in bodies ① and ②. These are to be worked out in the next few sections, by making use of the point force solution (*Green's function*) developed in **Chapter 4**. Here, let us give the explicit formulae associated with a point force first.

Consider the problem in cylindrical coordinates (r, α, z), as shown in Fig. 7.3. The concentrated force is applied at the surface point M whose coordinates are $(r, \alpha, z) = (r_0, \alpha_0, 0)$. The components of the force in x-, y- and z-directions are Q_x, Q_y and N, respectively. According to Eqs. (4.1.1) and (4.1.25), it is clear that the functions φ_i (i = 1, 2, 3) in the general solution of Eqs. (2.2.69) and (2.2.70) have the following explicit expressions:

(1) The half-space subjected to a normal force N only

In this case, we have

$$\varphi_1(r,\alpha,z;r_0,\alpha_0) = \frac{H}{(\alpha_1 - s_1)} N \ln(R_1 + z_1), \tag{7.1.35}$$

$$\varphi_2(r,\alpha,z;r_0,\alpha_0)=\frac{H}{(\alpha_2-s_2)}N\ln(R_2+z_2),$$

$$\varphi_3(r,\alpha,z;r_0,\alpha_0)=0,$$

where

$$R_j^2=r^2+r_0^2-2rr_0\cos(\alpha-\alpha_0)+z_j^2,\quad z_j=zs_j,\quad (j=1,2,3), \tag{7.1.36}$$

$$H=\frac{(s_1+s_2)c_{11}}{2\pi s_1 s_2(c_{11}c_{33}-c_{13}^2)}. \tag{7.1.37}$$

It is noted that, Eq. (7.1.35) can be derived by the substitution of Eq. (4.1.11) into Eq. (4.1.1) with P replaced by N, and the following identities should be employed

$$\frac{s_1 s_2}{2\pi c_{66}(s_2-s_1)k_{31}}=\frac{H}{\alpha_1-s_1},\quad \frac{s_1 s_2}{2\pi c_{66}(s_1-s_2)k_{32}}=\frac{H}{\alpha_2-s_2}.$$

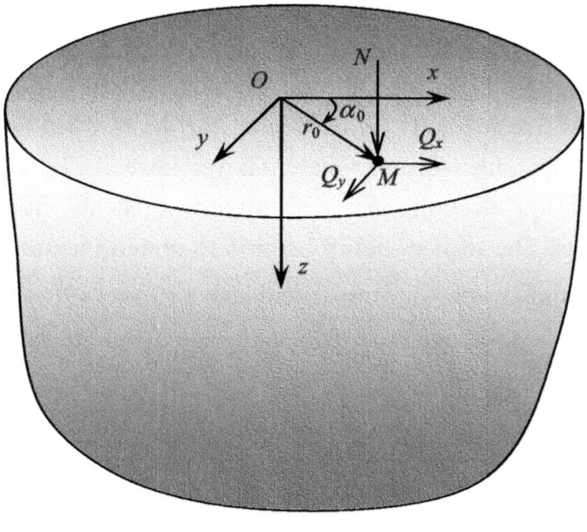

Fig. 7.3 Concentrated forces on a transversely isotropic half-space.

(2) The half-space subjected to tangential forces Q_x and Q_y only

Now we have[4]

$$
\begin{aligned}
\varphi_1(r,\alpha,z;r_0,\alpha_0) &= \frac{H}{2s_2(\alpha_1 - s_1)}(Q\bar{\Delta} + \bar{Q}\Delta)\chi(z_1),\\
\varphi_2(r,\alpha,z;r_0,\alpha_0) &= \frac{H}{2s_1(\alpha_2 - s_2)}(Q\bar{\Delta} + \bar{Q}\Delta)\chi(z_2),\\
\varphi_3(r,\alpha,z;r_0,\alpha_0) &= -\mathrm{i}\frac{s_3}{4\pi c_{66}}(Q\bar{\Delta} - \bar{Q}\Delta)\chi(z_3),
\end{aligned}
\tag{7.1.38}
$$

where $Q = Q_x + \mathrm{i}\,Q_y$, $\bar{Q}$ is the conjugate of Q, $\bar{\Delta}$ is the conjugate operator of Δ in Eq. (2.2.68) and

$$
\chi(z_j) = z_j \ln(R_j + z_j) - R_j, (j = 1, 2, 3). \tag{7.1.39}
$$

In the following sections, we will introduce a unified and more direct method to achieve the solutions of four fundamental frictional contact problems (Hanson, 1992*a*, 1992*b*, 1993, 1994; Hanson and Johnson, 1993). With the aid of the Green's function for a transversely isotropic half-space, the analytical solutions will be developed by the conventional semi-inverse method[5]. In particular, the stress distribution in the contact area follows exactly that for isotropic materials (Gladwell, 1980; Johnson, 1985) as illustrated earlier in this section. We will discuss in detail transversely isotropic materials with $s_1 \neq s_2$. The reader should be able to obtain the solutions for materials with $s_1 = s_2$ using a similar method. In particular, Fabrikant (1989) proposed a method to obtain solutions for an isotropic material directly from those for a transversely isotropic material by using *L'Hospital's rule.*

---

[4] The complex representation of Eq. (7.1.38) was first given by Fabrikant (1989). Note that Eq. (7.1.38) can be derived directly from Eqs. (4.1.25) and (4.1.39).

[5] The semi-inverse method was first introduced by Saint-Venant. Details can be found in the book by Zhang (2001). The reader interested in potential theory may consult the works by Fabrikant (1989, 1991).

7.2 CONTACT OF A SPHERE WITH A HALF-SPACE

Consider the contact sliding of a sphere and a half-space, Fig. 7.4, both made of transversely isotropic materials, with *Coulomb friction*. Assume that the deformation in the sphere and half-space is very small so that the principle of superposition applies. Hence, the displacements and stresses caused by normal and tangential loads can be studied individually.

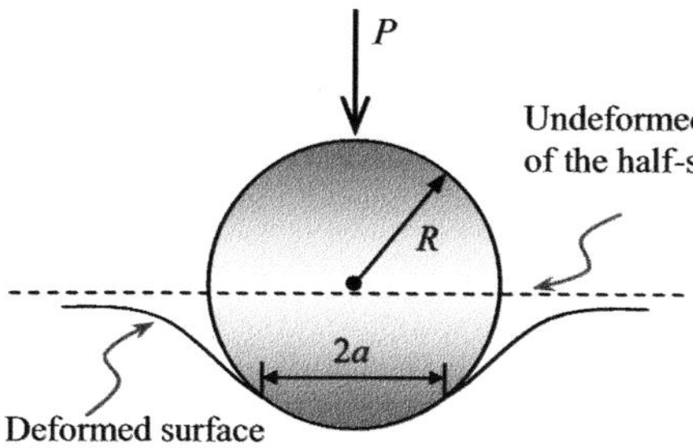

Fig. 7.4 A sphere contact with a half-space.

7.2.1 Contact with Normal Loading

From the discussion in the last section we know that the contact pressure for a system of transversely isotropic material has a distribution similar to that of an isotropic material. material. Thus for the half-space, the stress boundary conditions can be described as

$$\begin{aligned}
&\sigma_{ij}(r,\alpha,z)=0, \quad \text{when } r\to\infty, \\
&\tau_{zx}(r,\alpha,0)=\tau_{yz}(r,\alpha,0)=0, \\
&\sigma_z(r,\alpha,0)=-\frac{3P}{2\pi a^3}\sqrt{a^2-r^2}, \quad (0\le r\le a), \\
&\sigma_z(r,\alpha,0)=0, \quad (r>a),
\end{aligned} \tag{7.2.1}$$

where P is the resultant normal force and a is the radius of the contact area determined by Eq. (7.1.34).

To obtain the displacements and stresses in the half-space caused by this loading, we can make use of Eq. (7.1.35). In the present case, the normal force on an infinitesimal surface area can be considered as a concentrated force, which, according to the boundary condition, Eq. (7.2.1), can be expressed as

$$[3P\sqrt{a^2-r_0^2}/(2\pi a^3)]r_0\,\mathrm{d}r_0\,\mathrm{d}\alpha_0 .$$

Thus when the resultant force N in Eq. (7.1.35) is replaced by this force in the infinitesimal area, integration over the whole contact area ($0\le r_0\le a$, $0\le\alpha_0\le 2\pi$) leads to the displacement functions required, *i.e.*,

$$\begin{aligned}
\varphi_1(r,\alpha,z)&=\frac{3HP}{2\pi a^3(\alpha_1-s_1)}\psi(r,\alpha,z_1),\\
\varphi_2(r,\alpha,z)&=\frac{3HP}{2\pi a^3(\alpha_2-s_2)}\psi(r,\alpha,z_2),\\
\varphi_3(r,\alpha,z)&=0,
\end{aligned} \tag{7.2.2}$$

where

$$\psi(r,\alpha,z)=\int_0^{2\pi}\int_0^{a}\sqrt{a^2-r_0^2}\,\ln(R+z)r_0\,\mathrm{d}r_0\,\mathrm{d}\alpha_0, \tag{7.2.3}$$

and $R^2=r^2+r_0^2-2rr_0\cos(\alpha-\alpha_0)+z^2$.

The integral of Eq. (7.2.3) was studied by Fabrikant (1988); it can be expressed by elementary functions as

$$\begin{aligned}
\psi(r,\alpha,z)=\frac{\pi}{2}\Bigg[z\left(2a^2-r^2+\frac{2}{3}z^2\right)\sin^{-1}\left(\frac{l_1}{r}\right)+\frac{1}{3}\left(5r^2-\frac{10}{3}a^2-2l_2^2-\frac{11}{3}l_1^2\right)\\
\times\sqrt{a^2-l_1^2}+\frac{4}{3}a^3\ln(l_2+\sqrt{l_2^2-r^2})\Bigg],
\end{aligned} \tag{7.2.4}$$

where

$$\begin{aligned}
l_1=l_1(a)&=\frac{1}{2}[\sqrt{(r+a)^2+z^2}-\sqrt{(r-a)^2+z^2}],\\
l_2=l_2(a)&=\frac{1}{2}[\sqrt{(r+a)^2+z^2}+\sqrt{(r-a)^2+z^2}].
\end{aligned} \tag{7.2.5}$$

With these displacement functions, displacements and stresses can be obtained easily from Eqs. (2.2.69) and (2.2.70) by carrying out some differentiation operations[6]. For convenience, we use complex functions similar to Eqs. (2.2.56*a*) and (2.2.63*a*). Letting $U = u + \mathrm{i}v$, $\sigma_1 = \sigma_x + \sigma_y$, $\sigma_2 = \sigma_x - \sigma_y + 2\mathrm{i}\tau_{xy}$ and $\tau_z = \tau_{zx} + \mathrm{i}\tau_{yz}$ (Fabrikant, 1989), we get

$$U = \frac{3HP}{2a^3} r\,\mathrm{e}^{\mathrm{i}\alpha} \times \sum_{j=1}^{2} \frac{1}{\alpha_j - s_j}\left[-z_j \sin^{-1}\left(\frac{l_{1j}}{r}\right) + \sqrt{a^2 - l_{1j}^2}\left(1 - \frac{l_{1j}^2 + 2a^2}{3r^2}\right) + \frac{2a^3}{3r^2}\right], \tag{7.2.6a}$$

$$w = \frac{3HP}{4a^3}\sum_{j=1}^{2}\frac{\alpha_j}{\alpha_j - s_j}\left[(2a^2 + 2z_j^2 - r^2)\sin^{-1}\left(\frac{l_{1j}}{r}\right) + \frac{3l_{1j}^2 - 2a^2}{a}\sqrt{l_{2j}^2 - a^2}\right], \tag{7.2.6b}$$

$$\sigma_1 = \frac{6HPc_{66}}{a^3}\sum_{j=1}^{2}\frac{k_{1j} - 1}{\alpha_j - s_j}\left[z_j \sin^{-1}\left(\frac{l_{1j}}{r}\right) - \sqrt{a^2 - l_{1j}^2}\right], \tag{7.2.6c}$$

$$\sigma_2 = -\frac{2HPc_{66}}{a^3}\frac{\mathrm{e}^{\mathrm{i}2\alpha}}{r^2}\sum_{j=1}^{2}\frac{1}{\alpha_j - s_j}[2a^3 - (l_{1j}^2 + 2a^2)\sqrt{a^2 - l_{1j}^2}], \tag{7.2.6d}$$

$$\sigma_z = \frac{3Ps_1s_2}{2\pi a^3(s_2 - s_1)}\sum_{j=1}^{2}\frac{(-1)^{j+1}}{s_j}\left[z_j \sin^{-1}\left(\frac{l_{1j}}{r}\right) - \sqrt{a^2 - l_{1j}^2}\right], \tag{7.2.6e}$$

$$\tau_z = \frac{3Ps_1s_2}{4\pi a^3(s_2 - s_1)} r\,\mathrm{e}^{\mathrm{i}\alpha}\sum_{j=1}^{2}(-1)^{j+1}\left[-\sin^{-1}\left(\frac{l_{1j}}{r}\right) + \frac{a\sqrt{l_{2j}^2 - a^2}}{l_{2j}^2}\right], \tag{7.2.6f}$$

where k_{1j} is determined by Eq. (2.2.64) and

$$l_{1j} = \frac{1}{2}[\sqrt{(r+a)^2 + z_j^2} - \sqrt{(r-a)^2 + z_j^2}],$$
$$l_{2j} = \frac{1}{2}[\sqrt{(r+a)^2 + z_j^2} + \sqrt{(r-a)^2 + z_j^2}]. \tag{7.2.7}$$

To show the effect of transverse isotropy on the stress distribution, we consider two transversely isotropic materials, Cadmium and InSe, and an isotropic material with

[6] Fabrikant (1989, 1991) showed some derivatives of $\psi(r, \alpha, z)$.

Poisson's ratio $\nu = 0.3$. The material constants of the two transversely isotropic materials can be found in Table 1.4.

Given in Figs. 7.5-7.7 are the contours of the normalized hoop stress ($\sigma_0 = \sigma_y a^2 / P$ in the $y = 0$ plane), which is an important index for subsurface cracking due to indentation of brittle materials (Hanson, 1994). Although the shape of contours for Cadmium is similar to the one for the isotropic material, the magnitude is much different, and both the shape and magnitude become completely different for InSe. It can be seen that the maximum hoop stresses for Cadmium and the isotropic material both occur on the axisymmetric axis ($x = 0$), with $\sigma_{0\max} = 0.0167$ and $\sigma_{0\max} = 0.0028$, respectively. On the other hand, the hoop stress reaches its maximum of $\sigma_{0\max} = 0.0071$ at the point $(x/a, z/a) = (1.7447, 0.7663)$ for InSe. Thus, in the study of indentation induced subsurface fracture, it is very important to take the material anisotropy into consideration.

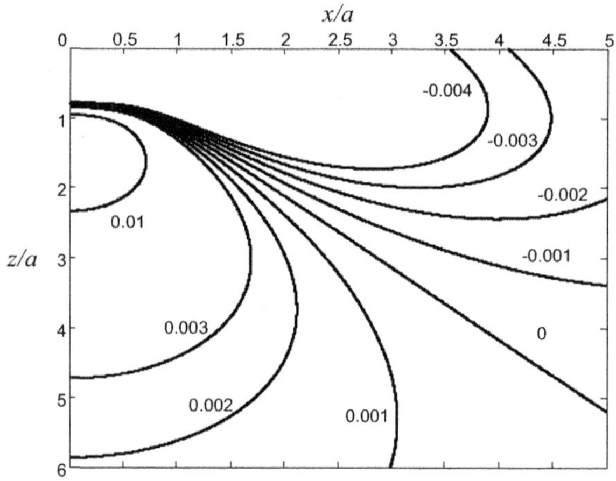

Fig. 7.5 Contours of $\sigma_0(x, 0, z)$ for Cadmium.

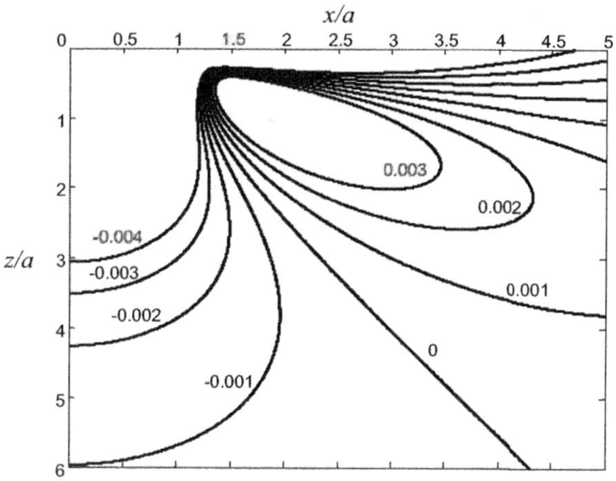

Fig. 7.6 Contours of $\sigma_0(x, 0, z)$ for InSe.

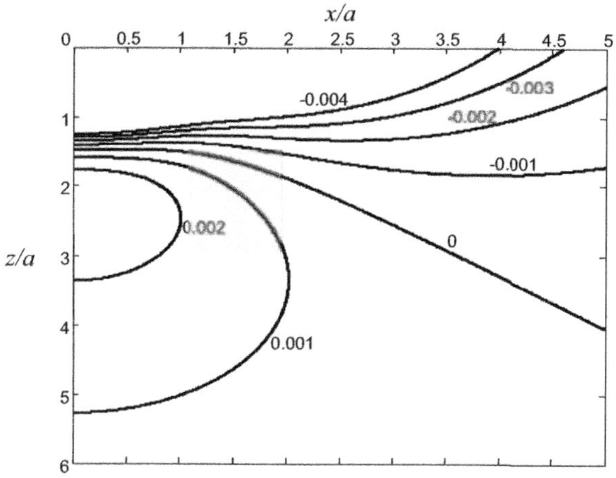

Fig. 7.7 Contours of $\sigma_0(x, 0, z)$ for isotropic material.

7.2.2 Contact with Tangential Loading

Since the contact sliding follows Coulomb friction law, the tangential force in the contact area on the half-space is simply the product of the friction coefficient and the normal force, *i.e.*,

$$Q_x = f_x N, \qquad Q_y = f_y N, \tag{7.2.8}$$

where f_x and f_y are the *friction coefficients* in *x*- and *y*-directions, respectively. Just as we obtained Eq. (7.2.2), we use the force acting on an infinitesimal element, $[3Pf/(2\pi a^3)]\sqrt{a^2 - r_0^2}\, r_0\, \mathrm{d}r_0\, \mathrm{d}\alpha_0$ where $f = f_x + \mathrm{i} f_y$, to replace the concentrated force Q of Eq. (7.1.38) and then integrate over the contact area ($0 \le r_0 \le a$, $0 \le \alpha_0 \le 2\pi$). This gives the following displacement functions:

$$\begin{aligned}
\varphi_1(r,\alpha,z) &= \frac{3HP}{4\pi a^3 s_2(\alpha_1 - s_1)}(f\bar{\Delta} + \bar{f}\Delta)[z_1\psi(r,\alpha,z_1) - \Phi(r,\alpha,z_1)], \\
\varphi_2(r,\alpha,z) &= \frac{3HP}{4\pi a^3 s_1(\alpha_2 - s_2)}(f\bar{\Delta} + \bar{f}\Delta)[z_2\psi(r,\alpha,z_2) - \Phi(r,\alpha,z_2)], \\
\varphi_3(r,\alpha,z) &= -\mathrm{i}\frac{3Ps_3}{8\pi^2 a^3 c_{66}}(f\bar{\Delta} - \bar{f}\Delta)[z_3\psi(r,\alpha,z_3) - \Phi(r,\alpha,z_3)],
\end{aligned} \tag{7.2.9}$$

where $\psi(r,\alpha,z)$ was given by Eq. (7.2.4) and $\Phi(r,\alpha,z)$ is defined as

$$\Phi(r,\alpha,z) = \int_0^{2\pi}\int_0^a \sqrt{a^2 - r_0^2}\, R r_0\, \mathrm{d}r_0\, \mathrm{d}\alpha_0 . \tag{7.2.10}$$

According to Hanson (1992*b*), we have

$$\begin{aligned}
\Phi(r,\alpha,z) = \frac{\pi}{32}\Bigg\{&[8(a^2+z^2)^2 + 8a^2r^2 - r^4 - 8r^2z^2]\sin^{-1}\left(\frac{l_1}{r}\right) \\
&+ [2a^2(4a^2 + r^2 - 4z^2) + l_1^2(6a^2 - r^2 + 14z^2)]\frac{z}{\sqrt{a^2 - l_1^2}}\Bigg\},
\end{aligned} \tag{7.2.11}$$

Similarly, the displacements and stresses can be obtained as

$$U = \frac{3HP}{4s_1 s_2 a^3} \sum_{j=1}^{2} \frac{s_j}{\alpha_j - s_j} \left\{ f \left[\left(\frac{1}{2} r^2 - a^2 - z_j^2 \right) \sin^{-1}\left(\frac{l_{1j}}{r} \right) + \frac{2a^2 - 3l_{1j}^2}{2l_{1j}} \sqrt{r^2 - l_{1j}^2} \right] \right.$$
$$+ \bar{f} \mathrm{e}^{\mathrm{i}2\alpha} \left[-\frac{4a^3 z_j}{3r^2} + \frac{1}{4} r^2 \sin^{-1}\left(\frac{l_{1j}}{r} \right) \right.$$
$$\left. \left. + [8a^4 + a^2 r^2 - l_{1j}^2 (z_j^2 + 5r^2/2 + 5a^2)] \frac{\sqrt{r^2 - l_{1j}^2}}{6r^2 l_{1j}} \right] \right\}$$
$$- \frac{3Ps_3}{8\pi a^3 c_{66}} \left\{ f \left[\left(\frac{1}{2} r^2 - a^2 - z_3^2 \right) \sin^{-1}\left(\frac{l_{13}}{r} \right) + \frac{2a^2 - 3l_{13}^2}{2l_{13}} \sqrt{r^2 - l_{13}^2} \right] \right.$$
$$- \bar{f} \mathrm{e}^{\mathrm{i}2\alpha} \left[-\frac{4a^3 z_3}{3r^2} + \frac{1}{4} r^2 \sin^{-1}\left(\frac{l_{13}}{r} \right) \right.$$
$$\left. \left. + [8a^4 + a^2 r^2 - l_{13}^2 (z_j^2 + 5r^2/2 + 5a^2)] \frac{\sqrt{r^2 - l_{13}^2}}{6r^2 l_{13}} \right] \right\}, \tag{7.2.12a}$$

$$w = \frac{3HP}{4a^3 s_1 s_2} (\bar{f} \mathrm{e}^{\mathrm{i}\alpha} + f \mathrm{e}^{-\mathrm{i}\alpha}) r \sum_{j=1}^{2} \frac{\alpha_j s_j}{\alpha_j - s_j} \left[-z_j \sin^{-1}\left(\frac{l_{1j}}{r} \right) \right.$$
$$\left. + \sqrt{a^2 - l_{1j}^2} \left(1 - \frac{l_{1j}^2 + 2a^2}{3r^2} \right) + \frac{2a^3}{3r^2} \right], \tag{7.2.12b}$$

$$\sigma_1 = \frac{3HPc_{66}}{2a^3 s_1 s_2} (\bar{f} \mathrm{e}^{\mathrm{i}\alpha} + f \mathrm{e}^{-\mathrm{i}\alpha}) r \sum_{j=1}^{2} \frac{s_j (k_{1j} - 1)}{\alpha_j - s_j} \left[-\sin^{-1}\left(\frac{l_{1j}}{r} \right) + \sqrt{r^2 - l_{1j}^2} \frac{l_{1j}}{r^2} \right], \tag{7.2.12c}$$

$$\sigma_2 = \frac{3HPc_{66}}{a^3 s_1 s_2} \sum_{j=1}^{2} \frac{s_j}{\alpha_j - s_j} \left\{ f \mathrm{e}^{\mathrm{i}\alpha} \left[\frac{1}{2} r \sin^{-1}\left(\frac{l_{1j}}{r} \right) - \frac{l_{1j}}{2r} \sqrt{r^2 - l_{1j}^2} \right] \right.$$
$$\left. - \bar{f} \mathrm{e}^{\mathrm{i}3\alpha} \left(4[(l_{1j}^2 + 2a^2)\sqrt{a^2 - l_{1j}^2} - 2a^3] \frac{z_j}{3r^3} + \frac{l_{1j}^3}{r^3} \sqrt{r^2 - l_{1j}^2} \right) \right\}$$
$$- \frac{3Ps_3}{2\pi a^3} \left\{ f \mathrm{e}^{\mathrm{i}\alpha} \left[\frac{1}{2} r \sin^{-1}\left(\frac{l_{13}}{r} \right) - \frac{l_{13}}{2r} \sqrt{r^2 - l_{13}^2} \right] \right.$$
$$\left. + \bar{f} \mathrm{e}^{\mathrm{i}3\alpha} \left(4[(l_{13}^2 + 2a^2)\sqrt{a^2 - l_{13}^2} - 2a^3] \frac{z_3}{3r^3} + \frac{l_{13}^3}{r^3} \sqrt{r^2 - l_{13}^2} \right) \right\}, \tag{7.2.12d}$$

$$\sigma_z = \frac{3P}{8\pi a^3 (s_2 - s_1)} (\bar{f} \mathrm{e}^{\mathrm{i}\alpha} + f \mathrm{e}^{-\mathrm{i}\alpha}) r \sum_{j=1}^{2} (-1)^{j+1} \left[-\sin^{-1}\left(\frac{l_{1j}}{r} \right) + \frac{l_{1j}}{r^2} \sqrt{r^2 - l_{1j}^2} \right], \tag{7.2.12e}$$

$$\tau_z = \frac{3P}{4\pi a^3(s_2 - s_1)} \sum_{j=1}^{2} (-1)^{j+1} s_j \left\{ f\left[-z_j \sin^{-1}\left(\frac{l_{1j}}{r}\right) + \sqrt{a^2 - l_{1j}^2} \right] \right.$$
$$\left. - \bar{f} \, e^{i2\alpha} \frac{2a^3 - (2a^2 + l_{1j}^2)\sqrt{a^2 - l_{1j}^2}}{3r^2} \right\} \tag{7.2.12f}$$
$$+ \frac{3P}{4\pi a^3} \left\{ f\left[-z_3 \sin^{-1}\left(\frac{l_{13}}{r}\right) + \sqrt{a^2 - l_{13}^2} \right] + \bar{f} \, e^{i2\alpha} \frac{2a^3 - (2a^2 + l_{13}^2)\sqrt{a^2 - l_{13}^2}}{3r^2} \right\}.$$

The resultant displacements and stresses due to the contact sliding are the sums of Eqs. (7.2.6) and (7.2.12)[7]. It must be pointed out that because of the surface deflection under the indentation of the sphere, the surface of the half-space inside and in the neighbourhood of the contact area is no longer flat. In addition, with the frictional force, the shape and size of the contact area generally will be different from a frictionless contact (Goodman and Keer, 1975). Thus in this sense, the solution obtained above is approximate because all these have been ignored in the formulation (Hanson, 1992*b*). However, for isotropic materials, Johnson (1985) concluded that the influence of frictional force on the shape and size of the contact area is minor when the coefficient of friction is less than one. Thus under the conditions of small deformation and small friction coefficients (f_x, $f_y < 1$), the solution above will be sufficiently accurate. This argument also applies to the other three contact problems to be considered in the following and will not be repeated for brevity. A detailed discussion on solutions of frictional contact will be presented in **Section 7.6** separately.

7.3 CONTACT OF A CYLINDRICAL PUNCH WITH A HALF-SPACE

A cylindrical punch of radius a slides on a transversely isotropic half-space, Fig. 7.8. The resultant normal force, P, acts through the central axis of the punch. Again, Coulomb friction law applies and the deformation in the contact system is small. Hence, as in the last section, the displacements and stresses induced by the normal and tangential contact forces can be studied separately following the principle of superposition.

---

[7] The solution in Eqs. (7.2.6) and (7.2.12) is consistent with that obtained by Chen (1969), except for a number of typos in Chen's formulation, as pointed out by Hanson (1992*b*).

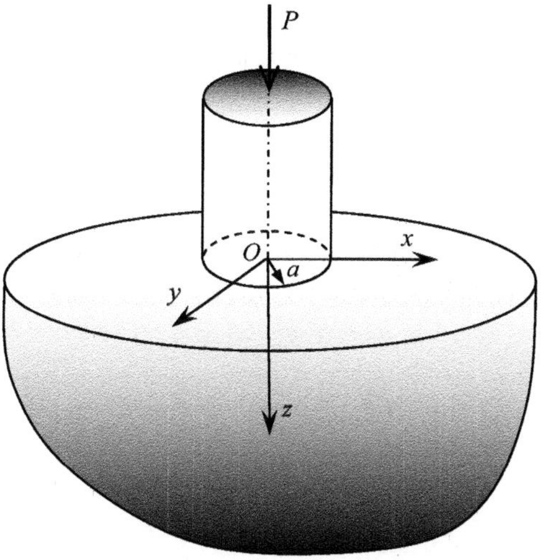

Fig. 7.8 An upright cylindrical punch contact with a half-space.

7.3.1 Contact with Normal Loading

The interface pressure within the contact area, $0 \le r \le a$, is similar to that for an isotropic half-space, *i.e.*,

$$\sigma_z(r,\alpha,0) = -\frac{P}{2\pi a}\frac{1}{\sqrt{a^2 - r^2}}, \quad (0 \le r \le a). \tag{7.3.1}$$

If we use $[P(a^2 - r_0^2)^{-1/2}/(2\pi a)]r_0\,\mathrm{d}r_0\,\mathrm{d}\alpha_0$ to replace the concentrated normal force N in Eq. (7.1.35) and then integrate over the contact area ($0 \le r_0 \le a, 0 \le \alpha_0 \le 2\pi$), we obtain the following displacement functions:

$$\begin{aligned}
\varphi_1(r,\alpha,z) &= \frac{HP}{a(\alpha_1 - s_1)}\Xi(r,\alpha,z_1), \\
\varphi_2(r,\alpha,z) &= \frac{HP}{a(\alpha_2 - s_2)}\Xi(r,\alpha,z_2), \\
\varphi_3(r,\alpha,z) &= 0,
\end{aligned} \tag{7.3.2}$$

where

$$\Xi(r,\alpha,z)=\int_0^{2\pi}\int_0^a (a^2-r_0^2)^{-1/2}\ln(R+z)r_0\,\mathrm{d}r_0\,\mathrm{d}\alpha_0 \\ =2\pi\left[z\sin^{-1}\left(\frac{l_1}{r}\right)-\sqrt{a^2-l_1^2}+a\ln(l_2+\sqrt{l_2^2-r^2})\right], \tag{7.3.3}$$

which is the partial derivative with respect to a of the potential ψ defined in Eq. (7.2.3) (Fabrikant, 1988). The displacements and stresses are therefore

$$U=\frac{PH\mathrm{e}^{\mathrm{i}\alpha}}{r}\sum_{j=1}^{2}\frac{1}{\alpha_j-s_j}\left(1-\frac{\sqrt{a^2-l_{1j}^2}}{a}\right), \tag{7.3.4a}$$

$$w=\frac{PH}{a}\sum_{j=1}^{2}\frac{\alpha_j}{\alpha_j-s_j}\sin^{-1}\left(\frac{l_{1j}}{r}\right), \tag{7.3.4b}$$

$$\sigma_1=-\frac{2PHc_{66}}{a}\sum_{j=1}^{2}\frac{k_{1j}-1}{\alpha_j-s_j}\frac{\sqrt{a^2-l_{1j}^2}}{l_{2j}^2-l_{1j}^2}, \tag{7.3.4c}$$

$$\sigma_2=\frac{2PHc_{66}}{a}\mathrm{e}^{\mathrm{i}2\alpha}\sum_{j=1}^{2}\frac{1}{\alpha_j-s_j}\left[\frac{\sqrt{a^2-l_{1j}^2}}{l_{2j}^2-l_{1j}^2}-\frac{2a}{r^2}\left(1-\frac{\sqrt{a^2-l_{1j}^2}}{a}\right)\right], \tag{7.3.4d}$$

$$\sigma_z=\frac{Ps_1s_2}{2\pi a(s_2-s_1)}\sum_{j=1}^{2}\frac{(-1)^j}{s_j}\frac{\sqrt{a^2-l_{1j}^2}}{l_{2j}^2-l_{1j}^2}, \tag{7.3.4e}$$

$$\tau_z=\frac{Ps_1s_2}{2\pi a(s_2-s_1)}\mathrm{e}^{\mathrm{i}\alpha}\sum_{j=1}^{2}(-1)^{j+1}\frac{l_{1j}\sqrt{l_{2j}^2-a^2}}{l_{2j}(l_{2j}^2-l_{1j}^2)}. \tag{7.3.4f}$$

This solution is consistent with that of Elliott (1949).

The asymptotic property of these stress and displacement fields near the punch edge is of importance to many engineering designs. By introducing the stress intensity factor,

$$k_1=\sqrt{a-r}\,\sigma_z,\quad \text{when}\quad z=0\quad\text{and}\quad r\to a, \tag{7.3.5}$$

and using Eq. (7.3.4e) to replace the σ_z in the above expression, we get

$$k_{\mathrm{I}} = -\frac{P}{2\pi a^2}\sqrt{\frac{a}{2}}. \tag{7.3.6}$$

Now, if we introduce a local spherical coordinate system, (R, θ, α), whose origin is at the edge of the contact area, we get the following equations and asymptotic expressions (Fabrikant, 1988):

$$\begin{aligned}
&r = a + R\sin\theta,\ z = R\cos\theta,\ l_{1j} \approx a - \frac{1}{2}RS_j^2,\ l_{2j} \approx a + \frac{1}{2}RT_j^2,\\
&l_{2j}^2 - l_{1j}^2 \approx 2aRP_j,\quad \sqrt{a^2 - l_{1j}^2} \approx \sqrt{aR}S_j,\\
&\sqrt{l_{2j}^2 - a^2} \approx \sqrt{aR}T_j,\quad \sin^{-1}\left(\frac{l_{1j}}{r}\right) \approx -\sqrt{\frac{R}{a}}T_j,
\end{aligned} \tag{7.3.7}$$

where

$$P_j = \sqrt{\sin^2\theta + s_j^2\cos^2\theta},\quad S_j = \sqrt{P_j - \sin\theta},\quad T_j = \sqrt{P_j + \sin\theta},\quad (j = 1, 2). \tag{7.3.8}$$

Hence, by substituting Eqs. (7.3.7) and (7.3.8) into Eq. (7.3.4), we can see the asymptotic nature of the stresses and displacements in the neighbourhood of the punch edge, *i.e.*,

$$\begin{aligned}
&U = 2\pi Hk_{\mathrm{I}}\sqrt{2R}\,\mathrm{e}^{\mathrm{i}\alpha}\left(\frac{S_1}{\alpha_1 - s_1} + \frac{S_2}{\alpha_2 - s_2}\right) + 0(1),\\
&w = -2\pi Hk_{\mathrm{I}}\sqrt{2R}\left(\frac{\alpha_1 T_1}{\alpha_1 - s_1} + \frac{\alpha_2 T_2}{\alpha_2 - s_2}\right) + 0(1),\\
&\sigma_1 = 2\pi c_{66}Hk_{\mathrm{I}}\sqrt{2/R}\sum_{j=1}^{2}\frac{(k_{1j} - 1)S_j}{(\alpha_j - s_j)P_j} + 0(1),\\
&\sigma_2 = -2\pi c_{66}Hk_{\mathrm{I}}\sqrt{2/R}\,\mathrm{e}^{\mathrm{i}2\alpha}\sum_{j=1}^{2}\frac{S_j}{(\alpha_j - s_j)P_j} + 0(\sqrt{R}),
\end{aligned} \tag{7.3.9}$$

$$\sigma_z = \frac{k_I s_1 s_2}{\sqrt{2R}(s_2 - s_1)}\left(\frac{S_1}{s_1 P_1} - \frac{S_2}{s_2 P_2}\right) + 0(1),$$

$$\tau_z = -\frac{k_I s_1 s_2}{\sqrt{2R}(s_2 - s_1)} \mathrm{e}^{\mathrm{i}\alpha}\left(\frac{T_1}{P_1} - \frac{T_2}{P_2}\right) + 0(\sqrt{R}).$$

As pointed out by Fabrikant (1988), the singular behaviour near the punch edge is the same as in a fracture mechanics problem. Thus mathematically the results in Eq. (7.3.9) are the same as those in Kassir and Sih (1975).

7.3.2 Contact with Tangential Loading

Considering the contact under friction, we can use

$$\frac{Pf}{2\pi a}\frac{1}{\sqrt{a^2 - r_0^2}} r_0 \, \mathrm{d} r_0 \, \mathrm{d}\alpha_0, \tag{7.3.10}$$

to replace the concentrated tangential force in Eq. (7.1.38) and integrate over the contact area ($0 \le r_0 \le a$, $0 \le \alpha_0 \le 2\pi$), where $f = f_x + \mathrm{i} f_y$ is the complex coefficient of friction. The displacement functions are therefore

$$\begin{aligned}
\varphi_1(r,\alpha,z) &= \frac{HP}{4\pi a s_2(\alpha_1 - s_1)}(f\bar{\Delta} + \bar{f}\Delta)[z_1 \Xi(r,\alpha,z_1) - \Theta(r,\alpha,z_1)], \\
\varphi_2(r,\alpha,z) &= \frac{HP}{4\pi a s_1(\alpha_2 - s_2)}(f\bar{\Delta} + \bar{f}\Delta)[z_2 \Xi(r,\alpha,z_2) - \Theta(r,\alpha,z_2)], \\
\varphi_3(r,\alpha,z) &= -\mathrm{i}\frac{P s_3}{8\pi^2 a c_{66}}(f\bar{\Delta} - \bar{f}\Delta)[z_3 \Xi(r,\alpha,z_3) - \Theta(r,\alpha,z_3)],
\end{aligned} \tag{7.3.11}$$

where $\Xi(r,\alpha,z)$ was given in Eq. (7.3.3), and $\Theta(r,\alpha,z)$ is defined as

$$\begin{aligned}
\Theta(r,\alpha,z) &= \int_0^{2\pi}\int_0^a (a^2 - r_0^2)^{-1/2} R r_0 \, \mathrm{d} r_0 \, \mathrm{d}\alpha_0 \\
&= \frac{\pi}{2}\left[(2a^2 + 2z^2 + r^2)\sin^{-1}\left(\frac{l_1}{r}\right) + \frac{1}{a}(2a^2 + l_1^2)\sqrt{l_2^2 - a^2}\right],
\end{aligned} \tag{7.3.12}$$

which can be obtained through differentiation with respect to a of Φ in Eq. (7.2.11) (Hanson, 1994). The displacements and stresses can thus be obtained by substituting Eq. (7.3.11) into Eq. (2.2.56*a*) and (2.2.63*a*), *i.e.*,

$$U=\frac{HP}{2as_1s_2}\sum_{j=1}^{2}\frac{s_j}{\alpha_j-s_j}\left[-f\sin^{-1}\left(\frac{l_{1j}}{r}\right)+\bar{f}\,\mathrm{e}^{\mathrm{i}2\alpha}\left(\frac{2z_j}{r^2}\sqrt{a^2-l_{1j}^2}-\frac{2az_j}{r^2}+\frac{l_{1j}}{r^2}\sqrt{r^2-l_{1j}^2}\right)\right]$$
$$+\frac{Ps_3}{4\pi ac_{66}}\left[f\sin^{-1}\left(\frac{l_{13}}{r}\right)+\bar{f}\,\mathrm{e}^{\mathrm{i}2\alpha}\left(\frac{2z_3}{r^2}\sqrt{a^2-l_{13}^2}-\frac{2az_3}{r^2}+\frac{l_{13}}{r^2}\sqrt{r^2-l_{13}^2}\right)\right],\qquad(7.3.13a)$$

$$w=\frac{HP}{2as_1s_2}(\bar{f}\,\mathrm{e}^{\mathrm{i}\alpha}+f\,\mathrm{e}^{-\mathrm{i}\alpha})\sum_{j=1}^{2}\frac{\alpha_js_j}{\alpha_j-s_j}\frac{a-\sqrt{a^2-l_{1j}^2}}{r},\qquad(7.3.13b)$$

$$\sigma_1=-\frac{HPc_{66}}{as_1s_2}(\bar{f}\,\mathrm{e}^{\mathrm{i}\alpha}+f\,\mathrm{e}^{-\mathrm{i}\alpha})\sum_{j=1}^{2}\frac{(k_{1j}-1)s_j}{\alpha_j-s_j}\frac{l_{1j}\sqrt{r^2-l_{1j}^2}}{r(l_{2j}^2-l_{1j}^2)},\qquad(7.3.13c)$$

$$\sigma_2=\frac{HPc_{66}}{as_1s_2}\sum_{j=1}^{2}\frac{s_j}{\alpha_j-s_j}\left[f\,\mathrm{e}^{\mathrm{i}\alpha}\frac{l_{1j}\sqrt{r^2-l_{1j}^2}}{r(l_{2j}^2-l_{1j}^2)}\right.$$
$$\left.+\bar{f}\,\mathrm{e}^{\mathrm{i}3\alpha}\left\{\frac{8az_j}{r^3}-\frac{\sqrt{r^2-l_{1j}^2}}{r^3l_{1j}(l_{2j}^2-l_{1j}^2)}[8a^2l_{2j}^2-8a^2r^2+(3r^2-4a^2+4z_j^2)l_{1j}^2]\right\}\right]$$
$$-\frac{Ps_3}{2\pi a}\left[f\,\mathrm{e}^{\mathrm{i}\alpha}\frac{l_{13}\sqrt{r^2-l_{13}^2}}{r(l_{23}^2-l_{13}^2)}\right.$$
$$\left.-\bar{f}\,\mathrm{e}^{\mathrm{i}3\alpha}\left\{\frac{8az_3}{r^3}-\frac{\sqrt{r^2-l_{13}^2}}{r^3l_{13}(l_{23}^2-l_{13}^2)}[8a^2l_{23}^2-8a^2r^2+(3r^2-4a^2+4z_3^2)l_{13}^2]\right\}\right],\qquad(7.3.13d)$$

$$\sigma_z=\frac{P}{4\pi a(s_2-s_1)}(\bar{f}\,\mathrm{e}^{\mathrm{i}\alpha}+f\,\mathrm{e}^{-\mathrm{i}\alpha})\sum_{j=1}^{2}(-1)^j\frac{l_{1j}\sqrt{r^2-l_{1j}^2}}{r(l_{2j}^2-l_{1j}^2)},\qquad(7.3.13e)$$

$$\tau_z = \frac{P}{4\pi a(s_2 - s_1)} \sum_{j=1}^{2} (-1)^{j+1} s_j \left\{ f \frac{\sqrt{a^2 - l_{1j}^2}}{l_{2j}^2 - l_{1j}^2} + \bar{f}\, \mathrm{e}^{\mathrm{i}2\alpha} \left[\frac{\sqrt{a^2 - l_{1j}^2}}{l_{2j}^2 - l_{1j}^2} - \frac{2a}{r^2} + \frac{2}{r^2}\sqrt{a^2 - l_{1j}^2} \right] \right\}$$
$$- \frac{P}{4\pi a} \left\{ f \frac{\sqrt{a^2 - l_{13}^2}}{l_{23}^2 - l_{13}^2} - \bar{f}\, \mathrm{e}^{\mathrm{i}2\alpha} \left[\frac{\sqrt{a^2 - l_{13}^2}}{l_{23}^2 - l_{13}^2} - \frac{2a}{r^2} + \frac{2}{r^2}\sqrt{a^2 - l_{13}^2} \right] \right\}. \tag{7.3.13f}$$

The total displacements and stresses due to the contact sliding of an upright cylindrical flat punch are the sums of Eqs. (7.3.4) and (7.3.13).

7.4 INDENTATION BY A CONE

A cone with a vertex angle of $2\phi_0$ indents a transversely isotropic half-space axisymmetrically about the central axis of the cone, as illustrated in Fig. 7.9. The resultant force is P, which is along the cone axis and is perpendicular to the half-space surface. During the indentation, sliding with Coulomb friction occurs between the cone and space surfaces, except at the apex. Again, we assume that the deformation is small and the principle of superposition applies.

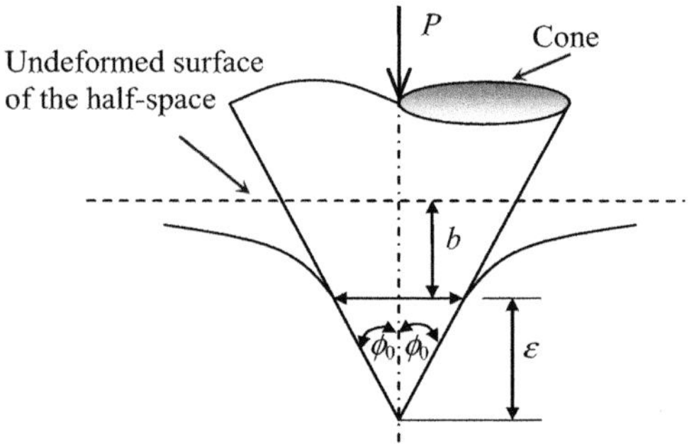

Fig. 7.9 The frictional indentation of a cone into a half-space.

7.4.1 Contact with Normal Loading

The contact pressure is the same as that for an isotropic half-space (Sneddon, 1948; Elliott, 1949) and can be described as

$$\sigma_z(r,\alpha,0) = -\frac{P}{\pi a^2}\cosh^{-1}\left(\frac{a}{r}\right), \quad 0 \le r \le a, \tag{7.4.1}$$

where a is the radius of the contact area to be determined. To obtain the displacements and stresses introduced by this pressure, we replace the concentrated force N in Eq. (7.1.35) by $[P\cosh^{-1}(a/r_0)/(\pi a)^2]r_0\,\mathrm{d}r_0\,\mathrm{d}\alpha_0$, and integrate over the whole contact area ($0 \le r_0 \le a, 0 \le \alpha_0 \le 2\pi$). The displacement functions are therefore

$$\begin{aligned}
\varphi_1(r,\alpha,z) &= \frac{HP}{\pi a^2(\alpha_1 - s_1)}\Omega(r,\alpha,z_1), \\
\varphi_2(r,\alpha,z) &= \frac{HP}{\pi a^2(\alpha_2 - s_2)}\Omega(r,\alpha,z_2), \\
\varphi_3(r,\alpha,z) &= 0,
\end{aligned} \tag{7.4.2}$$

where

$$\Omega(r,\alpha,z) = \int_0^{2\pi}\int_0^a \cosh^{-1}\left(\frac{a}{r_0}\right)\ln(R+z)r_0\,\mathrm{d}r_0\,\mathrm{d}\alpha_0\,. \tag{7.4.3}$$

This integral can be expressed in terms of elementary functions as (Hanson, 1992*a*)

$$\begin{aligned}
\Omega(r,\alpha,z) = 2\pi\Bigg[& az\sin^{-1}\left(\frac{l_1}{r}\right) + \frac{1}{4}(2a^2 + r^2 - 2z^2)\ln(l_2 + \sqrt{l_2^2 - r^2}) \\
&+ \frac{3zl_2(r^2 - 2l_1^2)}{4r\sqrt{r^2 - l_1^2}} - \frac{3z\sqrt{r^2+z^2}}{4} - \frac{1}{4}(r^2 - 2z^2)\ln(z + \sqrt{z^2 + r^2})\Bigg].
\end{aligned} \tag{7.4.4}$$

The displacements and stresses can then be derived by simple differentiations when using Eqs. (2.2.56*a*) and (2.2.63*a*), *i.e.*,

$$U = \frac{2HP}{a^2} e^{i\alpha} \sum_{j=1}^{2} \frac{1}{\alpha_j - s_j} \left[\frac{(rl_{2j} - 2al_{1j})\sqrt{l_{2j}^2 - r^2}}{2r^2} + \frac{r}{2}\ln(l_{2j} + \sqrt{l_{2j}^2 - r^2}) \right.$$
$$\left. - \frac{z_j\sqrt{r^2 + z_j^2}}{2r} - \frac{r}{2}\ln(z_j + \sqrt{r^2 + z_j^2}) + \frac{a^2}{2r} \right], \tag{7.4.5a}$$

$$w = \frac{2HP}{a^2} \sum_{j=1}^{2} \frac{\alpha_j}{\alpha_j - s_j} \left[a\sin^{-1}\left(\frac{l_{1j}}{r}\right) + \sqrt{l_{2j}^2 - a^2} - \sqrt{r^2 + z_j^2} \right.$$
$$\left. - z_j \ln(l_{2j} + \sqrt{l_{2j}^2 - r^2}) + z_j \ln(z_j + \sqrt{r^2 + z_j^2}) \right], \tag{7.4.5b}$$

$$\sigma_1 = -\frac{4HPc_{66}}{a^2} \sum_{j=1}^{2} \frac{k_{1j} - 1}{\alpha_j - s_j} [\ln(l_{2j} + \sqrt{l_{2j}^2 - r^2}) - \ln(z_j + \sqrt{r^2 + z_j^2})], \tag{7.4.5c}$$

$$\sigma_2 = \frac{4HPc_{66}}{a^2} e^{i2\alpha} \sum_{j=1}^{2} \frac{1}{\alpha_j - s_j} \left[\frac{1}{ar^2}(2a^2 - l_{2j}^2)\sqrt{a^2 - l_{1j}^2} + \frac{z_j\sqrt{r^2 + z_j^2}}{r^2} - \frac{a^2}{r^2} \right], \tag{7.4.5d}$$

$$\sigma_z = \frac{Ps_1s_2}{\pi a^2(s_2 - s_1)} \sum_{j=1}^{2} \frac{(-1)^j}{s_j} [\ln(l_{2j} + \sqrt{l_{2j}^2 - r^2}) - \ln(z_j + \sqrt{r^2 + z_j^2})], \tag{7.4.5e}$$

$$\tau_z = \frac{Ps_1s_2}{\pi a^2(s_2 - s_1)} e^{i\alpha} \sum_{j=1}^{2} (-1)^{j+1} \left(\frac{1}{r}\sqrt{l_{2j}^2 - a^2} - \frac{\sqrt{r^2 + z_j^2}}{r} \right). \tag{7.4.5f}$$

The radius of the contact area, a, in these expressions can be determined by the compatibility of deformation (Hanson, 1992*a*). According to Fig. 7.9, it is clear that the normal displacement at the half-space surface in the contact region can be written as

$$w = b + \varepsilon\left(1 - \frac{r}{a}\right), \quad (0 \le r \le a), \tag{7.4.6}$$

where $\varepsilon = a\cot\phi_0$. However, Eq. (7.4.5*b*) specifies that at the surface,

$$w = \frac{2HP}{a^2}\left(\frac{\pi}{2}a - r\right) = \frac{2HP}{a}\left(\frac{\pi}{2} - 1\right) + \frac{2HP}{a}\left(1 - \frac{r}{a}\right), \quad (0 \le r \le a). \tag{7.4.7}$$

Since the normal displacement at the contact interface must be continuous, the right hand sides of Eqs. (7.4.6) and (7.4.7) must be the same. This gives rise to

$$\varepsilon=\frac{2HP}{a}\,,\quad b=\varepsilon\left(\frac{\pi}{2}-1\right). \tag{7.4.8}$$

When the resultant force P is known, the contact parameters shown in Fig. 7.9 can be determined from

$$a=\sqrt{2HP\tan\phi_0}\,,\quad \varepsilon=\sqrt{2HP\cot\phi_0}\,,\quad b=\varepsilon\left(\frac{\pi}{2}-1\right). \tag{7.4.9}$$

If, instead of the indentation force, we know the indentation depth to be $d_p=b+\varepsilon$, then the radius of contact area, a, and the indentation force, P, can be calculated by

$$a=\frac{2}{\pi}d_p\tan\phi_0\,,\quad P=\frac{2}{\pi^2 H}d_p^2\tan\phi_0\,. \tag{7.4.10}$$

If the half-space is isotropic, $H=(1-\nu^2)/(\pi E)$ and the results reduce to those of Sneddon (1948) except for the incorrect definitions of two parameters, as pointed out by Hanson (1992*a*).

7.4.2 Contact with Tangential Loading

Now using

$$\frac{Pf}{\pi a^2}\cosh^{-1}\left(\frac{a}{r_0}\right)r_0\,\mathrm{d}r_0\,\mathrm{d}\alpha_0\,, \tag{7.4.11}$$

to replace the concentrated tangential force Q in Eq. (7.1.38) and integrating over the contact area, $(0\le r_0\le a\,,0\le\alpha_0\le 2\pi)$, we find the following displacement functions:

$$\varphi_1(r,\alpha,z)=\frac{HP}{2\pi a^2 s_2(\alpha_1-s_1)}(f\bar{\Delta}+\bar{f}\Delta)[z_1\Omega(r,\alpha,z_1)-\Gamma(r,\alpha,z_1)],$$
$$\varphi_2(r,\alpha,z)=\frac{HP}{2\pi a^2 s_1(\alpha_2-s_2)}(f\bar{\Delta}+\bar{f}\Delta)[z_2\Omega(r,\alpha,z_2)-\Gamma(r,\alpha,z_2)], \tag{7.4.12}$$
$$\varphi_3(r,\alpha,z)=-\mathrm{i}\frac{Ps_3}{4\pi^2 a^2 c_{66}}(f\bar{\Delta}-\bar{f}\Delta)[z_3\Omega(r,\alpha,z_3)-\Gamma(r,\alpha,z_3)],$$

where $\Omega(r,\alpha,z)$ was given by Eq. (7.4.4) and $\Gamma(r,\alpha,z)$ is defined by (Hanson, 1992*a*)

$$\begin{aligned}\Gamma(r,\alpha,z)&=\int_0^{2\pi}\int_0^a \cosh^{-1}\left(\frac{a}{r_0}\right)Rr_0\,\mathrm{d}r_0\,\mathrm{d}\alpha_0\\&=2\pi\left[\frac{a}{12}(2a^2+6z^2+3r^2)\sin^{-1}\left(\frac{l_1}{r}\right)+\frac{1}{3}z^3\ln(z+\sqrt{z^2+r^2})\right.\\&\quad-\frac{1}{3}z^3\ln(l_2+\sqrt{l_2^2-r^2})-\frac{1}{9}(4z^2+r^2)\sqrt{z^2+r^2}\\&\quad\left.+\frac{1}{36}(21l_1^2+16l_2^2-12r^2-10a^2)\sqrt{l_2^2-a^2}\right].\end{aligned} \tag{7.4.13}$$

The displacements and stresses are correspondingly:

$$\begin{aligned}U=&\frac{HP}{a^2 s_1 s_2}\sum_{j=1}^{2}\frac{s_j}{\alpha_j-s_j}\left\{f\left[-a\sin^{-1}\left(\frac{l_{1j}}{r}\right)-\sqrt{l_{2j}^2-a^2}+z_j\ln(l_{2j}+\sqrt{l_{2j}^2-r^2})\right.\right.\\&\left.-z_j\ln(z_j+\sqrt{r^2+z_j^2})+\sqrt{r^2+z_j^2}\right]\\&\left.+\bar{f}e^{\mathrm{i}2\alpha}\left[-\frac{a^2z_j}{r^2}+\frac{(r^2+z_j^2)^{3/2}}{3r^2}+\frac{4a^2-3l_{1j}^2-l_{2j}^2}{3r^2}\sqrt{l_{2j}^2-a^2}\right]\right\}\\&-\frac{Ps_3}{2\pi a^2 c_{66}}\left\{f\left[-a\sin^{-1}\left(\frac{l_{13}}{r}\right)-\sqrt{l_{23}^2-a^2}+z_3\ln(l_{23}+\sqrt{l_{23}^2-r^2})\right.\right.\\&\left.-z_3(z_3+\sqrt{r^2+z_3^2})+\sqrt{r^2+z_3^2}\right]\\&\left.-\bar{f}e^{\mathrm{i}2\alpha}\left[-\frac{a^2z_3}{r^2}+\frac{(r^2+z_3^2)^{3/2}}{3r^2}+\frac{4a^2-3l_{13}^2-l_{23}^2}{3r^2}\sqrt{l_{23}^2-a^2}\right]\right\},\end{aligned}$$

(7.4.14*a*)

$$w=\frac{HP}{a^2 s_1 s_2}(\bar{f}\,\mathrm{e}^{\mathrm{i}\alpha}+f\,\mathrm{e}^{-\mathrm{i}\alpha})\sum_{j=1}^{2}\frac{s_j\alpha_j}{\alpha_j-s_j}\left[\frac{r}{2}\ln(l_{2j}+\sqrt{l_{2j}^2-r^2})\right.$$

$$\left.-\frac{r}{2}\ln(z_j+\sqrt{z_j^2+r^2})+\frac{a^2}{2r}-\frac{z_j\sqrt{r^2+z_j^2}}{2r}+\frac{(rl_{2j}-2al_{1j})\sqrt{l_{2j}^2-r^2}}{2r^2}\right], \tag{7.4.14b}$$

$$\sigma_1=\frac{2HPc_{66}}{a^2 s_1 s_2}(\bar{f}\,\mathrm{e}^{\mathrm{i}\alpha}+f\,\mathrm{e}^{-\mathrm{i}\alpha})\sum_{j=1}^{2}\frac{(k_{1j}-1)s_j}{\alpha_j-s_j}\left(\frac{\sqrt{l_{2j}^2-a^2}}{r}-\frac{\sqrt{r^2+z_j^2}}{r}\right), \tag{7.4.14c}$$

$$\sigma_2=\frac{2HPc_{66}}{a^2 s_1 s_2}\sum_{j=1}^{2}\frac{s_j}{\alpha_j-s_j}\left\{f\,\mathrm{e}^{\mathrm{i}\alpha}\left(\frac{\sqrt{r^2+z_j^2}}{r}-\frac{\sqrt{l_{2j}^2-a^2}}{r}\right)\right.$$

$$\left.+\bar{f}\,\mathrm{e}^{\mathrm{i}3\alpha}\left[(12l_{1j}^2+4l_{2j}^2-3r^2-16a^2)\frac{\sqrt{l_{2j}^2-a^2}}{3r^3}+\frac{4a^2z_j}{r^3}-\frac{4z_j^4+r^4+5r^2z_j^2}{3r^3\sqrt{r^2+z_j^2}}\right]\right\}$$

$$-\frac{Ps_3}{\pi a^2}\left[f\,\mathrm{e}^{\mathrm{i}\alpha}\left(\frac{\sqrt{r^2+z_3^2}}{r}-\frac{\sqrt{l_{23}^2-a^2}}{r}\right)\right.$$

$$\left.-\bar{f}\,\mathrm{e}^{\mathrm{i}3\alpha}\left[(12l_{13}^2+4l_{23}^2-3r^2-16a^2)\frac{\sqrt{l_{23}^2-a^2}}{3r^3}+\frac{4a^2z_3}{r^3}-\frac{4z_3^4+r^4+5r^2z_3^2}{3r^3\sqrt{r^2+z_3^2}}\right]\right\}, \tag{7.4.14d}$$

$$\sigma_z=\frac{P}{2\pi a^2(s_2-s_1)}(\bar{f}\,\mathrm{e}^{\mathrm{i}\alpha}+f\,\mathrm{e}^{-\mathrm{i}\alpha})\sum_{j=1}^{2}(-1)^{j+1}\left(\frac{\sqrt{l_{2j}^2-a^2}}{r}-\frac{\sqrt{r^2+z_j^2}}{r}\right), \tag{7.4.14e}$$

$$\tau_z=\frac{P}{2\pi a^2(s_2-s_1)}\sum_{j=1}^{2}(-1)^{j+1}s_j\left\{f[\ln(l_{2j}+\sqrt{l_{2j}^2-r^2})-\ln(z_j+\sqrt{r^2+z_j^2})]\right.$$

$$\left.+\bar{f}\,\mathrm{e}^{\mathrm{i}2\alpha}\left[\frac{(2a^2-l_{2j}^2)\sqrt{a^2-l_{1j}^2}}{ar^2}+\frac{z_j\sqrt{r^2+z_j^2}}{r^2}-\frac{a^2}{r^2}\right]\right\}$$

$$-\frac{P}{2\pi a^2}\left\{f[\ln(l_{23}+\sqrt{l_{23}^2-r^2})-\ln(z_3+\sqrt{r^2+z_3^2})]\right.$$

$$\left.-\bar{f}\,\mathrm{e}^{\mathrm{i}2\alpha}\left[\frac{(2a^2-l_{23}^2)\sqrt{a^2-l_{13}^2}}{ar^2}+\frac{z_3\sqrt{r^2+z_3^2}}{r^2}-\frac{a^2}{r^2}\right]\right\}. \tag{7.4.14f}$$

The resultant displacements and stresses due to the contact sliding of a rigid circular cone are the sums of Eqs. (7.4.5) and (7.4.14).

7.5 INCLINED CONTACT OF A CYLINDRICAL PUNCH WITH A HALF-SPACE

As illustrated in Fig. 7.10, a cylindrical punch of radius a is in contact with a transversely isotropic half-space. The punch is subjected to moments M_x about x-axis and M_y about y-axis. The rotation angles caused by the moments are b_x and b_y, respectively.

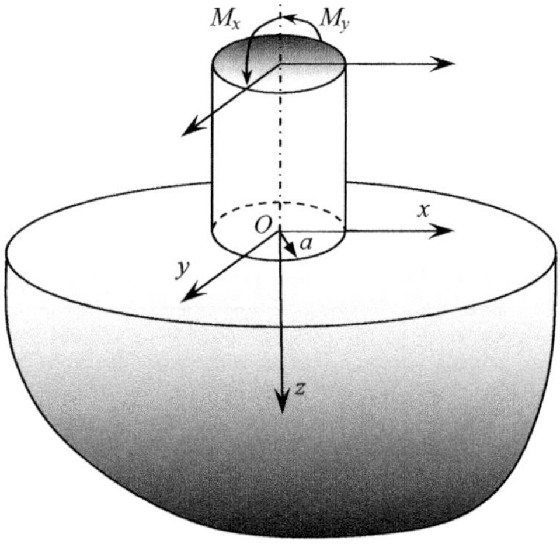

Fig. 7.7 A cylindrical punch in an inclined contact with a half-space.

For convenience, introduce the following complex moment and complex rotation angle:

$$M = M_x + \mathrm{i} M_y, \quad b = b_x + \mathrm{i} b_y. \tag{7.5.1}$$

M and b have a very simple relationship as shown by Fabrikant (1988), *i.e.*,

$$M = \frac{4a^3 b}{3\pi H}, \tag{7.5.2}$$

where H was defined by Eq. (7.1.37).

Assume that slip occurs between the punch and half-space surfaces with Coulomb friction. We will solve the problem in two steps, *i.e.*, by considering normal and tangential forces separately.

7.5.1 Contact with Normal Loading

The interface contact pressure in the present case can be described by (Fabrikant, 1988)

$$\sigma_z(r,\alpha,0) = -\frac{3r}{2\pi a^3}\frac{M_x \sin\alpha - M_y \cos\alpha}{\sqrt{a^2 - r^2}}, \quad 0 \le r \le a. \tag{7.5.3}$$

Again, by replacing the concentrated normal force N in Eq. (7.1.35) by

$$[3r_0(M_x \sin\alpha_0 - M_y \cos\alpha_0)/(2\pi a^3\sqrt{a^2 - r_0^2})]r_0\,\mathrm{d}r_0\,\mathrm{d}\alpha_0,$$

and carrying out integration over the contact area, $(0 \le r_0 \le a, 0 \le \alpha_0 \le 2\pi)$, we obtain the displacement functions as follows:

$$\begin{aligned}
\varphi_1(r,\alpha,z) &= \frac{3H}{2\pi a^3(\alpha_1 - s_1)}\mathrm{Y}(r,\alpha,z_1),\\
\varphi_2(r,\alpha,z) &= \frac{3H}{2\pi a^3(\alpha_2 - s_2)}\mathrm{Y}(r,\alpha,z_2),\\
\varphi_3(r,\alpha,z) &= 0,
\end{aligned} \tag{7.5.4}$$

where

$$\mathrm{Y}(r,\alpha,z) = \mathrm{Im}\left[\overline{M}\int_0^{2\pi}\int_0^a \frac{r_0\,\mathrm{e}^{\mathrm{i}\alpha_0}}{\sqrt{a^2 - r_0^2}}\ln(R+z)r_0\,\mathrm{d}r_0\,\mathrm{d}\alpha_0\right], \tag{7.5.5}$$

which can be expressed explicitly as (Fabrikant, 1988)

$$\Upsilon(r,\alpha,z)=\frac{\pi\mathrm{i}}{2}(Mr\,\mathrm{e}^{-\mathrm{i}\alpha}-\bar{M}r\,\mathrm{e}^{\mathrm{i}\alpha})\left[z\sin^{-1}\left(\frac{l_1}{r}\right)-\frac{2a^3}{3r^2}-\sqrt{a^2-l_1^2}\left(1-\frac{2a^2+l_1^2}{3r^2}\right)\right]. \tag{7.5.6}$$

Accordingly, the displacements and stresses can be obtained as

$$U=\frac{3H\mathrm{i}}{2a^3}\sum_{j=1}^{2}\frac{1}{\alpha_j-s_j}\left\{M\left[z_j\sin^{-1}\left(\frac{l_{1j}}{r}\right)-\sqrt{a^2-l_{1j}^2}\right]-\bar{M}\,\mathrm{e}^{\mathrm{i}2\alpha}\frac{2a^3-(l_{1j}^2+2a^2)\sqrt{a^2-l_{1j}^2}}{3r^2}\right\}, \tag{7.5.7a}$$

$$w=\frac{3H}{2a^3}(M_x y-M_y x)\sum_{j=1}^{2}\frac{\alpha_j}{\alpha_j-s_j}\left[\sin^{-1}\left(\frac{l_{1j}}{r}\right)-\frac{a\sqrt{l_{2j}^2-a^2}}{l_{2j}^2}\right], \tag{7.5.7b}$$

$$\sigma_1=-\frac{6Hc_{66}}{a}(M_x y-M_y x)\sum_{j=1}^{2}\frac{k_{1j}-1}{\alpha_j-s_j}\frac{\sqrt{a^2-l_{1j}^2}}{l_{2j}^2(l_{2j}^2-l_{1j}^2)}, \tag{7.5.7c}$$

$$\sigma_2=\frac{3Hc_{66}}{a^3}\mathrm{i}\mathrm{e}^{\mathrm{i}\alpha}\sum_{j=1}^{2}\frac{1}{\alpha_j-s_j}\left\{M\frac{al_{1j}\sqrt{a^2-l_{1j}^2}}{l_{2j}(l_{2j}^2-l_{1j}^2)}\right.$$
$$\left.-\bar{M}\,\mathrm{e}^{\mathrm{i}2\alpha}\left[\frac{(4l_{1j}^2+8a^2)\sqrt{a^2-l_{1j}^2}-8a^3}{3r^3}+\frac{al_{1j}\sqrt{a^2-l_{1j}^2}}{l_{2j}(l_{2j}^2-l_{1j}^2)}\right]\right\}, \tag{7.5.7d}$$

$$\sigma_z=\frac{3(M_x y-M_y x)s_1 s_2}{2\pi a^3(s_2-s_1)}\sum_{j=1}^{2}\frac{(-1)^j}{s_j}\frac{a^2\sqrt{a^2-l_{1j}^2}}{l_{2j}^2(l_{2j}^2-l_{1j}^2)}, \tag{7.5.7e}$$

$$\tau_z=\frac{3s_1 s_2\mathrm{i}}{4\pi a^3(s_2-s_1)}\sum_{j=1}^{2}(-1)^{j+1}\left\{\left[M\sin^{-1}\left(\frac{l_{1j}}{r}\right)-\frac{a\sqrt{l_{2j}^2-a^2}}{l_{2j}^2-l_{1j}^2}\right]+\bar{M}\,\mathrm{e}^{\mathrm{i}2\alpha}\frac{al_{1j}^2\sqrt{l_{2j}^2-a^2}}{l_{2j}^2(l_{2j}^2-l_{1j}^2)}\right\}. \tag{7.5.7f}$$

7.5.2 Contact with Tangential Loading

In this case, we use

$$\frac{3r_0(M_x \sin\alpha_0 - M_y \cos\alpha_0)f}{2\pi a^3\sqrt{a^2-r_0^2}} r_0\,\mathrm{d}r_0\,\mathrm{d}\alpha_0, \tag{7.5.8}$$

to replace the concentrated tangential force Q in Eq. (7.1.38) and then carry out an integration over the contact area, ($0 \le r_0 \le a, 0 \le \alpha_0 \le 2\pi$). The displacement functions can be found to be

$$\begin{aligned}
\varphi_1(r,\alpha,z) &= \frac{3H}{4\pi a^3 s_2(\alpha_1 - s_1)}(f\bar{\Delta} + \bar{f}\Delta)[z_1 \Upsilon(r,\alpha,z_1) - \Pi(r,\alpha,z_1)],\\
\varphi_2(r,\alpha,z) &= \frac{3H}{4\pi a^3 s_1(\alpha_2 - s_2)}(f\bar{\Delta} + \bar{f}\Delta)[z_2 \Upsilon(r,\alpha,z_2) - \Pi(r,\alpha,z_2)],\\
\varphi_3(r,\alpha,z) &= -\mathrm{i}\frac{3s_3}{8\pi^2 a^3 c_{66}}(f\bar{\Delta} - \bar{f}\Delta)[z_3 \Upsilon(r,\alpha,z_3) - \Pi(r,\alpha,z_3)],
\end{aligned} \tag{7.5.9}$$

where $\Upsilon(r,\alpha,z)$ was given by Eq. (7.5.6) and $\Pi(r,\alpha,z)$ is defined by

$$\Pi(r,\alpha,z) = \mathrm{Im}\left[\bar{M}\int_0^{2\pi}\int_0^a \frac{r_0\,\mathrm{e}^{\mathrm{i}\alpha_0}}{\sqrt{a^2-r_0^2}} R r_0\,\mathrm{d}r_0\,\mathrm{d}\alpha_0\right], \tag{7.5.10}$$

whose explicit form, when using the integration formula of Hanson (1994), can be written as

$$\begin{aligned}
\Pi(r,\alpha,z) = \frac{\pi\mathrm{i}}{2}(Mr\,\mathrm{e}^{-\mathrm{i}\alpha} - \bar{M}r\,\mathrm{e}^{\mathrm{i}\alpha})\Bigg[&\frac{1}{8}(r^2 + 4z^2 - 4a^2)\sin^{-1}\left(\frac{l_1}{r}\right)\\
&- (l_1^2 + 2l_2^2 - 3r^2/2)\frac{l_1\sqrt{r^2-l_1^2}}{4r^2}\Bigg].
\end{aligned} \tag{7.5.11}$$

The displacements and stresses are obtained by substituting the above displacement functions into Eqs. (2.2.56*a*) and (2.2.63*a*), *i.e.*,

$$U = \frac{3H\mathrm{i}}{8a^3 s_1 s_2} \sum_{j=1}^{2} \frac{s_j}{\alpha_j - s_j} \left((f\overline{M}\mathrm{e}^{\mathrm{i}\alpha} - fM\,\mathrm{e}^{-\mathrm{i}\alpha} - \bar{f}M\mathrm{e}^{\mathrm{i}\alpha}) \left[r\sin^{-1}\left(\frac{l_{1j}}{r}\right) - \frac{l_{1j}}{r}\sqrt{r^2 - l_{1j}^2} \right] \right.$$
$$\left. + 2\bar{f}\overline{M}\mathrm{e}^{\mathrm{i}3\alpha} \frac{1}{3r^3} \{4z_j[2a^3 - (2a^2 + l_{1j}^2)\sqrt{a^2 - l_{1j}^2}] - 3l_{1j}^3\sqrt{r^2 - l_{1j}^2}\} \right)$$
$$- \frac{3\mathrm{i}s_3}{16\pi a^3 c_{66}} \left((f\overline{M}\,\mathrm{e}^{\mathrm{i}\alpha} - fM\,\mathrm{e}^{-\mathrm{i}\alpha} - \bar{f}M\,\mathrm{e}^{\mathrm{i}\alpha}) \left[r\sin^{-1}\left(\frac{l_{13}}{r}\right) - \frac{l_{13}}{r}\sqrt{r^2 - l_{13}^2} \right] \right.$$
$$\left. - 2\bar{f}\overline{M}\mathrm{e}^{\mathrm{i}3\alpha} \frac{1}{3r^3} \{4z_3[2a^3 - (2a^2 + l_{13}^2)\sqrt{a^2 - l_{13}^2}] - 3l_{13}^3\sqrt{r^2 - l_{13}^2}\} \right), \tag{7.5.12a}$$

$$w = \frac{3H\mathrm{i}}{4a^3 s_1 s_2} \sum_{j=1}^{2} \frac{s_j\alpha_j}{\alpha_j - s_j} \left\{ (\bar{f}M - f\overline{M}) \left[z_j \sin^{-1}\left(\frac{l_{1j}}{r}\right) - \sqrt{a^2 - l_{1j}^2} \right] \right.$$
$$\left. + (fM\mathrm{e}^{-\mathrm{i}2\alpha} - \bar{f}\overline{M}\,\mathrm{e}^{\mathrm{i}2\alpha}) \frac{1}{3r^2} [2a^3 - (2a^2 + l_{1j}^2)\sqrt{a^2 - l_{1j}^2}\,] \right\}, \tag{7.5.12b}$$

$$\sigma_1 = \frac{3H\mathrm{i}c_{66}}{2a^3 s_1 s_2} \sum_{j=1}^{2} \frac{(k_{1j} - 1)s_j}{\alpha_j - s_j} \left\{ (\bar{f}M - f\overline{M}) \left[\sin^{-1}\left(\frac{l_{1j}}{r}\right) - \frac{a l_{2j}\sqrt{r^2 - l_{1j}^2}}{r(l_{2j}^2 - l_{1j}^2)} \right] \right.$$
$$\left. - (fM\,\mathrm{e}^{-\mathrm{i}2\alpha} - \bar{f}\overline{M}\,\mathrm{e}^{\mathrm{i}2\alpha}) \frac{l_{1j}^3\sqrt{r^2 - l_{1j}^2}}{r^2(l_{2j}^2 - l_{1j}^2)} \right\}, \tag{7.5.12c}$$

$$\sigma_2 = \frac{3H\mathrm{i}c_{66}}{2a^3 s_1 s_2} \sum_{j=1}^{2} \frac{s_j}{\alpha_j - s_j} \left((\bar{f}M\mathrm{e}^{\mathrm{i}2\alpha} - f\overline{M}\,\mathrm{e}^{\mathrm{i}2\alpha} - \bar{f}\overline{M}\,\mathrm{e}^{\mathrm{i}4\alpha}) \frac{l_{1j}^3\sqrt{r^2 - l_{1j}^2}}{r^2(l_{2j}^2 - l_{1j}^2)} \right.$$
$$- fM\left[\sin^{-1}\left(\frac{l_{1j}}{r}\right) - \frac{a l_{2j}\sqrt{r^2 - l_{1j}^2}}{r(l_{2j}^2 - l_{1j}^2)} \right]$$
$$\left. + \bar{f}\overline{M}\mathrm{e}^{\mathrm{i}4\alpha} \frac{1}{r^4} \{6l_{1j}^3\sqrt{r^2 - l_{1j}^2} - 8z_j[2a^3 - (2a^2 + l_{1j}^2)\sqrt{a^2 - l_{1j}^2}\,]\} \right)$$
$$+ \frac{3\mathrm{i}s_3}{4\pi a^3} \left((\bar{f}M\mathrm{e}^{\mathrm{i}2\alpha} + f\overline{M}\mathrm{e}^{\mathrm{i}2\alpha} - \bar{f}\overline{M}\,\mathrm{e}^{\mathrm{i}4\alpha}) \frac{l_{13}^3\sqrt{r^2 - l_{13}^2}}{r^2(l_{23}^2 - l_{13}^2)} \right.$$

$$+ fM\left[\sin^{-1}\left(\frac{l_{13}}{r}\right) - \frac{al_{23}\sqrt{r^2 - l_{13}^2}}{r(l_{23}^2 - l_{13}^2)}\right]$$

$$\left. + \bar{f}\bar{M}\,\mathrm{e}^{\mathrm{i}4\alpha}\frac{1}{r^4}\{6l_{13}^3\sqrt{r^2 - l_{13}^2} - 8z_3[2a^3 - (2a^2 + l_{13}^2)\sqrt{a^2 - l_{13}^2}\,]\}\right\}, \tag{7.5.12d}$$

$$\sigma_z = \frac{3\mathrm{i}}{8\pi a^3(s_2 - s_1)}\sum_{j=1}^{2}(-1)^{j+1}\left\{(\bar{f}M - f\bar{M})\left[\sin^{-1}\left(\frac{l_{1j}}{r}\right) - \frac{al_{2j}\sqrt{r^2 - l_{1j}^2}}{r(l_{2j}^2 - l_{1j}^2)}\right]\right.$$
$$\left. -(fM\,\mathrm{e}^{-\mathrm{i}2\alpha} - \bar{f}\bar{M}\,\mathrm{e}^{\mathrm{i}2\alpha})\frac{l_{1j}^3\sqrt{r^2 - l_{1j}^2}}{r^2(l_{2j}^2 - l_{1j}^2)}\right\}, \tag{7.5.12e}$$

$$\tau_z = \frac{3\mathrm{i}}{8\pi a^3(s_2 - s_1)}\sum_{j=1}^{2}(-1)^{j+1}s_j\left\{(fM\mathrm{e}^{-\mathrm{i}\alpha} - f\bar{M}\,\mathrm{e}^{\mathrm{i}\alpha} + \bar{f}M\,\mathrm{e}^{\mathrm{i}\alpha} - \bar{f}\bar{M}\,\mathrm{e}^{\mathrm{i}3\alpha})\frac{l_{1j}^2\sqrt{a^2 - l_{1j}^2}}{r(l_{2j}^2 - l_{1j}^2)}\right.$$
$$\left. + \bar{f}\bar{M}\,\mathrm{e}^{\mathrm{i}3\alpha}\frac{4}{3r^3}[2a^3 - (2a^2 + l_{1j}^2)\sqrt{a^2 - l_{1j}^2}\,]\right\}$$
$$+ \frac{3\mathrm{i}}{8\pi a^3}\left\{(f\bar{M}\mathrm{e}^{\mathrm{i}\alpha} - fM\mathrm{e}^{-\mathrm{i}\alpha} + \bar{f}M\,\mathrm{e}^{\mathrm{i}\alpha} - \bar{f}\bar{M}\,\mathrm{e}^{\mathrm{i}3\alpha})\frac{l_{13}^2\sqrt{a^2 - l_{13}^2}}{r(l_{23}^2 - l_{13}^2)}\right.$$
$$\left. + \bar{f}\bar{M}\,\mathrm{e}^{\mathrm{i}3\alpha}\frac{4}{3r^3}[2a^3 - (2a^2 + l_{13}^2)\sqrt{a^2 - l_{13}^2}\,]\right\}. \tag{7.5.12f}$$

The total displacements and stresses due to the contact sliding of a tilted cylindrical punch are the sums of Eqs. (7.5.7) and (7.5.12).

7.6 DISCUSSIONS ON SOLUTIONS FOR FRICTIONAL CONTACT

In the previous sections, we presented analytical solutions for four frictional contact problems. Coulomb friction law was adopted to measure the tangential force induced by the friction. The full-space field variables were all expressed in terms of elementary functions. These solutions are now discussed from a physically more rigor point of view.

In the end of **Section 7.2**, we pointed out that the frictional force, in general, alters both the size and shape of the contact area. As an exception, Johnson (1985, §7.1) showed that when the two isotropic elastic bodies in contact are of the same elastic properties, then any shear force transferring between the two bodies will not lead to any change of the distribution of contact pressure; and hence the shape and size of the contact area are only determined by the normal force and the shapes of the two bodies in contact themselves. For transversely isotropic materials, we readily see from Eq. (4.1.48) and the discussion of Johnson (1985) that if the material properties of the two bodies are the same, the frictional force also will not affect the size and shape of the contact area. There is another exception for isotropic materials, however, which was not mentioned by Johnson (1985). Equation (3.75*c*) of Johnson (1985) shows that when the materials of the two bodies are incompressible (Poisson's ratio $\nu = 0.5$), then a tangential surface force will not induce any normal displacement at $z = 0$, and hence the shape and size of the contact area will not vary with the frictional force. For transversely isotropic materials, the vanishing of surface normal displacement due to tangential force requires, from Eq. (4.1.48),

$$c_{11} - c_{13} s_1 s_2 = 0 . \tag{7.6.1}$$

Since $s_1 s_2 = \sqrt{c_{11} / c_{33}}$ (see **Section 2.2**), this equation is equivalent to

$$\sqrt{c_{11} c_{33}} - c_{13} = 0 . \tag{7.6.2}$$

Equation (7.6.2) conflicts with the inequality derived from the energy consideration as shown in **Section 2.2**. Note that, we can only obtain $c_{11} + c_{12} = 2c_{13}$ and $c_{11} = c_{33}$ by taking account of the incompressibility of transversely isotropic materials, *i.e.* $\varepsilon_x + \varepsilon_y + \varepsilon_z = 0$. Thus, for contact problems of rigid punches considered in **Sections 7.3, 7.4** and **7.5**, the frictional force will inevitably affect the size and shape of the contact region.

For isotropic materials, it also has been found that the tangential displacements on the half-space surface are generally not aligned with the direction of the frictional force (Johnson, 1985). Here we follow Hanson (1994) to examine the mechanism of such a frictional force effect when the materials are transversely isotropic.

Consider the frictional contact of a sphere. When $z \to 0$, *i.e.*, at the surface of the half-space, l_1 of Eq. (7.2.5) and l_{1j} of Eq. (7.2.7) approach min(*a*, *r*), while l_2 of Eq. (7.2.5) and l_{2j} of Eq. (7.2.7) approach max(*a*, *r*). Thus in the contact area, according to Eqs. (7.2.12*a*, *b*), the surface displacements can be obtained as

$$U=\frac{3\pi fP}{8a^3}G_1\left(a^2-\frac{r^2}{2}\right)+\frac{3\pi P\bar{f}}{32a^3}G_2r^2\,\mathrm{e}^{\mathrm{i}2\alpha},\quad (r\le a),\tag{7.6.3}$$

$$w=\frac{HP\vartheta}{a^3}(f_x\cos\alpha+f_y\sin\alpha)\frac{a^3-(a^2-r^2)^{3/2}}{r},\quad (r\le a),\tag{7.6.4}$$

where

$$G_1=\beta+\frac{H}{s_1s_2},\quad G_2=\beta-\frac{H}{s_1s_2},\quad \beta=\frac{s_3}{2\pi c_{66}},$$

$$\vartheta=\frac{1}{s_1s_2}\sum_{j=1}^{2}\frac{\alpha_js_j}{\alpha_j-s_j}=\frac{\sqrt{c_{11}c_{33}}-c_{13}}{c_{11}(s_1+s_2)}s_1s_2.\tag{7.6.5}$$

It is known from the discussion in **Section 2.2** that s_1s_2 and s_1+s_2 are both real and greater than zero, and H, defined in Eq. (7.1.37), is also real and greater than zero. Thus, G_1, β and ϑ are all positive real parameters, while G_2 is just a real parameter. Having these in mind, we know from Eq. (7.6.3) that at the contact interface, the direction of the tangential displacement is generally inconsistent with that of the tangential force, except for the special case of $G_2=0$. We can obtain similar formulations

$$U=\frac{G_1P}{a^2}f\left(\frac{\pi a}{2}-r\right)+\frac{G_2P}{3a^2}r\bar{f}\,\mathrm{e}^{\mathrm{i}2\alpha},\quad (0\le r\le a),\tag{7.6.6}$$

$$w=\frac{HP\vartheta}{2a^2}(\bar{f}\,\mathrm{e}^{\mathrm{i}\alpha}+f\,\mathrm{e}^{-\mathrm{i}\alpha})\left[r\ln\left(\frac{a+\sqrt{a^2-r^2}}{r}\right)+\frac{a^2}{r}-\frac{a\sqrt{a^2-r^2}}{r}\right],\quad (0\le r\le a),\tag{7.6.7}$$

for the frictional contact of a cone considered in **Section 7.4**, and

$$U=\frac{3G_1\pi}{8a^3}f(yM_x-xM_y)-\frac{3\mathrm{i}\pi G_2}{16a^3}\bar{f}M(x+\mathrm{i}y),\quad (0\le r\le a),\tag{7.6.8}$$

$$w=\frac{3H\mathrm{i}\vartheta}{4a^3}\left\{-(\bar{f}M-f\bar{M})\sqrt{a^2-r^2}+(fM\mathrm{e}^{-\mathrm{i}2\alpha}-\bar{f}\bar{M}\mathrm{e}^{\mathrm{i}2\alpha})\times\frac{1}{3r^2}[2a^3-(2a^2+r^2)\sqrt{a^2-r^2}]\right\},\quad 0\le r\le a, \tag{7.6.9}$$

for the frictional contact of an inclined cylindrical punch considered in **Section 7.5**.

This inconsistence means that the solutions for frictional contact problems derived in **Sections 7.2**, **7.4** and **7.5** somehow deviates from the Coulomb friction law and hence is not exact, except when $G_2=0$. This actually follows the fact that a surface tangential concentrated force, say T in the x-direction, leads to not only the surface tangential displacement in the x-direction but also the displacement in the y-direction, as seen from Eq. (4.1.43). In our analysis, the effect of the secondary frictional force (in the y-direction) due to the movement of surface particle in the y-direction induced by T has been neglected. Thus, for frictional contact problems, the tangential displacement is generally inconsistent with that of the tangential force. In certain particular cases, even when $G_2\neq 0$, however, the situation may change. For example, for the upright cylindrical flat punch considered in **Section 7.3**, we obtain from Eq. (7.3.13*a*,*b*)

$$U=\frac{fP\pi}{4a}G_1,\quad (r\le a), \tag{7.6.10}$$

$$w=\frac{HP\vartheta}{a}(f_x\cos\alpha+f_y\sin\alpha)\frac{a-\sqrt{a^2-r^2}}{r},\quad (r\le a). \tag{7.6.11}$$

It can be seen from Eq. (7.6.10) that the only complex number at the right hand side of the equation is f. Hence the complex tangential displacement can be written as $U=u+\mathrm{i}v=C(f_x+\mathrm{i}f_y)$, where C is a real constant, implying that the directions of the tangential displacements and tangential forces are the same. Thus the solution for the frictional contact of an upright cylindrical punch is consistent with the physical essence of the Coulomb friction. It is actually because the frictional force distributed in the contact region in the x-direction or y-direction yields the x-component or y-component displacement only at every point in the contact region due to the counteraction of displacement in the other direction by forces at all points.

The influence of the frictional force of contact on the surface normal displacement also can be seen from Eqs. (7.6.4), (7.6.7), (7.6.9) and (7.6.11). It is obvious that w vanishes everywhere for all four contact problems when $\vartheta=0$, *i.e.* when Eq. (7.6.2) holds true. However ϑ can not be zero for a general transversely isotropic material as discussed earlier. For an isotropic material, we have $\vartheta=(1-2\nu)/(2-2\nu)$, and hence $\vartheta=0$ corresponds to $\nu=1/2$, which implies an incompressible material.

8

BENDING, VIBRATION AND STABILITY OF PLATES

This chapter studies the bending, vibration and stability of transversely isotropic rectangular and circular plates using the general solution obtained in **Chapter 2** based on the three-dimensional theory of elasticity. There is a detailed discussion on the application of the state-space method to the free vibration of laminated rectangular and circular plates.

8.1 GENERAL SOLUTION METHOD

8.1.1 Rectangular Plates

Consider the bending, vibration and stability of a transversely isotropic rectangular plate of length a, width b and thickness h, as shown in Fig. 8.1. The plane of isotropy is parallel to the middle plane $z=0$. In the absence of body forces and temperature change, the equilibrium equations in terms of displacements, Eq. (2.2.1), can be re-written as

$$\begin{aligned}
&\left(c_{11}\frac{\partial^2}{\partial x^2}+c_{66}\frac{\partial^2}{\partial y^2}+c_{44}\frac{\partial^2}{\partial z^2}\right)u+(c_{12}+c_{66})\frac{\partial^2 v}{\partial x\partial y}+(c_{13}+c_{44})\frac{\partial^2 w}{\partial x\partial z}=Ku\,,\\
&(c_{12}+c_{66})\frac{\partial^2 u}{\partial x\partial y}+\left(c_{66}\frac{\partial^2}{\partial x^2}+c_{11}\frac{\partial^2}{\partial y^2}+c_{44}\frac{\partial^2}{\partial z^2}\right)v+(c_{13}+c_{44})\frac{\partial^2 w}{\partial y\partial z}=Kv\,,\\
&(c_{13}+c_{44})\frac{\partial^2 u}{\partial x\partial z}+(c_{13}+c_{44})\frac{\partial^2 v}{\partial y\partial z}+\left(c_{44}\Lambda+c_{33}\frac{\partial^2}{\partial z^2}\right)w=Kw\,,
\end{aligned}\tag{8.1.1}$$

where $c_{66}=(c_{11}-c_{12})/2$ and Λ is the two-dimensional Laplace operator defined by Eq. (2.2.9). For bending, vibration and stability problems, the operator K in Eq. (8.1.1) is 0,

$\rho\partial^2/\partial t^2$ and $T_1\partial^2/\partial x^2 + T_2\partial^2/\partial y^2$, respectively, where T_1 and T_2 are the uniform compressive stresses applied on the edges of the rectangular plate along x and y directions.

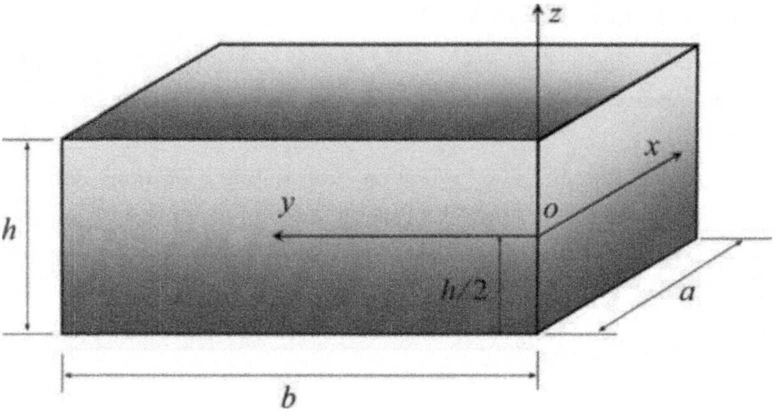

Fig. 8.1 The geometry of a rectangular plate and coordinates.

If the plate is simply supported on all its four edges, we have

$$\begin{aligned} &\sigma_x = v = w = 0, \quad \text{at } x = 0, a, \\ &\sigma_y = u = w = 0, \quad \text{at } y = 0, b. \end{aligned} \tag{8.1.2}$$

According to Eq. (2.2.34), we can take

$$\begin{aligned} u &= -\frac{\partial\varphi}{\partial y} - \frac{\partial^2 F}{\partial x \partial z}, \\ v &= \frac{\partial\varphi}{\partial x} - \frac{\partial^2 F}{\partial y \partial z}, \\ w &= \left(a_0\Lambda + b_0\frac{\partial^2}{\partial z^2} - \frac{b_0}{c_{44}}K \right) F, \end{aligned} \tag{8.1.3}$$

where $a_0 = c_{11}/(c_{13} + c_{44})$ and $b_0 = c_{44}/(c_{13} + c_{44})$. Then Eq. (8.1.1) yields

$$\left(c_{66}\Lambda + c_{44}\frac{\partial^2}{\partial z^2} - K \right)\varphi = 0, \tag{8.1.4}$$

$$\left\{ c_{11}c_{44}\Lambda^2 + [c_{11}c_{33} + c_{44}^2 - (c_{13} + c_{44})^2]\Lambda\frac{\partial^2}{\partial z^2} + c_{33}c_{44}\frac{\partial^4}{\partial z^4} - \left[(c_{11} + c_{44})\Lambda + (c_{33} + c_{44})\frac{\partial^2}{\partial z^2} \right] K + K^2 \right\} F = 0. \tag{8.1.5}$$

To satisfy the boundary conditions in Eq. (8.1.2), we assume[1]

$$\begin{aligned} \varphi(x,y,z,t) &= h^2 \sum_{m=1}^{\infty}\sum_{n=1}^{\infty} \psi_{mn}(\zeta)\cos(m\pi\xi)\cos(n\pi\eta)\exp(\mathrm{i}\,\omega t), \\ F(x,y,z,t) &= h^2 \sum_{m=1}^{\infty}\sum_{n=1}^{\infty} P_{mn}(\zeta)\sin(m\pi\xi)\sin(n\pi\eta)\exp(\mathrm{i}\,\omega t), \end{aligned} \tag{8.1.6}$$

where $\xi = x/a$, $\eta = y/b$, and $\zeta = z/h$ are dimensionless coordinates and m and n are integers. For free vibration, ω is the circular frequency, while for bending and stability, $\omega = 0$. Substituting this equation into Eqs. (8.1.4) and (8.1.5) and making use of the orthogonal property of trigonometric functions, for any pair of (m,n) we have

$$\psi''_{mn} - [(c_{66}/c_{44})k - \Omega^2]\psi_{mn} = 0, \tag{8.1.7}$$

$$P''''_{mn} - g_1 P''_{mn} + g_2 P_{mn} = 0, \tag{8.1.8}$$

where the prime denotes a differentiation with respect to ζ, $k = k_1^2 + k_2^2$, $k_1 = m\pi h/a$, $k_2 = n\pi h/b$, and

$$\begin{aligned} g_1 &= (c_{11}c_{33} - 2c_{13}c_{44} - c_{13}^2)k/(c_{33}c_{44}) - \Omega^2(c_{33} + c_{44})/c_{33}, \\ g_2 &= [c_{11}k^2 - (c_{11} + c_{44})k\Omega^2 + \Omega^4 c_{44}]/c_{33}. \end{aligned} \tag{8.1.9}$$

In these equations, $\Omega^2 = 0$ for bending, $\Omega^2 = \omega^2 h^2 \rho / c_{44}$ for free vibration and $\Omega^2 = (T_1 k_1^2 + T_2 k_2^2)/c_{44}$ for stability. The solutions to Eqs. (8.1.7) and (8.1.8) are

---

[1] In the expression for φ, either m or n can be zero, corresponding to a particular in-plane mode, as will be discussed later.

$$\psi_{mn}(\zeta)=A_3\cosh(\lambda_3\zeta)+B_3\sinh(\lambda_3\zeta), \tag{8.1.10}$$

$$P_{mn}(\zeta)=A_1\cosh(\lambda_1\zeta)+A_2\cosh(\lambda_2\zeta)+B_1\sinh(\lambda_1\zeta)+B_2\sinh(\lambda_2\zeta), \tag{8.1.11}$$

where

$$\lambda_3=\sqrt{(c_{66}/c_{44})k-\Omega^2},\ \lambda_{1,2}=\sqrt{(g_1\mp\sqrt{g_1^2-4g_2})/2}. \tag{8.1.12}$$

Hence, Eqs. (8.1.3) and (8.1.6) yield

$$\begin{aligned}
u&=h\sum_{m=1}^{\infty}\sum_{n=1}^{\infty}(k_2\psi_{mn}-k_1P'_{mn})\cos(m\pi\xi)\sin(n\pi\eta)\exp(\mathrm{i}\,\omega t),\\
v&=h\sum_{m=1}^{\infty}\sum_{n=1}^{\infty}(-k_1\psi_{mn}-k_2P'_{mn})\sin(m\pi\xi)\cos(n\pi\eta)\exp(\mathrm{i}\,\omega t),\\
w&=h\sum_{m=1}^{\infty}\sum_{n=1}^{\infty}[b_0P''_{mn}-(a_0k-b_0\Omega^2)P_{mn}]\sin(m\pi\xi)\sin(n\pi\eta)\exp(\mathrm{i}\,\omega t).
\end{aligned} \tag{8.1.13}$$

Using the constitutive relations for transversely isotropic materials, Eq. (1.2.33), we obtain the stresses as

$$\begin{aligned}
\sigma_x&=\sum_{m=1}^{\infty}\sum_{n=1}^{\infty}\{c_{13}b_0P'''_{mn}+[c_{13}(b_0\Omega^2-a_0k)+c_{11}k_1^2+c_{12}k_2^2]P'_{mn}+k_1k_2(c_{12}-c_{11})\psi_{mn}\}\\
&\qquad\times\sin(m\pi\xi)\sin(n\pi\eta)\exp(\mathrm{i}\,\omega t),\\
\sigma_y&=\sum_{m=1}^{\infty}\sum_{n=1}^{\infty}\{c_{13}b_0P'''_{mn}+[c_{13}(b_0\Omega^2-a_0k)+c_{12}k_1^2+c_{11}k_2^2]P'_{mn}-k_1k_2(c_{12}-c_{11})\psi_{mn}\}\\
&\qquad\times\sin(m\pi\xi)\sin(n\pi\eta)\exp(\mathrm{i}\,\omega t),\\
\sigma_z&=\sum_{m=1}^{\infty}\sum_{n=1}^{\infty}\{c_{33}b_0P'''_{mn}+[c_{33}b_0\Omega^2+(c_{13}-c_{33}a_0)k]P'_{mn}\}\sin(m\pi\xi)\sin(n\pi\eta)\exp(\mathrm{i}\,\omega t),\\
\tau_{yz}&=c_{44}\sum_{m=1}^{\infty}\sum_{n=1}^{\infty}[k_2(b_0-1)P''_{mn}+k_2(b_0\Omega^2-a_0k)P_{mn}-k_1\psi'_{mn}]\\
&\qquad\times\sin(m\pi\xi)\cos(n\pi\eta)\exp(\mathrm{i}\,\omega t),\\
\tau_{zx}&=c_{44}\sum_{m=1}^{\infty}\sum_{n=1}^{\infty}[k_1(b_0-1)P''_{mn}+k_1(b_0\Omega^2-a_0k)P_{mn}+k_2\psi'_{mn}]\\
&\qquad\times\cos(m\pi\xi)\sin(n\pi\eta)\exp(\mathrm{i}\,\omega t),
\end{aligned}$$

$$\tau_{xy}=c_{66}\sum_{m=1}^{\infty}\sum_{n=1}^{\infty}[(k_2^2-k_1^2)\psi_{mn}-2k_1k_2P'_{mn}]\cos(m\pi\xi)\cos(n\pi\eta)\exp(\mathrm{i}\,\omega t)\,. \tag{8.1.14}$$

(1) Bending

As mentioned earlier, we have $\Omega^2=0$ in this case and the eigenvalues λ_i $(i=1,2,3)$ can be determined by Eq. (8.1.12). Thus the six unknown constants, A_i and B_i $(i=1,2,3)$, can be determined by the boundary conditions at the upper and lower surfaces of the plate.

When the plate is subjected to normal loads $q_1(x,y)$ and $q_2(x,y)$ on its upper and lower surfaces, respectively, we have

$$\begin{aligned}&\sigma_z=-q_1(x,y)\,,\quad \tau_{zx}=\tau_{yz}=0\,,\quad (\zeta=-1/2),\\&\sigma_z=-q_2(x,y)\,,\quad \tau_{zx}=\tau_{yz}=0\,,\quad (\zeta=1/2).\end{aligned} \tag{8.1.15}$$

Expanding $q_1(x,y)$ and $q_2(x,y)$ into the series of double sine functions, *i.e.*,

$$\begin{aligned}&q_1(x,y)=c_{44}\sum_{m=1}^{\infty}\sum_{n=1}^{\infty}a_{mn}^1\sin(m\pi\xi)\sin(n\pi\eta),\\&q_2(x,y)=c_{44}\sum_{m=1}^{\infty}\sum_{n=1}^{\infty}a_{mn}^2\sin(m\pi\xi)\sin(n\pi\eta),\end{aligned} \tag{8.1.16}$$

where

$$a_{mn}^i=\frac{4}{c_{44}}\int_0^1\int_0^1 q_i(\xi,\eta)\sin(m\pi\xi)\sin(n\pi\eta)\,\mathrm{d}\xi\,\mathrm{d}\eta\,,\quad (i=1,2), \tag{8.1.17}$$

and substituting the third, fourth and fifth equations of Eq. (8.1.14) into Eq. (8.1.15), then for any pair of (m,n), we get

$$\begin{aligned}&c_{33}b_0P'''_{mn}(-1/2)+(c_{13}-c_{33}a_0)kP'_{mn}(-1/2)=-c_{44}a_{mn}^1\,,\\&k_2(b_0-1)P''_{mn}(-1/2)-a_0kk_2P_{mn}(-1/2)-k_1\psi'_{mn}(-1/2)=0\,,\\&k_1(b_0-1)P''_{mn}(-1/2)-a_0kk_1P_{mn}(-1/2)+k_2\psi'_{mn}(-1/2)=0\,,\\&c_{33}b_0P'''_{mn}(1/2)+(c_{13}-c_{33}a_0)kP'_{mn}(1/2)=-c_{44}a_{mn}^2\,,\end{aligned} \tag{8.1.18}$$

$$k_2(b_0-1)P''_{mn}(1/2)-a_0kk_2P_{mn}(1/2)-k_1\psi'_{mn}(1/2)=0\,,$$
$$k_1(b_0-1)P''_{mn}(1/2)-a_0kk_1P_{mn}(1/2)+k_2\psi'_{mn}(1/2)=0\,.$$

With these, the second, third, fifth and sixth equations of Eq. (8.1.18) immediately lead to

$$\psi'_{mn}(-1/2)=\psi'_{mn}(1/2)=0\,, \tag{8.1.19}$$

$$\begin{aligned}&(b_0-1)P''_{mn}(-1/2)-a_0kP_{mn}(-1/2)=0,\\&(b_0-1)P''_{mn}(1/2)-a_0kP_{mn}(1/2)=0.\end{aligned} \tag{8.1.20}$$

Using Eq. (8.1.10) and the property of the hyperbolic functions, Eq. (8.1.19) leads to

$$A_3=B_3=0\,, \tag{8.1.21}$$

which implies that for bending we always have $\varphi=0$. The substitution of P_{mn} in Eq. (8.1.11) into Eq. (8.1.20) and the first and fourth equations of Eq. (8.1.18) gives us the four inhomogeneous, linear algebraic equations with four unknowns A_i, B_i $(i=1,2)$ to be determined.

(2) Free vibration [2]

In this case, the upper and lower surfaces are traction-free, thus

$$\sigma_z=\tau_{zx}=\tau_{yz}=0\,,\qquad(\zeta=\pm1/2)\,. \tag{8.1.22}$$

Equation (8.1.14) then gives rise to

$$\begin{aligned}&b_0P'''_{mn}(-1/2)+[b_0\Omega^2+(c_{13}/c_{33}-a_0)k]P'_{mn}(-1/2)=0\,,\\&k_2(b_0-1)P''_{mn}(-1/2)+k_2(b_0\Omega^2-a_0k)P_{mn}(-1/2)-k_1\psi'_{mn}(-1/2)=0\,,\\&k_1(b_0-1)P''_{mn}(-1/2)+k_1(b_0\Omega^2-a_0k)P_{mn}(-1/2)+k_2\psi'_{mn}(-1/2)=0\,,\\&b_0P'''_{mn}(1/2)+[b_0\Omega^2+(c_{13}/c_{33}-a_0)k]P'_{mn}(1/2)=0\,,\end{aligned} \tag{8.1.23}$$

[2] The analysis mainly follows the work of Wang and He (1986). Nosier *et al.* (1993) presented a slightly different derivation by introducing a function $\Phi=\partial u/\partial y-\partial v/\partial x$ to simplify Eq. (8.1.1) directly. It can be seen from Eq. (8.1.3) that $\Phi=-\Lambda\varphi$ so that Φ also satisfies Eq. (8.1.4).

$$k_2(b_0-1)P''_{mn}(1/2)+k_2(b_0\Omega^2-a_0k)P_{mn}(1/2)-k_1\psi'_{mn}(1/2)=0,$$
$$k_1(b_0-1)P''_{mn}(1/2)+k_1(b_0\Omega^2-a_0k)P_{mn}(1/2)+k_2\psi'_{mn}(1/2)=0.$$

Hence,

$$\psi'_{mn}(\zeta_i)=A_3\lambda_3\sinh(\lambda_3\zeta_i)+B_3\lambda_3\cosh(\lambda_3\zeta_i)=0, \tag{8.1.24}$$

$$\begin{aligned}&b_0P'''_{mn}(\zeta_i)+[b_0\Omega^2+(c_{13}/c_{33}-a_0)k]P'_{mn}(\zeta_i)\\&\quad=G_1(\lambda_1)[A_1\sinh(\lambda_1\zeta_i)+B_1\cosh(\lambda_1\zeta_i)]+G_1(\lambda_2)[A_2\sinh(\lambda_2\zeta_i)+B_2\cosh(\lambda_2\zeta_i)]=0,\\&(b_0-1)P''_{mn}(\zeta_i)+(b_0\Omega^2-a_0k)P_{mn}(\zeta_i)\\&\quad=G_2(\lambda_1)[A_1\cosh(\lambda_1\zeta_i)+B_1\sinh(\lambda_1\zeta_i)]+G_2(\lambda_2)[A_2\cosh(\lambda_2\zeta_i)+B_2\sinh(\lambda_2\zeta_i)]=0,\end{aligned} \tag{8.1.25}$$

where $i=1,\ 2$, $\zeta_1=-1/2$, $\zeta_2=1/2$, and

$$\begin{aligned}G_1(\lambda)&=b_0\lambda^3+[b_0\Omega^2+(c_{13}/c_{33}-a_0)k]\lambda,\\G_2(\lambda)&=(b_0-1)\lambda^2+b_0\Omega^2-a_0k.\end{aligned} \tag{8.1.26}$$

To have a nontrivial solution, the coefficient determinants of Eqs. (8.1.24) and (8.1.25) should vanish, which gives rise to

$$\lambda_3^2\sinh(\lambda_3)=0, \tag{8.1.27}$$

$$\begin{aligned}&[G_{12}\sinh(\lambda_1/2)\cosh(\lambda_2/2)-G_{21}\sinh(\lambda_2/2)\cosh(\lambda_1/2)]\\&\quad\times[G_{12}\cosh(\lambda_1/2)\sinh(\lambda_2/2)-G_{21}\cosh(\lambda_2/2)\sinh(\lambda_1/2)]=0,\end{aligned} \tag{8.1.28}$$

where $G_{12}=G_1(\lambda_1)G_2(\lambda_2)$ and $G_{21}=G_1(\lambda_2)G_2(\lambda_1)$. Clearly, $\lambda_3=0$ is a root of $\sinh(\lambda_3)=0$, thus these two eigen-equations contain three independent factors, the mechanics meanings of which are to be discussed in the following.

(*a*) For

$$\sinh(\lambda_3)=0, \tag{8.1.29}$$

we obtain

$$\lambda_3 = p\pi \mathrm{i} , \tag{8.1.30}$$

where p is an integer. This equation gives

$$\Omega = \sqrt{(c_{66} / c_{44})k + p^2\pi^2} , \tag{8.1.31}$$

and the corresponding vibration mode[3]

$$\begin{aligned} u &= hk_2\psi_{mn} \cos(m\pi\xi)\sin(n\pi\eta)\exp(\mathrm{i}\omega t), \\ v &= -hk_1\psi_{mn} \sin(m\pi\xi)\cos(n\pi\eta)\exp(\mathrm{i}\omega t), \\ w &= 0, \end{aligned} \tag{8.1.32}$$

in which only a typical term is given, and

$$\psi_{mn} = \begin{cases} A_3 \cosh(\lambda_3\zeta) , & p \text{ is even,} \\ B_3 \sinh(\lambda_3\zeta) , & p \text{ is odd.} \end{cases} \tag{8.1.33}$$

It can be seen that the volume strain $e = \partial u / \partial x + \partial v / \partial y + \partial w / \partial z = 0$. This kind of vibration corresponds to the *thickness-shear* motion described by Mindlin (1951*b*). The frequency obtained by the Mindlin plate theory is

$$\Omega = \sqrt{(c_{66} / c_{44})k + 12k_t^2} , \tag{8.1.34}$$

where k_t^2 is the *modified shear factor*. If $k_t^2 = \pi^2 / 12$ and $p = 1$, Eq. (8.1.34) becomes identical to Eq. (8.1.31).

(*b*) For

$$G_{12} \sinh(\lambda_1 / 2)\cosh(\lambda_2 / 2) - G_{21} \sinh(\lambda_2 / 2)\cosh(\lambda_1 / 2) = 0 , \tag{8.1.35}$$

$A_3 = B_3 = B_1 = B_2 = 0$, *i.e.*

[3] As mentioned before, m or n can be zero (not both, otherwise no deformation occurs). In fact, the case $m = 0$ corresponds to $v = 0$ and u dependent on y only, while $n = 0$ corresponds to $u = 0$ and v dependent on x only.

$$\psi_{mn}=0,\qquad P_{mn}=A_1\cosh(\lambda_1\zeta)+A_2\cosh(\lambda_2\zeta). \tag{8.1.36}$$

Substituting this equation into Eq. (8.1.13), we find

$$u(\zeta)=-u(-\zeta),\quad v(\zeta)=-v(-\zeta),\quad w(\zeta)=w(-\zeta), \tag{8.1.37}$$

which indicates that the vibration is anti-symmetric with respect to the middle plane of the plate, $\zeta=0$.

(*c*) For

$$G_{21}\sinh(\lambda_1/2)\cosh(\lambda_2/2)-G_{12}\sinh(\lambda_2/2)\cosh(\lambda_1/2)=0, \tag{8.1.38}$$

we have $A_3=B_3=A_1=A_2=0$, *i.e.*

$$\psi_{mn}=0,\qquad P_{mn}=B_1\sinh(\lambda_1\zeta)+B_2\sinh(\lambda_2\zeta). \tag{8.1.39}$$

Thus Eq. (8.1.13) leads to

$$u(\zeta)=u(-\zeta),\quad v(\zeta)=v(-\zeta),\quad w(\zeta)=-w(-\zeta). \tag{8.1.40}$$

In this case, the vibration is symmetric with respect to the middle plane. This type of vibration cannot be predicted by any plate bending theories since we have $w(0)=0$, *i.e.* the middle plane doesn't bend during vibration.

We refer to the vibration described by Eq. (8.1.33) as the *first class of vibration*, while that by Eq. (8.1.36) or Eq. (8.1.39) as the *second class of vibration*[4]. The first class is characterized by the absence of transverse displacement and the volume strain. For the second class of vibration, though all three components of displacement do not vanish, the z-component of rotation does vanish, *i.e.* $(\partial v/\partial x-\partial u/\partial y)/2=0$.

The numerical investigation by Wang and He (1986) showed that the effect of the ratio of in-plane to transverse moduli, $\kappa_1=E/E'$, on the frequency is negligible when

[4] Lamb (1882) found that the free oscillation of isotropic spheres could be classified into two classes: the first class corresponds to an equivoluminal motion and is characterized by the absence of a radial displacement component, while, for the second class, the displacement has in general, both transverse and radial components, but the rotation has no radial component. When the radius of the sphere tends to infinity, we obtain two corresponding classes of vibrations of a plate. The transverse isotropy of materials does not affect this classification as shown here.

k is small (a large k means a higher-order vibration mode and *vice versa*). Hence, the assumption that the transverse normal stress has no effect on the deformation, which has been widely adopted in most plate theories, is reasonable for small wave numbers n and m [cf. Eq. (1.2.35)]. When k is large, however, the effect of κ_1 becomes considerable, indicating that the transverse normal stress becomes significant in the calculation of deformation of the plate and its effect should be taken into consideration. Moreover, when k is large but the ratio of the in-plane and transverse shear moduli, $\kappa_2 = G/G'$, is less than 0.75, the lowest natural frequency of the plate is almost equal to that of the thickness-shear vibration mode. This phenomenon was also observed in the study of vibration of anisotropic beams (Sayir, 1980).

(3) Stability

We have $\Omega^2 = (T_1 k_1^2 + T_2 k_2^2)/c_{44}$ in the stability analysis of a plate subjected to uniform compressive stresses on edges. The detailed solution procedure is the same as that for vibration (Wang, 1986).

The three-dimensional exact analyses available are usually for rectangular plates simply supported on all the four edges (Pagano, 1969, 1970, 1972; Srinivas and Rao, 1970; Srinivas *et al.*, 1970; Wang, 1986; Wang and He, 1986). Let us find the three-dimensional solution for a plate with the *rigidly slipping conditions* described by

$$\begin{aligned} &\tau_{xy} = \tau_{zx} = u = 0, \qquad (x = 0, a), \\ &\tau_{xy} = \tau_{yz} = v = 0, \qquad (y = 0, b). \end{aligned} \tag{8.1.41}$$

This type of boundary conditions is also referred to the guided or rigidly smooth contact conditions (Messina and Soldatos, 2002). To satisfy Eq. (8.1.41), we can take

$$\begin{aligned} \varphi(x,y,z,t) &= h^2 \sum_{m=1}^{\infty}\sum_{n=1}^{\infty} \psi_{mn}(\zeta)\sin(m\pi\xi)\sin(n\pi\eta)\exp(\mathrm{i}\omega t), \\ F(x,y,z,t) &= h^2 \sum_{m=1}^{\infty}\sum_{n=1}^{\infty} P_{mn}(\zeta)\cos(m\pi\xi)\cos(n\pi\eta)\exp(\mathrm{i}\omega t). \end{aligned} \tag{8.1.42}$$

Then the remaining solution steps are the same as those for the simply supported plates, and thus are omitted here.

We can actually deal with any combination of simply supported and rigidly slipping conditions. For example, if we have the following combined conditions

$$\begin{aligned}
&\sigma_x = v = w = 0\,, \quad (x=0),\\
&\tau_{xy} = \tau_{zx} = u = 0\,, \quad (x=a),\\
&\sigma_y = u = w = 0\,, \quad (y=0,b),
\end{aligned} \tag{8.1.43}$$

we can take

$$\begin{aligned}
\varphi(x,y,z,t) &= h^2 \sum_{m=1}^{\infty}\sum_{n=1}^{\infty} \psi_{mn}(\zeta)\sin\left(\frac{2m+1}{2}\pi\xi\right)\sin(n\pi\eta)\exp(\mathrm{i}\,\omega t),\\
F(x,y,z,t) &= h^2 \sum_{m=1}^{\infty}\sum_{n=1}^{\infty} P_{mn}(\zeta)\cos\left(\frac{2m+1}{2}\pi\xi\right)\cos(n\pi\eta)\exp(\mathrm{i}\,\omega t).
\end{aligned} \tag{8.1.44}$$

8.1.2 Circular Plates

In cylindrical coordinates (r,α,z), the general solution of transversely isotropic elasticity, Eq. (2.2.34), is

$$\begin{aligned}
u_r &= -\frac{1}{r}\frac{\partial\varphi}{\partial\alpha} - \frac{\partial^2 F}{\partial r\partial z},\\
u_\alpha &= \frac{\partial\varphi}{\partial r} - \frac{1}{r}\frac{\partial^2 F}{\partial\alpha\partial z},\\
w &= \left(a_0\Lambda + b_0\frac{\partial^2}{\partial z^2} - \frac{b_0}{c_{44}}\rho\frac{\partial^2}{\partial t^2}\right)F,
\end{aligned} \tag{8.1.45}$$

where $a_0 = c_{11}/(c_{13}+c_{44})$, $b_0 = c_{44}/(c_{13}+c_{44})$, and the displacement functions φ and F satisfy Eqs. (2.2.36) and (2.2.37), respectively. The two-dimensional Laplace operator, Λ, in cylindrical coordinates was defined in Eq. (2.2.73).

Consider the transversely isotropic circular plate of radius a and thickness h, as illustrated in Fig. 8.2. The plane of isotropy is assumed to be parallel to the middle plane $z=0$. We can therefore take

$$\begin{aligned}
\varphi &= h^2\psi_n(\zeta)J_n(\kappa\xi)\sin(n\alpha)\exp(\mathrm{i}\omega t),\\
F &= h^3 P_n(\zeta)J_n(\kappa\xi)\cos(n\alpha)\exp(\mathrm{i}\omega t),
\end{aligned} \tag{8.1.46}$$

where $\xi = r/a$ and $\zeta = z/h$ are dimensionless coordinates, $J_n(x)$ are Bessel functions of the first kind, n is an integer, $\psi_n(\zeta)$ and $P_n(\zeta)$ are functions to be determined and κ is to be determined by the boundary conditions at $r = a$. As in the formulae for rectangular plates, ω is the circular frequency in vibration and $\omega = 0$ in bending. Strictly, Eq. (8.1.46) should be the sum of all the series terms, *i.e.*, from $n = 0$ to $n = \infty$. However, the orthogonality of the trigonometric functions allows us to consider a term individually.

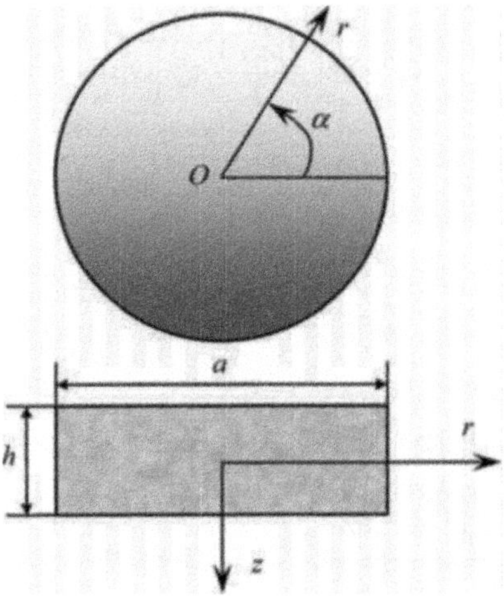

Fig. 8.2 A circular plate and the coordinates.

Using the properties of Bessel functions, we have

$$\begin{aligned}\Lambda\varphi &= -kJ_n(\kappa\xi)\psi_n(\zeta)\sin(n\alpha)\exp(\mathrm{i}\,\omega t),\\ \Lambda F &= -hkJ_n(\kappa\xi)P_n(\zeta)\cos(n\alpha)\exp(\mathrm{i}\,\omega t),\end{aligned} \tag{8.1.47}$$

where $k = t_0^2\kappa^2$, and $t_0 = h/a$ is the thickness-to-radius ratio. Using Eq. (8.1.47) and substituting Eq. (8.1.46) into Eqs. (2.2.36) and (2.2.37), we get

$$\psi_n''(\zeta) - [(c_{66}/c_{44})k - \Omega^2]\psi_n(\zeta) = 0, \tag{8.1.48}$$

$$P_n''''(\zeta) - g_1 P_n''(\zeta) + g_2 P_n(\zeta) = 0, \tag{8.1.49}$$

where the prime denotes differentiation with respect to ζ; $\Omega = \omega h\sqrt{\rho / c_{44}}$ is the non-dimensional frequency; g_1 and g_2 were given in Eq. (8.1.9). The solutions to Eqs. (8.1.48) and (8.1.49) are therefore

$$\psi_n(\zeta) = A_3\cosh(\lambda_3\zeta) + B_3\sinh(\lambda_3\zeta), \tag{8.1.50}$$

$$P_n(\zeta) = A_1\cosh(\lambda_1\zeta) + A_2\cosh(\lambda_2\zeta) + B_1\sinh(\lambda_1\zeta) + B_2\sinh(\lambda_2\zeta), \tag{8.1.51}$$

where λ_i $(i = 1,2,3)$ were defined in Eq. (8.1.12).

With Eqs. (8.1.45) and (8.1.46), the displacements can be written as

$$\begin{aligned} u_r &= -ht_0\left[\frac{n}{\xi}\psi_n(\zeta)J_n(\kappa\xi) + P_n'(\zeta)J_n^{\xi}(\kappa\xi)\right]\cos(n\alpha)\exp(\mathrm{i}\omega t), \\ u_\alpha &= ht_0\left[\psi_n(\zeta)J_n^{\xi}(\kappa\xi) + \frac{n}{\xi}P_n'(\zeta)J_n(\kappa\xi)\right]\sin(n\alpha)\exp(\mathrm{i}\omega t), \\ w &= h[b_0P_n''(\zeta) - (a_0k - b_0\Omega^2)P_n(\zeta)]J_n(\kappa\xi)\cos(n\alpha)\exp(\mathrm{i}\omega t), \end{aligned} \tag{8.1.52}$$

where the superscript ξ denotes differentiation with respect to ξ. By virtue of the constitutive relations, Eq. (1.2.37), and the geometric relations, Eq. (1.2.2), the stresses can be obtained as

$$\begin{aligned} \sigma_r &= \left\{-2c_{66}\frac{nt_0^2}{\xi}\left(J_n^{\xi} - \frac{1}{\xi}J_n\right)\psi_n - 2c_{66}t_0^2P_n'J_n^{\xi\xi} + c_{13}[b_0P_n''' + (b_0\Omega^2 - a_0k + c_{12}k/c_{13})P_n']J_n\right\} \\ &\quad \times\cos(n\alpha)\exp(\mathrm{i}\omega t), \\ \sigma_\alpha &= \left\{2c_{66}\frac{nt_0^2}{\xi}\left[J_n^{\xi} - \frac{1}{\xi}J_n\right]\psi_n + 2c_{66}t_0^2P_n'J_n^{\xi\xi} + c_{13}[b_0P_n''' + (b_0\Omega^2 - a_0k + (c_{11}/c_{13})k]P_n']J_n\right\} \\ &\quad \times\cos(n\alpha)\exp(\mathrm{i}\omega t), \\ \sigma_z &= c_{33}\{b_0P_n''' + [b_0\Omega^2 - a_0k + (c_{13}/c_{33})k]P_n'\}J_n\cos(n\alpha)\exp(\mathrm{i}\omega t), \\ \tau_{\alpha z} &= c_{44}\left\{t_0\psi_n'J_n^{\xi} - \frac{nt_0}{\xi}[(b_0 - 1)P_n'' + (b_0\Omega^2 - a_0k)P_n]J_n\right\}\sin(n\alpha)\exp(\mathrm{i}\omega t), \\ \tau_{zr} &= c_{44}\left\{-\frac{nt_0}{\xi}\psi_n'J_n + t_0[(b_0 - 1)P_n'' + (b_0\Omega^2 - a_0k)P_n]J_n^{\xi}\right\}\cos(n\alpha)\exp(\mathrm{i}\omega t), \\ \tau_{r\alpha} &= c_{66}\left[t_0^2(2J_n^{\xi\xi} + \kappa^2J_n)\psi_n + \frac{2nt_0^2}{\xi}\left(J_n^{\xi} - \frac{1}{\xi}J_n\right)P_n'\right]\sin(n\alpha)\exp(\mathrm{i}\omega t), \end{aligned} \tag{8.1.53}$$

where $c_{66}=(c_{11}-c_{12})/2$. For simplicity, we have used J_n, ψ_n and P_n to represent $J_n(\kappa\xi)$, $\psi_n(\zeta)$ and $P_n(\zeta)$, respectively.

(1) Bending

In this case, $\Omega^2=0$ and the eigenvalues λ_i $(i=1,2,3)$ can be directly determined by Eq. (8.1.12). The six unknown constants, A_i, B_i $(i=1,2,3)$, can be determined by the boundary conditions at the upper and lower surfaces of the circular plate.

Assume that the plate is subjected to normal loads at the upper and lower surfaces, $q_1(r,\alpha)$ and $q_2(r,\alpha)$, respectively, *i.e.*

$$\begin{aligned}&\sigma_z=-q_1(r,\alpha),\quad \tau_{zr}=\tau_{\alpha z}=0\ ,\quad (\zeta=-1/2),\\&\sigma_z=-q_2(r,\alpha),\quad \tau_{zr}=\tau_{\alpha z}=0\ ,\quad (\zeta=1/2).\end{aligned}\tag{8.1.54}$$

With zero shear stresses, the fourth and fifth equations of Eq. (8.1.53) lead to

$$\begin{aligned}&t_0\psi_n'(-1/2)J_n^{\xi}-\frac{nt_0}{\xi}[(b_0-1)P_n''(-1/2)-a_0kP_n(-1/2)]J_n=0\,,\\&-\frac{nt_0}{\xi}\psi_n'(-1/2)J_n+t_0[(b_0-1)P_n''(-1/2)-a_0kP_n(-1/2)]J_n^{\xi}=0\,,\\&t_0\psi_n'(1/2)J_n^{\xi}-\frac{nt_0}{\xi}[(b_0-1)P_n''(1/2)-a_0kP_n(1/2)]J_n=0\,,\\&-\frac{nt_0}{\xi}\psi_n'(1/2)J_n+t_0[(b_0-1)P_n''(1/2)-a_0kP_n(1/2)]J_n^{\xi}=0\,,\end{aligned}\tag{8.1.55}$$

which, in turn, gives rise to

$$\psi_n'(-1/2)=\psi_n'(1/2)=0\,,\tag{8.1.56}$$

$$\begin{aligned}&(b_0-1)P_n''(-1/2)-a_0kP_n(-1/2)=0,\\&(b_0-1)P_n''(1/2)-a_0kP_n(1/2)=0.\end{aligned}\tag{8.1.57}$$

By making use of Eq. (8.1.50) and the property of hyperbolic functions, we obtain, from Eq. (8.1.56),

$$A_3 = B_3 = 0, \tag{8.1.58}$$

showing that we always have $\varphi = 0$ for bending problems.

Now we can determine parameter κ. From Eqs. (8.1.52) and (8.1.53), we can see that if κ satisfies

$$J_n(\kappa) = 0, \tag{8.1.59}$$

then at the boundary $r = a$ $(\xi = 1)$

$$w = 0, \quad u_\alpha = 0, \quad 2c_{66}u_r + a\sigma_r = 0. \tag{8.1.60}$$

Equation (8.1.60) corresponds to the *elastically supported boundary conditions* as discussed by Ding, Xu and Guo (1999). On the other hand, if κ satisfies

$$J_n^\xi(\kappa) = 0, \tag{8.1.61}$$

then at $r = a$ $(\xi = 1)$, we have

$$u_r = 0, \quad \tau_{zr} = 0, \quad 2c_{66}u_\alpha + a\tau_{r\alpha} = 0. \tag{8.1.62}$$

Expanding the loads, $q_1(r,\alpha)$ and $q_2(r,\alpha)$, in a series form, we can obtain a typical term corresponding to $J_n(\kappa\xi)$ as

$$\begin{aligned} q_1(r,\alpha) &= c_{44}a_n^1 J_n(\kappa\xi)\cos(n\alpha), \\ q_2(r,\alpha) &= c_{44}a_n^2 J_n(\kappa\xi)\cos(n\alpha), \end{aligned} \tag{8.1.63}$$

where

$$a_n^i = \frac{C}{\pi c_{44}} \int_0^{2\pi} \int_0^1 q_i(\xi,\alpha) J_n(\kappa\xi)\cos(n\alpha)\xi \,\mathrm{d}\xi \,\mathrm{d}\alpha, \quad (i = 1, 2), \tag{8.1.64}$$

where constant C is (Watson, 1922)

$$C = \frac{2}{[J_{n+1}(\kappa)]^2}, \tag{8.1.65}$$

when κ satisfies Eq. (8.1.59), or

$$C = \frac{2\kappa^2}{(\kappa^2 - n^2)[J_n(\kappa)]^2}, \tag{8.1.66}$$

when κ satisfies Eq. (8.1.61).

With Eq. (8.1.63), Eqs. (8.1.53) and (8.1.54) lead to,

$$\begin{aligned} &c_{33}[b_0 P_n'''(-1/2) - (a_0 - c_{13}/c_{33})kP_n'(-1/2)] = -c_{44}a_n^1, \\ &c_{33}[b_0 P_n'''(1/2) - (a_0 - c_{13}/c_{33})kP_n'(1/2)] = -c_{44}a_n^2. \end{aligned} \tag{8.1.67}$$

By substituting $P_n(\zeta)$ in Eq. (8.1.51) into Eqs. (8.1.57) and (8.1.67), we obtain four inhomogeneous, linear algebraic equations, which determine the four unknowns, A_i and B_i $(i = 1, 2)$.

(2) Free vibration

Since the upper and lower surfaces are traction-free, *i.e.*

$$\sigma_z = \tau_{zr} = \tau_{\alpha z} = 0, \quad (\zeta = \pm 1/2), \tag{8.1.68}$$

Eq. (8.1.53) gives rise to

$$\begin{aligned} &b_0 P_n'''(-1/2) + (b_0\Omega^2 - a_0 k + c_{13}k/c_{33})P_n'(-1/2) = 0, \\ &t_0\psi_n'(-1/2)J_n^\xi - \frac{nt_0}{\xi}[(b_0 - 1)P_n''(-1/2) + (b_0\Omega^2 - a_0 k)P_n(-1/2)]J_n = 0, \\ &-\frac{nt_0}{\xi}\psi_n'(-1/2)J_n + t_0[(b_0 - 1)P_n''(-1/2) + (b_0\Omega^2 - a_0 k)P_n(-1/2)]J_n^\xi = 0, \\ &b_0 P_n'''(1/2) + (b_0\Omega^2 - a_0 k + c_{13}k/c_{33})P_n'(1/2) = 0, \\ &t_0\psi_n'(1/2)J_n^\xi - \frac{nt_0}{\xi}[(b_0 - 1)P_n''(1/2) + (b_0\Omega^2 - a_0 k)P_n(1/2)]J_n = 0, \\ &-\frac{nt_0}{\xi}\psi_n'(1/2)J_n + t_0[(b_0 - 1)P_n''(1/2) + (b_0\Omega^2 - a_0 k)P_n(1/2)]J_n^\xi = 0, \end{aligned} \tag{8.1.69}$$

from which, one gets

$$\psi'_n(\zeta_i) = A_3\lambda_3 \sinh(\lambda_3\zeta_i) + B_3\lambda_3 \cosh(\lambda_3\zeta_i) = 0\,, \tag{8.1.70}$$

$$\begin{aligned}
&b_0 P'''_n(\zeta_i) + [b_0\Omega^2 - (a_0 - c_{13}/c_{33})k]P'_n(\zeta_i) \\
&\quad = G_1(\lambda_1)[A_1 \sinh(\lambda_1\zeta_i) + B_1 \cosh(\lambda_1\zeta_i)] + G_1(\lambda_2)[A_2 \sinh(\lambda_2\zeta_i) + B_2 \cosh(\lambda_2\zeta_i)] = 0, \\
&(b_0 - 1)P''_n(\zeta_i) + (b_0\Omega^2 - a_0 k)P_n(\zeta_i) \\
&\quad = G_2(\lambda_1)[A_1 \cosh(\lambda_1\zeta_i) + B_1 \sinh(\lambda_1\zeta_i)] + G_2(\lambda_2)[A_2 \cosh(\lambda_2\zeta_i) + B_2 \sinh(\lambda_2\zeta_i)] = 0,
\end{aligned} \tag{8.1.71}$$

where $i = 1,\ 2$, $\zeta_1 = -1/2$, $\zeta_2 = 1/2$, and $G_1(\lambda)$ and $G_2(\lambda)$ were defined in Eq. (8.1.26).

Again, to have a nontrivial solution, the determinants of Eqs. (8.1.70) and (8.1.71) must vanish, giving rise to the eigen-equations as shown in Eqs. (8.1.27) and (8.1.28), respectively.

(*a*) For

$$\sinh(\lambda_3) = 0\,, \tag{8.1.72}$$

we obtain

$$\Omega = \sqrt{(c_{66}/c_{44})k + p^2\pi^2}\,, \tag{8.1.73}$$

where p is an integer and $A_1 = B_1 = A_2 = B_2 = 0$. According to Eq. (8.1.51), we have $P_n(\zeta) = 0$, and the vibrational mode can be obtained as

$$\begin{aligned}
u_r &= -ht_0 \frac{n}{\xi}\psi_n(\zeta)J_n(\kappa\xi)\cos(n\alpha)\exp(\mathrm{i}\,\omega t), \\
u_\alpha &= ht_0\psi_n(\zeta)J_n^{\xi}(\kappa\xi)\sin(n\alpha)\exp(\mathrm{i}\,\omega t), \\
w &= 0,
\end{aligned} \tag{8.1.74}$$

where $\psi_n(\zeta)$ is the same as $\psi_{mn}(\zeta)$ in Eq. (8.1.33). These equations show that the volume strain $e = \partial u_r/\partial r + (1/r)\partial u_\alpha/\partial\alpha + u_r/r + \partial w/\partial z = 0$, indicating that the vibration corresponds to a thickness-shear motion[5]. Furthermore, when $k_t^2 = \pi^2/12$ and $p = 1$, Eq. (8.1.34) becomes the same as Eq. (8.1.73).

Let us have a further look at parameter κ. Equations (8.1.52) and (8.1.53) show that when κ satisfies Eq. (8.1.59), Eq. (8.1.62) holds, but when it satisfies Eq. (8.1.61), Eq. (8.1.60) applies. This is different from the case in bending.

(*b*) For

$$G_{12}\sinh(\lambda_1/2)\cosh(\lambda_2/2) - G_{21}\sinh(\lambda_2/2)\cosh(\lambda_1/2) = 0, \tag{8.1.75}$$

we have $A_3 = B_3 = B_1 = B_2 = 0$, *i.e.*

$$\psi_n(\zeta) = 0, \quad P_n(\zeta) = A_1\cosh(\lambda_1\zeta) + A_2\cosh(\lambda_2\zeta). \tag{8.1.76}$$

Equation (8.1.52) then yields

$$u_r(\zeta) = -u_r(-\zeta), \quad u_\alpha(\zeta) = -u_\alpha(-\zeta), \quad w(\zeta) = w(-\zeta). \tag{8.1.77}$$

This implies that the vibration is anti-symmetric with respect to the middle plane of the circular plate, $\zeta = 0$. In this case, parameter κ and the boundary conditions are the same as in bending.

(*c*) For

$$G_{21}\sinh(\lambda_1/2)\cosh(\lambda_2/2) - G_{12}\sinh(\lambda_2/2)\cosh(\lambda_1/2) = 0, \tag{8.1.78}$$

we have $A_3 = B_3 = A_1 = A_2 = 0$, *i.e.*

$$\psi_n = 0, \quad P_n = B_1\sinh(\lambda_1\zeta) + B_2\sinh(\lambda_2\zeta). \tag{8.1.79}$$

Equation (8.1.52) then brings about

$$u_r(\zeta) = u_r(-\zeta), \quad u_\alpha(\zeta) = u_\alpha(-\zeta), \quad w(\zeta) = -w(-\zeta). \tag{8.1.80}$$

---

[5] Using the Mindlin plate theory, Xu (1999) obtained the same frequencies of a circular plate as shown in Eq. (8.1.34).

This means that now the vibration is symmetric with respect to the middle plane. As with the previous rectangular plate, this type of vibration cannot be predicted by any plate bending theories. In the present case, parameter κ and the boundary conditions are also the same as in bending.

The fundamental (lowest) frequencies of a transversely isotropic circular PZT-4 plate are given in Tables 8.1 and 8.2 for boundary conditions in Eqs. (8.1.60) and (8.1.62), respectively. The material properties of ceramic PZT-4 were given in Table 1.4.

It is shown that the fundamental frequencies of the *in-plane vibration* (thickness-shear motion) for boundary conditions (8.1.60) are always smaller than those for boundary conditions (8.1.62), which however is contrary to the case of the *out-of-plane vibration* (antisymmetric or symmetric motion). This follows that boundary conditions (8.1.60) impose a constraint on the transverse displacement, yielding a relative larger rigidity for the out-of-plane motion, while boundary conditions (8.1.62) impose a constraint on the radial displacement which leads to a relative larger stiffness for the in-plane vibration.

Table 81 The fundamental frequencies (Ω)corresponding to boundary conditions (8160)

t_0	in-plane			anti-symmetric			Symmetric		
	$n=1$	$n=2$	$n=3$	$n=1$	$n=2$	$n=3$	$n=1$	$n=2$	$n=3$
01	02013	03339	04593	00776	01364	02052	07205	09637	11941
02	04026	06678	09186	02874	04823	06937	14294	18980	23289
03	06039	10018	13780	05837	09362	12970	21123	27621	33118
04	08052	13357	18373	09286	14359	19369	27494	34950	40293
05	10065	16695	22966	12988	19529	25853	33143	40433	45000
06	12078	20035	27559	16812	24748	32325	37803	44307	48733
07	14091	23375	32152	20690	29964	38757	41409	47402	52464
08	16104	26714	36745	24585	35158	45145	44202	50345	56507
09	18117	30053	41339	28479	40323	51490	46555	53419	60908
10	20130	30092	45932	32362	45459	57798	48754	56716	65627

Table 8.2 The fundamental frequencies (Ω) corresponding to boundary conditions (8.1.62).

t_0	in-plane			anti-symmetric			symmetric		
	$n=1$	$n=2$	$n=3$	$n=1$	$n=2$	$n=3$	$n=1$	$n=2$	$n=3$
0.1	0.4189	0.5615	0.6975	0.0183	0.0498	0.0927	0.3469	0.5749	0.7896
0.2	0.8378	1.1230	1.3951	0.0718	0.1892	0.3392	0.6926	1.1440	1.5636
0.3	1.2568	1.6844	2.0926	0.1566	0.3955	0.6798	1.0357	1.7006	2.3023
0.4	1.6757	2.2459	2.7902	0.2673	0.6459	1.0693	1.3749	2.2365	2.9775
0.5	2.0946	2.8074	3.4877	0.3988	0.9234	1.4821	1.7084	2.7406	3.5525
0.6	2.5135	3.3689	4.1853	0.5460	1.2169	1.9052	2.0343	3.1991	4.0008
0.7	2.9325	3.9304	4.8828	0.7050	1.5195	2.3318	2.3505	3.5976	4.3360
0.8	3.3514	4.4918	5.5804	0.8728	1.8269	2.7589	2.6541	3.9279	4.6039
0.9	3.7703	5.0533	6.2779	1.0468	2.1366	3.1849	2.9421	4.1942	4.8465
1.0	4.1892	5.6148	6.9755	1.2254	2.4472	3.6092	3.2113	4.4129	5.0886

When n is not an integer, the method described above can be extended to bending and vibration problems of sectorial plates with straight boundaries (Xu, 1999)[6]. It is worthwhile to emphasise that usually only the lowest natural frequency is of interest, although there are an infinity of frequencies Ω for any κ determined by Eq. (8.1.59) or Eq. (8.1.61).

8.2 THE STATE-SPACE METHOD FOR LAMINATED PLATES

As shown in the last section, the analyses for bending, vibration and stability are similar to each other. Thus in this section, we will present only the solution procedure for the free vibration of laminated plates, using the state-space method.

8.2.1 Laminated Rectangular Plates

In this section, new state-space formulations for transversely isotropic elasticity will be derived and used to analyse the free vibration of rectangular plates. But before so doing, we will present the conventional state-space formulations for comparison purpose.

[6] Xu (1999) also made a comprehensive comparison between the three-dimensional theory and some plate theories.

The state-space formulations for orthotropic elasticity have been discussed in detail in **Section 2.1**. By virtue of Eq. (1.4.12), we can obtain the state-space formulation for transversely isotropic elasticity with thermal effects[7]

$$\frac{\partial}{\partial z}\begin{Bmatrix}\sigma_z\\ u\\ v\\ w\\ \tau_{zx}\\ \tau_{yz}\end{Bmatrix}=\begin{bmatrix}\mathbf{0} & \mathbf{A}_1\\ \mathbf{A}_2 & \mathbf{0}\end{bmatrix}\begin{Bmatrix}\sigma_z\\ u\\ v\\ w\\ \tau_{zx}\\ \tau_{yz}\end{Bmatrix}+\begin{Bmatrix}-F_z\\ 0\\ 0\\ -\beta_3 T/c_{33}\\ -F_x-(\beta_1-c_{13}\beta_3/c_{33})\partial T/\partial x\\ -F_y-(\beta_1-c_{13}\beta_3/c_{33})\partial T/\partial y\end{Bmatrix}, \tag{8.2.1}$$

and

$$\begin{aligned}\sigma_x&=\frac{c_{13}}{c_{33}}\sigma_z+\left(c_{11}-\frac{c_{13}^2}{c_{33}}\right)\frac{\partial u}{\partial x}+\left(c_{12}-\frac{c_{13}^2}{c_{33}}\right)\frac{\partial v}{\partial y}+\left(\beta_1-\frac{c_{13}}{c_{33}}\beta_3\right)T,\\ \sigma_y&=\frac{c_{13}}{c_{33}}\sigma_z+\left(c_{12}-\frac{c_{13}^2}{c_{33}}\right)\frac{\partial u}{\partial x}+\left(c_{11}-\frac{c_{13}^2}{c_{33}}\right)\frac{\partial v}{\partial y}+\left(\beta_1-\frac{c_{13}}{c_{33}}\beta_3\right)T,\\ \tau_{xy}&=c_{66}\left(\frac{\partial u}{\partial y}+\frac{\partial v}{\partial x}\right),\end{aligned} \tag{8.2.2}$$

where

$$\mathbf{A}_1=\begin{bmatrix}\rho\dfrac{\partial^2}{\partial t^2} & -\dfrac{\partial}{\partial x} & -\dfrac{\partial}{\partial y}\\ -\dfrac{\partial}{\partial x} & \dfrac{1}{c_{44}} & 0\\ -\dfrac{\partial}{\partial y} & 0 & \dfrac{1}{c_{44}}\end{bmatrix},$$

[7] If buckling problem is considered, the applied normal edge stresses T_1 and T_2, as in Eq. (8.1.1), should vary from ply to ply to keep the laminate at an initial state of uniform strains ε_x, ε_y and $\sigma_z = 0$. For more detailed discussion, see Srinivas and Rao (1970).

$$\mathbf{A}_2 = \begin{bmatrix} \dfrac{1}{c_{33}} & -\dfrac{c_{13}}{c_{33}}\dfrac{\partial}{\partial x} & -\dfrac{c_{13}}{c_{33}}\dfrac{\partial}{\partial y} \\ -\dfrac{c_{13}}{c_{33}}\dfrac{\partial}{\partial x} & \rho\dfrac{\partial^2}{\partial t^2} - \left(c_{11} - \dfrac{c_{13}^2}{c_{33}}\right)\dfrac{\partial^2}{\partial x^2} - c_{66}\dfrac{\partial^2}{\partial y^2} & -\left(c_{66} + c_{12} - \dfrac{c_{13}^2}{c_{33}}\right)\dfrac{\partial^2}{\partial x \partial y} \\ -\dfrac{c_{13}}{c_{33}}\dfrac{\partial}{\partial y} & -\left(c_{66} + c_{12} - \dfrac{c_{13}^2}{c_{33}}\right)\dfrac{\partial^2}{\partial x \partial y} & \rho\dfrac{\partial^2}{\partial t^2} - c_{66}\dfrac{\partial^2}{\partial x^2} - \left(c_{22} - \dfrac{c_{13}^2}{c_{33}}\right)\dfrac{\partial^2}{\partial y^2} \end{bmatrix}. \tag{8.2.3}$$

In these equations, σ_z, u, v, w, τ_{zx} and τ_{yz} are usually known as the state variables or basic variables, while σ_x, σ_y and τ_{xy} are the induced variables. For a simply supported thick or laminated rectangular plate, these state variables can be expanded in terms of double sine functions (Fan and Ye, 1990*a*, *b*). Then, Eq. (8.2.1) can be transformed to a state equation with constant coefficients, which can be solved for the state variables, using matrix theory. With the state variables obtained, the induced variables can be determined using Eq. (8.2.2).

Now we will introduce a state-space method based on the new equations for transversely isotropic elasticity (Ding, Chen and Xu, 2001). Using the geometric relations, Eq. (1.2.1), and considering the thermal effect, we find that the stresses are

$$\begin{aligned} \sigma_x &= c_{11}\frac{\partial u}{\partial x} + c_{12}\frac{\partial v}{\partial y} + c_{13}\frac{\partial w}{\partial z} + \beta_1 T, \\ \sigma_y &= c_{12}\frac{\partial u}{\partial x} + c_{11}\frac{\partial v}{\partial y} + c_{13}\frac{\partial w}{\partial z} + \beta_1 T, \\ \sigma_z &= c_{13}\frac{\partial u}{\partial x} + c_{13}\frac{\partial v}{\partial y} + c_{33}\frac{\partial w}{\partial z} + \beta_3 T, \\ \tau_{yz} &= c_{44}\left(\frac{\partial w}{\partial y} + \frac{\partial v}{\partial z}\right), \quad \tau_{zx} = c_{44}\left(\frac{\partial w}{\partial x} + \frac{\partial u}{\partial z}\right), \quad \tau_{xy} = c_{66}\left(\frac{\partial u}{\partial y} + \frac{\partial v}{\partial x}\right). \end{aligned} \tag{8.2.4}$$

We introduce the following separation formulae

$$u = \frac{\partial \psi}{\partial y} - \frac{\partial G}{\partial x}, \quad v = -\frac{\partial \psi}{\partial x} - \frac{\partial G}{\partial y}, \tag{8.2.5}$$

$$\tau_{zx} = \frac{\partial \tau_1}{\partial y} - \frac{\partial \tau_2}{\partial x}, \quad \tau_{yz} = -\frac{\partial \tau_1}{\partial x} - \frac{\partial \tau_2}{\partial y}, \tag{8.2.6}$$

$$F_x = \frac{\partial F_1}{\partial y} - \frac{\partial F_2}{\partial x}, \quad F_y = -\frac{\partial F_1}{\partial x} - \frac{\partial F_2}{\partial y}. \tag{8.2.7}$$

With Eqs. (8.2.5) and (8.2.6), the fourth and fifth equations in Eq. (8.2.4) yield

$$\begin{aligned} &\frac{\partial}{\partial y}\left(\tau_1 - c_{44}\frac{\partial \psi}{\partial z}\right) - \frac{\partial}{\partial x}\left(\tau_2 + c_{44}w - c_{44}\frac{\partial G}{\partial z}\right) = 0, \\ &-\frac{\partial}{\partial x}\left(\tau_1 - c_{44}\frac{\partial \psi}{\partial z}\right) - \frac{\partial}{\partial y}\left(\tau_2 + c_{44}w - c_{44}\frac{\partial G}{\partial z}\right) = 0. \end{aligned} \tag{8.2.8}$$

These two equations will be satisfied if we take

$$\tau_1 - c_{44}\frac{\partial \psi}{\partial z} = 0, \tag{8.2.9}$$

$$\tau_2 + c_{44}w - c_{44}\frac{\partial G}{\partial z} = 0. \tag{8.2.10}$$

By substituting Eqs. (8.2.4) to (8.2.7) into the first two equations in Eq. (1.2.10), as in the derivation presented in **Section 2.2**, we obtain

$$c_{66}\Lambda\psi + \frac{\partial \tau_1}{\partial z} + F_1 - \rho\frac{\partial^2 \psi}{\partial t^2} = 0, \tag{8.2.11}$$

$$c_{11}\Lambda G - c_{13}\frac{\partial w}{\partial z} - \beta_1 T + \frac{\partial \tau_2}{\partial z} + F_2 - \rho\frac{\partial^2 G}{\partial t^2} = 0. \tag{8.2.12}$$

With the aid of Eqs. (8.2.5) and (8.2.6), Eq. (1.2.10) and the third equation of Eq. (8.2.4) yield

$$\frac{\partial \sigma_z}{\partial z} = \rho\frac{\partial^2 w}{\partial t^2} + \Lambda\tau_2 - F_z, \tag{8.2.13}$$

$$\frac{\partial w}{\partial z} = \frac{c_{13}}{c_{33}}\Lambda G + \frac{1}{c_{33}}(\sigma_z - \beta_3 T). \tag{8.2.14}$$

In addition, the expressions of $\partial \tau_1 / \partial z$ and $\partial \tau_2 / \partial z$ are given by Eqs. (8.2.11) and (8.2.12) when Eq. (8.2.14) is used. Finally, we get

$$\frac{\partial}{\partial z}\begin{Bmatrix} \psi \\ \tau_1 \end{Bmatrix} = \begin{bmatrix} 0 & \dfrac{1}{c_{44}} \\ \rho \dfrac{\partial^2}{\partial t^2} - c_{66}\Lambda & 0 \end{bmatrix} \begin{Bmatrix} \psi \\ \tau_1 \end{Bmatrix} + \begin{Bmatrix} 0 \\ -F_1 \end{Bmatrix}, \tag{8.2.15}$$

$$\frac{\partial}{\partial z}\begin{Bmatrix} G \\ \sigma_z \\ \tau_2 \\ w \end{Bmatrix} = \begin{bmatrix} 0 & 0 & \dfrac{1}{c_{44}} & 1 \\ 0 & 0 & \Lambda & \rho \dfrac{\partial^2}{\partial t^2} \\ \rho \dfrac{\partial^2}{\partial t^2} - \left(c_{11} - \dfrac{c_{13}^2}{c_{33}} \right)\Lambda & \dfrac{c_{13}}{c_{33}} & 0 & 0 \\ \dfrac{c_{13}}{c_{33}}\Lambda & \dfrac{1}{c_{33}} & 0 & 0 \end{bmatrix} \begin{Bmatrix} G \\ \sigma_z \\ \tau_2 \\ w \end{Bmatrix} + \begin{Bmatrix} 0 \\ -F_z \\ \left(\beta_1 - \dfrac{c_{13}}{c_{33}}\beta_3 T \right) - F_2 \\ -\dfrac{\beta_3}{c_{33}} T \end{Bmatrix}. \tag{8.2.16}$$

By virtue of Eq. (8.2.4), the three induced variables can be obtained as

$$\begin{aligned} \sigma_x + \sigma_y &= -\left(c_{11} + c_{12} - 2\frac{c_{13}^2}{c_{33}} \right)\Lambda G + 2\frac{c_{13}}{c_{33}}\sigma_z + 2\left(\beta_1 - \frac{c_{13}}{c_{33}}\beta_3 \right)T, \\ \sigma_x - \sigma_y &= -2c_{66}\left[2\frac{\partial^2 \psi}{\partial x \partial y} + \left(\frac{\partial^2}{\partial x^2} - \frac{\partial^2}{\partial y^2} \right)G \right], \\ \tau_{xy} &= c_{66}\left[\left(\frac{\partial^2}{\partial x^2} - \frac{\partial^2}{\partial y^2} \right)\psi - 2\frac{\partial^2 G}{\partial x \partial y} \right]. \end{aligned} \tag{8.2.17}$$

It can be seen that the six state variables, ψ, τ_1, G, σ_z, τ_2 and w, have been divided into two uncoupled groups, satisfying Eqs. (8.2.15) and (8.2.16), respectively. Compared with Eq. (8.2.1), the order of either Eq. (8.2.15) or Eq. (8.2.16) is smaller. This will facilitate, for example, the solution of a plate bending problem because only Eq. (8.2.16) needs to be solved.

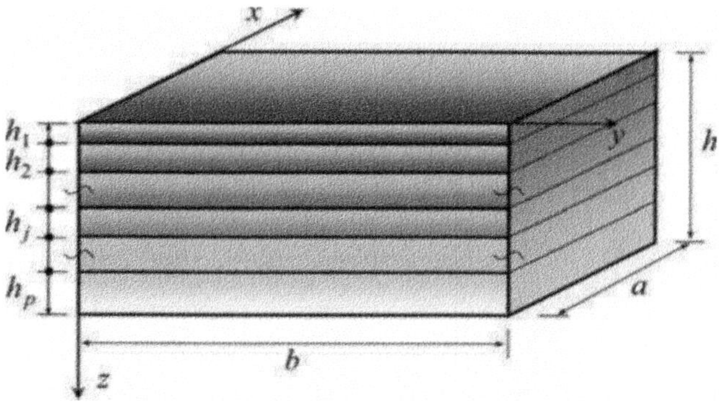

Fig. 8.3 A laminated rectangular plate.

Consider the free vibration of a p-ply simply supported transversely isotropic laminated rectangular plate illustrated in Fig. 8.3, whose isotropic plane is parallel to its middle plane. We can take

$$\begin{Bmatrix} \psi \\ \tau_1 \\ G \\ \sigma_z \\ \tau_2 \\ w \end{Bmatrix} = \sum_{m=0}^{\infty}\sum_{n=0}^{\infty} \begin{bmatrix} h^2\overline{\psi}(\zeta)\cos(m\pi\xi)\cos(n\pi\eta) \\ hc_{44}^{(1)}\overline{\tau}_1(\zeta)\cos(m\pi\xi)\cos(n\pi\eta) \\ h^2\overline{G}(\zeta)\sin(m\pi\xi)\sin(n\pi\eta) \\ c_{44}^{(1)}\overline{\sigma}_z(\zeta)\sin(m\pi\xi)\sin(n\pi\eta) \\ hc_{44}^{(1)}\overline{\tau}_2(\zeta)\sin(m\pi\xi)\sin(n\pi\eta) \\ h\overline{w}(\zeta)\sin(m\pi\xi)\sin(n\pi\eta) \end{bmatrix} \exp(i\omega t), \tag{8.2.18}$$

where $\zeta = z/h$, $\xi = x/a$ and $\eta = y/b$ are dimensionless coordinates, ω is the circular frequency and $c_{44}^{(1)}$ represents the material constant of the first layer. Equation (8.2.18) satisfies the simply supported conditions specified by Eq. (8.1.2). Substituting Eq. (8.2.18) into Eqs. (8.2.15) and (8.2.16) and neglecting body forces and any possible temperature change, for any pair of (m, n) we obtain

$$\frac{\mathrm{d}}{\mathrm{d}\zeta}\begin{Bmatrix} \overline{\psi} \\ \overline{\tau}_1 \end{Bmatrix} = \mathbf{M}_1 \begin{Bmatrix} \overline{\psi} \\ \overline{\tau}_1 \end{Bmatrix}, \tag{8.2.19}$$

$$\frac{\mathrm{d}}{\mathrm{d}\zeta}\begin{Bmatrix} \overline{G} \\ \overline{\sigma}_z \\ \overline{\tau}_2 \\ \overline{w} \end{Bmatrix} = \mathbf{M}_2 \begin{Bmatrix} \overline{G} \\ \overline{\sigma}_z \\ \overline{\tau}_2 \\ \overline{w} \end{Bmatrix}, \tag{8.2.20}$$

where

$$\mathbf{M}_1 = \begin{bmatrix} 0 & \dfrac{c_{44}^{(1)}}{c_{44}} \\ \dfrac{c_{66}k}{c_{44}^{(1)}} - f & 0 \end{bmatrix}, \quad \mathbf{M}_2 = \begin{bmatrix} 0 & 0 & \dfrac{c_{44}^{(1)}}{c_{44}} & 1 \\ 0 & 0 & -k & -f \\ \dfrac{(c_{33}c_{11} - c_{13}^2)k}{c_{33}c_{44}^{(1)}} - f & \dfrac{c_{13}}{c_{33}} & 0 & 0 \\ -\dfrac{c_{13}k}{c_{33}} & \dfrac{c_{44}^{(1)}}{c_{33}} & 0 & 0 \end{bmatrix}. \tag{8.2.21}$$

in which $k = k_1^2 + k_2^2$, $k_1 = m\pi h/a$, $k_2 = n\pi h/b$, $f = \Omega^2 \rho/\rho^{(1)}$, $\Omega = \omega h\sqrt{\rho^{(1)}/c_{44}^{(1)}}$ is the dimensionless frequency, and $\rho^{(1)}$ is the density of the first layer. The solutions to Eqs. (8.2.19) and (8.2.20) can now be obtained by matrix theory as

$$\begin{Bmatrix} \overline{\psi}(\zeta) \\ \overline{\tau}_1(\zeta) \end{Bmatrix} = \exp[\mathbf{M}_1(\zeta - \zeta_{j-1})] \begin{Bmatrix} \overline{\psi}(\zeta_{j-1}) \\ \overline{\tau}_1(\zeta_{j-1}) \end{Bmatrix}, \quad (\zeta_{j-1} \le \zeta \le \zeta_j,\ j = 1,2,\cdots,p), \tag{8.2.22}$$

$$\begin{Bmatrix} \overline{G}(\zeta) \\ \overline{\sigma}_z(\zeta) \\ \overline{\tau}_2(\zeta) \\ \overline{w}(\zeta) \end{Bmatrix} = \exp[\mathbf{M}_2(\zeta - \zeta_{j-1})] \begin{Bmatrix} \overline{G}(\zeta_{j-1}) \\ \overline{\sigma}_z(\zeta_{j-1}) \\ \overline{\tau}_2(\zeta_{j-1}) \\ \overline{w}(\zeta_{j-1}) \end{Bmatrix}, \quad (\zeta_{j-1} \le \zeta \le \zeta_j,\ j = 1,2,\cdots,p), \tag{8.2.23}$$

where $\zeta_0 = 0$ and $\zeta_j = \sum_{i=1}^{j} h_i/h$, with h_i being the thickness of the i-th layer. The exponential matrix $\exp[\mathbf{M}_k(\zeta - \zeta_{j-1})]$ is known as the *transfer matrix*; it can be expressed as a polynomial in the matrix $\mathbf{M}_k$, according to the *Cayley-Hamilton theorem* (Bellman, 1970)

$$\exp[\mathbf{M}_1(\zeta - \zeta_{j-1})] = c_0(\zeta - \zeta_{j-1})\mathbf{I}_{2\times2} + c_1(\zeta - \zeta_{j-1})\mathbf{M}_1, \tag{8.2.24}$$

$$\exp[\mathbf{M}_2(\zeta - \zeta_{j-1})] = d_0(\zeta - \zeta_{j-1})\mathbf{I}_{4\times4} + \sum_{k=1}^{3} d_k(\zeta - \zeta_{j-1})(\mathbf{M}_2)^k, \tag{8.2.25}$$

where $\mathbf{I}_{q\times q}$ is the qth order identity matrix and

$$\begin{Bmatrix} c_0(\zeta-\zeta_{j-1}) \\ c_1(\zeta-\zeta_{j-1}) \end{Bmatrix} = \begin{bmatrix} 1 & \lambda_1 \\ 1 & \lambda_2 \end{bmatrix}^{-1} \begin{Bmatrix} \exp[\lambda_1(\zeta-\zeta_{j-1}) \\ \exp[\lambda_2(\zeta-\zeta_{j-1}) \end{Bmatrix}, \tag{8.2.26}$$

$$\begin{Bmatrix} d_0(\zeta-\zeta_{j-1}) \\ d_1(\zeta-\zeta_{j-1}) \\ d_2(\zeta-\zeta_{j-1}) \\ d_3(\zeta-\zeta_{j-1}) \end{Bmatrix} = \begin{bmatrix} 1 & \delta_1 & \delta_1^2 & \delta_1^3 \\ 1 & \delta_2 & \delta_2^2 & \delta_2^3 \\ 1 & \delta_3 & \delta_3^2 & \delta_3^3 \\ 1 & \delta_4 & \delta_4^2 & \delta_4^3 \end{bmatrix}^{-1} \begin{Bmatrix} \exp[\delta_1(\zeta-\zeta_{j-1}) \\ \exp[\delta_2(\zeta-\zeta_{j-1}) \\ \exp[\delta_3(\zeta-\zeta_{j-1}) \\ \exp[\delta_4(\zeta-\zeta_{j-1}) \end{Bmatrix}, \tag{8.2.27}$$

in which λ_i and δ_i are the eigenvalues of matrices $\mathbf{M}_1$ and $\mathbf{M}_2$, respectively. We can derive the explicit expressions of c_i and d_i to avoid performing the inverse of the matrices in these two equations (Chen, Ding and Yu, 1996). It should be pointed out that Eqs. (8.2.26) or (8.2.27) is applicable to the cases that have different eigenvalues, λ_i or δ_i. When the eigenvalues are equal, different forms should be employed (Bellman, 1970).

For a perfectly bonded laminated plate, u, v, w, σ_z, τ_{zx} and τ_{yz} must be continuous at the interfaces of the laminates, which, in turn, requires that the six state variables must be continuous at the interfaces. Thus Eqs. (8.2.22) and (8.2.23) give rise to

$$\begin{Bmatrix} \overline{\psi}(1) \\ \overline{\tau}_1(1) \end{Bmatrix} = \mathbf{T}_1 \begin{Bmatrix} \overline{\psi}(0) \\ \overline{\tau}_1(0) \end{Bmatrix}, \tag{8.2.28}$$

$$[\overline{G}(1), \overline{\sigma}_z(1), \overline{\tau}_2(1), \overline{w}(1)]^{\mathrm{T}} = \mathbf{T}_2[\overline{G}(0), \overline{\sigma}_z(0), \overline{\tau}_2(0), \overline{w}(0)]^{\mathrm{T}}, \tag{8.2.29}$$

where $\mathbf{T}_1 = \prod_{j=p}^{1} \exp[\mathbf{M}_1(\zeta_j - \zeta_{j-1})]$ and $\mathbf{T}_2 = \prod_{j=p}^{1} \exp[\mathbf{M}_2(\zeta_j - \zeta_{j-1})]$ are square matrices of order 2 and 4, respectively.

For free vibration problems, the upper and lower surfaces of the plate are traction-free, *i.e.*

$$\overline{\sigma}_z = \overline{\tau}_1 = \overline{\tau}_2 = 0, \quad (\zeta = 0, 1), \tag{8.2.30}$$

from which and Eqs. (8.2.28) and (8.2.29), we can get two frequency equations corresponding to two different classes of vibration, respectively,

$$T_{121}=0, \tag{8.2.31}$$

$$\begin{vmatrix} T_{221} & T_{224} \\ T_{231} & T_{234} \end{vmatrix}=0, \tag{8.2.32}$$

where T_{1ij} and T_{2ij} are the elements of matrices $\mathbf{T}_1$ and $\mathbf{T}_2$, respectively. Equation (8.2.31) corresponds to in-plane vibration (the first class of vibration), while Eq. (8.2.32) to general bending vibration (the second class of vibration)[8]. The dimensionless frequency Ω can be obtained from Eq. (8.2.31) or Eq. (8.2.32) numerically, and then the vibrational modes of the state variables can be determined using Eqs. (8.2.28), (8.2.29), (8.2.22) and (8.2.23). The vibrational modes of the induced variables are given by Eq. (8.2.17).

As a numerical example, consider the free vibration of a simply supported three-layered rectangular plate. The thickness ratio of the three layers is 0.4:0.2:0.4, the length-to-thickness ratio of the plate is $a/h=10$ and the width-to-thickness ratio is $b/h=5$. The top and bottom layers of the plate are of the same transversely isotropic material with

$$c_{11}=20\times10^{10}\,\text{Pa},\quad c_{12}=12\times10^{10}\,\text{Pa},\quad c_{13}=c_{33}=2\times10^{10}\,\text{Pa},\quad c_{44}=1\times10^{10}\,\text{Pa}.$$

The intermediate layer is an isotropic material with Poisson's ratio $\nu=0.3$ and Young's modulus $E=21\times10^{10}$ Pa. Note that for isotropic materials, we have

$$c_{11}=c_{33}=\frac{E(1-\nu)}{(1+\nu)(1-2\nu)},\quad c_{12}=c_{13}=\frac{E\nu}{(1+\nu)(1-2\nu)},\quad c_{44}=c_{66}=\frac{E}{2(1+\nu)}. \tag{8.2.33}$$

The density ratio between the transversely isotropic material and isotropic material is 0.5. The variations of the lowest non-dimensional frequency versus mode numbers m and n are shown in Figs. 8.4 and 8.5 for the first and second classes of vibration, respectively. It should be pointed out that for the first class of vibration, we can take either m or n zero (they should not be zero at the same time), but for the second kind

[8] In **Section 8.1**, we showed that for a transversely isotropic homogeneous plate, the vibration can be divided into two different classes. Here we obtain the same conclusion for a laminated transversely isotropic plates based on state-space formulations. A noticeable difference is that for a generally laminated plate, it is usually difficult to further classify the second class of vibration into symmetric and antisymmetric modes as for a homogeneous plate.

vibration, both m and n should be greater than zero, as shown directly by Eqs. (8.2.5), (8.2.6) and (8.2.18).

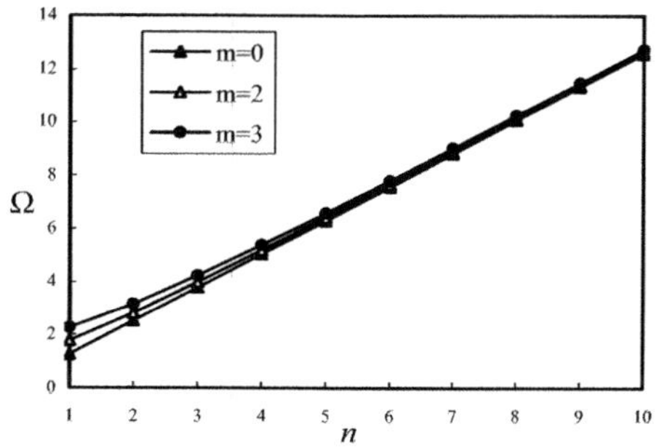

Fig. 8.4 The lowest frequency spectra of the first class of vibration.

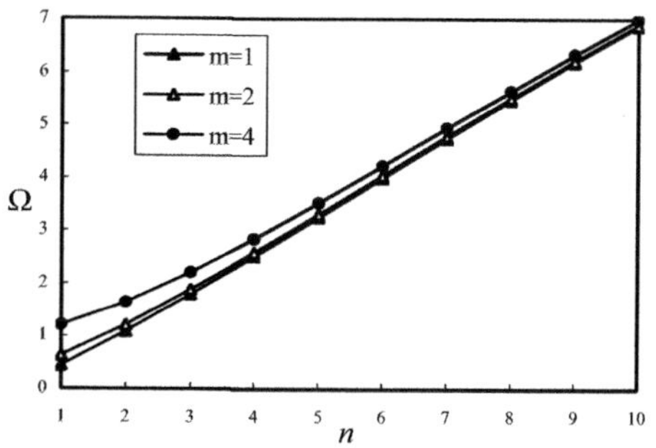

Fig. 8.5 The lowest frequency spectra of the second class of vibration.

This method is also applicable to the analyses of the bending and stability of simply supported laminated rectangular plates (Ding, Chen and Xu, 2001).

8.2.2 Laminated Circular Plates[9]

Under axisymmetric deformation, the equations of motion without body forces, referring to Eq. (1.2.11), become

$$\begin{aligned} &\frac{\partial \sigma_r}{\partial r}+\frac{\partial \tau_{zr}}{\partial z}+\frac{\sigma_r-\sigma_\alpha}{r}=\rho\frac{\partial^2 u_r}{\partial t^2}, \\ &\frac{\partial \tau_{zr}}{\partial r}+\frac{\partial \sigma_z}{\partial z}+\frac{\tau_{zr}}{r}=\rho\frac{\partial^2 w}{\partial t^2}. \end{aligned} \tag{8.2.34}$$

[9] Celep (1978, 1980) studied the axisymmetric free vibration of an isotropic circular plate using the state-space method, which was later extended to transversely isotropic materials by Fan and Ye (1990*c*). However, Ding, Xu and Chen *et al.* (1998) and Xu(1999) recently found that there was a significant error in these studies.

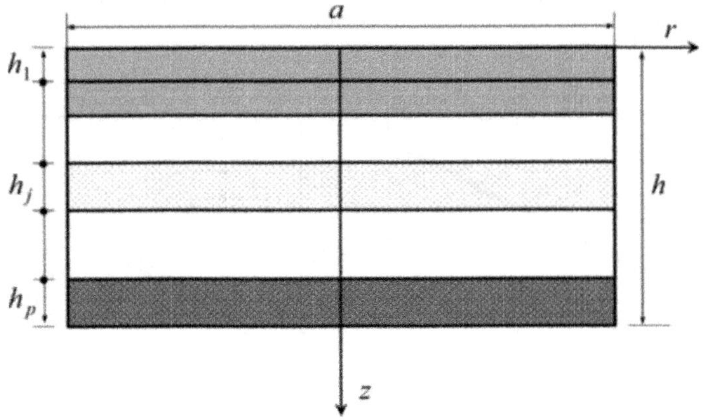

Fig. 8.6 A laminated circular plate.

The stress-displacement relations for a transversely isotropic body are

$$\begin{aligned}
\sigma_r &= c_{11}\frac{\partial u_r}{\partial r} + c_{12}\frac{u_r}{r} + c_{13}\frac{\partial w}{\partial z},\\
\sigma_\alpha &= c_{12}\frac{\partial u_r}{\partial r} + c_{11}\frac{u_r}{r} + c_{13}\frac{\partial w}{\partial z},\\
\sigma_z &= c_{13}\frac{\partial u_r}{\partial r} + c_{13}\frac{u_r}{r} + c_{33}\frac{\partial w}{\partial z},\\
\tau_{zr} &= c_{44}\left(\frac{\partial u_r}{\partial z} + \frac{\partial w}{\partial r}\right).
\end{aligned} \tag{8.2.35}$$

Rearranging Eqs. (8.2.34) and (8.2.35), we get

$$\frac{\partial}{\partial z}\begin{Bmatrix} u_r \\ \sigma_z \\ \tau_{zr} \\ w \end{Bmatrix} = \begin{bmatrix} 0 & 0 & \dfrac{1}{c_{44}} & -\dfrac{\partial}{\partial r} \\ 0 & 0 & -\left(\dfrac{\partial}{\partial r} + \dfrac{1}{r}\right) & \rho\dfrac{\partial^2}{\partial t^2} \\ \rho\dfrac{\partial^2}{\partial t^2} - \left(c_{11} - \dfrac{c_{13}^2}{c_{33}}\right)\left(\dfrac{\partial^2}{\partial r^2} + \dfrac{1}{r}\dfrac{\partial}{\partial r} - \dfrac{1}{r^2}\right) & -\dfrac{c_{13}}{c_{33}}\dfrac{\partial}{\partial r} & 0 & 0 \\ -\dfrac{c_{13}}{c_{33}}\left(\dfrac{\partial}{\partial r} + \dfrac{1}{r}\right) & \dfrac{1}{c_{33}} & 0 & 0 \end{bmatrix}\begin{Bmatrix} u_r \\ \sigma_z \\ \tau_{zr} \\ w \end{Bmatrix}, \tag{8.2.36}$$

$$\sigma_r = \left(c_{11} - \frac{c_{13}^2}{c_{33}}\right)\frac{\partial u_r}{\partial r} + \left(c_{12} - \frac{c_{13}^2}{c_{33}}\right)\frac{u_r}{r} + \frac{c_{13}}{c_{33}}\sigma_z,$$
$$\sigma_\alpha = \left(c_{12} - \frac{c_{13}^2}{c_{33}}\right)\frac{\partial u_r}{\partial r} + \left(c_{11} - \frac{c_{13}^2}{c_{33}}\right)\frac{u_r}{r} + \frac{c_{13}}{c_{33}}\sigma_z. \qquad (8.2.37)$$

Consider the vibration of the p-ply circular plate of radius a and thickness h, as illustrated in Fig. 8.6. For the j-th ply we can take

$$\begin{aligned} u_r &= h\bar{u}_r(\xi,\zeta)\exp(\mathrm{i}\omega t), \\ \sigma_z &= c_{44}^{(1)}\bar{\sigma}_z(\xi,\zeta)\exp(\mathrm{i}\omega t), \\ \tau_{zr} &= c_{44}^{(1)}\bar{\tau}_{zr}(\xi,\zeta)\exp(\mathrm{i}\omega t), \\ w &= h\bar{w}(\xi,\zeta)\exp(\mathrm{i}\omega t), \end{aligned} \qquad (8.2.38)$$

where ω is the circular frequency and $\xi = r/a$ and $\zeta = z/h$ are dimensionless coordinates. Substituting the above equation into Eq. (8.2.36), we have

$$\frac{\partial}{\partial \zeta}\begin{Bmatrix} \bar{u}_r \\ \bar{\sigma}_z \\ \bar{\tau}_{zr} \\ \bar{w} \end{Bmatrix} = \begin{bmatrix} 0 & 0 & \dfrac{c_{44}^{(1)}}{c_{44}} & -t_0\dfrac{\partial}{\partial \xi} \\ 0 & 0 & -t_0\left(\dfrac{\partial}{\partial \xi} + \dfrac{1}{\xi}\right) & f \\ f - \dfrac{c_{11}c_{33} - c_{13}^2}{c_{33}c_{44}^{(1)}}t_0^2\left(\dfrac{\partial^2}{\partial \xi^2} + \dfrac{1}{\xi}\dfrac{\partial}{\partial \xi} - \dfrac{1}{\xi^2}\right) & -\dfrac{c_{13}}{c_{33}}t_0\dfrac{\partial}{\partial \xi} & 0 & 0 \\ -\dfrac{c_{13}}{c_{33}}t_0\left(\dfrac{\partial}{\partial \xi} + \dfrac{1}{\xi}\right) & \dfrac{c_{44}^{(1)}}{c_{33}} & 0 & 0 \end{bmatrix}\begin{Bmatrix} \bar{u}_r \\ \bar{\sigma}_z \\ \bar{\tau}_{zr} \\ \bar{w} \end{Bmatrix}, \qquad (8.2.39)$$

where $t_0 = h/a$, $f = -\Omega^2\rho/\rho^{(1)}$ and $\Omega = \omega h\sqrt{\rho^{(1)}/c_{44}^{(1)}}$ is the dimensionless frequency.

Bearing in mind that the *finite Hankel transform* is defined by (Sneddon, 1951)

$$H_u[f] = \int_0^1 f(\xi)J_\mu(\kappa\xi)\xi\,\mathrm{d}\xi, \qquad (8.2.40)$$

where $J_\mu(x)$ is the Bessel function of the first kind of order μ, we can take

$$\begin{aligned} &U(\kappa,\zeta)=H_1[\bar{u}_r], \quad \sigma(\kappa,\zeta)=H_0[\bar{\sigma}_z], \\ &\tau(\kappa,\zeta)=H_1[\bar{\tau}_{zr}], \quad W(\kappa,\zeta)=H_0[\bar{w}]. \end{aligned} \tag{8.2.41}$$

Hence the transformed equation of Eq. (8.2.39) is

$$\frac{\partial}{\partial\zeta}\mathbf{R}(\kappa,\zeta)=\mathbf{M}\mathbf{R}(\kappa,\zeta)+\mathbf{Q}, \tag{8.2.42}$$

where $\mathbf{R}=[U(\kappa,\zeta),\sigma(\kappa,\zeta),-\tau(\kappa,\zeta),W(\kappa,\zeta)]^{\mathrm{T}}$, and

$$\mathbf{M}=\begin{bmatrix} 0 & 0 & -\dfrac{c_{44}^{(1)}}{c_{44}} & \kappa t_0 \\ 0 & 0 & \kappa t_0 & f \\ -f-\dfrac{c_{11}c_{33}-c_{13}^2}{c_{33}c_{44}^{(1)}}\kappa^2 t_0^2 & -\dfrac{c_{13}}{c_{33}}\kappa t_0 & 0 & 0 \\ -\dfrac{c_{13}}{c_{33}}\kappa t_0 & \dfrac{c_{44}^{(1)}}{c_{33}} & 0 & 0 \end{bmatrix}, \tag{8.2.43}$$

$$\mathbf{Q}=\begin{Bmatrix} -t_0\bar{w}(1,\zeta)J_1(\kappa) \\ -t_0\bar{\tau}_{zr}(1,\zeta)J_0(\kappa) \\ \left[\dfrac{c_{11}-c_{12}}{c_{44}^{(1)}}t_0^2\bar{u}_r(1,\zeta)+t_0\bar{\sigma}_r(1,\zeta)\right]J_1(\kappa)-\dfrac{c_{11}c_{33}-c_{13}^2}{c_{33}c_{44}^{(1)}}t_0^2\kappa\bar{u}_r(1,\zeta)J_0(\kappa) \\ -\dfrac{c_{13}}{c_{33}}t_0\bar{u}_r(1,\zeta)J_0(\kappa) \end{Bmatrix}. \tag{8.2.44}$$

When the circular plate has one of the following two boundary conditions at $r=a$ ($\xi=1$), we can make $\mathbf{Q}=\mathbf{0}$.

(*a*) Elastically supported

$$\bar{w}(1,\zeta)=0, \quad \frac{c_{11}-c_{12}}{c_{44}^{(1)}}t_0^2\bar{u}_r(1,\zeta)+t_0\bar{\sigma}_r(1,\zeta)=0, \quad \text{and } J_0(\kappa)=0. \tag{8.2.45}$$

(*b*) Rigidly slipping

$$\bar{u}_r(1,\zeta)=0, \quad \bar{\tau}_{zr}(1,\zeta)=0, \quad \text{and } J_1(\kappa)=0. \tag{8.2.46}$$

Under these two boundary conditions, Eq. (8.2.42) becomes homogeneous, and the solution is

$$\mathbf{R}(\kappa,\zeta)=\exp[\mathbf{M}(\zeta-\zeta_{j-1})]\mathbf{R}(\kappa,\zeta_{j-1}),\quad(\zeta_{j-1}\le\zeta\le\zeta_j,\ j=1,2,\cdots,p),\tag{8.2.47}$$

where $\zeta_0=0,\ \zeta_j=\sum_{i=1}^{j}h_i/h$ and the exponential matrix $\exp[\mathbf{M}(\zeta-\zeta_{j-1})]$ is the transfer matrix that can be expressed in terms of a polynomial about $\mathbf{M}$, see **Section 8.2.1**.

Since the state variables u_r, σ_z, τ_{zr} and w are continuous at the interface of two adjacent layers, Eq. (8.2.47) yields

$$\mathbf{R}(\kappa,1)=\mathbf{T}\mathbf{R}(\kappa,0),\tag{8.2.48}$$

where $\mathbf{T}=\prod_{j=p}^{1}\exp[\mathbf{M}(\zeta_j-\zeta_{j-1})]$ is a fourth-order square matrix.

If a circulate plate is in free vibration, the upper and lower surfaces of the plate are traction-free, *i.e.*

$$\sigma(1)=\tau(1)=\sigma(0)=\tau(0)=0.\tag{8.2.49}$$

Then Eq. (8.2.48) will give rise to a set of linear, homogeneous algebraic equations. Again, to have a nonzero solution, the determinant of the equations must be zero, so that

$$\begin{vmatrix}T_{21} & T_{24}\\ T_{31} & T_{34}\end{vmatrix}=0.\tag{8.2.50}$$

Equation (8.2.50) is the frequency equation and is a transcendental equation for the dimensionless frequency Ω.

With the elastically supported condition, parameter κ satisfies

$$J_0(\kappa)=0,\tag{8.2.51}$$

from which we can obtain a series of positive roots κ_m $(m=1,2,\cdots)$. Individual non-dimensional frequencies Ω corresponding to every κ_m can then be calculated by Eq. (8.2.50). Note that for every κ_m, there are infinity of Ω. To obtain the mode shapes, we need to use the inverse Hankel transform (Sneddon, 1951), which leads to

$$\begin{aligned}\overline{u}_r(\xi,\zeta)&=2U(\kappa_m,\zeta)\frac{J_1(\kappa_m\xi)}{[J_1(\kappa_m)]^2},\\ \overline{w}(\xi,\zeta)&=2W(\kappa_m,\zeta)\frac{J_0(\kappa_m\xi)}{[J_1(\kappa_m)]^2},\end{aligned} \tag{8.2.52}$$

for a particular κ_m.

For the rigidly slipping condition, parameter κ satisfies

$$J_1(\kappa)=0, \tag{8.2.53}$$

and the formulae for calculating vibrational mode corresponding to a particular κ_m are

$$\begin{aligned}\overline{u}_r(\xi,\zeta)&=2U(\kappa_m,\zeta)\frac{J_1(\kappa_m\xi)}{[J_0(\kappa_m)]^2},\\ \overline{w}(\xi,\zeta)&=2W(\kappa_m,\zeta)\frac{J_0(\kappa_m\xi)}{[J_0(\kappa_m)]^2}.\end{aligned} \tag{8.2.54}$$

These solutions can be used to analyse individual problems easily. Consider a circular plate composed of three laminates with thickness ratio 0.25:0.5:0.25. The top and bottom layers are of the same isotropic material (steel) with $E=21\times10^{10}$ Pa, $\nu=0.3$ and $\rho=7.8\times10^3\,\text{kg/m}^3$, while the intermediate layer is ceramic PZT-4, which is transversely isotropic and the material constants can be found in Table 1.4. Table 8.3 lists the lowest natural frequencies of the plate and compares the results from the current state-space method (SSM) with those calculated by the finite element method (FEM). They are in very good agreement[10].

[10] Xu (1999) made a comprehensive comparison of the exact solution with some plate theories and obtained some useful observations. This method is applicable also to the static and dynamic analyses of annular plates (Xu, 1999; Ding and Xu, 2000*b*).

Table 8.3 The lowest natural frequency Ω of a three-layered circular plate

t_0	Elastically supported		Rigidly Slipping	
	SSM	FEM	SSM	FEM
0.1	0.0333	0.0333	0.0823	0.0823
0.2	0.1261	0.1263	0.2908	0.2910
0.3	0.2621	0.2621	0.5637	0.5639
0.4	0.4250	0.4253	0.8636	0.8638
0.5	0.6040	0.6042	1.1753	1.1756
0.6	0.7920	0.7920	1.4936	1.4939
0.7	0.9852	0.9852	1.8171	1.8173
0.8	1.1821	1.1821	2.4159	2.4163
0.9	1.3811	1.3813	2.4805	2.4809
1.0	1.5824	1.5827	2.8214	2.8218

9

VIBRATIONS OF CYLINDERS AND CYLINDRICAL SHELLS OF TRANSVERSELY ISOTROPIC MATERIALS

This chapter analyses the free vibration of transversely isotropic cylinders or cylindrical shells based on the three-dimensional theory of elasticity. The aim of the chapter is two fold. One is to promote the physical understanding of the fundamentals of vibration, which will be realized by introducing first the three simple vibration modes, *i.e.*, the axisymmetric torsional vibration, breathing mode vibration and thickness-shear vibration and then the general asymmetric vibration. The second is to solve some problems closely related to engineering applications. Thus the coupled vibrations of cylindrical shells interacting with fluid and elastic media will be discussed. Most published studies are based on the two-dimensional shell theories or numerical methods such as the finite element method and the boundary element method. The three-dimensional analysis presented in this chapter provides a benchmark for determining the validity and suitability of these approximate shell theories and numerical methods.

9.1 THREE SIMPLE MODES OF VIBRATION

The basic equations of a transversely isotropic elastic body in cylindrical coordinates (r,α,z), have been discussed in **Chapter 1**. With the aid of the geometric and constitutive relations, Eqs. (1.2.2) and (1.2.37), the stresses can be expressed in terms of displacements as

$$\sigma_r = c_{11}\frac{\partial u_r}{\partial r} + c_{12}\left(\frac{1}{r}\frac{\partial u_\alpha}{\partial \alpha} + \frac{u_r}{r}\right) + c_{13}\frac{\partial w}{\partial z},$$

$$\sigma_\alpha = c_{12}\frac{\partial u_r}{\partial r} + c_{11}\left(\frac{1}{r}\frac{\partial u_\alpha}{\partial \alpha} + \frac{u_r}{r}\right) + c_{13}\frac{\partial w}{\partial z},$$

$$\sigma_z = c_{13}\frac{\partial u_r}{\partial r} + c_{13}\left(\frac{1}{r}\frac{\partial u_\alpha}{\partial \alpha} + \frac{u_r}{r}\right) + c_{33}\frac{\partial w}{\partial z},$$

$$\tau_{\alpha z} = c_{44}\left(\frac{\partial u_\alpha}{\partial z} + \frac{1}{r}\frac{\partial w}{\partial \alpha}\right), \quad \tau_{zr} = c_{44}\left(\frac{\partial w}{\partial r} + \frac{\partial u_r}{\partial z}\right), \tag{9.1.1}$$

$$\tau_{r\alpha} = c_{66}\left(\frac{1}{r}\frac{\partial u_r}{\partial \alpha} + \frac{\partial u_\alpha}{\partial r} - \frac{u_\alpha}{r}\right),$$

where $c_{66} = (c_{11} - c_{12})/2$. Hence, the equations of motion in absence of body forces can be written as

$$\left[c_{11}\left(\frac{\partial^2}{\partial r^2} + \frac{1}{r}\frac{\partial}{\partial r} - \frac{1}{r^2}\right) + c_{66}\frac{1}{r^2}\frac{\partial^2}{\partial \alpha^2} + c_{44}\frac{\partial^2}{\partial z^2}\right]u_r$$
$$+\left[(c_{12}+c_{66})\frac{1}{r}\frac{\partial^2}{\partial \alpha \partial r} - (c_{11}+c_{66})\frac{1}{r^2}\frac{\partial}{\partial \alpha}\right]u_\alpha + (c_{13}+c_{44})\frac{\partial^2}{\partial r \partial z}w = \rho\frac{\partial^2 u_r}{\partial t^2},$$

$$\left[(c_{12}+c_{66})\frac{1}{r}\frac{\partial^2}{\partial \alpha \partial r} + (c_{11}+c_{66})\frac{1}{r^2}\frac{\partial}{\partial \alpha}\right]u_r$$
$$+\left[c_{66}\left(\frac{\partial^2}{\partial r^2} + \frac{1}{r}\frac{\partial}{\partial r} - \frac{1}{r^2}\right) + c_{11}\frac{1}{r^2}\frac{\partial^2}{\partial \alpha^2} + c_{44}\frac{\partial^2}{\partial z^2}\right]u_\alpha + (c_{13}+c_{44})\frac{1}{r}\frac{\partial^2 w}{\partial \alpha \partial z} = \rho\frac{\partial^2 u_\alpha}{\partial t^2},$$

$$(c_{13}+c_{44})\left(\frac{\partial^2}{\partial r \partial z} + \frac{1}{r}\frac{\partial}{\partial z}\right)u_r + (c_{13}+c_{44})\frac{1}{r}\frac{\partial^2 u_\alpha}{\partial \alpha \partial z}$$
$$+\left[c_{44}\left(\frac{\partial^2}{\partial r^2} + \frac{1}{r}\frac{\partial}{\partial r} + \frac{1}{r^2}\frac{\partial^2}{\partial \alpha^2}\right) + c_{33}\frac{\partial^2}{\partial z^2}\right]w = \rho\frac{\partial^2 w}{\partial t^2}.$$

$$\tag{9.1.2}$$

These equations can be also obtained directly from Eq. (2.1.5), by taking $c_{22} = c_{11}$, $c_{23} = c_{13}$, $c_{55} = c_{44}$ and neglecting the terms of body forces and temperature change.

Let a, b be the inner and outer radii of the cylindrical shell, $h = b - a$ the thickness, $R = (a+b)/2$ the mean radius, and L the length. We will first consider three simple vibration modes of the cylinder; all are of practical importance in the design of resonators. For example, the breathing mode vibration can be used to generate sound radiation (Sittig and Coquin, 1970), while the torsional or thickness-shear mode resonator microsensor is capable of *in situ* monitoring of physical and chemical properties of fluids (Rosler *et al.*, 1998).

9.1.1 Axisymmetric Torsional Vibration

Considering the axisymmetric torsional free vibration of a transversely isotropic cylinder (or a cylindrical shell), we have

$$u_r = w = 0, \quad u_\alpha = V(r,z,t), \quad \frac{\partial}{\partial \alpha}[\cdots] = 0. \tag{9.1.3}$$

The first and third equations in Eq. (9.1.2) are satisfied, but the second equation requires that

$$\left[c_{66}\left(\frac{\partial^2}{\partial r^2} + \frac{1}{r}\frac{\partial}{\partial r} - \frac{1}{r^2}\right) + c_{44}\frac{\partial^2}{\partial z^2}\right]V = \rho\frac{\partial^2 V}{\partial t^2}. \tag{9.1.4}$$

Clearly, the solution of this equation takes the form of[1]

$$V(r,z,t) = f_1(r)\cos(m\pi z / L)\exp(\mathrm{i}\omega t), \tag{9.1.5}$$

where ω is the circular frequency and m is the axial half-wave number. It can be shown that this displacement solution satisfies the rigidly slipping condition (Lusher and Hardy, 1988; Chau, 1994) described by

$$w = \tau_{zr} = \tau_{\alpha z} = 0, \quad (z = 0,\, L). \tag{9.1.6}$$

In torsional vibration, this corresponds to the traction-free condition at the two ends since w is identically zero. Substituting Eq. (9.1.5) into Eq. (9.1.4), we can obtain the equation for the unknown function $f_1(r)$, *i.e.*,

$$\frac{\mathrm{d}^2 f_1(r)}{\mathrm{d}r^2} + \frac{1}{r}\frac{\mathrm{d}f_1(r)}{\mathrm{d}r} + \left(p^2 - \frac{1}{r^2}\right)f_1(r) = 0, \tag{9.1.7}$$

---

[1] If the factor cos($m\pi z/L$) in Eq. (9.1.5) is replaced by sin ($m\pi z/L$), the solution corresponds to a cylinder simply supported at both ends. On the other hand, for an infinite cylinder or cylindrical shell, this factor can be replaced by exp(ikz), where k is the axial wave number. In this case, Eq. (9.1.5) represents a wave motion along the z-axis.

where

$$p^2 = \omega^2 / v_3^2 - k^2 / s_3^2, \quad k = m\pi / L, \quad v_3^2 = c_{66} / \rho, \quad s_3^2 = c_{66} / c_{44}. \tag{9.1.8}$$

Equation (9.1.7) is the *Bessel's equation* of the first-order [cf. **Appendix B**], whose solution is

$$f_1(r) = \begin{cases} A_1 J_1(pr) + B_1 Y_1(pr), & p^2 > 0, \\ A_1 r + B_1 \dfrac{1}{r}, & p^2 = 0, \\ A_1 I_1(\tilde{p}r) + B_1 K_1(\tilde{p}r), & \tilde{p}^2 = -p^2 > 0, \end{cases} \tag{9.1.9}$$

where J_1 and Y_1 are the Bessel functions of the first and second kinds, respectively, I_1 and K_1 are the modified Bessel functions of the first and second kinds, respectively, and A_1 and B_1 are arbitrary constants.

According to Eqs. (9.1.1), (9.1.3) and (9.1.5), the nonzero stress components are

$$\begin{aligned} \tau_{\alpha z} &= -(m\pi / L) c_{44} f_1(r) \sin(m\pi z / L) \exp(\mathrm{i}\omega t), \\ \tau_{r\alpha} &= c_{66} \left[\frac{\mathrm{d} f_1(r)}{\mathrm{d} r} - \frac{1}{r} \right] \cos(m\pi z / L) \exp(\mathrm{i}\omega t). \end{aligned} \tag{9.1.10}$$

In the following, we will discuss the free torsional vibrations of cylindrical shells and cylinders separately.

(*a*) Cylinder

The value of $f_1(r)$ at $r = 0$ must be finite, which requires that $B_1 = 0$ in Eq. (9.1.9). Hence, the nonzero stress component at the cylindrical surface is

$$\tau_{r\alpha} = c_{66} A_1 \left\{ \begin{array}{ll} pJ_0(pr) - \dfrac{2}{r} J_1(pr), & p^2 > 0 \\ 0, & p^2 = 0 \\ \tilde{p} I_0(\tilde{p}r) - \dfrac{2}{r} I_1(\tilde{p}r), & p^2 < 0 \end{array} \right\} \cos(m\pi z / L) \exp(\mathrm{i}\omega t). \tag{9.1.11}$$

However, the surface is traction-free, *i.e.*, $\tau_{r\alpha} = 0 \ (r = b)$, thus the following frequency equation must hold

$$pbJ_0(pb) - 2J_1(pb) = 0, \qquad p^2 > 0, \tag{9.1.12a}$$

$$p = 0, \tag{9.1.12b}$$

$$\tilde{p}bI_0(\tilde{p}b) - 2I_1(\tilde{p}b) = 0, \qquad p^2 < 0. \tag{9.1.12c}$$

Function $y = xI_0(x) - 2I_1(x)$ is anti-symmetric about y-axis, as shown in Fig. 9.1, indicating that Eq. (9.1.12c) can only have a zero real root. Also, $p = 0$ is a root of Eq. (9.1.12a). Thus the frequency equation for torsional vibration is given by Eq. (9.1.12a), whose first 10 positive roots are listed in Table 9.1. There are several powerful mathematical manipulation codes, such as MATHEMATICA, MAPLE and MATHCAD, that can be readily used to perform the calculation involving special functions.

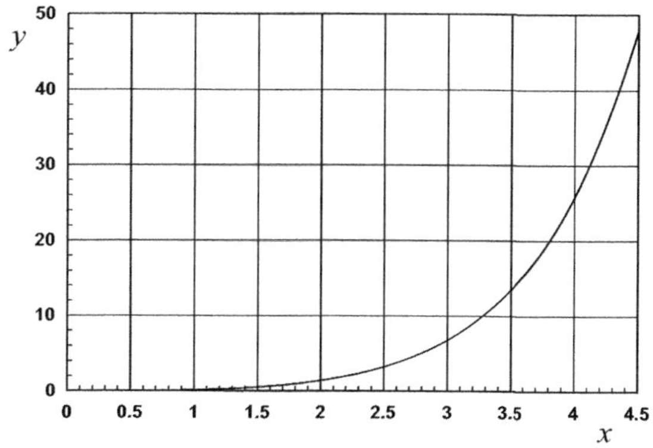

Fig. 9.1 Function $y = xI_0(x) - 2I_1(x)$.

Table 9.1 The first 10 positive roots of Eq. (9.1.12a).

j	1	2	3	4	5
pb	5.13562	8.41724	11.6198	14.7960	17.9598
j	6	7	8	9	10
pb	21.1170	24.2701	27.4206	30.5692	33.7165

When pb is large, we can also use the *asymptotic expansions* of Bessel functions in Eq. (B.4.5) to derive a simpler form of Eq. (9.1.12a). According to Eq. (B.4.5) in **Appendix B**, the following asymptotic expansions are adopted here

$$J_0(x) = \sqrt{\frac{2}{\pi x}}\left[\cos\left(x - \frac{\pi}{4}\right) + \frac{1}{8x}\sin\left(x - \frac{\pi}{4}\right)\right],$$

$$J_1(x)=\sqrt{\frac{2}{\pi x}}\left[\sin\left(x-\frac{\pi}{4}\right)\left(1+\frac{15}{128x^2}\right)+\frac{3}{8x}\cos\left(x-\frac{\pi}{4}\right)\right]. \tag{9.1.13}$$

Applying Eq. (9.1.13) to Eq. (9.1.12*a*) and discarding proper higher-order terms, we get

$$pb\cos\left(pb-\frac{\pi}{4}\right)-\frac{15}{8}\sin\left(pb-\frac{\pi}{4}\right)=0. \tag{9.1.14a}$$

The second and third roots of this equation are 5.14853 and 8.42028, respectively; these two roots are very close to the first and second roots of Eq. (9.1.12*a*). The larger the value, the closer the roots of Eqs. (9.1.14*a*) and (9.1.12*a*). We can further delete the second term in Eq. (9.1.14*a*) if pb is very large; the solution of the resulting equation is obvious:

$$pb=\frac{(2n-1)\pi}{2}+\frac{\pi}{4}, \tag{9.1.14b}$$

where n is an integer. When $n=3$, we get $pb=8.63938$, which is close to the second root of Eq. (9.1.12*a*), and when $n=11$, we get $pb=33.7721$, which is almost the same as the tenth root of Eq. (9.1.12*a*). Thus, the value of pb calculated from Eq. (9.1.14*b*) for a large n can serve as a good approximation of the root of Eq. (9.1.12*a*).

Table 9.2 The first 40 positive roots of Bessel function $J_1(x)$.

j	1	2	3	4	5	6	7	8	9	10
x	3.83171	7.01559	10.1735	13.3237	16.4706	19.6159	22.7601	25.9037	29.0468	32.1897
j	11	12	13	14	15	16	17	18	19	20
x	35.3323	38.4748	41.6171	44.7593	47.9015	51.0435	54.1856	57.3275	60.4695	63.6114
j	21	22	23	24	25	26	27	28	29	30
x	66.7532	69.8951	73.0369	76.1787	79.3205	82.4623	85.6040	88.7458	91.8875	95.0292
j	31	32	33	34	35	36	37	38	39	40
x	98.1710	101.313	104.454	107.596	110.738	113.879	117.021	120.163	123.304	126.446

The discussion presented above focuses on the free torsional vibration of a cylinder with a traction-free surface. It is also straightforward to use Eqs. (9.1.5) and (9.1.9) to obtain the frequency equation for a cylinder with a fixed surface (*i.e.*, $u_\alpha=0$ at $r=b$) as $J_1(pb)=0$. Table 9.2 lists the first 40 positive roots of function $J_1(x)$, from which the corresponding frequencies can be obtained. The function $J_1(x)$ exhibits a nearly periodic behaviour like the sine function, as can be seen from the asymptotic series in Eq. (B.4.5). With the asymptotic expansions, we can also obtain the higher-order frequencies approximately according to $x = n\pi + \pi/4$, here n is an integer.

(*b*) Cylindrical shell

In this case, the nonzero stress component at the cylindrical surfaces is

$$\tau_{r\alpha} = c_{66}\begin{Bmatrix} A_1[pJ_0(pr) - \frac{2}{r}J_1(pr)] + B_1[pY_0(pr) - \frac{2}{r}Y_1(pr)], & p^2 > 0 \\ -\frac{2}{r^2}B_1, & p^2 = 0 \\ A_1[\tilde{p}I_0(\tilde{p}r) - \frac{2}{r}I_1(\tilde{p}r)] - B_1[\tilde{p}K_0(\tilde{p}r) + \frac{2}{r}K_1(\tilde{p}r)], & p^2 < 0 \end{Bmatrix} \times \cos(m\pi z/L)\exp(\mathrm{i}\omega t). \tag{9.1.15}$$

When the inner and outer surfaces of a cylindrical shell are traction-free, *i.e.*, $\tau_{r\alpha} = 0\ (r = a, b)$, the frequency equation becomes

$$\begin{vmatrix} paJ_0(pa) - 2J_1(pa) & paY_0(pa) - 2Y_1(pa) \\ pbJ_0(pb) - 2J_1(pb) & pbY_0(pb) - 2Y_1(pb) \end{vmatrix} = 0, \quad p^2 > 0, \tag{9.1.16a}$$

$$p = 0, \tag{9.1.16b}$$

$$\begin{vmatrix} \tilde{p}aI_0(\tilde{p}a) - 2I_1(\tilde{p}a) & \tilde{p}aK_0(\tilde{p}a) + 2K_1(\tilde{p}a) \\ \tilde{p}bI_0(\tilde{p}b) - 2I_1(\tilde{p}b) & \tilde{p}bK_0(\tilde{p}b) + 2K_1(\tilde{p}b) \end{vmatrix} = 0, \quad p^2 < 0. \tag{9.1.16c}$$

Introducing the following dimensionless quantities

$$\begin{gathered} t_r = h/R, \quad t_1 = a/R = 1 - t_r/2, \quad t_2 = b/R = 1 + t_r/2, \\ \Omega^2 = \omega^2 R^2 \rho / c_{66}, \quad t_L = m\pi R/L, \quad m_1 = c_{44}/c_{66}, \end{gathered} \tag{9.1.17}$$

and defining that $x = \Omega^2 - m_1 t_L^2$ and $x_1 = m_1 t_L^2 - \Omega^2$, we can transfer Eq. (9.1.16) to

$$\begin{vmatrix} xt_1 J_0(xt_1) - 2J_1(xt_1) & xt_1 Y_0(xt_1) - 2Y_1(xt_1) \\ xt_2 J_0(xt_2) - 2J_1(xt_2) & xt_2 Y_0(xt_2) - 2Y_1(xt_2) \end{vmatrix} = T_1(x, t_r) = 0, \tag{9.1.18a}$$

$$\Omega = \sqrt{m_1}\, t_L, \tag{9.1.18b}$$

$$\begin{vmatrix} x_1 t_1 I_0(x_1 t_1) - 2I_1(x_1 t_1) & x_1 t_1 K_0(x_1 t_1) + 2K_1(x_1 t_1) \\ x_1 t_2 I_0(x_1 t_2) - 2I_1(x_1 t_2) & x_1 t_2 K_0(x_1 t_2) + 2K_1(x_1 t_2) \end{vmatrix} = T_2(x_1, t_r) = 0. \tag{9.1.18c}$$

Equation (9.1.18*c*) has no positive real root, as shown by the typical curves of $T_2(x_1, t_r)$ ($x_1 > 0$) in Fig. 9.2, indicating that Eqs. (9.1.18*a*) and (9.1.18*b*) are the frequency equations of the free torsional vibration of a cylindrical shell. Table 9.3 lists the first 10 positive real roots of equation $T_1(x, t_r) = 0$ for $t_r = 0.1$, 0.5, 1.0 and 1.5. The corresponding dimensionless frequencies can then be obtained.

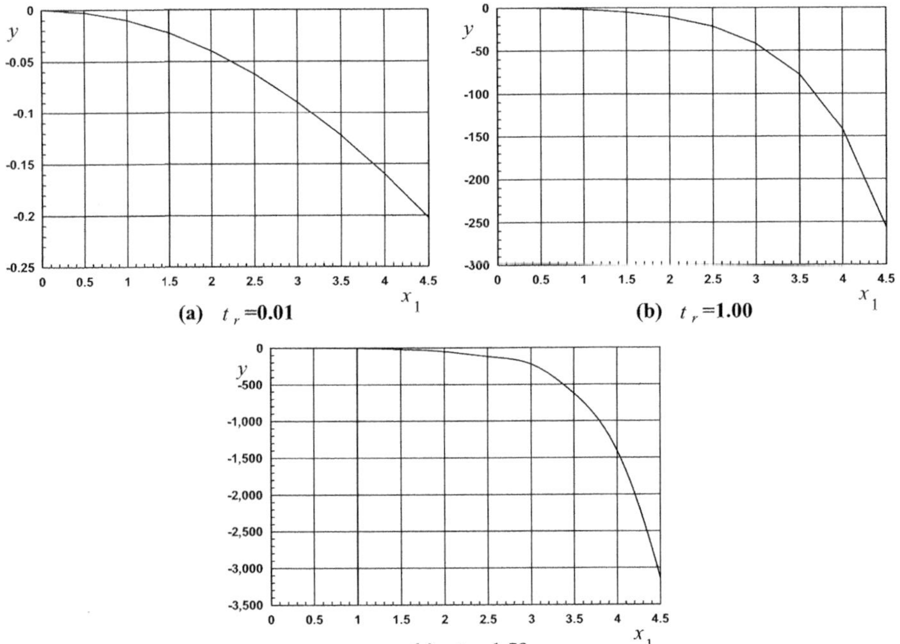

(a) t_r**=0.01** **(b)** t_r**=1.00**

(c) t_r**=1.50**

Fig. 9.2 Function $y = T_2(x_1, t_r)$.

Again, we can use the asymptotic properties listed in Eq. (B.4.5). In addition to Eq. (9.1.13), we shall employ the following formulae for Bessel functions of the second kind for large x:

$$Y_0(x) = \sqrt{\frac{2}{\pi x}}\left[\sin\left(x-\frac{\pi}{4}\right)-\frac{1}{8x}\cos\left(x-\frac{\pi}{4}\right)\right],$$
$$Y_1(x) = \sqrt{\frac{2}{\pi x}}\left[-\cos\left(x-\frac{\pi}{4}\right)\left(1+\frac{15}{128x^2}\right)+\frac{3}{8x}\sin\left(x-\frac{\pi}{4}\right)\right]. \tag{9.1.19}$$

Table 9.3 The first 10 positive roots (x) of $T_1(x,t_r)=0$.

t_r	1	2	3	4	5	6	7	8	9	10
0.1	31.4756	62.8618	94.2677	125.679	157.092	188.506	219.920	251.335	282.750	314.165
0.5	6.58174	12.7229	18.9549	25.2120	31.4794	37.7521	44.0277	50.3052	56.5840	62.8637
1.0	3.73599	6.64765	9.67962	12.7609	15.8648	18.9809	22.1040	25.2317	28.3624	31.4952
1.5	2.94936	4.88793	6.84232	8.83265	10.8515	12.8900	14.9421	17.0036	19.0718	21.1449

Applying Eqs. (9.1.13) and (9.1.19) to Eq. (9.1.18*a*), we can obtain, by omitting proper higher-order terms,

$$120xt_r\cos(xt_r)-\left[64x^2t_1t_2+225-\frac{48}{t_1t_2}(t_1^2+t_2^2)\right]\sin(xt_r)=0. \tag{9.1.20a}$$

This equation approximates Eq. (9.1.18*a*) accurately for large x. For example, the first two roots of Eq. (9.1.20*a*) for $t_r=0.5$ are 6.57240 and 12.7214, respectively, which agree well with those in Table 9.3. We can further simplify Eq. (9.1.20*a*) by retaining the dominant term only, which gives the following evaluation of the higher-order frequencies:

$$xt_r = n\pi, \tag{9.1.20b}$$

where n is an integer.

When the shell is very thin, $t_r \to 0$. Since $xt_r \neq 0$, xt_1 and xt_2 can be regarded as large numbers so that Eq. (9.1.20*a*) holds. Moreover, we have $t_1, t_2 \approx 1$ and $\tan(xt_r)/(xt_r)\to 1$. Thus, from Eq. (9.1.20*a*), we can deduce

$$64x^2+9=0, \tag{9.1.21}$$

which has no real root. On the other hand, Eq. (9.1.18*b*) is always valid, which predicts the frequency for the torsional vibration of a cylindrical shell as

$$\omega^2 = \frac{c_{44}}{\rho}\frac{m^2\pi^2}{L^2}. \tag{9.1.22}$$

When the material is isotropic, this is consistent with that presented on page 546 of Love (1927).

The reader should now be able to modify the preceding analysis to obtain the frequency equation of a cylindrical shell with fixed surfaces.

9.1.2 Breathing Mode Vibration

The breathing mode vibration is a purely radial vibration so that $u_r \neq 0$ and

$$w = u_\alpha = 0, \quad u_r = U(r,t), \quad \frac{\partial}{\partial\alpha}[\cdots] = \frac{\partial}{\partial z}[\cdots] = 0. \tag{9.1.23}$$

It can be shown that the second and third equations in Eq. (9.1.2) are satisfied automatically and the first one becomes

$$c_{11}\left(\frac{\partial^2}{\partial r^2} + \frac{1}{r}\frac{\partial}{\partial r} - \frac{1}{r^2}\right)U = \rho\frac{\partial^2 U}{\partial t^2}. \tag{9.1.24}$$

Because of the special nature of the breathing mode vibration, the aspect ratio of the cylinder, L/R, should be large enough so that the cylinder can be treated as infinite. Assuming that

$$U(r,t) = f_2(r)\exp(\mathrm{i}\omega t), \tag{9.1.25}$$

the substitution of this into Eq. (9.1.24) yields

$$\frac{\mathrm{d}^2 f_2(r)}{\mathrm{d}r^2} + \frac{1}{r}\frac{\mathrm{d}f_2(r)}{\mathrm{d}r} + \left(q^2 - \frac{1}{r^2}\right)f_2(r) = 0, \tag{9.1.26}$$

where $q^2 = \omega^2/v_1^2$ and $v_1^2 = c_{11}/\rho$. The solution to this equation is

$$f_2(r)=\begin{cases} A_2J_1(qr)+B_2Y_1(qr), & q>0, \\ A_2r+B_2\dfrac{1}{r}, & q=0, \end{cases} \tag{9.1.27}$$

where A_2 and B_2 are arbitrary constants. Therefore, Eqs. (9.1.1), (9.1.23) and (9.1.25) yield

$$\begin{aligned} \sigma_r &= \left[c_{11}\frac{\mathrm{d}f_2(r)}{\mathrm{d}r}+c_{12}\frac{f_2(r)}{r}\right]\exp(\mathrm{i}\omega t), \\ \sigma_\alpha &= \left[c_{12}\frac{\mathrm{d}f_2(r)}{\mathrm{d}r}+c_{11}\frac{f_2(r)}{r}\right]\exp(\mathrm{i}\omega t), \\ \sigma_z &= c_{13}\left[\frac{\mathrm{d}f_2(r)}{\mathrm{d}r}+\frac{f_2(r)}{r}\right]\exp(\mathrm{i}\omega t). \end{aligned} \tag{9.1.28}$$

(*a*) Cylinder

Similarly, $f_2(r)$ must be finite at $r=0$, which demands that $B_2=0$. Hence the nonzero stress component at the cylindrical surfaces is

$$\sigma_r=\left\{\begin{matrix} A_2\left\{c_{11}\left[qJ_0(qr)-\dfrac{1}{r}J_1(qr)\right]+c_{12}\dfrac{1}{r}J_1(qr)\right\}, & q>0 \\ (c_{11}+c_{12})A_2, & q=0 \end{matrix}\right\}\exp(\mathrm{i}\omega t). \tag{9.1.29}$$

As the surface is traction-free, *i.e.*, $\sigma_r=0$ $(r=b)$, we get $A_2=0$ when $q=0$. When $q>0$, we get the frequency equation

$$qbJ_0(qb)-(1-m_2)J_1(qb)=0, \tag{9.1.30}$$

where $m_2=c_{12}/c_{11}$. The above equation can be rewritten in a dimensionless form as

$$xt_2J_0(xt_2)-(1-m_2)J_1(xt_2)=0, \tag{9.1.31}$$

where $x=qR$ and $t_2=b/R$. Table 9.4 gives the first 10 positive roots of the above equation for $m_2=0.1$, 0.3, 0.5 and 0.7. Again, corresponding frequencies can be easily obtained using these roots.

Table 9.4 The first 10 positive roots (xt_2) of Eq. (9.1.31) .

m_2	1	2	3	4	5	6	7	8	9	10
0.1	1.91539	5.35084	8.54818	11.7146	14.8703	18.0211	21.1691	24.3154	27.4607	30.6052
0.3	2.04885	5.38936	8.57186	11.7318	14.8838	18.0322	21.1786	24.3237	27.4680	30.6117
0.5	2.16587	5.42743	8.59543	11.7489	14.8973	18.0433	21.1880	24.3319	27.4753	30.6183
0.7	2.26955	5.46496	8.61886	11.7660	14.9108	18.0544	21.1975	24.3401	27.4826	30.6248

Using the asymptotic expansions in Eq. (9.1.13), we get from Eq. (9.1.31) for large xt_2

$$xt_2 \cos\left(xt_2 - \frac{\pi}{4}\right) - \left(1 - m_2 - \frac{1}{8}\right)\sin\left(xt_2 - \frac{\pi}{4}\right) = 0 . \tag{9.1.32a}$$

We can consider only the dominant term if xt_2 is very large; this yields:

$$xt_2 = \frac{(2n-1)\pi}{2} + \frac{\pi}{4}, \tag{9.1.32b}$$

where n is an integer. It is then clear that the ratio of elastic constants $m_2 = c_{12} / c_{11}$ almost has no influence on large xt_2, as evidenced by the data in Table 9.4.

If the cylinder has a fixed surface, the boundary condition $u_r = 0, (r = b)$ applies. In this case, the frequency equation is determined from Eqs. (9.1.25) and (9.1.27) as $J_1(qb) = 0$. Table 9.2 has listed the first 40 positive roots of the Bessel function of the first kind of order one, which allows the calculation of the corresponding frequencies.

(*b*) Cylindrical shell

The nonzero stress component at the cylindrical surfaces, according to Eqs. (9.1.27) and (9.1.28), is given by

$$\sigma_r = \begin{cases} A_2\left\{c_{11}\left[qJ_0(qr) - \frac{1}{r}J_1(qr)\right] + c_{12}\frac{1}{r}J_1(qr)\right\} \\ \quad + B_2\left\{c_{11}\left[qY_0(qr) - \frac{1}{r}Y_1(qr)\right] + c_{12}\frac{1}{r}Y_1(qr)\right\}\exp(i\omega t), \quad q > 0, \\ \left[(c_{11} + c_{12})A_2 + (c_{12} - c_{11})B_2\frac{1}{r^2}\right]\exp(i\omega t), \quad q = 0. \end{cases} \tag{9.1.33}$$

Again, since the shell surfaces are traction-free, *i.e.*, $\sigma_r = 0$ $(r = a, b)$, $A_2 = 0$ and $B_2 = 0$ when $q = 0$, *i.e.*, when the shell is static. For $q > 0$, we get the frequency equation

$$\begin{vmatrix} qaJ_0(qa) - (1-m_2)J_1(qa) & qaY_0(qa) - (1-m_2)Y_1(qa) \\ qbJ_0(qb) - (1-m_2)J_1(qb) & qbY_0(qb) - (1-m_2)Y_1(qb) \end{vmatrix} = 0, \tag{9.1.34}$$

that has the non-dimensional form of

$$\begin{vmatrix} xt_1J_0(xt_1) - (1-m_2)J_1(xt_1) & xt_1Y_0(xt_1) - (1-m_2)Y_1(xt_1) \\ xt_2J_0(xt_2) - (1-m_2)J_1(xt_2) & xt_2Y_0(xt_2) - (1-m_2)Y_1(xt_2) \end{vmatrix} = T_3(x, m_2, t_r) = 0. \tag{9.1.35}$$

Table 9.5 lists the first 10 positive roots of $T_3(x, m_2, t_r) = 0$ at $t_r = 0.03$, 0.1, 0.5 and 1.0 $(m_2 = 0.3)$, which can be used for the corresponding frequency calculations.

Table 9.5 The first 10 positive roots (x) of $T_3(x, m_2, t_r) = 0$ $(m_2 = 0.3)$.

t_r	1	2	3	4	5	6	7	8	9	10
0.03	0.954018	104.725	209.442	314.161	418.880	523.600	628.319	733.039	837.759	942.478
0.10	0.954814	31.4343	62.8410	94.2539	125.668	157.083	188.499	219.914	251.330	282.745
0.50	0.975732	6.38307	12.6155	18.8822	25.1572	31.4355	43.9963	50.2777	56.5595	62.8416
1.00	1.03692	3.40809	6.41021	9.50781	12.6281	15.7572	18.8905	22.0262	25.1633	28.3015

Equation (9.1.35) is similar to Eq. (9.1.18*a*), and the following equation can be derived for large x

$$(56 - 64m_2)xt_r \cos(xt_r) - \left[64x^2t_1t_2 + (8 - 8m_2 - 1)^2 - \frac{24(1-m_2)}{t_1t_2}(t_1^2 + t_2^2)\right]\sin(xt_r) = 0. \tag{9.1.36a}$$

When $m_2 = -1$[2], Eq. (9.1.35) becomes identical to Eq. (9.1.18*a*), and Eq. (9.1.36*a*) also is the same as Eq. (9.1.20*a*).

Considering the dominant term only in Eq. (9.1.36a) gives approximately the higher-order frequencies:

[2] This is not possible from the physical point of view as $c_{11} > |c_{12}|$ [cf. **Section 2.2** or Payton (1983, page 4)].

$$xt_r = n\pi \, , \tag{9.1.36b}$$

where n is an integer. Hence, the parameter m_2 has little influence on the higher-order frequencies for the breathing mode vibration of a cylindrical shell.

When the shell is very thin, we can obtain similarly from Eq. (9.1.36*a*)

$$64x^2 + 64(1-m_2)^2 - 128(1-m_2) + 9 = 0 \, . \tag{9.1.37}$$

It can be shown that when $-\sqrt{55/64} < m_2 < \sqrt{55/64}$, which is generally occupied by most engineering materials as those listed in Table 1.4, this equation has a real positive root and governs the breathing mode free vibration of thin cylindrical shells. If m_2 of a particular material does not satisfy this condition, higher-order terms should be added to Eq. (9.1.36*a*) to obtain the corresponding frequency equation. This is however omitted for brevity.

Similarly, the breathing mode vibration of the cylindrical shell with fixed surfaces can be easily analysed.

9.1.3 Thickness-Shear Vibration

Under certain external stimuli, a cylinder or a cylindrical shell can have an axial displacement (w) only that is independent of coordinate z. Such vibration is known as the thickness-shear vibration (Baltrukonis and Gottenberg, 1959). In this case, we have

$$u_r = u_\alpha = 0 \, , \; w = W(r,\alpha,t) \, , \; \frac{\partial}{\partial z}[\cdots] = 0 \, . \tag{9.1.38}$$

Just as the breathing mode, the thickness-shear mode is also z-independent; the frequencies of both modes are thus the limiting frequencies for waves that can propagate along the cylinder/cylindrical shell, namely, these z-independent resonances are cut-offs that pin down one end of a dispersion curve for a mode (Thurston, 1992). As such it is expected that the thickness-shear modes would be contained within the framework of certain approximate theories. Thus the thickness-shear modes are available for correction of or comparison with these approximate theories. For example, Mindlin and Deresiewicz (1955) used the thickness-shear modes to calculate the shear coefficient of Timoshenko beam theory.

The first two equations in Eq. (9.1.2) are satisfied, but the third becomes

$$c_{44}\left(\frac{\partial^2}{\partial r^2}+\frac{1}{r}\frac{\partial}{\partial r}+\frac{1}{r^2}\frac{\partial^2}{\partial \alpha^2}\right)W=\rho\frac{\partial^2 W}{\partial t^2}. \tag{9.1.39}$$

If

$$W(r,\alpha,t)=f_3(r)\exp(\mathrm{i}n\alpha)\exp(\mathrm{i}\omega t), \tag{9.1.40}$$

the substitution of the above equation into Eq. (9.1.39) yields

$$\frac{\mathrm{d}^2 f_3(r)}{\mathrm{d}r^2}+\frac{1}{r}\frac{\mathrm{d}f_3(r)}{\mathrm{d}r}+\left(s^2-\frac{n^2}{r^2}\right)f_3(r)=0, \tag{9.1.41}$$

where $s^2=\omega^2/v_2^2$ and $v_2^2=c_{44}/\rho$. The solution of Eq. (9.1.41) is

$$f_3(r)=\begin{cases} A_3J_n(sr)+B_3Y_n(sr), & s>0,\\ A_3r^n+B_3r^{-n}, & s=0 \text{ and } n>0,\\ A_3+B_3\ln r, & s=0 \text{ and } n=0,\end{cases} \tag{9.1.42}$$

where A_3 and B_3 are two arbitrary constants. Hence, Eqs. (9.1.1), (9.1.38) and (9.1.40) give rise to the following nonzero stress components

$$\begin{aligned}\tau_{zr}&=c_{44}\frac{\mathrm{d}f_3(r)}{\mathrm{d}r}\exp(\mathrm{i}n\alpha)\exp(\mathrm{i}\omega t),\\ \tau_{\alpha z}&=c_{44}\frac{\mathrm{i}n}{r}f_3(r)\exp(\mathrm{i}n\alpha)\exp(\mathrm{i}\omega t).\end{aligned} \tag{9.1.43}$$

For a finite cylinder or a finite cylindrical shell, the natural boundary condition at the two ends for the thickness-shear vibration is

$$u_r=u_\alpha=\sigma_z=0, \quad (z=0,L), \tag{9.1.44}$$

which are just the simply supported conditions discussed by Hoff and Soong (1965) and Hoff and Rehfield (1965).

(*a*) Cylinder

Because of the requirement of the finite value of $f_3(r)$ at $r=0$, $B_3=0$. Hence, the nonzero stress component at the cylindrical surfaces is

$$\tau_{zr} = \begin{Bmatrix} c_{44}A_3\left[\dfrac{n}{r}J_n(sr) - sJ_{n+1}(sr)\right], & s>0 \\ c_{44}A_3 nr^{n-1}, & s=0 \end{Bmatrix} \exp(\mathrm{i}n\alpha)\exp(\mathrm{i}\omega t), \tag{9.1.45}$$

For $s=0$, by making use of the free surface condition, $\tau_{zr}=0 \quad (r=b)$, we get $n=0$, for which $w=A_3\exp(\mathrm{i}\omega t)$ representing a rigid body motion along z-axis, or $A_3=0$ corresponding to the static state.

When $s>0$, we get the following frequency equation

$$nJ_n(sb) - sbJ_{n+1}(sb) = 0, \tag{9.1.46}$$

or in a dimensionless form,

$$nJ_n(xt_2) - xt_2 J_{n+1}(xt_2) = 0, \tag{9.1.47}$$

where $x=sR$. Table 9.6 lists the first 5 positive roots of this equation for $n=0$-4 for the natural frequency calculations of the thickness-shear vibration.

Table 9.6 The first 5 positive roots (xt_2) of Eq. (9.1.47).

n	1	2	3	4	5
0	3.8317	7.0156	10.1735	13.3237	16.4706
1	1.8412	5.3314	8.5363	11.7060	14.8636
2	3.0542	6.7061	9.9695	13.1704	16.3475
3	4.2012	8.0152	11.3459	14.5859	17.7888
4	5.3175	9.2824	12.6819	15.9641	19.1960

For the asymptotic analysis, we adopt the following formulae from Eq. (B.4.5)

$$J_n(x) = \sqrt{\frac{2}{\pi x}}\left\{\cos\left(x - \frac{n\pi}{2} - \frac{\pi}{4}\right) - \sin\left(x - \frac{n\pi}{2} - \frac{\pi}{4}\right)\frac{4n^2-1}{8x}\right\},$$

$$J_{n+1}(x) = \sqrt{\frac{2}{\pi x}}\left\{\sin\left(x - \frac{n\pi}{2} - \frac{\pi}{4}\right)\left[1 - \frac{16(n+1)^4 - 40(n+1)^2 + 9}{128x^2}\right] + \cos\left(x - \frac{n\pi}{2} - \frac{\pi}{4}\right)\frac{4(n+1)^2-1}{8x}\right\}. \tag{9.1.48}$$

We then obtain from Eq. (9.1.47) for large x

$$\frac{4n^2+3}{8}\cos\left(xt_2-\frac{n\pi}{2}-\frac{\pi}{4}\right)+xt_2\sin\left(xt_2-\frac{n\pi}{2}-\frac{\pi}{4}\right)=0. \tag{9.1.49a}$$

When $n=2$, for example, the second and third roots of this equation are 6.7293 and 9.9765, respectively, which agree with the accurate ones in Table 9.6. Further considering the dominant term only in Eq. (9.1.49*a*) gives higher-order frequencies approximately:

$$xt_2 = m\pi+\frac{n\pi}{2}+\frac{\pi}{4}, \tag{9.1.49b}$$

where m is an integer.

For the fixed surface, the boundary condition becomes $w=0$ $(r=b)$. Hence Eqs. (9.1.40) and (9.1.42) lead to $J_n(sb)=0$ whose first 5 positive roots for n= 0-4 are given in Table 9.7. The asymptotic analysis gives approximately higher-order frequencies as $sb=(4m+2n-1)\pi/4$, with m being an integer.

Table 9.7 The first 5 positive roots (sb) of $J_n(sb)=0$.

n	1	2	3	4	5
0	2.0448	5.5201	8.6537	11.7915	14.9309
1	3.8317	7.0156	10.1735	13.3237	16.4706
2	5.1356	8.4172	11.6198	14.7960	17.9598
3	6.3802	9.7610	13.0152	16.2235	19.4094
4	7.5883	11.6047	14.3725	17.6160	20.8269

(*b*) Cylindrical shell

From Eqs. (9.1.42) and (9.1.43), the nonzero stress component at the cylindrical surfaces is

$$\tau_{zr}=\left\{\begin{array}{ll} c_{44}A_3\left[\frac{n}{r}J_n(sr)-sJ_{n+1}(sr)\right]+c_{44}B_3\left[\frac{n}{r}Y_n(sr)-sY_{n+1}(sr)\right], & s>0 \\ c_{44}A_3nr^{n-1}-c_{44}B_3nr^{-n-1}, & s=0 \text{ and } n>0 \\ c_{44}B_3\frac{1}{r}, & s=0 \text{ and } n=0 \end{array}\right\} \times\exp(\mathrm{i}n\alpha)\exp(\mathrm{i}\omega t). \tag{9.1.50}$$

The traction-free condition, $\tau_{zr}=0$ $(r=a,b)$, specifies that when $s=0$, $w=A_3\exp(\mathrm{i}\omega t)$ for $n=0$ (a rigid body motion along z-axis) or $A_3=B_3=0$ for a static state. If $s>0$, the following frequency equation holds

$$\begin{vmatrix} nJ_n(sa)-saJ_{n+1}(sa) & nY_n(sa)-saY_{n+1}(sa) \\ nJ_n(sb)-sbJ_{n+1}(sb) & nY_n(sb)-sbY_{n+1}(sb) \end{vmatrix}=0, \tag{9.1.51}$$

or in a non-dimensional form,

$$\begin{vmatrix} nJ_n(xt_1)-xt_1J_{n+1}(xt_1) & nY_n(xt_1)-xt_1Y_{n+1}(xt_1) \\ nJ_n(xt_2)-xt_2J_1(xt_2) & nY_n(xt_2)-xt_2Y_{n+1}(xt_2) \end{vmatrix}=T_4(x,n,t_r)=0. \tag{9.1.52}$$

Table 9.8 gives the first 5 positive roots of the equation $T_4(x,n,t_r)=0$ for $t_r=0.03$, 0.1, 0.5 and 1.0 ($n=0,1,2$) which are needed for frequency calculations.

Table 9.8 The first 5 positive roots (x) of $T_4(x,n,t_r)=0$.

n	t_r	1	2	3	4	5
0	0.03	104.723	209.441	314.160	418.880	523.599
	0.10	31.4279	62.8378	94.2518	125.667	157.082
	0.50	6.34407	12.5978	18.8707	25.1486	31.4286
	1.00	3.27123	6.35768	9.47618	12.6054	15.7394
1	0.03	104.728	209.444	314.162	418.881	523.600
	0.10	31.4439	62.8458	94.2571	125.671	157.085
	0.50	1.00966	6.43287	12.6408	18.8991	25.1699
	1.00	1.02724	3.51553	6.47222	9.54988	12.6598
2	0.03	104.742	209.451	314.167	418.885	523.603
	0.10	31.4918	62.8698	94.2730	125.683	157.095
	0.50	2.01276	6.69366	12.7692	18.9843	25.2337
	1.00	1.95499	4.18021	6.81335	9.77013	12.8225

For the asymptotic analysis, in addition to Eq. (9.1.48), we shall employ

$$Y_n(x)=\sqrt{\frac{2}{\pi x}}\left\{\sin\left(x-\frac{n\pi}{2}-\frac{\pi}{4}\right)+\cos\left(x-\frac{n\pi}{2}-\frac{\pi}{4}\right)\frac{4n^2-1}{8x}\right\},$$

$$Y_{n+1}(x)=\sqrt{\frac{2}{\pi x}}\left\{-\cos\left(x-\frac{n\pi}{2}-\frac{\pi}{4}\right)\left[1-\frac{16(n+1)^4-40(n+1)^2+9}{128x^2}\right]\right. \tag{9.1.53}$$

$$\left.+\sin\left(x-\frac{n\pi}{2}-\frac{\pi}{4}\right)\frac{4(n+1)^2-1}{8x}\right\}.$$

Then we obtain from Eq. (9.1.52) by omitting proper higher-order terms

$$\left\{x^2 t_1 t_2 + n^2 + \frac{[4(n+1)^2-1]^2}{64} - \frac{16(n+1)^4 - 40(n+1)^2 + 9}{128 t_1 t_2}(t_1^2 + t_2^2) + \frac{n(4n^2-1)}{8 t_1 t_2}(t_1^2 + t_2^2) - \frac{n[4(n+1)^2-1]}{4}\right\} \sin(x t_r) - \frac{4n^2+3}{8} x t_r \cos(x t_r) = 0. \tag{9.1.54a}$$

If we consider only the leading term in this equation, then we get the approximation of higher-order frequencies

$$x t_r = m\pi, \quad (m \text{ is an integer}). \tag{9.1.54b}$$

It becomes obvious that the circumferential wave number n almost has no influence on the higher-order frequencies for the thickness-shear vibration of a cylindrical shell, as also seen from the data in Table 9.8.

As with the discussion of breathing mode vibration, we obtain from Eq. (9.1.54*a*) for thin cylindrical shells

$$x = \omega R \sqrt{\frac{\rho}{c_{44}}} = n. \tag{9.1.55}$$

We can also derive this formula from the thin shell theory directly [cf. Love (1927, page 545) for isotropic materials].

9.2 ASYMMETRIC VIBRATION[3]

For the general asymmetric free vibration, it is more convenient to use the displacement functions introduced in **Chapter 2**. Equation (2.2.2*a*) can be rewritten as

$$u_r = \frac{1}{r}\frac{\partial \psi}{\partial \alpha} - \frac{\partial G}{\partial r}, \quad u_\alpha = -\frac{\partial \psi}{\partial r} - \frac{1}{r}\frac{\partial G}{\partial \alpha}, \quad w = W. \tag{9.2.1}$$

[3] The three-dimensional solution for an isotropic cylinder was published by Pochhammer (1876), and later elaborated by Chree (1889), see Love (1927).

Hence, Eq. (9.1.2) gives

$$\begin{aligned} &\left(c_{66}\Lambda + c_{44}\frac{\partial^2}{\partial z^2} - \rho\frac{\partial^2}{\partial t^2}\right)\psi = 0, \\ &\left(c_{11}\Lambda + c_{44}\frac{\partial^2}{\partial z^2} - \rho\frac{\partial^2}{\partial t^2}\right)G - (c_{13}+c_{44})\frac{\partial W}{\partial z} = 0, \\ &-(c_{13}+c_{44})\Lambda\frac{\partial G}{\partial z} + \left(c_{44}\Lambda + c_{33}\frac{\partial^2}{\partial z^2} - \rho\frac{\partial^2}{\partial t^2}\right)W = 0, \end{aligned} \tag{9.2.2}$$

where $\Lambda = (1/r)\partial/\partial r(r\partial/\partial r) + (1/r^2)\partial^2/\partial\alpha^2$. This equation can also be obtained directly from Eqs. (2.2.6), (2.2.14) and (2.2.18). If the cylindrical shell has rigidly slipping conditions at the two ends as shown in Eq. (9.1.6), we can assume

$$\begin{aligned} \psi(r,\alpha,z,t) &= R^2\overline{\psi}(\xi)\cos(m\pi\zeta)\sin(n\alpha)\exp(\mathrm{i}\omega t), \\ G(r,\alpha,z,t) &= R^2\overline{G}(\xi)\cos(m\pi\zeta)\cos(n\alpha)\exp(\mathrm{i}\omega t), \\ W(r,\alpha,z,t) &= R\overline{W}(\xi)\sin(m\pi\zeta)\cos(n\alpha)\exp(\mathrm{i}\omega t), \end{aligned} \tag{9.2.3}$$

where $\xi = r/R$ and $\zeta = z/L$ are two dimensionless variables. It can be shown that when $n = 0$, the shell vibrates axisymmetrically and $\psi = 0$. Substituting Eq. (9.2.3) into Eq. (9.2.2), we obtain

$$(\Lambda_1 + k_1^2)\overline{\psi} = 0, \tag{9.2.4}$$

$$\begin{aligned} (\Lambda_1 + g_1)\overline{G} + g_2\overline{W} &= 0, \\ (\Lambda_1 + g_3)\overline{W} + g_4\overline{G} &= 0, \end{aligned} \tag{9.2.5}$$

where $\Lambda_1 = (1/\xi)(\mathrm{d}/\mathrm{d}\xi)(\xi\,\mathrm{d}/\mathrm{d}\xi) - n^2/\xi^2$, and

$$\begin{aligned} &k_1^2 = 2(\Omega^2 - t_L^2)/(p_1 - p_2), \quad \Omega = \omega/\omega_s, \quad \omega_s = v_2/R, \\ &p_1 = c_{11}/c_{44}, \quad p_2 = c_{12}/c_{44}, \quad p_3 = c_{13}/c_{44}, \quad p_4 = c_{33}/c_{44}, \\ &g_1 = (\Omega^2 - t_L^2)/p_1, \quad g_2 = -(p_3+1)t_L/p_1, \\ &g_3 = [(p_3+1)^2 - p_1 p_4]t_L^2/p_1 + \Omega^2, \quad g_4 = -(p_3+1)t_L(\Omega^2 - t_L^2)/p_1, \end{aligned} \tag{9.2.6}$$

where $v_2 = \sqrt{c_{44}/\rho}$ is the velocity of the transverse wave in the shell. Note that the non-dimensional frequency defined here is different from that in Eq. (9.1.17).

Equation (9.2.4) is a Bessel's equation of order n, whose solution is

$$\overline{\psi} = \begin{cases} A_1 J_n(k_1\xi) + B_1 Y_n(k_1\xi), & k_1^2 > 0, \\ A_1\xi^n + B_1\xi^{-n}, & k_1^2 = 0, \\ A_1 I_n(\tilde{k}_1\xi) + B_1 K_n(\tilde{k}_1\xi), & \tilde{k}_1^2 = -k_1^2 > 0, \end{cases} \tag{9.2.7}$$

where J_n and Y_n are the Bessel functions of the first and second kinds of order n, respectively, I_n and K_n are the modified Bessel functions of the first and second kinds of order n, respectively, and A_1 and B_1 are arbitrary constants.

Equation (9.2.5) can be rewritten as

$$\overline{W} = -\frac{1}{g_2}(\Lambda_1 + g_1)\overline{G}, \tag{9.2.8a}$$

$$(\Lambda_1 - k_2^2)(\Lambda_1 - k_3^2)\overline{G} = 0, \tag{9.2.8b}$$

where k_2 and k_3 (assume that $\mathrm{Re}[k_2] \geq \mathrm{Re}[k_3] \geq 0$) are two roots of the following eigen-equation

$$k^4 + \overline{B}k^2 + \overline{C} = 0, \tag{9.2.9}$$

where $\overline{B} = g_1 + g_3$ and $\overline{C} = g_1 g_3 - g_2 g_4$. Equation (9.2.9) is identical to that obtained by Mirsky (1965) and Chau (1994) using different methods.

The solution to Eq. (9.2.8*b*) depends on the nature of the roots of Eq. (9.2.9) that, in turn, relies on the range of the frequencies and the properties of the particular material under consideration. Mirsky (1965) and Chau (1994) pointed out that there are four possible cases associated with the roots of Eq. (9.2.9):

(a) all the roots are complex when $\overline{B}^2 - 4\overline{C} < 0$;

(b) all are imaginary when $\overline{B}^2 - 4\overline{C} > 0$, $\overline{B} > 0$ and $\overline{C} > 0$;

(c) two are real and the other two are imaginary when $\overline{C} < 0$; and

(d) all are real when $\overline{B}^2 - 4\overline{C} > 0$, $\overline{B} < 0$ and $\overline{C} > 0$.

They presented a solution using Bessel functions with real arguments. However, as we will show later, for zinc, the arguments may be complex as the frequency can be very low. Mirsky (1965) and Prasad and Jain (1966) did not obtain the solution with complex roots but Mirsky (1965) suggested that shell theory should be employed in this case. On the other hand, Chau (1994) considered a specific transversely isotropic material, sapphire, that allows only the latter three cases, *i.e.*, (b) to (d).

As a matter of fact, the solution to the first case may be obtained directly using the modified Bessel functions with complex arguments. Since k_2 and k_3 are conjugated in this case and the terms with k_2 always appear with k_3, the final expressions of displacements and stresses as well as the frequency equation will all be real. Hence, the operator theory can give rise to the solution of Eq. (9.2.8*b*), *i.e.*,

$$\overline{G}(\xi) = \overline{G}_1(\xi) + \overline{G}_2(\xi). \tag{9.2.10}$$

Table 9.9 summarizes the forms of the functions $\overline{G}_i(\xi)$ $(i = 1,2)$ for the four cases, where $(\widetilde{k}_j)^2 = -k_j^2, (j = 2, 3)$.

Table 9.9 Different forms of $\overline{G}_i(\xi)$.

Case (a)		Case (b)		Case (c)		Case (d)	
$\overline{G}_1(\xi)$	$\overline{G}_2(\xi)$	$\overline{G}_1(\xi)$	$\overline{G}_2(\xi)$	$\overline{G}_1(\xi)$	$\overline{G}_2(\xi)$	$\overline{G}_1(\xi)$	$\overline{G}_2(\xi)$
$I_n(k_2\xi)$, $K_n(k_2\xi)$	$I_n(k_3\xi)$, $K_n(k_3\xi)$	$J_n(\widetilde{k}_2\xi)$, $Y_n(\widetilde{k}_2\xi)$	$J_n(\widetilde{k}_3\xi)$, $Y_n(\widetilde{k}_3\xi)$	$I_n(k_2\xi)$, $K_n(k_2\xi)$	$J_n(\widetilde{k}_3\xi)$, $Y_n(\widetilde{k}_3\xi)$	$I_n(k_2\xi)$, $K_n(k_2\xi)$	$I_n(k_3\xi)$, $K_n(k_3\xi)$

In the following, let us consider Case (a) as an example of derivation. In this case, functions $\overline{G}_1(\xi)$ and $\overline{G}_2(\xi)$ in Eq. (9.2.10) are

$$\begin{aligned} \overline{G}_1(\xi) &= A_2 I_n(k_2\xi) + B_2 K_n(k_2\xi), \\ \overline{G}_2(\xi) &= A_3 I_n(k_3\xi) + B_3 K_n(k_3\xi). \end{aligned} \tag{9.2.11}$$

The substitution of Eqs. (9.2.10) and (9.2.11) into Eq. (9.2.8*a*) gives

$$\overline{W}(\xi) = -[(k_2^2 + g_1)\overline{G}_1(\xi) + (k_3^2 + g_1)\overline{G}_2(\xi)]/g_2. \tag{9.2.12}$$

With the aid of Eqs. (9.1.1) and (9.2.1), the displacements and stresses at the cylindrical surfaces in terms of displacement functions can be obtained as

$$\begin{aligned}
&\bar{u}_r = (n\bar{\psi}/\xi - \bar{G}')\cos(m\pi\zeta)\cos(n\alpha)\exp(\mathrm{i}\,\omega t), \\
&\bar{u}_\alpha = (-\bar{\psi}' + n\bar{G}/\xi)\cos(m\pi\zeta)\sin(n\alpha)\exp(\mathrm{i}\,\omega t), \\
&\bar{w} = \bar{W}\sin(m\pi\zeta)\cos(n\alpha)\exp(\mathrm{i}\,\omega t), \\
&\bar{\sigma}_r = [-p_1\bar{G}'' - p_2\bar{G}'/\xi + p_2 n^2\bar{G}/\xi^2 + (p_1 - p_2)n\bar{\psi}'/\xi - (p_1 - p_2)n\bar{\psi}/\xi^2 \\
&\qquad + p_3 t_L\bar{W}]\cos(m\pi\zeta)\cos(n\alpha)\exp(\mathrm{i}\,\omega t), \\
&\bar{\tau}_{zr} = (t_L\bar{G}' - t_L n\bar{\psi}/\xi + \bar{W}')\sin(m\pi\zeta)\cos(n\alpha)\exp(\mathrm{i}\,\omega t), \\
&\bar{\tau}_{r\alpha} = (2n\bar{G}'/\xi - 2n\bar{G}/\xi^2 - \bar{\psi}'' + \bar{\psi}'/\xi - n^2\bar{\psi}/\xi^2) \\
&\qquad \times\cos(m\pi\zeta)\sin(n\alpha)\exp(\mathrm{i}\,\omega t),
\end{aligned} \tag{9.2.13}$$

where $\bar{u}_r = u_r / R$, $\bar{u}_\alpha = u_\alpha / R$, $\bar{w} = w/R$, $\bar{\sigma}_r = \sigma_r / c_{44}$, $\bar{\tau}_{zr} = \tau_{zr}/c_{44}$ and $\bar{\tau}_{r\alpha} = \tau_{r\alpha}/c_{44}$ are dimensionless displacement and stress components, and the prime represents differentiation with respect to ξ.

If a cylindrical shell or a cylinder is subjected to free vibration, its surface is traction-free, *i.e.*,

$$\sigma_r = \tau_{zr} = \tau_{r\alpha} = 0, \quad (\xi = t_1, t_2). \tag{9.2.14}$$

In the following we will consider a cylinder and a cylindrical shell separately.

(*a*) Cylinder

Since the solution should be finite at $r = 0$, we know that $B_i = 0$ $(i = 1, 2, 3)$. Further using the traction-free condition at the surface $(\xi = t_2)$, we can derive three linear homogeneous algebraic equations about A_i $(i = 1, 2, 3)$ from Eqs. (9.2.7), (9.2.10), (9.2.12) and (9.2.13). The coefficient determinant of these homogeneous equations must be zero, thus

$$|E_{ij}^1| = 0, \quad (i, j = 1, 2, 3). \tag{9.2.15}$$

For $k_1^2 > 0$ and Case (a) listed in Table 9.9, the elements in the frequency equation (9.2.15) have the following forms

$$
\begin{aligned}
&E_{11}^1 = (p_1 - p_2)nJ_n'(k_1t_2)/t_2 - (p_1 - p_2)nJ_n(k_1t_2)/t_2^2, \\
&E_{12}^1 = -p_1I_n''(k_2t_2) - p_2I_n'(k_2t_2)/t_2 + (p_2n^2/t_2^2 + p_3t_Lq_1)I_n(k_2t_2), \\
&E_{13}^1 = -p_1I_n''(k_3t_2) - p_2I_n'(k_3t_2)/t_2 + (p_2n^2/t_2^2 + p_3t_Lq_2)I_n(k_3t_2), \\
&E_{21}^1 = -t_LnJ_n(k_1t_2)/t_2, \quad E_{22}^1 = (t_L + q_1)I_n'(k_2t_2), \quad E_{23}^1 = (t_L + q_2)I_n'(k_3t_2), \qquad (9.2.16) \\
&E_{31}^1 = -J_n''(k_1t_2) + J_n'(k_1t_2)/t_2 - n^2J_n(k_1t_2)/t_2^2, \\
&E_{32}^1 = 2nI_n'(k_2t_2)/t_2 - 2nI_n(k_2t_2)/t_2^2, \\
&E_{33}^1 = 2nI_n'(k_3t_2)/t_2 - 2nI_n(k_3t_2)/t_2^2,
\end{aligned}
$$

where $q_1 = -(k_2^2 + g_1)/g_2$ and $q_2 = -(k_3^2 + g_1)/g_2$. The solution for a cylinder with simply supported ends, as specified by Eq. (9.1.44), can be obtained by simply interchanging the terms $\cos(m\pi\zeta)$ and $\sin(m\pi\zeta)$ in Eq. (9.2.3). The resulting frequency equation still has the same form as Eq. (9.2.15) (Chen, Ying and Yang, 1998). Therefore, the numerical results below are valid for both these two boundary conditions.

Table 9.10 lists the lowest natural frequencies of a sapphire cylinder for the axisymmetric mode ($n = 0$) and the bending mode ($n = 1$) for $m = 1 - 4$. The elastic constants and density of sapphire are available in Table 1.4. We also take $L = 2.97815$ cm and $b = 1.4986$ cm. The experimental study by Lusher and Hardy (1988) showed that the frequency range of axisymmetric modes for the sapphire cylinder is in the interval of 150 kHz to 450 kHz with the lowest frequency (axisymmetric vibration) being 175.1 or 175.2 kHz. Table 9.10 shows that the theoretical prediction agrees well with the experimental observation. However, it is interesting to note that the lowest frequency of the cylinder, according to the theoretical prediction, is associated with a non-axisymmetric vibration mode and is out of the frequency range considered by Lusher and Hardy (1988).

Table 9.10 The lowest natural frequencies of the cylinder ($\omega/2\pi$, kHz).

m	1	2	3	4
$n = 0$	173.57	253.22	321.63	404.34
$n = 1$	77.09	179.71	280.28	379.84

(*b*) Cylindrical shell

In this case, six linear homogeneous algebraic equations with six unknowns, A_i and B_i $(i = 1, 2, 3)$ can be obtained using the boundary conditions, Eq. (9.2.14), and the solution, Eqs. (9.2.7) and (9.2.10) to (9.2.13). Again, the coefficient determinant of the homogeneous equations should vanish for a nontrivial solution. This leads to

$$\left|E_{ij}^{2}\right|=0, \quad (i,j=1,2,\cdots,6). \tag{9.2.17}$$

For $k_1^2 > 0$ and Case (a) in Table 9.9, we have $E_{ij}^2 = E_{ij}^1$ $(i, j = 1, 2, 3)$. E_{ij}^2 $(j = 4, 5, 6)$ can be obtained by replacing the (modified) Bessel functions of the first kind in E_{ij}^2 $(j = 1, 2, 3)$ with the ones of the second kind, while E_{ij}^2 $(i = 4, 5, 6)$ can be obtained by replacing t_2 in E_{ij}^2 $(i = 1, 2, 3)$ with t_1.

As before, we can perform an asymptotic analysis of the frequency equations given in Eqs. (9.2.15) and (9.2.17) for large arguments. This is, however, omitted for brevity. It is emphasized here that it is very easy to perform numerical calculation directly from Eqs. (9.2.15) and (9.2.17) by virtue of softwares such as MATHEMATICA, MATLAB, MATHCAD, *etc*.

In the following, we consider a transversely isotropic cylindrical shell made of zinc, whose elastic constants and density were given in Table 1.4. The dimensionless elastic constants are calculated as $p_1 = 3.9563$, $p_2 = 0.7883$, $p_3 = 1.1860$, and $p_4 = 1.5400$.

First, let us see the nature of the roots of Eq. (9.2.9) of the zinc shell. There are three possible situations:

1) $0 < \eta < 0.8876$, where $\eta = \Omega / t_L$, which leads to four complex roots;

2) $0.8876 < \eta < 1$ or $\eta > 1.2410$, which yields four pure imaginary roots; and

3) $1.0 < \eta < 1.2410$, which gives rise to two real and two pure imaginary roots.

The modified Bessel functions will have complex arguments in case 1).

The lowest natural frequencies calculated by the present method are in good agreement with the shear shell theory predictions (Jain, 1974), as compared in Table 9.11. It is noted that most lower frequencies fall in $0 < \eta < 0.8876$, which demonstrates the reliability of the present method. The lowest natural frequency increases with t_r slowly when mR/L is small but more rapidly when the parameter becomes greater. This indicates that the effect of thickness on frequency is more pronounced for a cylindrical shell with a large mR/L. On the other hand, Jain's results are in better agreement with the present predictions when $n=1$. This is because Jain (1974) introduced two modified shear factors ($\kappa_r = \kappa_\theta = \pi^2/12$) with the condition that both n and t_r are small. This certainly led to an error for large n or t_r, as pointed out by Mirsky (1964).

Table 9.11 The dimensionless frequencies of a simply supported zinc cylindrical shell.

t_r	mR/L	$n=1$			$n=2$		
		Ω Present	Ω Jain (1974)	η	Ω Present	Ω Jain (1974)	η
0.01	0.1	0.0644	0.0643	0.205	0.0262	0.0262	0.083
	0.2	0.2268	0.2269	0.361	0.0844	0.0845	0.134
	0.4	0.6761	0.6764	0.538	0.3028	0.3017	0.241
	0.6	1.1291	1.1289	0.599	0.5900	0.5899	0.313
	1.0	1.5488	1.5430	0.493	1.1247	1.1267	0.358
	2.0	1.7027	1.7012	0.271	1.5959	1.5990	0.254
0.10	0.1	0.0644	0.0644	0.205	0.1541	0.1543	0.490
	0.2	0.2275	0.2272	0.362	0.1810	0.1809	0.288
	0.4	0.6773	0.6775	0.539	0.3566	0.3582	0.284
	0.6	1.1329	1.1327	0.601	0.6409	0.6419	0.340
	1.0	1.5802	1.5781	0.503	1.2189	1.2198	0.388
	2.0	2.0797	2.0747	0.331	2.1237	2.1164	0.338
0.25	0.1	0.0647	0.0649	0.206	0.3700	0.3723	1.178
	0.2	0.2287	0.2288	0.364	0.3974	0.3985	0.633
	0.4	0.6824	0.6832	0.543	0.5504	0.5542	0.438
	0.6	1.1517	1.1515	0.611	0.8369	0.8412	0.444
	1.0	1.7122	1.7090	0.545	1.5519	1.5560	0.494
	2.0	3.0890	3.0651	0.492	3.3615	3.3266	0.535
0.50	0.1	0.0660	0.0664	0.210	0.6940	0.6843	2.209
	0.2	0.2331	0.2341	0.371	0.6875	0.7155	1.094
	0.4	0.6999	0.7012	0.557	0.8719	0.8753	0.694
	0.6	1.2026	1.2037	0.638	1.1706	1.1772	0.621
	1.0	2.0012	1.9917	0.637	2.0420	2.0404	0.650
	2.0	4.2097	4.1258	0.670	4.5742	4.4843	0.728

The proposed method can be easily applied to the free vibration of a laminated cylindrical shell. Assume that the shell consists of p layers of transversely isotropic materials. The stress and displacement continuity at the interface of the two adjacent layers can be described as

$$\begin{aligned}&(\sigma_r)_j=(\sigma_r)_{j+1},\quad (\tau_{zr})_j=(\tau_{zr})_{j+1},\quad (\tau_{r\alpha})_j=(\tau_{r\alpha})_{j+1},\\&(u_r)_j=(u_r)_{j+1},\quad (u_\alpha)_j=(u_\alpha)_{j+1},\quad (w)_j=(w)_{j+1},\\&\qquad\qquad (r=r_j,\ \ j=1,2,\cdots,p-1),\end{aligned}\tag{9.2.18}$$

where r_j is the radius at the interface between the jth and $(j+1)^{\text{th}}$ layers. Using Eq. (9.2.13) and noting that the inner and outer surfaces of the shell ($r = a, b$) are stress-free as specified by Eq. (9.2.14), we obtain, in connection with the continuity conditions in Eq. (9.2.18), $6p$ linear homogeneous algebraic equations with $6p$ arbitrary constants. The coefficient determinant of the equation should vanish to guarantee a nontrivial solution and leads to the frequency equation.

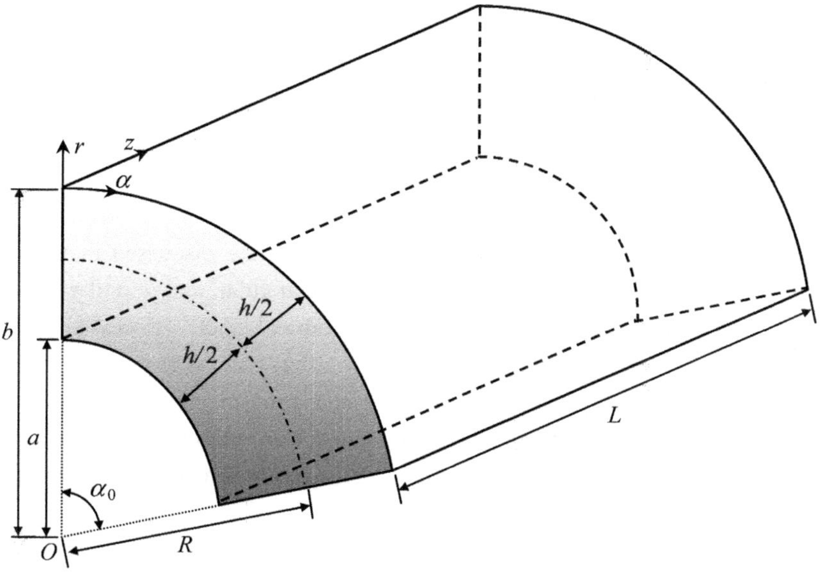

Fig. 9.3 Geometry of a cylindrical panel.

This analysis can be further applied to the free vibration of a *cylindrical panel*, as illustrated in Fig. 9.3, with all four edges simply supported[4] (Soldatos and Hadjigeorgiou, 1990)

$$\begin{aligned} &u_r = \sigma_z = u_\alpha = 0, \quad (z = 0,\ L), \\ &u_r = \sigma_\alpha = w = 0, \quad (\alpha = 0,\ \alpha_0). \end{aligned} \tag{9.2.19}$$

[4] We can treat an arbitrary combination of simply supported and rigidly slipping conditions by choosing the axial number or the circumferential number as an integer or a half integer (see **Section 8.1** for details).

We can take

$$\begin{aligned}\psi(r,\alpha,z,t)&=R^2\overline{\psi}(\xi)\sin(m\pi\zeta)\cos(n\alpha/\alpha_0)\exp(\mathrm{i}\omega t),\\ G(r,\alpha,z,t)&=R^2\overline{G}(\xi)\sin(m\pi\zeta)\sin(n\alpha/\alpha_0)\exp(\mathrm{i}\omega t),\\ W(r,\alpha,z,t)&=R\overline{W}(\xi)\cos(m\pi\zeta)\sin(n\alpha/\alpha_0)\exp(\mathrm{i}\omega t).\end{aligned} \tag{9.2.20}$$

The detailed analysis as well as numerical examples can be found in Chen, Cai and Ye, *et al.* (1998).

9.3 VIBRATION OF A LAYERED CYLINDRICAL SHELL

9.3.1 State-Space Formulations

The state-space formulations in Cartesian coordinates were discussed in **Section 2.1**. The relevant solution procedure for laminated rectangular and circular plates was described in **Section 8.2**. For a laminated cylindrical shell, however, it is more convenient to use $\mathbf{A}=[u_r, u_\alpha, w, \sigma_r, \tau_{r\alpha}, \tau_{zr}]^{\mathrm{T}}$ as the state vector (Fan and Zhang, 1992). The state equation for a transversely isotropic medium can be derived from Eqs. (9.1.1) and (1.2.11), or directly obtained from that for an orthotropic medium (Fan and Zhang, 1992). The state equation without considering body forces can be written as

$$\frac{\partial}{\partial r}\mathbf{A}=\mathbf{P}(r)\mathbf{A}, \tag{9.3.1}$$

where

$$\mathbf{P}(r)=\begin{bmatrix}\mathbf{P}_1(r) & \mathbf{P}_2(r)\\ \mathbf{P}_3(r) & \mathbf{P}_4(r)\end{bmatrix},$$

$$\mathbf{P}_1(r)=\begin{bmatrix}-\dfrac{c_{12}}{c_{11}}\dfrac{1}{r} & -\dfrac{c_{12}}{c_{11}}\dfrac{1}{r}\dfrac{\partial}{\partial\alpha} & -\dfrac{c_{13}}{c_{11}}\dfrac{\partial}{\partial z}\\ -\dfrac{1}{r}\dfrac{\partial}{\partial\alpha} & \dfrac{1}{r} & 0\\ -\dfrac{\partial}{\partial z} & 0 & 0\end{bmatrix},\quad \mathbf{P}_2(r)=\begin{bmatrix}\dfrac{1}{c_{11}} & 0 & 0\\ 0 & \dfrac{1}{c_{66}} & 0\\ 0 & 0 & \dfrac{1}{c_{44}}\end{bmatrix},$$

$$\mathbf{P}_3(r)=\begin{bmatrix} \rho\dfrac{\partial^2}{\partial t^2}+k_1\dfrac{1}{r^2} & k_1\dfrac{1}{r^2}\dfrac{\partial}{\partial\alpha} & k_2\dfrac{1}{r}\dfrac{\partial}{\partial z} \\ -k_1\dfrac{1}{r^2}\dfrac{\partial}{\partial\alpha} & \rho\dfrac{\partial^2}{\partial t^2}-k_1\dfrac{1}{r^2}\dfrac{\partial^2}{\partial\alpha^2}-c_{44}\dfrac{\partial^2}{\partial z^2} & -(c_{44}+k_2)\dfrac{1}{r}\dfrac{\partial^2}{\partial\alpha\partial z} \\ -k_2\dfrac{1}{r}\dfrac{\partial}{\partial z} & -(c_{44}+k_2)\dfrac{1}{r}\dfrac{\partial^2}{\partial\alpha\partial z} & \rho\dfrac{\partial^2}{\partial t^2}-c_{44}\dfrac{1}{r^2}\dfrac{\partial^2}{\partial\alpha^2}-k_3\dfrac{\partial^2}{\partial z^2} \end{bmatrix},$$

$$\mathbf{P}_4(r)=\begin{bmatrix} \left(\dfrac{c_{12}}{c_{11}}-1\right)\dfrac{1}{r} & -\dfrac{1}{r}\dfrac{\partial}{\partial\alpha} & -\dfrac{\partial}{\partial z} \\ -\dfrac{c_{12}}{c_{11}}\dfrac{1}{r}\dfrac{\partial}{\partial\alpha} & -\dfrac{2}{r} & 0 \\ -\dfrac{c_{13}}{c_{11}}\dfrac{\partial}{\partial z} & 0 & -\dfrac{1}{r} \end{bmatrix}, \tag{9.3.2}$$

in which

$$k_1=c_{11}-c_{12}^2/c_{11},\quad k_2=c_{13}-c_{12}c_{13}/c_{11},\quad k_3=c_{33}-c_{13}^2/c_{11}.$$

Meanwhile, we have

$$\begin{aligned} \sigma_\alpha &= k_1\frac{u_r}{r}+k_1\frac{1}{r}\frac{\partial u_\alpha}{\partial\alpha}+k_2\frac{\partial w}{\partial z}+\frac{c_{12}}{c_{11}}\sigma_r, \\ \sigma_z &= k_2\frac{u_r}{r}+k_2\frac{1}{r}\frac{\partial u_\alpha}{\partial\alpha}+k_3\frac{\partial w}{\partial z}+\frac{c_{13}}{c_{11}}\sigma_r, \\ \tau_{\alpha z} &= c_{44}\frac{\partial u_\alpha}{\partial z}+c_{44}\frac{1}{r}\frac{\partial w}{\partial\alpha}. \end{aligned} \tag{9.3.3}$$

Consider the free vibration of a simply supported laminated cylindrical panel illustrated in Fig. 9.4. To satisfy the boundary conditions in Eq. (9.2.19), we can take

$$\begin{aligned} u_r &= aU(\xi)\sin(m\pi\zeta)\sin(n\pi\alpha/\alpha_0)\exp(\mathrm{i}\omega t), \\ u_\alpha &= aV(\xi)\sin(m\pi\zeta)\cos(n\pi\alpha/\alpha_0)\exp(\mathrm{i}\omega t), \\ w &= aW(\xi)\cos(m\pi\zeta)\sin(n\pi\alpha/\alpha_0)\exp(\mathrm{i}\omega t), \\ \sigma_r &= c_{44}^{(1)}S(\xi)\sin(m\pi\zeta)\sin(n\pi\alpha/\alpha_0)\exp(\mathrm{i}\omega t), \\ \tau_{r\alpha} &= c_{44}^{(1)}Y(\xi)\sin(m\pi\zeta)\cos(n\pi\alpha/\alpha_0)\exp(\mathrm{i}\omega t), \\ \tau_{zr} &= c_{44}^{(1)}Z(\xi)\cos(m\pi\zeta)\sin(n\pi\alpha/\alpha_0)\exp(\mathrm{i}\omega t), \end{aligned} \tag{9.3.4}$$

where $\xi = r/a$, $\zeta = z/L$, ω is the circular frequency, m and n are integers, and the superscript (1) in $c_{44}^{(1)}$ denotes the material constants in the first layer. We will focus on the bending vibration and hence will not consider $m = 0$ or $n = 0$ (this corresponds to $u_r = \sigma_r = 0$, *i.e.*, the shell has only in-surface displacements).

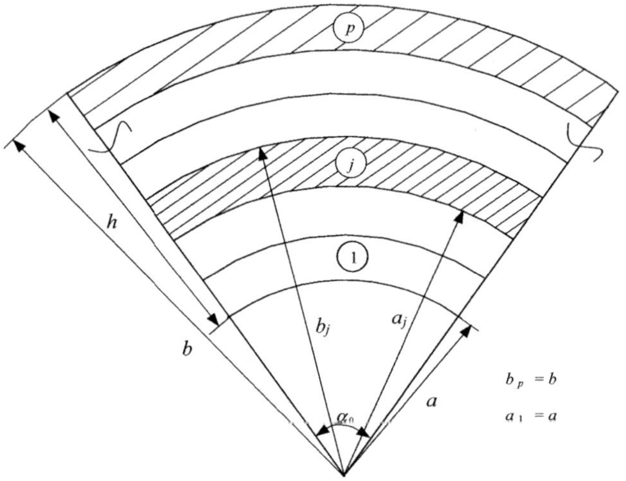

Fig. 9.4 A laminated cylindrical panel.

Therefore, the substitution of Eq. (9.3.4) into Eq. (9.3.1) leads to

$$\frac{\mathrm{d}}{\mathrm{d}\xi}\mathbf{N}(\xi) = \mathbf{M}(\xi)\mathbf{N}(\xi), \tag{9.3.5}$$

for an arbitrary couple of (m,n), where

$$\mathbf{N}(\xi) = [U(\xi), V(\xi), W(\xi), S(\xi), Y(\xi), Z(\xi)]^{\mathrm{T}},$$

is a dimensionless state vector and

$$\mathbf{M}(\xi) = \begin{bmatrix} \mathbf{M}_1(\xi) & \mathbf{M}_2(\xi) \\ \mathbf{M}_3(\xi) & \mathbf{M}_4(\xi) \end{bmatrix},$$

$$\mathbf{M}_1(\xi)=\begin{bmatrix} -\frac{c_{12}}{c_{11}}\frac{1}{\xi} & \frac{c_{12}}{c_{11}}\frac{t_\alpha}{\xi} & \frac{c_{13}}{c_{11}}t_L \\ -\frac{t_\alpha}{\xi} & \frac{1}{\xi} & 0 \\ -t_L & 0 & 0 \end{bmatrix},\quad \mathbf{M}_2(\xi)=\begin{bmatrix} \frac{c_{44}^{(1)}}{c_{11}} & 0 & 0 \\ 0 & \frac{c_{44}^{(1)}}{c_{66}} & 0 \\ 0 & 0 & \frac{c_{44}^{(1)}}{c_{44}} \end{bmatrix},$$

$$\mathbf{M}_3(\xi)=\begin{bmatrix} -\frac{\rho}{\rho^{(1)}}\Omega^2+\frac{k_1}{c_{44}^{(1)}}\frac{1}{\xi^2} & -\frac{k_1}{c_{44}^{(1)}}\frac{t_\alpha}{\xi^2} & -\frac{k_2}{c_{44}^{(1)}}\frac{t_L}{\xi} \\ -\frac{k_1}{c_{44}^{(1)}}\frac{t_\alpha}{\xi^2} & -\frac{\rho}{\rho^{(1)}}\Omega^2+\frac{k_1}{c_{44}^{(1)}}\frac{t_\alpha^2}{\xi^2}+\frac{c_{44}}{c_{44}^{(1)}}t_L^2 & \frac{c_{44}+k_2}{c_{44}^{(1)}}\frac{t_\alpha t_L}{\xi} \\ -\frac{k_2}{c_{44}^{(1)}}\frac{t_L}{\xi} & \frac{c_{44}+k_2}{c_{44}^{(1)}}\frac{t_\alpha t_L}{\xi} & -\frac{\rho}{\rho^{(1)}}\Omega^2+\frac{c_{44}}{c_{44}^{(1)}}\frac{t_\alpha^2}{\xi^2}+\frac{k_3}{c_{44}^{(1)}}t_L^2 \end{bmatrix},$$

$$\mathbf{M}_4(\xi)=\begin{bmatrix} \left(\frac{c_{12}}{c_{11}}-1\right)\frac{1}{\xi} & \frac{t_\alpha}{\xi} & t_L \\ -\frac{c_{12}}{c_{11}}\frac{t_\alpha}{\xi} & -\frac{2}{\xi} & 0 \\ -\frac{c_{13}}{c_{11}}t_L & 0 & -\frac{1}{\xi} \end{bmatrix},$$

$$t_L=m\pi a/L,\quad t_\alpha=n\pi/\alpha_0,\quad \Omega^2=\omega^2a^2\rho^{(1)}/c_{44}^{(1)}. \tag{9.3.6}$$

Since the coefficient matrix contains variable ξ, Eq. (9.3.5) is a state equation with variable coefficients.

9.3.2 Layerwise Method and State Vector Solution

It is usually difficult to solve a state equation with variable coefficients. Chen and Ding (2001*b*) solved such an equation of the vibration of a multi-layered hollow sphere based on Taylor's series expansion theorem. Here, we will use the *layerwise method* (Pestel and Leckie, 1963; Fan and Zhang, 1992). In this method, any layer of the laminated cylindrical shell is divided into a number of sub-layers of a very small thickness so that the variable ξ can be considered as a constant in such a sub-layer. Consequently, the coefficient matrix in Eq. (9.3.5) also becomes constant and hence the method described above will not produce a significant error. When the number of sub-layers is great enough, the solution will approach the exact one.

For convenience, we divide the jth layer, whose thickness is h_j, into j_q equal sub-layers so that the thickness of each sub-layer is $h_j / j_q = (b_j - a_j)/ j_q$. As variable ξ in a sub-layer is regarded as constant, we can simply use its value at the middle surface of the sub-layer. In the j_k^{th} ($j_k = 1, 2, \cdots, j_q$) sub-layer, we can set $\xi = \xi_{j0} + (2j_k - 1) \times (\xi_{j1} - \xi_{j0})/(2j_q) = \xi_{jj_k}$ in Eq. (9.3.5), where $\xi_{j0} = a_j / a$ and $\xi_{j1} = b_j / a$ are the inner and outer radii of the j-th layer. In so doing, the coefficient matrix $\mathbf{M}(\xi)$ in Eq. (9.3.5) for the j_k^{th} sub-layer of the jth layer becomes a constant matrix $\mathbf{M}(\xi_{jj_k})$. The solution of Eq. (9.3.5) therefore becomes (Bellman, 1970)

$$\begin{aligned} \mathbf{N}(\xi) &= \exp[(\xi - \xi^0_{jj_k})\mathbf{M}(\xi_{jj_k})]\mathbf{N}(\xi^0_{jj_k}), \\ &\quad (\xi^0_{jj_k} \le \xi \le \xi^1_{jj_k},\ j_k = 1, 2, \cdots, j_q;\ j = 1, 2, \cdots, p), \end{aligned} \tag{9.3.7}$$

where $\xi^0_{jj_k} = \xi_{j0} + (j_k - 1)(\xi_{j1} - \xi_{j0})/ j_q$ and $\xi^1_{jj} = \xi_{j0} + j_k(\xi_{j1} - \xi_{j0})/ j_q$. The exponential matrix $\exp[(\xi - \xi^0_{jj_k})\mathbf{M}(\xi_{jj_k})]$ is known as the transfer matrix (Fan and Ye, 1990*a*,*b*); it can be expressed in terms of a polynomial of the matrix $\mathbf{M}(\xi_{jj_k})$ using the Cayley-Hamilton theorem (Bellman, 1970) as discussed in **Section 8.2**.

Since the state vector should be continuous at the interface between two sub-layers, we can obtain from Eq. (9.3.7) that

$$\mathbf{N}(\xi_{j1}) = \mathbf{T}_j \mathbf{N}(\xi_{j0}), \qquad (j = 1, 2, \cdots, p), \tag{9.3.8}$$

where $\mathbf{T}_j = \prod_{j_k = j_q}^{1} \exp[(\xi_{j1} - \xi_{j0})\mathbf{M}(\xi_{jj_k})/ j_q]$ is a sixth-order square matrix.

The continuity conditions between the natural layers yield

$$\mathbf{N}(\xi_1) = \mathbf{T}\mathbf{N}(1), \tag{9.3.9}$$

where $\mathbf{T} = \prod_{j=p}^{1} \mathbf{T}_j$ is a sixth-order square matrix, and $\xi_1 = b / a$ is the ratio between the outer radius and inner radius.

9.3.3 Free Vibration Analysis and Numerical Results

Considering the traction-free condition under free vibration, we have

$$S(\xi) = Z(\xi) = Y(\xi) = 0\,, \quad (\xi = 1,\, \xi_1)\,. \tag{9.3.10}$$

Then Eq. (9.3.9) gives rise to

$$[U(\xi_1), V(\xi_1), W(\xi_1), 0, 0, 0]^{\mathrm{T}} = \mathbf{T}[U(1), V(1), W(1), 0, 0, 0]^{\mathrm{T}}\,. \tag{9.3.11}$$

The fourth, fifth and sixth equations above are

$$\begin{bmatrix} T_{41} & T_{42} & T_{43} \\ T_{51} & T_{52} & T_{53} \\ T_{61} & T_{62} & T_{63} \end{bmatrix} \begin{Bmatrix} U(1) \\ V(1) \\ W(1) \end{Bmatrix} = \{0\}, \tag{9.3.12}$$

where T_{ij} is the element on the i-th row and j-th column of matrix $\mathbf{T}$. The frequency equation is then given by

$$\begin{vmatrix} T_{41} & T_{42} & T_{43} \\ T_{51} & T_{52} & T_{53} \\ T_{61} & T_{62} & T_{63} \end{vmatrix} = 0\,. \tag{9.3.13}$$

The frequencies of a cylindrical zinc shell calculated with MATHEMATICA show that the state-space method gives identical results to those listed in Table 9.11.

Table 9.12 The lowest natural frequencies (Ω) of a three-layered cylindrical panel.

(t_L, t_α)	$j_q = 10$	$j_q = 15$
(0.5, 0.5)	0.34139	0.34139
(1.0, 0.5)	0.70990	0.70990
(3.0, 0.5)	1.84283	1.84283
(0.5, 1.0)	0.22554	0.22554
(0.5, 2.0)	0.40660	0.40661
(0.5, 3.0)	0.92853	0.92858

Now consider the free vibration of a three-layered cylindrical panel. Assume that the middle layer is a transversely isotropic material, sapphire, (see **Section 9.2** for its material constants) and that the outer and inner layers are of an isotropic material with $\nu = 0.3$, $E = 2.0\times10^{11}$ N/m^2 and $\rho = 6.8$ g/cm^3. The geometry of the panel is $a : b_1 : b_2 : b = 1 : 1.2 : 1.4 : 1.6$. Table 9.12 gives the lowest natural frequencies when t_L

and t_α vary. The numerical results from two division schemes of 10 and 15 sub-layers are almost identical, indicating that the present method has a very rapid convergence rate.

9.4 VIBRATION OF A CYLINDRICAL SHELL COUPLED WITH FLUID

9.4.1 Coupling Effect of Fluid

Consider a non-viscous, compressible fluid medium, whose *velocity potential* satisfies the following *wave equation*

$$\left(\frac{1}{c_f^2}\frac{\partial^2}{\partial t^2}-\nabla^2\right)\Phi=0, \tag{9.4.1}$$

where c_f is the *velocity of sound in fluid*, ∇^2 is the three-dimensional Laplace operator and defined in the cylindrical coordinates as follows

$$\nabla^2=\frac{1}{r}\frac{\partial}{\partial r}\left(r\frac{\partial}{\partial r}\right)+\frac{1}{r^2}\frac{\partial^2}{\partial\alpha^2}+\frac{\partial^2}{\partial z^2}. \tag{9.4.2}$$

The *separation of variables* technique, as outlined in Appendix B, can be used to obtain the following solution of Eq. (9.4.1)

$$\Phi(r,\alpha,z,t)=R^2\phi(\xi)\cos(m\pi\zeta)\cos(n\alpha)\exp(\mathrm{i}\,\omega t). \tag{9.4.3}$$

Substituting Eq. (9.4.3) into Eq. (9.4.1) gives

$$\frac{\mathrm{d}^2\phi}{\mathrm{d}\xi^2}+\frac{1}{\xi}\frac{\mathrm{d}\phi}{\mathrm{d}\xi}+\left(\kappa^2-\frac{n^2}{\xi^2}\right)\phi=0, \tag{9.4.4}$$

where $\kappa^2=\Omega^2/c_0^2-t_L^2$ and $c_0=c_f/v_2$. The solution of this equation depends on the boundary conditions, *i.e.*, with the fluid internally or externally, and the nature of κ^2 as shown in Table 9.13.

Table 9.13 The solution of Eq. (9.4.4).

	Internal	External
$\kappa^2 > 0$	$AJ_n(\kappa\xi)$	$AH_n^{(2)}(\kappa\xi)$
$\kappa^2 = 0$	$A\xi^n$	$A\xi^{-n}$
$\kappa^2 < 0$	$AI_n(\tilde{\kappa}\xi)$	$AK_n(\tilde{\kappa}\xi)$

In Table 9.13, we have $(\tilde{\kappa})^2 = -\kappa^2$, and A is an arbitrary constant. In the solution, we have used the condition that the function is finite at $\xi = 0$ for the fluid inside the shell. If the fluid is outside the shell, the function should be finite at infinity for $\kappa^2 < 0$, while for $\kappa^2 > 0$, the solution should represent an outgoing wave (Morse and Feshbach, 1953). For convenience, we write the solution of Eq. (9.4.4) in a unified form as

$$\phi(\xi) = AZ_n(\kappa\xi). \tag{9.4.5}$$

For a small perturbation, by neglecting the effect of static pressure, we obtain the dynamic pressure and velocity from the *Bernoulli equation* as follows

$$p_f = \mathrm{i}A\frac{\Omega\rho_0}{v_2 R}Z_n(\kappa\xi)\cos(m\pi\zeta)\cos(n\alpha)\exp(\mathrm{i}\omega t), \tag{9.4.6}$$

$$v_f = \frac{A}{R}Z_n'(\kappa\xi)\cos(m\pi\zeta)\cos(n\alpha)\exp(\mathrm{i}\omega t), \tag{9.4.7}$$

where ρ_f is the fluid density, $\rho_0 = \rho_f / \rho$ and the prime denotes differentiation with respect to ξ. Neglecting the viscosity of fluid, we have the following conditions at the *fluid-structure interface*

$$\bar{\sigma}_r = -p_f / c_{44}, \quad \bar{\tau}_{r\alpha} = \bar{\tau}_{zr} = 0, \quad v_r = v_f. \tag{9.4.8}$$

Since $v_r = \partial u_r / \partial t$, we obtain from Eqs. (9.4.6) to (9.4.8)

$$\bar{\sigma}_r = -\Omega^2\rho_0 T(\kappa\xi)\bar{u}_r, \quad (\xi = t_1,\ t_2), \tag{9.4.9}$$

where

$$T(\kappa\xi) = Z_n(\kappa\xi) / Z_n'(\kappa\xi). \quad (9.4.10)$$

In the following, we use the subscripts i and o to denote respectively the quantities associated with internal and external fluids. In addition, parameters $P_1(t_1)$ and $P_2(t_2)$ are defined by

$$P_1(t_1) = \Omega^2 \rho_{0i} T_1(\kappa_i t_1), \quad P_2(t_2) = \Omega^2 \rho_{0o} T_2(\kappa_o t_2). \quad (9.4.11)$$

9.4.2 Free Vibration of Submerged and/or Fluid-Filled Cylinders and Cylindrical Shells

(1) Free vibration of submerged cylinders

When a cylinder is submerged in a compressible but non-viscous fluid, we have the following boundary conditions at its outer surface

$$\bar{\sigma}_r = -\Omega^2 \rho_{0o} T(\kappa_o \xi) \bar{u}_r, \quad \bar{\tau}_{r\alpha} = \bar{\tau}_{zr} = 0, \quad (\xi = t_2). \quad (9.4.12)$$

Using Eq. (9.2.13), we can get three linear homogeneous algebraic equations about A_1, A_2 and A_3. Again, the frequency equation is specified by

$$\left| E_{ij}^3 \right| = 0, \quad (i, j = 1, 2, 3). \quad (9.4.13)$$

For $k_1^2 > 0$ and Case (a) listed in Table 9.9, the explicit expressions of the elements in Eq. (9.4.13) can be obtained as

$$\begin{aligned}
E_{11}^3 &= (p_1 - p_2) n J_n'(k_1 t_2) / t_2 - (p_1 - p_2) n J_n(k_1 t_2) / t_2^2 \\
&\quad + P_2(t_2) n J_n(k_1 t_2) / t_2, \\
E_{12}^3 &= -p_1 I_n''(k_2 t_2) - p_2 I_n'(k_2 t_2) / t_2 + (p_2 n^2 / t_2^2 + p_3 t_L q_1) I_n(k_2 t_2) \\
&\quad - P_2(t_2) I_n'(k_2 t_2), \\
E_{13}^3 &= -p_1 I_n''(k_3 t_2) - p_2 I_n'(k_3 t_2) / t_2 + (p_2 n^2 / t_2^2 + p_3 t_L q_2) I_n(k_3 t_2) \\
&\quad - P_2(t_2) I_n'(k_3 t_2), \\
E_{21}^3 &= -t_L n J_n(k_1 t_2) / t_2,
\end{aligned}$$

$$
\begin{aligned}
E_{22}^3 &= (t_L + q_1) I_n'(k_2 t_2), \\
E_{23}^3 &= (t_L + q_2) I_n'(k_3 t_2), \\
E_{31}^3 &= -J_n''(k_1 t_2) + J_n'(k_1 t_2)/t_2 - n^2 J_n(k_1 t_2)/t_2^2, \\
E_{32}^3 &= 2nI_n'(k_2 t_2)/t_2 - 2nI_n(k_2 t_2)/t_2^2, \\
E_{33}^3 &= 2nI_n'(k_3 t_2)/t_2 - 2nI_n(k_3 t_2)/t_2^2.
\end{aligned} \tag{9.4.14}
$$

(2) Submerged and fluid-filled cylindrical shell

Consider the free vibration of a fluid-filled transversely isotropic cylindrical shell submerged in another liquid. In this case, we have the following boundary conditions

$$
\begin{aligned}
&\bar{\sigma}_r = -\Omega^2 \rho_{0i} T(\kappa_i \xi) \bar{u}_r, \quad \bar{\tau}_{r\alpha} = \bar{\tau}_{zr} = 0, \quad (\xi = t_1), \\
&\bar{\sigma}_r = -\Omega^2 \rho_{0o} T(\kappa_o \xi) \bar{u}_r, \quad \bar{\tau}_{r\alpha} = \bar{\tau}_{zr} = 0, \quad (\xi = t_2),
\end{aligned} \tag{9.4.15}
$$

Using the results presented in **Section 9.2**, we can get six linear homogeneous algebraic equations with six unknown constants. For having a nontrivial solution, the coefficient determinant should be zero and hence

$$
\left| E_{ij}^4 \right| = 0, \quad (i, j = 1, 2, \cdots, 6), \tag{9.4.16}
$$

where $E_{ij}^4 = E_{ij}^3\ (i, j = 1, 2, 3)$. For $k_1^2 > 0$ and Case (a) in Table 9.9, $E_{ij}^4\ (j = 4, 5, 6)$ can be obtained by replacing J_n and I_n in $E_{ij}^4\ (j = 1, 2, 3)$ with Y_n and K_n, respectively, while $E_{ij}^4\ (i = 4, 5, 6)$ can be obtained by replacing t_2 in $E_{ij}^4\ (i = 1, 2, 3)$ with t_1 along with replacing $P_2(t_2)$ with $P_1(t_1)$.

When the shell is filled with fluid only internally, $P_2(t_2) = 0$ in Eq. (9.4.16), but if it is only submerged in a compressible fluid medium, $P_1(t_1) = 0$ in Eq. (9.4.16).

9.4.3 Numerical Results of a Fluid-Filled Cylindrical Shell

Table 9.14 shows the variation of the natural frequency of a cylindrical zinc shell filled with non-viscous compressible fluid with the thickness-to-mean radius ratio t_r. Jain (1974) considered a similar problem using a shear shell theory but did not include the effect of liquid compressibility. For a qualitative comparison, Jain's results are also listed in the table.

Table 9.14 The lowest natural frequencies (Ω) of a cylindrical fluid-filled zinc shell. ($\rho_{0i} = 0.09,\ c_{0i} = 0.3$)

t_r	mR/L	$n=1$			$n=2$		
		Ω Present	Ω Jain (1974)	η	Ω Present	Ω Jain (1974)	η
0.01	0.1	0.0286	0.0609	0.091	0.0155	0.0250	0.049
	0.2	0.1068	0.2068	0.170	0.0425	0.0773	0.066
	0.4	0.3116	0.5833	0.248	0.1532	0.2579	0.122
	0.6	0.4673	0.8946	0.248	0.3016	0.4832	0.160
	1.0	0.6815	1.1025	0.217	0.5874	0.8609	0.187
	2.0	1.0178	1.2244	0.162	0.9799	1.1634	0.156
0.10	0.1	0.0553	0.0631	0.176	0.1347	0.1535	0.429
	0.2	0.1960	0.2253	0.312	0.1582	0.1796	0.262
	0.4	0.5269	0.6682	0.428	0.3141	0.3537	0.250
	0.6	0.7350	1.2059	0.390	0.5597	0.6301	0.297
	1.0	1.1182	1.5226	0.356	1.0271	1.1863	0.327
	2.0	1.7652	2.0014	0.281	1.8405	2.0450	0.293
0.25	0.1	0.0563	0.0642	0.196	0.3537	0.3711	1.126
	0.2	0.2286	0.2275	0.364	0.3831	0.3983	0.610
	0.4	0.6810	0.6813	0.542	0.5498	0.5529	0.437
	0.6	1.1363	1.1460	0.603	0.8235	0.8380	0.437
	1.0	1.7055	1.6956	0.542	1.5392	1.5465	0.494
	2.0	3.0842	3.0408	0.491	3.3480	3.3020	0.553
0.50	0.1	0.0648	0.0653	0.206	0.6772	0.6835	2.156
	0.2	0.2292	0.2340	0.365	0.7048	0.7150	1.122
	0.4	0.6708	0.6998	0.534	0.8517	0.8739	0.678
	0.6	1.1993	1.2026	0.637	1.1307	1.1754	0.600
	1.0	1.9912	1.9973	0.634	2.0342	2.0374	0.649
	2.0	4.1962	4.1172	0.668	4.5702	4.4763	0.727

Comparing the results in Table 9.14 and those in Table 9.11, we find that when t_r is small (*e.g.*, 0.1 or 0.01), the filled fluid lowers the natural frequency significantly due to the *added-mass-effect*. Jain (1974) observed a similar phenomenon with incompressible fluid. The comparison of Table 9.14 with Table 9.11 also shows that the effect becomes more significant when the fluid is compressible. On the other hand, the extent of the effect decreases with t_r and becomes really minor, less than 4%, when $t_r = 0.5$. Chen and Ding (1999*b*) made a further comparison with both the classical and an improved shell theory and discussed the effect of normal strains.

9.5 VIBRATION OF A CYLINDRICAL SHELL COUPLED WITH THE SURROUNDING ELASTIC MEDIUM

9.5.1 Elastic Waves in an Isotropic Elastic Medium

The displacements in the surrounding elastic medium can be described by the three displacement functions according to the *Helmholtz separation theorem* (Pao and Mow, 1971), *i.e.*,

$$\begin{aligned} u_r &= \frac{\partial \varphi}{\partial r} + \frac{1}{r}\frac{\partial \vartheta}{\partial \alpha} - \frac{\partial^2 \chi}{\partial r \partial z}, \\ u_\alpha &= \frac{1}{r}\frac{\partial \varphi}{\partial \alpha} - \frac{\partial \vartheta}{\partial r} - \frac{1}{r}\frac{\partial^2 \chi}{\partial \alpha \partial z}, \\ w &= \frac{\partial \varphi}{\partial z} + \left[\frac{1}{r}\frac{\partial}{\partial r}\left(r\frac{\partial \chi}{\partial r}\right) + \frac{1}{r^2}\frac{\partial^2 \chi}{\partial \alpha^2}\right], \end{aligned} \tag{9.5.1}$$

where the displacement functions, φ, ϑ and χ, satisfy the following equations

$$\nabla^2 \varphi = \frac{1}{c_1^2}\frac{\partial^2 \varphi}{\partial t^2}, \quad \nabla^2 \vartheta = \frac{1}{c_2^2}\frac{\partial^2 \vartheta}{\partial t^2}, \quad \nabla^2 \chi = \frac{1}{c_2^2}\frac{\partial^2 \chi}{\partial t^2}, \tag{9.5.2}$$

in which $c_1 = \sqrt{(\lambda + 2\mu)/\rho_m}$ and $c_2 = \sqrt{\mu/\rho_m}$ are the velocities of the longitudinal and transverse waves in the elastic medium respectively, λ and μ are Lamé constants and ρ_m is the density of the elastic medium. Substituting Eq. (9.5.1) into the constitutive relations of an isotropic elastic medium, Eq. (1.2.43), we obtain the following stresses in terms of the displacement functions

$$\begin{aligned} \sigma_r &= \lambda \nabla^2 \varphi + 2\mu\left[\frac{\partial^2 \varphi}{\partial r^2} + \frac{\partial}{\partial r}\left(\frac{1}{r}\frac{\partial \vartheta}{\partial \alpha}\right) - \frac{\partial^3 \chi}{\partial r^2 \partial z}\right], \\ \sigma_\alpha &= \lambda \nabla^2 \varphi + 2\mu\left[\frac{1}{r}\left(\frac{\partial \varphi}{\partial r} + \frac{1}{r}\frac{\partial^2 \varphi}{\partial \alpha^2}\right) + \frac{1}{r}\left(\frac{1}{r}\frac{\partial \vartheta}{\partial \alpha} - \frac{\partial^2 \vartheta}{\partial r \partial \alpha}\right) - \frac{1}{r}\left(\frac{\partial^2 \chi}{\partial r \partial z} + \frac{1}{r}\frac{\partial^3 \chi}{\partial \alpha^2 \partial z}\right)\right], \\ \sigma_z &= \lambda \nabla^2 \varphi + 2\mu\left[\frac{\partial^2 \varphi}{\partial z^2} + \left(\frac{\partial}{\partial z}\nabla^2 \chi - \frac{\partial^3 \chi}{\partial z^3}\right)\right], \end{aligned}$$

$$\begin{aligned}
\tau_{r\alpha} &= \mu\left\{2\left[\frac{1}{r}\frac{\partial^2\varphi}{\partial\alpha\partial r}-\frac{1}{r^2}\frac{\partial\varphi}{\partial\alpha}\right]+\left[\frac{1}{r^2}\frac{\partial^2\vartheta}{\partial\alpha^2}-r\frac{\partial}{\partial r}\left(\frac{1}{r}\frac{\partial\vartheta}{\partial r}\right)\right]\right. \\
&\quad \left.-2\left[\frac{1}{r}\frac{\partial^3\chi}{\partial r\partial\alpha\partial z}-\frac{1}{r^2}\frac{\partial^2\chi}{\partial\alpha\partial z}\right]\right\}, \\
\tau_{zr} &= \mu\left\{2\frac{\partial^2\varphi}{\partial r\partial z}+\frac{1}{r}\frac{\partial^2\vartheta}{\partial\alpha\partial z}-\left[\frac{\partial^3\chi}{\partial r\partial z^2}-\frac{\partial}{\partial r}(\nabla^2\chi)\right]\right\}, \\
\tau_{\alpha z} &= \mu\left\{\frac{1}{r}\frac{\partial^2\varphi}{\partial\alpha\partial z}-\frac{\partial^2\vartheta}{\partial r\partial z}-\left[\frac{2}{r}\frac{\partial^3\chi}{\partial\alpha\partial z^2}-\frac{1}{r}\frac{\partial}{\partial\alpha}(\nabla^2\chi)\right]\right\}.
\end{aligned} \tag{9.5.3}$$

Now let us use this result to solve a transversely isotropic cylindrical shell of infinite length embedded in a linear, homogeneous and isotropic elastic medium and subjected to a combination of P-, SV- and SH-waves. Assume that the wave in the shell propagates in the axial direction and the wave length is $\Pi = 2\pi / k$, where k is the wave number. The wave velocity is $c = \omega / k$ and ω is the circular frequency. Since both the displacements and stresses of the surrounding medium consist of two parts induced by the incident and scattered waves, we have

$$u_i^e = u_i^{(i)} + u_i^{(s)}, \qquad \sigma_{ij}^e = \sigma_{ij}^{(i)} + \sigma_{ij}^{(s)}, \tag{9.5.4}$$

where the superscripts (i) and (s) stand for the incident and scattered waves, respectively, and the superscript e denotes the surrounding elastic medium. According to the wave propagation characteristics and the wave equation, we obtain

$$\begin{aligned}
\varphi^{(i)} &= d_1 R^2 I_n(\gamma\xi)\cos(n\alpha)\cos(kz-\omega t), \\
\vartheta^{(i)} &= d_2 R^2 I_n(\delta\xi)\sin(n\alpha)\cos(kz-\omega t), \\
\chi^{(i)} &= d_3 R^3 I_n(\delta\xi)\cos(n\alpha)\sin(kz-\omega t),
\end{aligned} \tag{9.5.5}$$

$$\begin{aligned}
\varphi^{(s)} &= AR^2 K_n(\gamma\xi)\cos(n\alpha)\cos(kz-\omega t), \\
\vartheta^{(s)} &= BR^2 K_n(\delta\xi)\sin(n\alpha)\cos(kz-\omega t), \\
\chi^{(s)} &= CR^3 K_n(\delta\xi)\cos(n\alpha)\sin(kz-\omega t),
\end{aligned} \tag{9.5.6}$$

where $d_j\,(j = 1, 2, 3)$ are the intensities of the incident P-, SV- and SH-waves, respectively, A, B and C are three undetermined constants, $\gamma = \beta\sqrt{1-\bar{c}_1^2}$,

$\delta = \beta\sqrt{1-\bar{c}_2^2}$, $\beta = kR = 2\pi R/\Pi$, $\bar{c}_1 = c/c_1$, and $\bar{c}_2 = c/c_2$. Substituting these two equations into Eqs. (9.5.1) and (9.5.3), we get

$$\begin{aligned}
u_r^{(i)} &= R\{[\gamma I_n'(\gamma\xi)]d_1 + [(n/\xi)I_n(\delta\xi)]d_2 - [\beta\delta I_n'(\delta\xi)]d_3\}\cos(n\alpha)\cos(kz-\omega t),\\
u_\alpha^{(i)} &= R\{[-(n/\xi)I_n(\gamma\xi)]d_1 - [\delta I_n(\delta\xi)]d_2 + [(n/\xi)\beta I_n(\delta\xi)]d_3\}\\
&\quad\times\sin(n\alpha)\cos(kz-\omega t),\\
w^{(i)} &= R\{[-\beta I_n(\gamma\xi)]d_1 + [\delta^2 I_n(\delta\xi)]d_3\}\cos(n\alpha)\sin(kz-\omega t),
\end{aligned} \tag{9.5.7}$$

$$\begin{aligned}
u_r^{(s)} &= R\{[\gamma I_n'(\gamma\xi)]A + [(n/\xi)I_n(\delta\xi)]B - [\beta\delta I_n'(\delta\xi)]C\}\cos(n\alpha)\cos(kz-\omega t),\\
u_\alpha^{(s)} &= R\{[-(n/\xi)I_n(\gamma\xi)]A - [\delta I_n(\delta\xi)]B + [(n/\xi)\beta I_n(\delta\xi)]C\}\sin(n\alpha)\cos(kz-\omega t),\\
w^{(s)} &= R\{[-\beta I_n(\gamma\xi)]A + [\delta^2 I_n(\delta\xi)]C\}\cos(n\alpha)\sin(kz-\omega t),
\end{aligned} \tag{9.5.8}$$

$$\begin{aligned}
\sigma_r^{(i)} &= 2\mu\{d_1[-\eta\beta^2\bar{c}_1^2 I_n(\gamma\xi) + \gamma^2 I_n''(\gamma\xi)] + d_2[-(n/\xi^2)I_n(\delta\xi) + (n\delta/\xi)I_n'(\delta\xi)]\\
&\qquad - d_3[\delta^2\beta I_n''(\delta\xi)]\}\cos(n\alpha)\cos(kz-\omega t),\\
\sigma_\alpha^{(i)} &= 2\mu\{d_1[-\eta\beta^2\bar{c}_1^2 I_n(\gamma\xi) + (\gamma/\xi)I_n'(\gamma\xi) - (n^2/\xi^2)I_n(\gamma\xi)] + d_2[(n/\xi^2)I_n(\delta\xi)\\
&\quad - (n\delta/\xi)I_n'(\delta\xi)] - d_3[(\delta\beta/\xi)I_n'(\delta\xi) - (n^2\beta/\xi^2)I_n(\delta\xi)]\\
&\quad\times\cos(n\alpha)\cos(kz-\omega t),\\
\sigma_z^{(i)} &= 2\mu\{d_1[-\eta\beta^2\bar{c}_1^2 I_n(\gamma\xi) - \beta^2 I_n(\gamma\xi)] + d_3[(-\bar{c}_2^2\beta^3 + \beta^3)I_n(\delta\xi)]\}\\
&\quad\times\cos(n\alpha)\cos(kz-\omega t),\\
\tau_{r\alpha}^{(i)} &= 2\mu\{d_1[-(n\gamma/\xi)I_n'(\gamma\xi) + (n/\xi^2)I_n(\gamma\xi)] + (1/2)d_2[(-n^2/\xi^2)I_n(\delta\xi)\\
&\quad + (\delta/\xi)I_n'(\delta\xi) - \delta^2 I_n''(\delta\xi)] - d_3[-(n\delta\beta/\xi)I_n'(\delta\xi) + (n\beta/\xi^2)I_n(\delta\beta)]\}\\
&\quad\times\sin(n\alpha)\cos(kz-\omega t),\\
\tau_{zr}^{(i)} &= 2\mu\{d_1[-\gamma\beta I_n'(\gamma\xi)] + (1/2)d_2[-(n\beta/\xi)I_n(\delta\xi)] - (1/2)d_3[-\delta\beta^2 I_n(\delta\xi)\\
&\quad + \bar{c}_2^2\beta^2\delta I_n'(\delta\xi)]\}\cos(n\alpha)\sin(kz-\omega t),\\
\tau_{\alpha z}^{(i)} &= \mu\{d_1[(n\beta/\xi)I_n(\gamma\xi)] + d_2[\delta\beta I_n'(\delta\xi)] - d_3[(2n\beta^2/\xi)I_n(\delta\xi)\\
&\quad - (n\bar{c}_2^2\beta^2/\xi)I_n(\delta\xi)]\}\sin(n\alpha)\sin(kz-\omega t),
\end{aligned} \tag{9.5.9}$$

$$\begin{aligned}
\sigma_r^{(s)} &= 2\mu\{A[-\eta\beta^2\bar{c}_1^2 I_n(\gamma\xi) + \gamma^2 I_n''(\gamma\xi)] + B[-(n/\xi^2)I_n(\delta\xi) + (n\delta/\xi)I_n'(\delta\xi)]\\
&\qquad - C[\delta^2\beta I_n''(\delta\xi)]\}\cos(n\alpha)\cos(kz-\omega t),\\
\sigma_\alpha^{(s)} &= 2\mu\{A[-\eta\beta^2\bar{c}_1^2 I_n(\gamma\xi) + (\gamma/\xi)I_n'(\gamma\xi) - (n^2/\xi^2)I_n(\gamma\xi)] + B[(n/\xi^2)I_n(\delta\xi)\\
&\quad - (n\delta/\xi)I_n'(\delta\xi)] - C[(\delta\beta/\xi)I_n'(\delta\xi) - (n^2\beta/\xi^2)I_n(\delta\xi)]\}\\
&\quad\times\cos(n\alpha)\cos(kz-\omega t),
\end{aligned}$$

$$\begin{aligned}
\sigma_z^{(s)} &= 2\mu\{A[-\eta\beta^2\bar{c}_1^2 I_n(\gamma\xi) - \beta^2 I_n(\gamma\xi)] + B[(-\bar{c}_2^2\beta^3 + \beta^3)I_n(\delta\xi)]\} \\
&\quad \times\cos(n\alpha)\cos(kz-\omega t), \\
\tau_{r\alpha}^{(s)} &= 2\mu\{A[-(n\gamma/\xi)I_n'(\gamma\xi) + (n/\xi^2)I_n(\gamma\xi)] + (1/2)B[(-n^2/\xi^2)I_n(\delta\xi) \\
&\quad + (\delta/\xi)I_n'(\delta\xi) - \delta^2 I_n''(\delta\xi)] - B[-(n\delta\beta/\xi)I_n'(\delta\xi) + (n\beta/\xi^2)I_n(\delta\beta)]\} \\
&\quad \times\sin(n\alpha)\cos(kz-\omega t), \\
\tau_{zr}^{(s)} &= 2\mu\{A[-\gamma\beta I_n'(\gamma\xi)] + (1/2)B[-(n\beta/\xi)I_n(\delta\xi)] - (1/2)C[-\delta\beta^2 I_n(\delta\xi) \\
&\quad + \bar{c}_2^2\beta^2\delta I_n'(\delta\xi)]\}\cos(n\alpha)\sin(kz-\omega t), \\
\tau_{\alpha z}^{(s)} &= \mu\{A(n\beta/\xi)I_n(\gamma\xi) + B[\delta\beta I_n'(\delta\xi)] - C[(2n\beta^2/\xi)I_n(\delta\xi) \\
&\quad - (n\bar{c}_2^2\beta^2/\xi)I_n(\delta\xi)]\}\sin(n\alpha)\sin(kz-\omega t),
\end{aligned} \tag{9.5.10}$$

where the prime denotes the differentiation with respect to ξ, $\eta = \lambda/(2\mu) = \nu_m/(1-2\nu_m)$ and ν_m is the Poisson's ratio of the surrounding elastic medium. According to Eq. (9.5.4), it is known that the total displacements are the sum of Eqs. (9.5.7) and (9.5.8) and the total stresses are the sum of Eqs. (9.5.9) and (9.5.10).

9.5.2 Displacements and Stresses in the Shell

Now the displacement functions in **Section 9.2** can take the following form

$$\begin{aligned}
\psi(r,\alpha,z,t) &= R^2\bar{\psi}(\xi)\sin(n\alpha)\cos(kz-\omega t), \\
G(r,\alpha,z,t) &= R^2\bar{G}(\xi)\cos(n\alpha)\cos(kz-\omega t), \\
W(r,\alpha,z,t) &= R\bar{W}(\xi)\cos(n\alpha)\sin(kz-\omega t).
\end{aligned} \tag{9.5.11}$$

Then similarly the displacements and stresses at the cylindrical surfaces can be written as

$$
\begin{aligned}
\overline{u}_r &= (n\overline{\psi}/\xi - \overline{G}')\cos(n\alpha)\cos(kz-\omega t),\\
\overline{u}_\alpha &= (-\overline{\psi}' + n\overline{G}/\xi)\sin(n\alpha)\cos(kz-\omega t),\\
\overline{w} &= \overline{W}\cos(n\alpha)\sin(kz-\omega t),\\
\overline{\sigma}_r &= [-p_1\overline{G}'' - p_2\overline{G}'/\xi + p_2 n^2\overline{G}/\xi^2 + (p_1-p_2)n\overline{\psi}' - (p_1-p_2)n\overline{\psi}/\xi^2\\
&\qquad + p_3\beta\overline{W}]\cos(n\alpha)\cos(kz-\omega t),\\
\overline{\tau}_{zr} &= (\beta\overline{G}' - \beta n\overline{\psi}/\xi + \overline{W}')\cos(n\alpha)\sin(kz-\omega t),\\
\overline{\tau}_{r\alpha} &= (2n\overline{G}'/\xi - 2n\overline{G}/\xi^2 - \overline{\psi}'' + \overline{\psi}'/\xi - n^2\overline{\psi}/\xi^2)\sin(n\alpha)\cos(kz-\omega t),
\end{aligned}
\tag{9.5.12}
$$

where $\overline{\psi}$, $\overline{G}$ and $\overline{W}$ are the same as the corresponding ones in **Section 9.2** provided that t_L in those expressions is replaced by β. For a cylindrical shell, Eq. (9.5.12) contains six unknown constants A_i and B_i $(i=1, 2, 3)$.

9.5.3 Vibration of the Shell

Now we consider the responses of the cylindrical shell due to seismic waves. At the interface between the shell and the elastic surrounding, the displacements and stresses should be continuous, while at the inner surface the stresses should vanish, *i.e.*,

$$\sigma_r^s = \tau_{r\alpha}^s = \tau_{zr}^s = 0, \quad (\xi = t_1), \tag{9.5.13a}$$

$$
\left.
\begin{aligned}
&\sigma_r^s = \sigma_r^e, \quad \tau_{r\alpha}^s = \tau_{r\alpha}^e, \quad \tau_{zr}^s = \tau_{zr}^e,\\
&u_r^s = u_r^e, \quad u_\alpha^s = u_{r\alpha}^e, \quad w^s = w^e,
\end{aligned}
\right\} \quad (\xi = t_2), \tag{9.5.13b}
$$

where the superscript s denotes the quantity associated with the shell. According to these nine conditions and the results presented in the last two sub-sections, we can obtain the following nine linear inhomogeneous algebraic equations with A_i, $B_i\,(i=1, 2, 3)$, A, B and C as the unknowns:

$$\mathbf{Ha} = \mathbf{P}, \tag{9.5.14}$$

where $\mathbf{H}$ is a square matrix of order nine, $\mathbf{P}$ is the effective load vector and

$$\mathbf{a} = [A_1, \;\; B_1, \;\; A_2, \;\; B_2, \;\; A_3, \;\; B_3, \;\; A, \;\; B, \;\; C]^{\mathrm{T}}.$$

If the shell is not under any external load, we must have

$$\mathbf{Ha} = 0. \tag{9.5.15}$$

To have a nontrivial solution, the coefficient determinant must be zero, which gives rise to the following frequency equation

$$|\mathbf{H}| = 0. \tag{9.5.16}$$

Since the scattered wave brings away energy, the free vibration of the embedded cylindrical shell shall be damped (Gonçalves and Batista, 1988; Men and Yuan, 1990). Thus its final form reads as

$$H_1 + \mathrm{i} H_2 = 0. \tag{9.5.17}$$

where H_1 and H_2 are real functions of related parameters as well as the frequency. The solution to this equation is not straight since the frequencies will be complex with damping coefficients. For simplicity, it is often to obtain approximate frequencies by finding the roots of the following equation (Men and Yuan, 1990)

$$H_1 = 0. \tag{9.5.18}$$

This method of analysis can be easily extended to the investigation of embedded laminated cylindrical shells or embedded cylindrical shells simultaneously filled with fluid.

10

SPHERICAL SHELLS OF SPHERICALLY ISOTROPIC MATERIALS

This chapter presents a three-dimensional analysis for the free vibration of spherically isotropic spherical shells, including single-layered and multi-layered ones. Coupled vibrations of shells interacting with fluid and elastic media are also studied. A modified two-parameter elastic foundation is derived using the energy principle to analyse a hollow sphere embedded in an infinite elastic medium. Finally, the state-space formulation in spherical coordinates is derived to produce the free vibration solution for laminated spherical shells.

10.1 FREE VIBRATION

10.1.1 Basic Equations and Solution

In **Section 2.4**, we introduced three displacement functions, ψ, w and G, to simplify and decouple the basic equations of a spherically isotropic elastic body, as given in Eqs. (2.4.11) to (2.4.13). For a free vibration problem without body forces and temperature change, these equations become

$$[c_{44}\nabla_3^2 + 2(c_{66} - c_{44}) + c_{66}\nabla_1^2]\psi - R^2\rho\frac{\partial^2\psi}{\partial t^2} = 0, \tag{10.1.1}$$

$$L_1 w - L_2 G + R^2\rho\frac{\partial^2 G}{\partial t^2} = 0, \tag{10.1.2}$$

$$L_3 w - L_4\nabla_1^2 G - R^2\rho\frac{\partial^2 w}{\partial t^2} = 0. \tag{10.1.3}$$

We can take

$$\begin{aligned}\psi &= \psi_n(R)S_n^m(\theta,\alpha)\exp(\mathrm{i}\omega t),\\ w &= w_n(R)S_n^m(\theta,\alpha)\exp(\mathrm{i}\omega t),\\ G &= G_n(R)S_n^m(\theta,\alpha)\exp(\mathrm{i}\omega t),\end{aligned} \tag{10.1.4}$$

to further simplify these equations, where $S_n^m(\theta,\alpha) = P_n^m(\cos\theta)\exp(im\alpha)$ are spherical harmonics, $P_n^m(\cos\theta)$ are the associated Legendre functions, n and m are integers, ω is the circular frequency, and ψ_n, G_n and w_n are unknown functions of variable R.

For convenience, introduce the following dimensionless notation

$$\begin{aligned}&U_n = \psi_n / R_0,\quad V_n = G_n / R_0,\quad W_n = w_n / R_0,\\ &\xi = R / R_0,\quad \Omega = \omega R_0 / v_2,\\ &f_1 = c_{11}/c_{44},\quad f_2 = c_{12}/c_{44},\quad f_3 = c_{13}/c_{44},\quad f_4 = c_{33}/c_{44},\end{aligned} \tag{10.1.5}$$

where $v_2 = \sqrt{c_{44}/\rho}$ is the velocity of the elastic wave; R_0 is a characteristic length, and will be taken as the mean radius of the spherical shell. By substituting Eq. (10.1.4) into Eqs. (10.1.1) to (10.1.3) and noticing the following identity [see Eq. (B.3.1)]

$$\nabla_1^2 S_n^m(\theta,\alpha) + n(n+1)S_n^m(\theta,\alpha) = 0, \tag{10.1.6}$$

we get

$$\xi^2 U_n'' + 2\xi U_n' + \{\Omega^2\xi^2 - [2 + (n^2 + n - 2)(f_1 - f_2)/2]\}U_n = 0, \tag{10.1.7}$$

$$\begin{aligned}&\xi^2 W_n'' + 2\xi W_n' + (\Omega^2\xi^2 / f_4 + p_1)W_n - p_2\xi V_n' - p_3 V_n = 0,\\ &\xi^2 V_n'' + 2\xi V_n' + (\Omega^2\xi^2 + p_4)V_n - p_5\xi W_n' - p_6 W_n = 0,\end{aligned} \tag{10.1.8}$$

where the prime denotes differentiation with respect to ξ and

$$\begin{aligned}p_1 &= [2(f_3 - f_1 - f_2) - n(n+1)]/f_4,\\ p_2 &= -n(n+1)(f_3 + 1)/f_4,\\ p_3 &= n(n+1)(f_1 + f_2 + 1 - f_3)/f_4,\\ p_4 &= f_1 - f_2 - n(n+1)f_1 - 2,\\ p_5 &= f_3 + 1,\\ p_6 &= f_1 + f_2 + 2.\end{aligned} \tag{10.1.9}$$

Equations (10.1.7) and (10.1.8) are identical to those obtained by Cohen *et al.* (1972)[1].

Equation (10.1.7) can be transformed into Bessel's differential equation by variable substitution. According to **Appendix B**, the solution to Eq. (10.1.7) is

$$U_n(\xi)=\begin{cases} B_{11}j_1(\Omega\xi)+B_{12}y_1(\Omega\xi), & (n=1), \\ \xi^{-1/2}[B_{n1}J_\eta(\Omega\xi)+B_{n2}Y_\eta(\Omega\xi)], & (n>1), \end{cases} \tag{10.1.10}$$

where j_1 and y_1 are the *spherical Bessel functions* of the first and second kinds, J_η and Y_η are the Bessel functions of the first and second kinds, B_{ij} $(i=1,2,\cdots;\ j=1,2)$ are arbitrary constants, and

$$\eta^2=\frac{1}{4}[9+2(n^2+n-2)(f_1-f_2)]>0. \tag{10.1.11}$$

The inequality is obvious since $c_{11}-c_{12}>0$ and $n\geq 1$.

The two equations in Eq. (10.1.8) are coupled and a constraint on the elastic constants is often needed to decouple them (Cohen *et al.*, 1972). In the special case of $n=0$, however, an independent second-order ordinary differential equation for W_0 will be obtained when substituting $u_R=RW_0(\xi)\exp(\mathrm{i}\omega t)$ and $u_\theta=u_\alpha=0$ into Eq. (2.4.1) and using Eq. (10.1.5). This gives rise to

$$\xi^2W_0''+2\xi W_0'+[\Omega^2\xi^2/f_4+2(f_3-f_1-f_2)/f_4]W_0=0. \tag{10.1.12}$$

This equation can also be derived directly from the first equation in Eq. (10.1.8) by omitting V_0 since there is only radial displacement for $n=0$. The solution to Eq. (10.1.12) is

$$W_0(\xi)=\xi^{-1/2}[C_{01}J_\zeta(\Omega f_4^{-1/2}\xi)+C_{02}Y_\zeta(\Omega f_4^{-1/2}\xi)], \tag{10.1.13}$$

[1] Instead of Eq. (2.4.3), Cohen *et al.* (1972) introduced two auxiliary variables
$\Delta=\left[\frac{\partial}{\partial\theta}(\sin\theta u_\theta)+\frac{\partial u_\alpha}{\partial\alpha}\right]/\sin\theta$, and $\Lambda=\left[\frac{\partial}{\partial\theta}(\sin\theta u_\alpha)-\frac{\partial u_\theta}{\partial\alpha}\right]/\sin\theta$
to simplify the basic equations. With expressions for Δ and Λ in hand, it is still a tricky task to obtain expressions for displacements as well as stresses. In our formulae, however, this can be finished by simple differentiation from ψ and G.

where C_{0i} $(i=1, 2)$ are two arbitrary constants and

$$\zeta^2 = 1/4 + 2(f_1 + f_2 - f_3)/f_4 > 0 . \tag{10.1.14}$$

This inequality can be obtained directly from Eq. (3.4.17).

When $n > 0$, V_n and W_n in the two equations of Eq. (10.1.8) are coupled. However, the two equations are similar and have only one regular singular point at $\xi = 0$, thus the *Frobenius power series method* can be applied to solve this set of ordinary differential equations (Wang and Guo, 2000; Ding, Chen and Liu, 1995). As shown in Ding, Chen and Liu (1995) and Chen (1995), the general solution of Eq. (10.1.8) can be expressed as a linear combination of four independent solutions, *i.e.*,

$$V_n(\xi) = \sum_{j=1}^{4} C_{nj} V_{nj}(\xi), \quad W_n(\xi) = \sum_{j=1}^{4} C_{nj} W_{nj}(\xi), \qquad (n > 0), \tag{10.1.15}$$

where C_{nj} $(j = 1, 2, \cdots, 4)$ are arbitrary constants and V_{nj} and W_{nj} are convergent and infinite series in ξ; they can also be treated as functions of the dimensionless circular frequency Ω once ξ is known. With the U_n, V_n and W_n thus obtained, the displacement functions can be derived from Eq. (10.1.4). The displacement and stress components are then fully defined by Eqs. (2.4.3), (1.2.3) and (1.2.40*a*), respectively. The stresses on a spherical surface can therefore be written as

$$\begin{aligned}
\sigma_R &= c_{13}\sigma_\xi = c_{13}\left[2W_n/\xi + n(n+1)V_n/\xi + (f_4/f_3)W_n'\right] S_n^m(\theta,\alpha)\exp(\mathrm{i}\omega t), \\
\tau_{R\theta} &= c_{44}\tau_{\xi\theta} = c_{44}\left[(W_n/\xi + V_n/\xi - V_n')\frac{\partial S_n^m(\theta,\alpha)}{\partial\theta} - (U_n' - U_n/\xi)\csc\theta\frac{\partial S_n^m(\theta,\alpha)}{\partial\alpha}\right] \\
&\quad \times\exp(\mathrm{i}\omega t), \\
\tau_{\alpha R} &= c_{44}\tau_{\alpha\xi} = c_{44}\left[(W_n/\xi + V_n/\xi - V_n')\csc\theta\frac{\partial S_n^m(\theta,\alpha)}{\partial\alpha} + (U_n' - U_n/\xi)\frac{\partial S_n^m(\theta,\alpha)}{\partial\theta}\right] \\
&\quad \times\exp(\mathrm{i}\omega t),
\end{aligned} \tag{10.1.16}$$

where σ_ξ, $\tau_{\xi\theta}$ and $\tau_{\alpha\xi}$ are dimensionless stress components on the spherical surface.

10.1.2 Free Vibration Analysis

(1) Single-layered shell

Consider the free vibration of a single-layered spherical shell of inner radius a and outer radius b. The traction-free conditions on the shell surfaces are

$$\sigma_\xi = \tau_{\xi\theta} = \tau_{\alpha\xi} = 0, \quad (\xi = 1 \pm t_R/2), \tag{10.1.17}$$

where $t_R = h/R_0$ is the thickness-to-mean radius ratio, $h = b - a$ is the shell thickness, and $R_0 = (a+b)/2$ is the mean radius. According to Eq. (10.1.16), the satisfaction of these boundary conditions requires that

$$2W_n/\xi + n(n+1)V_n/\xi + (f_4/f_3)W_n' = 0, \quad (\xi = 1 \pm t_R/2, n \ge 0), \tag{10.1.18}$$

$$\begin{aligned} &(W_n/\xi + V_n/\xi - V_n')\frac{\partial S_n}{\partial \theta} - (U_n' - U_n/\xi)\frac{\partial S_n}{\partial \alpha}\csc\theta = 0, \\ &(W_n/\xi + V_n/\xi - V_n')\frac{\partial S_n}{\partial \alpha}\csc\theta + (U_n' - U_n/\xi)\frac{\partial S_n}{\partial \theta} = 0, \end{aligned} \quad (\xi = 1 \pm t_R/2, \ n \ge 1), \tag{10.1.19}$$

where $S_n = S_n^m(\theta, \alpha)$. Since

$$\sin\theta \frac{\mathrm{d}\, P_n^m(\cos\theta)}{\mathrm{d}\theta} = \frac{1}{2n+1}\left[n(n-m+1)P_{n+1}^m(\cos\theta) - (n+1)(n+m)P_{n-1}^m(\cos\theta)\right], \tag{10.1.20}$$

and the associated Legendre functions are orthogonal with respect to the weight $\sin\theta$ (see **Appendix B, Section B.2**), we get

$$W_n/\xi + V_n/\xi - V_n' = 0, \quad (\xi = 1 \pm t_R/2, \ n \ge 1), \tag{10.1.21}$$

$$U_n' - U_n/\xi = 0, \quad (\xi = 1 \pm t_R/2, \ n \ge 1). \tag{10.1.22}$$

As in the free vibration analysis of an isotropic elastic sphere (Love, 1927), the free vibration of a spherically isotropic hollow sphere can be divided into two classes according to the boundary conditions and governing differential equations. The first class ($n \ge 1$), defined by Eqs. (10.1.7) and (10.1.22), corresponds to an equivoluminal motion of the shell without radial displacement. The second class ($n \geq 0$), specified by Eqs. (10.1.8), (10.1.12), (10.1.18) and (10.1.21), possesses in general both transverse and radial components except the case of $n = 0$.

(2) Multi-layered spherical shell

For the free vibration of a multi-layered spherical shell, as illustrated in Fig. 10.1, in addition to the boundary conditions at the inner and outer surfaces, the continuity of stresses and displacements at the layer interfaces must also be satisfied.

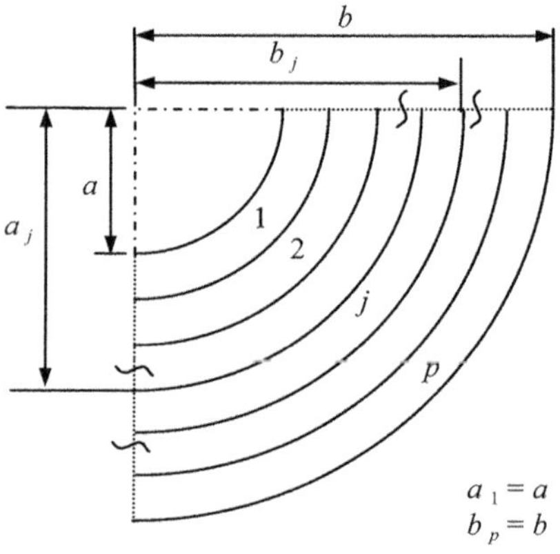

Fig. 10.1 A multi-layered spherical shell.

(a) Displacement continuity conditions

When two adjacent layers of a spherical shell, *e.g.*, the $(j-1)^{\text{th}}$ and j^{th} layers, are rigidly joined, the displacement components at the layer interface, $\xi = \xi_j = a_j / R_0$, must be continuous, *i.e.*,

$$u_R^{(j-1)} = u_R^{(j)}, \quad u_\theta^{(j-1)} = u_\theta^{(j)}, \quad u_\alpha^{(j-1)} = u_\alpha^{(j)}, \quad (\xi = \xi_j). \tag{10.1.23}$$

Applying Eqs. (2.4.3), (10.1.4) and (10.1.5), we get

$$W_n^{(j-1)}(\xi_j) - W_n^{(j)}(\xi_j) = 0, \tag{10.1.24}$$

$$[U_n^{(j-1)}(\xi_j)-U_n^{(j)}(\xi_j)]\sin\theta\frac{\mathrm{d}P_n^m(\cos\theta)}{\mathrm{d}\theta}-\mathrm{i}m[V_n^{(j-1)}(\xi_j)-V_n^{(j)}(\xi_j)]P_n^m(\cos\theta)=0,$$
$$\mathrm{i}m[U_n^{(j-1)}(\xi_j)-U_n^{(j)}(\xi_j)]P_n^m(\cos\theta)+[V_n^{(j-1)}(\xi_j)-V_n^{(j)}(\xi_j)]\sin\theta\frac{\mathrm{d}P_n^m(\cos\theta)}{\mathrm{d}\theta}=0. \tag{10.1.25}$$

Due to Eq. (10.1.20) and the orthogonality of the Legendre functions, Eq. (10.1.25) leads to

$$U_n^{(j-1)}(\xi_j)-U_n^{(j)}(\xi_j)=0, \tag{10.1.26}$$
$$V_n^{(j-1)}(\xi_j)-V_n^{(j)}(\xi_j)=0. \tag{10.1.27}$$

Equations (10.1.24), (10.1.26) and (10.1.27) describe the displacement continuity conditions between the $(j-1)^{\text{th}}$ and j^{th} layers.

(*b*) Stress continuity conditions

Similarly, when the $(j-1)^{\text{th}}$ and j^{th} layers of the shell are rigidly joined, the stresses at the layer interface $\xi=\xi_j$ must be continuous. This requires that

$$\sigma_R^{(j-1)}=\sigma_R^{(j)},\quad \tau_{R\theta}^{(j-1)}=\tau_{R\theta}^{(j)},\quad \tau_{\alpha R}^{(j-1)}=\tau_{\alpha R}^{(j)},\quad (\xi=\xi_j). \tag{10.1.28}$$

Again, by making use of the properties of Legendre functions, we obtain

$$\xi_j U_n'^{(j-1)}(\xi_j)-U_n^{(j-1)}(\xi_j)-g_2[\xi_j U_n'^{(j)}(\xi_j)-U_n^{(j)}(\xi_j)]=0, \tag{10.1.29}$$

$$2W_n^{(j-1)}(\xi_j)+n(n+1)V_n^{(j-1)}(\xi_j)+(f_4^{(j-1)}/f_3^{(j-1)})\xi_j W_n'^{(j-1)}(\xi_j)$$
$$-g_1[2W_n^{(j)}(\xi_j)+n(n+1)V_n^{(j)}(\xi_j)+(f_4^{(j)}/f_3^{(j)})\xi_j W_n'^{(j)}(\xi_j)]=0,$$
$$W_n^{(j-1)}(\xi_j)+V_n^{(j-1)}(\xi_j)-\xi_j V_n'^{(j-1)}(\xi_j)-g_2[W_n^{(j)}(\xi_j)+V_n^{(j)}(\xi_j)-\xi_j V_n'^{(j)}(\xi_j)]=0, \tag{10.1.30}$$

where $g_1=c_{13}^{(j)}/c_{13}^{(j-1)}$ and $g_2=c_{44}^{(j)}/c_{44}^{(j-1)}$.

It is interesting to note from the governing equations and the above boundary and continuity conditions that the free vibration of a multi-layered spherical shell of spherically isotropic materials can also be divided into two independent classes, with the first defined by Eqs. (10.1.7), (10.1.22), (10.1.26) and (10.1.29), and the second by Eqs. (10.1.8), (10.1.18), (10.1.21), (10.1.24), (10.1.27) and (10.1.30).

10.2 FREQUENCY EQUATIONS AND NUMERICAL RESULTS

10.2.1 Frequency Equations of a Single-Layered Hollow Sphere

We will focus on the frequency equations for a single-layered spherical shell. The same process applies to shells with multi-layers.

In the last section, we obtained the solutions to the governing equations, where the unknown constants can be determined by satisfying the boundary conditions. Thus the procedures are routine, *i.e.*, (1) substituting the corresponding solutions into the relevant boundary conditions to get a set of linear homogeneous algebraic equations about the unknown constants, and (2) obtaining the required frequency equations from the coefficient determinant of the equation set because to have a nontrivial solution of the unknown constants, the determinant must vanish.

(1) The first class

(a) $n=1$

This is a purely torsional or rotary vibration and the frequency equation can be obtained from the first equation in Eq. (10.1.10) and Eq. (10.1.22) as

$$\left|E_{ij}^1\right|=0, \quad (i,j=1,2), \tag{10.2.1}$$

where

$$\begin{aligned}
E_{11}^1 &= 3\Omega t_1\cos(\Omega t_1)-3\sin(\Omega t_1)+(\Omega t_1)^2\sin(\Omega t_1),\\
E_{12}^1 &= 3\Omega t_1\sin(\Omega t_1)+3\cos(\Omega t_1)-(\Omega t_1)^2\cos(\Omega t_1),\\
E_{21}^1 &= 3\Omega t_2\cos(\Omega t_2)-3\sin(\Omega t_2)+(\Omega t_2)^2\sin(\Omega t_2),\\
E_{22}^1 &= 3\Omega t_2\sin(\Omega t_2)+3\cos(\Omega t_2)-(\Omega t_2)^2\cos(\Omega t_2),
\end{aligned} \tag{10.2.2}$$

with $t_1=1-t_R/2$ and $t_2=1+t_R/2$.

(b) $n>1$

In this case, the second equation of Eq. (10.1.10) and Eq. (10.1.22) lead to the frequency equation

$$\left|E_{ij}^{2}\right| = 0, \quad (i, j = 1, 2), \tag{10.2.3}$$

where

$$\begin{aligned}
E_{11}^{2} &= (\eta - 3/2)J_{\eta}(\Omega t_1) - \Omega t_1 J_{\eta+1}(\Omega t_1), \\
E_{12}^{2} &= (\eta - 3/2)Y_{\eta}(\Omega t_1) - \Omega t_1 Y_{\eta+1}(\Omega t_1), \\
E_{21}^{2} &= (\eta - 3/2)J_{\eta}(\Omega t_2) - \Omega t_2 J_{\eta+1}(\Omega t_2), \\
E_{22}^{2} &= (\eta - 3/2)Y_{\eta}(\Omega t_2) - \Omega t_2 Y_{\eta+1}(\Omega t_2).
\end{aligned} \tag{10.2.4}$$

(2) The second class

(a) $n = 0$

This is a radial vibration and the frequency equation can be obtained from Eqs. (10.1.13) and (10.1.18), *i.e.*,

$$\left|E_{ij}^{3}\right| = 0, \quad (i, j = 1, 2), \tag{10.2.5}$$

where

$$\begin{aligned}
E_{11}^{3} &= [2 + (\zeta - 1/2)(f_4/f_3)]J_{\zeta}(\gamma t_1) - (f_4/f_3)\gamma t_1 J_{\zeta+1}(\gamma t_1), \\
E_{12}^{3} &= [2 + (\zeta - 1/2)(f_4/f_3)]Y_{\zeta}(\gamma t_1) - (f_4/f_3)\gamma t_1 Y_{\zeta+1}(\gamma t_1), \\
E_{21}^{3} &= [2 + (\zeta - 1/2)(f_4/f_3)]J_{\zeta}(\gamma t_2) - (f_4/f_3)\gamma t_2 J_{\zeta+1}(\gamma t_2), \\
E_{22}^{3} &= [2 + (\zeta - 1/2)(f_4/f_3)]Y_{\zeta}(\gamma t_2) - (f_4/f_3)\gamma t_2 Y_{\zeta+1}(\gamma t_2),
\end{aligned} \tag{10.2.6}$$

in which $\gamma = \Omega/\sqrt{f_4}$.

(b) $n > 0$

The frequency equation can be obtained from Eqs. (10.1.15), (10.1.18) and (10.1.21) as

$$\left|E_{ij}^{4}\right| = 0, \quad (i, j = 1, 2, 3, 4), \tag{10.2.7}$$

where

$$\begin{aligned}
E_{1i}^4 &= 2W_{ni}(t_1)/t_1 + n(n+1)V_{ni}(t_1)/t_1 + (f_4/f_3)W'_{ni}(t_1), \\
E_{2i}^4 &= W_{ni}(t_1)/t_1 + V_{ni}(t_1)/t_1 - V'_{ni}(t_1), \\
E_{3i}^4 &= W_{ni}(t_2)/t_2 + V_{ni}(t_2)/t_2 - V'_{ni}(t_2), \\
E_{4i}^4 &= 2W_{ni}(t_2)/t_2 + n(n+1)V_{ni}(t_2)/t_2 + (f_4/f_3)W'_{ni}(t_2),
\end{aligned} \quad (i = 1, 2, 3, 4) . \tag{10.2.8}$$

It can be seen that the integer m does not appear in the frequency equations. This means that the frequency of an asymmetric mode is the same as that of an axisymmetric mode, as it is for an isotropic spherical shell (Silbiger, 1962). It is understandable because a spherically isotropic material still possesses spherical symmetry.

For the vibration of a multi-layered spherical shell, the continuity conditions of stresses and displacements at the interfaces of adjacent layers must also be used in deriving the frequency equations. We can obtain a set of linear homogeneous algebraic equations for the unknown constants. For instance, for the first class of vibration of a shell with p layers, there will be $2p$ unknown constants.

This derivation is completely based on the three-dimensional theory of elasticity. Thus, the frequency equations will have an infinite number of roots. We will consider mainly the lowest natural frequency. It must be pointed out that there are two zero frequencies when $n = 1$, of which one is due to a *rigid-body rotation* in the first class vibration and the other is caused by a *rigid-body translation* in the second class vibration.

10.2.2 Some Special Cases

(1) Isotropic material

If the shell material is isotropic, Eq. (1.4.13) holds and $\zeta = 3/2$ according to Eq. (10.1.14). In particular, Eq. (10.2.5) becomes

$$\frac{\tan(\gamma t_R)}{\gamma t_R} = \frac{4(4 + \Omega^2 t_1 t_2)}{16\gamma^2 t_1 t_2 + (\Omega^2 t_1^2 - 4)(\Omega^2 t_2^2 - 4)} . \tag{10.2.9}$$

(2) Thin spherical shell

If the spherical shell is very thin, $t_R/(1 + t_R/2) \to 0$. Moreover, since $\Omega t_R \neq 0$, $\Omega t_i \to \infty\ (i = 1, 2)$. Thus the asymptotic expansions of Bessel functions can be used (Watson, 1922). Since Eq. (10.2.1) can be obtained directly from Eq. (10.2.3) by letting $n = 1$, we will just discuss Eq. (10.2.3).

When $x \to \infty$, the following asymptotic expansions [cf. **Appendix B**, Eq. (B.4.5)] are employed

$$\begin{aligned}
J_\eta(x) &= \sqrt{\frac{2}{\pi x}}\left\{\cos\left(x-\frac{\eta\pi}{2}-\frac{\pi}{4}\right)-\sin\left(x-\frac{\eta\pi}{2}-\frac{\pi}{4}\right)\frac{4\eta^2-1}{8x}\right\},\\
J_{\eta+1}(x) &= \sqrt{\frac{2}{\pi x}}\left\{\sin\left(x-\frac{\eta\pi}{2}-\frac{\pi}{4}\right)\left[1-\frac{16(\eta+1)^4-40(\eta+1)^2+9}{128x^2}\right]\right.\\
&\qquad\left.+\cos\left(x-\frac{\eta\pi}{2}-\frac{\pi}{4}\right)\frac{4(\eta+1)^2-1}{8x}\right\},\\
Y_\eta(x) &= \sqrt{\frac{2}{\pi x}}\left\{\sin\left(x-\frac{\eta\pi}{2}-\frac{\pi}{4}\right)+\cos\left(x-\frac{\eta\pi}{2}-\frac{\pi}{4}\right)\frac{4\eta^2-1}{8x}\right\},\\
Y_{\eta+1}(x) &= \sqrt{\frac{2}{\pi x}}\left\{-\cos\left(x-\frac{\eta\pi}{2}-\frac{\pi}{4}\right)\left[1-\frac{16(\eta+1)^4-40(\eta+1)^2+9}{128x^2}\right]\right.\\
&\qquad\left.+\sin\left(x-\frac{\eta\pi}{2}-\frac{\pi}{4}\right)\frac{4(\eta+1)^2-1}{8x}\right\}.
\end{aligned} \tag{10.2.10}$$

The substitution of Eq. (10.2.10) into Eq. (10.2.3), yields (Cohen *et al.*, 1972)

$$\frac{\tan(\Omega t_R)}{\Omega t_R}=\frac{4\eta^2+15}{8t_1t_2\Omega^2-4\eta^2+33}. \tag{10.2.11}$$

Now, because $t_R \to 0$, the left-hand side tends to unity, and $t_1, t_2 \approx 1$, giving

$$\Omega^2=\eta^2-9/4=\frac{1}{2}(n^2+n-2)(f_1-f_2)=0, \quad (n=1,2,\cdots). \tag{10.2.12}$$

This frequency equation represents a membrane state. For isotropic materials, since $f_1-f_2=2$, it is obtained from Eq. (10.2.12) that

$$\Omega^2=(n-1)(n+2), \quad (n=1,2,\cdots), \tag{10.2.13}$$

which is identical to that obtained by Love (1927) [page 549, Eq. (54)]. Similarly, when $t_R \to 0$, Eq. (10.2.9) becomes

$$\Omega = 2\sqrt{(1+\nu)/(1-\nu)}\,, \tag{10.2.14}$$

where ν is the Poisson's ratio of the isotropic material. It can be seen that Eq. (10.2.14) is identical to that obtained by Love (1927, p. 287) as well.

(3) Isotropic solid sphere

For an isotropic solid sphere, $t_2 = t_R = 2$. In addition, the displacements at the origin must be finite. Thus we can obtain the frequency equations of purely torsional and radial vibrations from Eqs. (10.2.1) and (10.2.9) respectively, as follows

$$\tan(2\Omega) = \frac{6\Omega}{3-4\Omega^2}, \quad (n=1)\,, \tag{10.2.15}$$

$$\frac{\tan(\gamma_R)}{\gamma_R} = \frac{1}{1-\Omega^2}, \quad (n=0)\,, \tag{10.2.16}$$

which are what Love obtained (Love, 1927). The frequency equation of the first class for $n \geq 1$ is also obtained from Eq. (10.2.3)

$$(\eta - 3/2)J_\eta(2\Omega) - 2\Omega J_{\eta+1}(2\Omega) = 0\,, \quad (n > 1)\,, \tag{10.2.17}$$

where for isotropic materials $\eta = n + 1/2$. It can be shown that Eq. (10.2.17) is also identical to the Love's equation (Love, 1927) [page 284, Eq. (38)].

10.2.3 An Example

As an example, let us analyse the free vibration of an isotropic solid sphere of Material (1) surrounded by a spherically isotropic spherical shell of Material (2), as illustrated in Fig. 10.2. The radius of the sphere is a and the outer radius of the shell is b. The bonding between them is perfect so that at the interface $R = a$ all the displacement and stress components are continuous. The density ratio of the two materials is $\bar{\rho} = \rho^{(2)}/\rho^{(1)} = 1.2$ and their elastic constants are listed in Table 10.1. Note that Material (2) is a hypothetical material exhibiting substantial anisotropy (Ramakrishnan and Shah, 1970).

The dimensionless material constants of the two-layered solid sphere can be obtained as

$$f_1^{(1)} = f_4^{(1)} = 3.38, f_2^{(1)} = f_3^{(1)} = 1.38,$$
$$f_1^{(2)} = 20, f_2^{(2)} = 12, f_3^{(2)} = f_4^{(2)} = 2, \quad (10.2.18)$$

where the superscript denotes the corresponding layer, *i.e.* the inner sphere or the outer spherical shell.

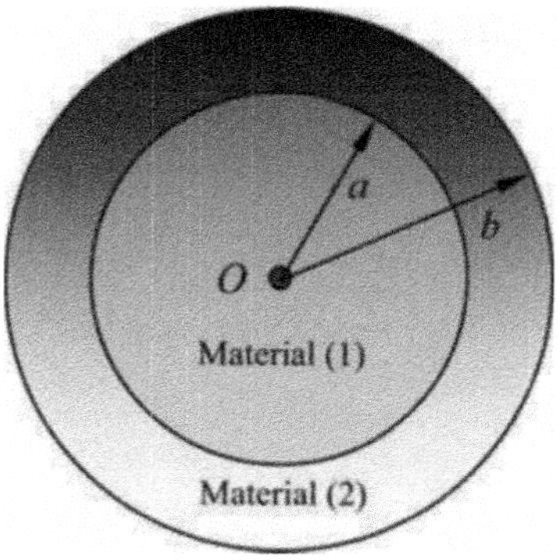

Fig. 10.2 A two-layered solid sphere.

Table 10.1 Material constants of three different materials.

Material	Material Constants
(1)	$E = 20.7\times10^{10}$ Pa, $\nu = 0.29$
(2)	$c_{11} = 20.0\times10^{10}$ Pa, $c_{12} = 12.0\times10^{10}$ Pa, $c_{13} = c_{33} = 2.0\times10^{10}$ Pa, $c_{44} = 1.0\times10^{10}$ Pa
(3)	$c_{11} = 5.97\times10^{10}$ Pa, $c_{12} = 2.62\times10^{10}$ Pa, $c_{13} = 2.17\times10^{10}$ Pa, $c_{33} = 6.17\times10^{10}$ Pa, $c_{44} = 1.64\times10^{10}$ Pa

According to the results presented in previous sections, we can have the frequency equations for calculation easily.

(1) Frequency equation

(*a*) The first class of vibration

When $n = 1$, the two-layered sphere is subjected to a torsional vibration and the frequency equation becomes

$$\left|E_{ij}^{5}\right| = 0, \quad (i, j = 1, 2, 3), \tag{10.2.19}$$

where

$$\begin{aligned}
&E_{11}^{5} = \lambda^2[\sin(\Omega\xi_1) - \Omega\xi_1\cos(\Omega\xi_1)], \quad E_{12}^{5} = \sin(\lambda\Omega\xi_1) - \lambda\Omega\xi_1\cos(\lambda\Omega\xi_1), \\
&E_{13}^{5} = -\cos(\lambda\Omega\xi_1) - \lambda\Omega\xi_1\sin(\lambda\Omega\xi_1), \\
&E_{21}^{5} = \lambda^2[3\Omega\xi_1\cos(\Omega\xi_1) - 3\sin(\Omega\xi_1) + (\Omega\xi_1)^2\sin(\Omega\xi_1)], \\
&E_{22}^{5} = g_2[3\lambda\Omega\xi_1\cos(\lambda\Omega\xi_1) - 3\sin(\lambda\Omega\xi_1) + (\lambda\Omega\xi_1)^2\sin(\lambda\Omega\xi_1)], \\
&E_{23}^{5} = g_2[3\lambda\Omega\xi_1\sin(\lambda\Omega\xi_1) + 3\cos(\lambda\Omega\xi_1) - (\lambda\Omega\xi_1)^2\cos(\lambda\Omega\xi_1)], \\
&E_{31}^{5} = 0, \quad E_{32}^{5} = 3\lambda\Omega\xi_2\cos(\lambda\Omega\xi_2) - 3\sin(\lambda\Omega\xi_2) + (\lambda\Omega\xi_2)^2\sin(\lambda\Omega\xi_2), \\
&E_{23}^{5} = 3\lambda\Omega\xi_2\sin(\lambda\Omega\xi_2) + 3\cos(\lambda\Omega\xi_2) - (\lambda\Omega\xi_2)^2\cos(\lambda\Omega\xi_2),
\end{aligned} \tag{10.2.20}$$

in which

$$\begin{aligned}
&g_2 = c_{44}^{(2)}/c_{44}^{(1)} = 2c_{44}^{(2)}(1+\nu)/E, \quad \lambda = (\bar{\rho}/g_2)^{1/2}, \quad \xi_1 = a/R_0, \\
&\xi_2 = b/R_0 = 2, \quad R_0 = b/2, \quad \Omega = \omega R_0\sqrt{\rho_1/c_{44}^{(1)}} = \omega R_0\sqrt{2\rho_1(1+\nu)/E}.
\end{aligned} \tag{10.2.21}$$

When $n > 1$, the frequency equation becomes

$$\left|E_{ij}^{6}\right| = 0, \quad (i, j = 1, 2, 3), \tag{10.2.22}$$

where

$$\begin{aligned}
&E_{11}^{6} = J_{\eta_1}(\Omega\xi_1), \; E_{12}^{6} = J_{\eta_2}(\lambda\Omega\xi_1), \; E_{13}^{6} = Y_{\eta_2}(\lambda\Omega\xi_1), \\
&E_{21}^{6} = (\eta_1 - 3/2)J_{\eta_1}(\Omega\xi_1) - \Omega\xi_1 J_{\eta_1+1}(\Omega\xi_1), \\
&E_{22}^{6} = g_2[(\eta_2 - 3/2)J_{\eta_2}(\lambda\Omega\xi_1) - \lambda\Omega\xi_1 J_{\eta_2+1}(\lambda\Omega\xi_1)], \\
&E_{23}^{6} = g_2[(\eta_2 - 3/2)Y_{\eta_2}(\lambda\Omega\xi_1) - \lambda\Omega\xi_1 Y_{\eta_2+1}(\lambda\Omega\xi_1)], \\
&E_{31}^{6} = 0, \; E_{32}^{6} = (\eta_2 - 3/2)J_{\eta_2}(\lambda\Omega\xi_2) - \lambda\Omega\xi_2 J_{\eta_2+1}(\lambda\Omega\xi_2), \\
&E_{33}^{6} = (\eta_2 - 3/2)Y_{\eta_2}(\lambda\Omega\xi_2) - \lambda\Omega\xi_2 Y_{\eta_2+1}(\lambda\Omega\xi_2),
\end{aligned} \tag{10.2.23}$$

in which

$$\eta_1 = \sqrt{n(n+1)+1/4}\,, \quad \eta_2 = \sqrt{9+2(n^2+n-2)(f_1^{(2)}-f_2^{(2)})/2}\,. \tag{10.2.24}$$

(*b*) The second class of vibration

When $n=0$, a radial vibration occurs. In this case, the frequency equation is given by

$$\left|E_{ij}^7\right| = 0\,, \quad (i,j=1,2,3)\,, \tag{10.2.25}$$

where

$$\begin{aligned}
&E_{11}^7 = J_{\zeta_1}(\gamma_1\Omega\xi_1), \quad E_{12}^7 = J_{\zeta_2}(\lambda\gamma_2\Omega\xi_1), \quad E_{13}^7 = Y_{\zeta_2}(\lambda\gamma_2\Omega\xi_1)\,,\\
&E_{21}^7 = [2+(\zeta_1-1/2)\beta_1]J_{\zeta_1}(\gamma_1\Omega\xi_1) - \gamma_1\Omega\xi_1\beta_1 J_{\zeta_1+1}(\gamma_1\Omega\xi_1),\\
&E_{22}^7 = g_1\{[2+(\zeta_2-1/2)\beta_2]J_{\zeta_2}(\lambda\gamma_2\Omega\xi_1) - \lambda\gamma_2\Omega\xi_1\beta_2 J_{\zeta_2+1}(\lambda\gamma_2\Omega\xi_1)\},\\
&E_{23}^7 = g_1\{[2+(\zeta_2-1/2)\beta_2]Y_{\zeta_2}(\lambda\gamma_2\Omega\xi_1) - \lambda\gamma_2\Omega\xi_1\beta_2 Y_{\zeta_2+1}(\lambda\gamma_2\Omega\xi_1)\},\\
&E_{31}^7 = 0,\\
&E_{32}^7 = [2+(\zeta_2-1/2)\beta_2]J_{\zeta_2}(\lambda\gamma_2\Omega\xi_2) - \lambda\gamma_2\Omega\xi_2\beta_2 J_{\zeta_2+1}(\lambda\gamma_2\Omega\xi_2),\\
&E_{33}^7 = [2+(\zeta_2-1/2)\beta_2]Y_{\zeta_2}(\lambda\gamma_2\Omega\xi_2) - \lambda\gamma_2\Omega\xi_2\beta_2 Y_{\zeta_2+1}(\lambda\gamma_2\Omega\xi_2),
\end{aligned} \tag{10.2.26}$$

in which

$$\begin{aligned}
&\zeta_1 = 3/2\,, \quad \gamma_1 = 1/\sqrt{f_4^{(1)}}\,, \quad \gamma_2 = 1/\sqrt{f_4^{(2)}}\,, \quad g_1 = c_{13}^{(2)}/c_{13}^{(1)}\,,\\
&\zeta_2 = \sqrt{1/4+2(f_1^{(2)}+f_2^{(2)}-f_3^{(2)})/f_4^{(2)}}\,,\\
&\beta_1 = f_4^{(1)}/f_3^{(1)}\,, \quad \beta_2 = f_4^{(2)}/f_3^{(2)}\,.
\end{aligned} \tag{10.2.27}$$

However, the frequency equation for $n>0$ becomes

$$\left|E_{ij}^8\right| = 0\,, \quad (i,j=1,2,3)\,, \tag{10.2.28}$$

where

$$
\begin{aligned}
&E_{1l}^{8} = W_{nl}^{1}(\xi_1), \quad E_{1j}^{8} = W_{ni}^{2}(\xi_1),\\
&E_{2l}^{8} = V_{nl}^{1}(\xi_1), \quad E_{2j}^{8} = V_{ni}^{2}(\xi_1),\\
&E_{3l}^{8} = 2W_{nl}^{1}(\xi_1)/\xi_1 + n(n+1)V_{nl}^{1}(\xi_1)/\xi_1 + (f_4^1/f_3^1)W_{nl}^{\prime 1}(\xi_1),\\
&E_{3j}^{8} = g_1[2W_{ni}^{2}(\xi_1)/\xi_1 + n(n+1)V_{ni}^{2}(\xi_1)/\xi_1 + (f_4^2/f_3^2)W_{ni}^{\prime 2}(\xi_1)],\\
&E_{4l}^{8} = W_{nl}^{1}(\xi_1)/\xi_1 + V_{nl}^{1}(\xi_1)/\xi_1 - V_{nl}^{\prime 1}(\xi_1),\\
&E_{4j}^{8} = g_2[W_{ni}^{2}(\xi_1)/\xi_1 + V_{ni}^{2}(\xi_1)/\xi_1 - V_{ni}^{\prime 2}(\xi_1)],\\
&E_{5l}^{8} = E_{6l}^{8} = 0,\\
&E_{5j}^{8} = 2W_{ni}^{2}(\xi_2)/\xi_2 + n(n+1)V_{ni}^{2}(\xi_2)/\xi_2 + (f_4^2/f_3^2)W_{ni}^{\prime 2}(\xi_2),\\
&E_{6j}^{8} = W_{ni}^{2}(\xi_2)/\xi_2 + V_{ni}^{2}(\xi_2)/\xi_2 - V_{ni}^{\prime 2}(\xi_2),\\
&\qquad\qquad (l = 1, 2; j = i+2;\ i = 1, 2, 3, 4).
\end{aligned}
\tag{10.2.29}
$$

(2) Numerical results

The following dimensionless parameters can be obtained according to the material constants of the two-layered solid sphere specified above:

$$
\begin{aligned}
&\lambda = 3.1029, \quad g_1 = 0.1805, \quad g_2 = 0.1246, \quad \gamma_1 = 0.5439,\\
&\gamma_2 = 0.7071, \quad \zeta_2 = 5.5, \quad \beta_2 = 0.4085, \quad \beta_2 = 1.
\end{aligned}
$$

Using Eqs. (10.2.19), (10.2.22), (10.2.25) and (10.2.28), we can easily calculate the natural frequencies of the sphere. Figure 10.3 shows the first five nonzero frequencies when the sphere is under torsional vibration. Clearly the frequencies vary with the non-dimensionless parameter ξ_1. As pointed out earlier, the first class of vibration contains a rigid-body rotation associated with a zero frequency when $n=1$. When $\xi_1 = 2$ the layered-sphere becomes an isotropic solid sphere of Material (1).

It can be seen from Fig.10.3 that the dimensionless frequency (Ω) of torsional vibration increases with ξ_1. However, the increment is minor when ξ_1 is small.

Figure 10.4 demonstrates the spectra of the lowest frequency of the first class of vibration of non-torsional modes with $n = 2, 3, 4$ and 5. It can be seen that when $0 < \xi_1 \le 1$, the dimensionless frequency keeps almost constant and the value is very close to that of a spherically isotropic hollow sphere.

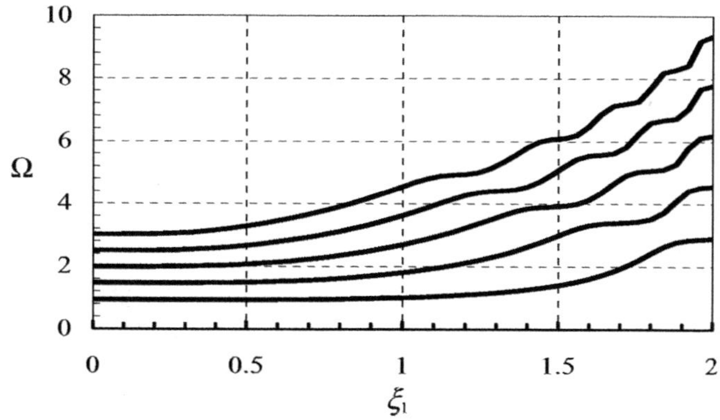

Fig. 10.3 Dimensionless frequencies under torsional vibration.

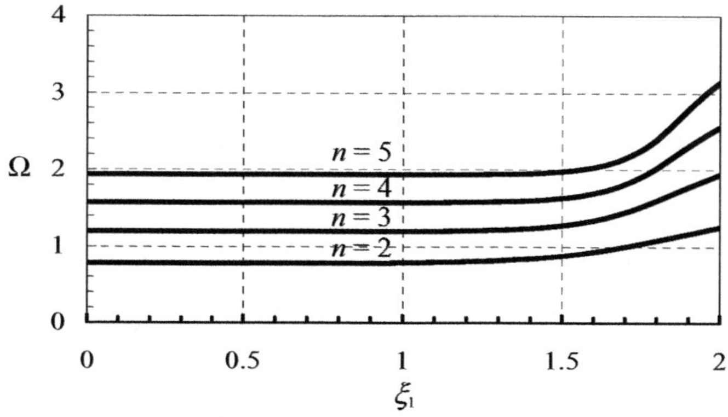

Fig. 10.4 Frequency spectra for other cases of the first class vibration.

Figure 10.5 shows the variations of the first five frequencies of a radial vibration of the two-layered solid sphere. The curves are similar to those in Fig. 10.3, but the values are greater. Note that when $\xi_1 = 2$, the frequencies can be directly calculated from Eq. (10.2.16), instead of Eq. (10.2.25); the two give the same result. Table 10.2 lists the lowest dimensionless frequencies of the modes $n = 2, \cdots, 5$ and 6 of the second class of vibration (non-breathing mode)[2].

The results presented in Figs. 10.3-10.5 and Table 10.2 all show that the frequency increases with ξ_1. This is reasonable since Material (1) is more stiffer than Material (2),

---

[2] The zero frequency associated with the rigid-body translation at $n = 1$ is not included here.

and with increasing ξ_1, the two-layered sphere, as shown in Fig. 10.2, will contain a bigger isotropic solid sphere. Note that, when ξ_1 is small, the contribution to the rigidity of the whole system from the isotropic solid sphere is also small, which leads to an inconspicuous variation of frequency, as shown in Figs. 10.3-10.5.

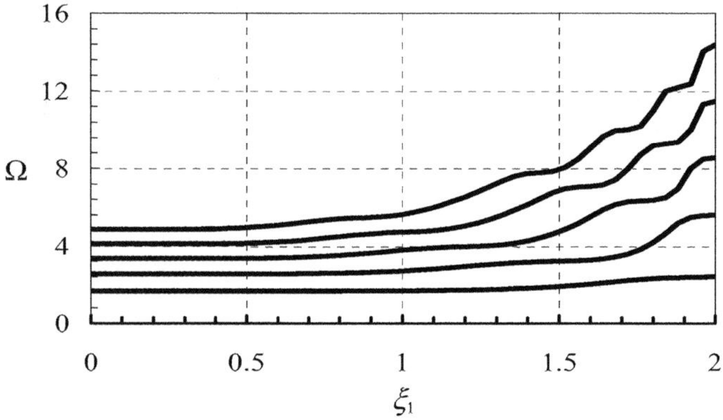

Fig. 10.5 Frequency spectra for purely radial vibration.

Table 10.2 The lowest non-dimensional frequency (Ω) for non-breathing mode.

n	2	3	4	5	6
$\xi_1 = 0.5$	0.655740	0.876974	1.053884	1.218613	1.379598
$\xi_1 = 1.0$	0.713541	0.916906	1.078925	1.233544	1.388211
$\xi_1 = 1.5$	0.981592	1.240201	1.392323	1.515428	1.633446
$\xi_1 = 2.0$	1.322503	1.966689	2.519444	3.036829	3.207820

To clarify the applicability of various approximate shell theories, Chen (1995) compared this three-dimensional analysis with three typical shell theories, *i.e.* the *classical thin shell theory*, the *five-mode shell theory* that includes the effects of shear deformation and rotatory inertia and the *six-mode shell theory* that takes into account the transverse normal strain in addition to the shear and inertia considerations. The governing equations of all these shell theories can be derived using *Hamilton's principle* (Ramakrishnan and Shah, 1970; Ding and Chen, 1995, 1996*a*; Chen, 1995). It was found that the applicability of these theories becomes more limited when the material anisotropy increases. In a wide range of thickness-to-mean radius ratio, the six-mode shell theory always agrees well with the exact three-dimensional theory, while the classical theory fails when the thickness-to-mean radius ratio increases. As expected, the five-mode shell theory has a wider range of application than the classical thin shell theory.

10.3 VIBRATION COUPLED WITH FLUID

This section will discuss the free vibration of a submerged and/or fluid-filled spherically isotropic hollow sphere using the three-dimensional method developed in **Section 10.1**.

10.3.1 Effect of Fluid

Consider a non-viscous, compressible fluid whose velocity potential Φ satisfies the wave equation as in Eq. (9.4.1), *i.e.*

$$\left(\frac{1}{c_f^2}\frac{\partial^2}{\partial t^2}-\nabla^2\right)\Phi=0, \tag{10.3.1}$$

where the Laplace operator ∇^2 in spherical coordinates is

$$\nabla^2=\frac{1}{R^2}\frac{\partial}{\partial R}\left(R^2\frac{\partial}{\partial R}\right)+\frac{1}{R^2\sin\theta}\frac{\partial}{\partial\theta}\left(\sin\theta\frac{\partial}{\partial\theta}\right)+\frac{1}{R^2\sin^2\theta}\frac{\partial^2}{\partial\alpha^2}. \tag{10.3.2}$$

The separation of variables technique can be used (see **Appendix B**). For the outer and inner fluids, the solutions are respectively

$$\Phi_o=[B_1h_n^{(1)}(kR)+B_2h_n^{(2)}(kR)]S_n^m(\theta,\alpha)\exp(\mathrm{i}\omega t), \tag{10.3.3a}$$

$$\Phi_i=[B_3j_n(kR)+B_4y_n(kR)]S_n^m(\theta,\alpha)\exp(\mathrm{i}\omega t), \tag{10.3.3b}$$

where the subscripts i and o indicate the inner and outer fluids, $B_i\,(i=1,2,3,4)$ are arbitrary constants, $h_n^{(1)}(x)$ and $h_n^{(2)}(x)$ are *spherical Hankel functions* of the first and second kinds, $j_n(x)$ and $y_n(x)$ are spherical Bessel functions of the first and second kinds, and k is the wave number defined by

$$k = \omega / c_f \,. \tag{10.3.4}$$

For the outer fluid, the scattering condition at the infinity must be satisfied. This requires that $B_1 = 0$. Similarly, the finite value condition applies to the inner fluid and makes $B_4 = 0$. Thus we have

$$\Phi_o = B_2 h_n^{(2)}(kR) S_n^m(\theta, \alpha) \exp(\mathrm{i}\omega t) , \tag{10.3.5a}$$

$$\Phi_i = B_3 j_n(kR) S_n^m(\theta, \alpha) \exp(\mathrm{i}\omega t) . \tag{10.3.5b}$$

With a small perturbation, the dynamic fluid pressure can be obtained from the Bernoulli equation as

$$P_o = -\rho_f \frac{\partial \Phi_o}{\partial t} = -B_2 \,\mathrm{i}\omega \rho_f h_n^{(2)}(kR) S_n^m(\theta, \alpha) \exp(\mathrm{i}\omega t) , \tag{10.3.6a}$$

$$P_i = -\rho_f \frac{\partial \Phi_i}{\partial t} = -B_3 \,\mathrm{i}\omega \rho_f j_n(kR) S_n^m(\theta, \alpha) \exp(\mathrm{i}\omega t) , \tag{10.3.6b}$$

where ρ_f is the fluid density. In this derivation, the effect of static pressure of the fluid has been omitted.

The radial velocity of the fluid can be obtained as

$$v_o = \frac{\partial \Phi_o}{\partial R} = B_2 \frac{\mathrm{d}}{\mathrm{d}R} [h_n^{(2)}(kR)] S_n^m(\theta, \alpha) \exp(\mathrm{i}\omega t) , \tag{10.3.7a}$$

$$v_i = \frac{\partial \Phi_i}{\partial R} = B_3 \frac{\mathrm{d}}{\mathrm{d}R} [j_n(kR)] S_n^m(\theta, \alpha) \exp(\mathrm{i}\omega t) . \tag{10.3.7b}$$

At the solid-fluid interface, $R = c$, we have the following boundary conditions

$$\tau_{R\theta} = \tau_{\alpha R} = 0 , \tag{10.3.8}$$

$$P_i \,,\ P_o = -\sigma_R , \tag{10.3.9}$$

$$v_i \,,\ v_o = v_s , \tag{10.3.10}$$

where v_s is the radial velocity of the spherical shell:

$$v_s = \frac{\partial u_R}{\partial t} = \mathrm{i}\,\omega u_R = \mathrm{i}\,\omega w. \tag{10.3.11}$$

Thus from Eqs. (10.3.6), (10.3.7), (10.3.10) and (10.3.11), we can obtain the following relations

$$\begin{Bmatrix} P_i \\ P_o \end{Bmatrix} = \rho_f \omega c_f \begin{Bmatrix} g_n(kR) \\ h_n(kR) \end{Bmatrix} w, \quad (R = c), \tag{10.3.12}$$

where

$$g_n(x) = \frac{j_n(x)}{\frac{\mathrm{d}}{\mathrm{d}x}[j_n(x)]} = \frac{x j_n(x)}{n j_n(x) - x j_{n+1}(x)},$$
$$h_n(x) = \frac{h_n^{(2)}(x)}{\frac{\mathrm{d}}{\mathrm{d}x}[h_n^{(2)}(x)]} = \frac{x h_n^{(2)}(x)}{n h_n^{(2)}(x) - x h_{n+1}^{(2)}(x)}. \tag{10.3.13}$$

Equation (10.3.12) can be written in a dimensionless form for an arbitrary independent mode number n, *i.e.*,

$$\begin{Bmatrix} P_i \\ P_o \end{Bmatrix} = c_{44}\Omega\rho_0 c_0 \begin{Bmatrix} g_n(\Omega\xi_0/c_0) \\ h_n(\Omega\xi_0/c_0) \end{Bmatrix} W_n(\xi_0), \quad (\xi = \xi_0 = c/R_0), \tag{10.3.14}$$

where

$$\rho_0 = \rho_f/\rho, \quad c_0 = c_f/v_2. \tag{10.3.15}$$

Eq. (10.3.14) is equivalent to the boundary conditions described by Eq. (10.3.10) and will be used in our analysis below.

If the fluid is incompressible, then the velocity potential satisfies Laplace's equation, *i.e.*

$$\nabla^2\Phi = 0. \tag{10.3.16}$$

The solution can be obtained by separation of variables (see **Appendix B**), which gives rise to

$$\Phi = (B_5 R^n + B_6 R^{-n-1}) S_n^m(\theta, \alpha) \exp(\mathrm{i}\omega t), \tag{10.3.17}$$

where B_5 and B_6 are two arbitrary constants. Similarly, we have $B_6 = 0$ for the inner fluid, and $B_5 = 0$ for the outer fluid. At the interface $R = c$,

$$\begin{Bmatrix} P_i \\ P_o \end{Bmatrix} = \rho_f \omega^2 c \begin{Bmatrix} 1/n \\ -1/(n+1) \end{Bmatrix} w, \quad (R = c). \tag{10.3.18}$$

The dimensionless form of this equation is

$$\begin{Bmatrix} P_i \\ P_o \end{Bmatrix} = c_{44} \rho_0 \Omega^2 \xi_0 \begin{Bmatrix} 1/n \\ -1/(n+1) \end{Bmatrix} W_n(\xi_0), \quad (\xi = \xi_0). \tag{10.3.19}$$

By comparing this equation with Eq. (10.3.14), we see that for the inner fluid, Eq. (10.3.19) can be obtained by simply replacing $c_0 g_n(\Omega\xi_0 / c_0)$ by $\Omega\xi_0 / n$ in Eq. (10.3.14), and for the outer fluid, this can be done by replacing $c_0 h_n(\Omega\xi_0 / c_0)$ by $-\Omega\xi_0 / (n+1)$ in Eq. (10.3.14).

10.3.2 Frequency Equations

Consider the free vibration of a submerged and fluid-filled spherically isotropic hollow sphere. The boundary conditions at the inner and outer surfaces are

$$\begin{aligned} &\tau_{\xi\theta} = \tau_{\alpha\xi} = 0, \quad (\xi = 1 \pm t_R / 2), \\ &\sigma_\xi = -P_i / c_{13}, \quad (\xi = 1 - t_R / 2), \\ &\sigma_\xi = -P_o / c_{13}, \quad (\xi = 1 + t_R / 2). \end{aligned} \tag{10.3.20}$$

Similar to **Section 10.1**, we can obtain two independent classes of vibrations. Since the effect of viscosity of the fluid is neglected and the first class of vibration has no radial displacement component, the corresponding frequency equation is identical to that of the uncoupled vibration. The second class of vibration is affected by the fluid and is discussed below.

(1) Breathing mode ($n = 0$)

The frequency equation is

$$\left|E_{ij}^{9}\right| = 0, \quad (i, j = 1, 2), \tag{10.3.21}$$

where

$$\begin{aligned}
E_{11}^{9} &= [2 + (\zeta - 1/2)(f_4 / f_3)]J_{\zeta}(\gamma t_1) - (f_4 / f_3)\gamma t_1 J_{\zeta+1}(\gamma t_1) + G_0(t_1) t_1 J_{\zeta}(\gamma t_1) / f_3, \\
E_{12}^{9} &= [2 + (\zeta - 1/2)(f_4 / f_3)]Y_{\zeta}(\gamma t_1) - (f_4 / f_3)\gamma t_1 Y_{\zeta+1}(\gamma t_1) + G_0(t_1) t_1 Y_{\zeta}(\gamma t_1) / f_3, \\
E_{21}^{9} &= [2 + (\zeta - 1/2)(f_4 / f_3)]J_{\zeta}(\gamma t_2) - (f_4 / f_3)\gamma t_2 J_{\zeta+1}(\gamma t_2) + H_0(t_2) t_2 J_{\zeta}(\gamma t_2) / f_3, \\
E_{22}^{9} &= [2 + (\zeta - 1/2)(f_4 / f_3)]Y_{\zeta}(\gamma t_2) - (f_4 / f_3)\gamma t_2 Y_{\zeta+1}(\gamma t_2) + H_0(t_2) t_2 Y_{\zeta}(\gamma t_2) / f_3,
\end{aligned} \tag{10.3.22}$$

in which

$$G_n(x) = \rho_{0i} c_{0i} \Omega g_n(\Omega x / c_{0i}), \quad H_n(x) = \rho_{0o} c_{0o} \Omega h_n(\Omega x / c_{0o}), \tag{10.3.23}$$

where the second subscript i denotes the parameter of the inner fluid and o denotes that of the outer fluid.

Equation (10.3.19) shows that when the inner fluid is incompressible, the dynamic pressure of the fluid becomes infinite. This means that a hollow sphere filled with incompressible fluid cannot vibrate in a breathing mode.

(2) Non-breathing mode ($n > 0$)

The frequency equation is

$$\left|E_{ij}^{10}\right| = 0, \quad (i, j = 1, 2, 3, 4), \tag{10.3.24}$$

where

$$
\begin{aligned}
E_{1i}^{10} &= 2W_{ni}(t_1)/t_1 + n(n+1)V_{ni}(t_1)/t_1 + (f_4/f_3)W'_{ni}(t_1) + G_n(t_1)W_{ni}(t_1)/f_3, \\
E_{2i}^{10} &= W_{ni}(t_1)/t_1 + V_{ni}(t_1)/t_1 - V'_{ni}(t_1), \\
E_{3i}^{10} &= W_{ni}(t_2)/t_2 + V_{ni}(t_2)/t_2 - V'_{ni}(t_2), \\
E_{4i}^{10} &= 2W_{ni}(t_2)/t_2 + n(n+1)V_{ni}(t_2)/t_2 + (f_4/f_3)W'_{ni}(t_2) + H_n(t_2)W_{ni}(t_2)/f_3,
\end{aligned}
\tag{10.3.25}
$$

in which $i = 1, 2, 3$, and 4.

Since the problem is spherically symmetrical, like an uncoupled case, integer m does not appear in the frequency equations. In addition, because the viscosity and static pressure of the fluid are ignored, there exists a rigid-body translation associated with the second class of vibration for $n = 1$, at which the frequency is zero.

10.3.3 Numerical Results

(1) Numerical method

As one can see, the frequency equations of a hollow sphere submerged in compressible fluid contain spherical Hankel functions of the second kind, and the displacement functions are also obtained in a complicated series form. Thus both Eqs. (10.3.21) and (10.3.24) are complex transcendental equations. Lou and Su (1978) pointed out that the free vibration of a submerged thin spherical shell is usually attenuated. The motion is damped because energy is dissipated by waves moving outwards into the fluid and not returning. Ding and Chen (1998) confirmed Lou and Su's opinion by using the classical thin shell theory. Generally, the roots of the two three-dimensional frequency equations are also complex, and so difficult to obtain. On the other hand, only roots with small damping components are needed for free vibrations of submerged shells in practical engineering; roots with large imaginary parts correspond to highly attenuated vibrations. Therefore, we adopt the *method of small damping coefficient* suggested by Lou and Su (1978) to seek such roots in the form

$$
\Omega = \overline{\Omega}(1 + \mathrm{i}\,\varepsilon), \tag{10.3.26}
$$

where $\overline{\Omega} \geq 0$ is the real part of the complex frequency and $0 \leq \varepsilon << 1$. Applying Eq. (10.3.26) to the frequency equations, expressing each function in terms of its Taylor's series expansion and then ignoring all the high order terms in ε, we obtain

$$
(D_{11} + D_{12}\varepsilon) + \mathrm{i}(D_{21} + D_{22}\varepsilon) = 0, \tag{10.3.27}
$$

where $D_{ij}\,(i, j = 1, 2)$ are real functions of $\overline{\Omega}$. Hence, we have

$$D_{11} + D_{12}\varepsilon = 0\,, \quad D_{21} + D_{22}\varepsilon = 0\,, \tag{10.3.28}$$

and

$$F(\overline{\Omega}) = D_{11}D_{22} - D_{12}D_{21} = 0\,. \tag{10.3.29}$$

In general, Eq. (10.3.29) can be solved by a conventional iterative algorithm such as the Newton-Raphson method. Because Eq. (10.3.29) is still a transcendental equation, there are an infinite number of roots for each n. In the following discussion, we will concentrate on the smallest root.

When $\overline{\Omega}$ is obtained, the damping factor ε can be calculated by

$$\varepsilon = -\frac{D_{21}}{D_{22}} = -\frac{D_{11}}{D_{12}}\,, \tag{10.3.30}$$

and the damping component of the complex frequency is given by

$$\underline{\Omega} = \varepsilon\overline{\Omega}\,. \tag{10.3.31}$$

(2) Numerical results and discussions

For convenience, we call $\overline{\Omega}$ the real part of the complex frequency or simply the (natural) frequency and $\underline{\Omega}$ the imaginary part or damping component.

(*a*) Submerged spherical shell[3]

① Breathing mode ($n = 0$)

The complex frequencies of an isotropic solid sphere submerged in compressible fluid are given in Table 10.3 to be compared with those of the sphere in incompressible fluid. The frequencies of the uncoupled vibration are also listed. In the calculation, the material constants are taken as $(\lambda + \mu)/\mu = 3$, $\rho_{0o} = 0.127$ and $c_{0o} = 0.279$. The frequencies of the uncoupled vibration are identical to those in Love (1927). It can be seen that the frequency of a submerged hollow sphere is lower than the corresponding uncoupled one, due to the added-mass-effect (Junger and Feit, 1986). The effect of incompressible fluid on the breathing mode frequency is more pronounced than that of a

[3] Ding and Chen (1996*b*).

compressible fluid. This is true because the incompressibility of fluid leads to a stronger interaction between fluid and shell for this particular vibration mode.

Table 10.3 Frequencies of an isotropic solid sphere ($n = 0$).

Order		1	2	3	4	5	6
Unc.		2.220000	5.246962	8.036596	10.789673	13.529357	16.262580
Inc.		1.862313	4.708344	7.326833	9.956271	12.606949	15.274670
Com.	$\overline{\Omega}$	2.218855	5.246573	8.036334	10.789489	13.529212	16.262453
	$\underline{\Omega}$	3.299E-2	2.183E-2	2.089E-2	2.058E-2	2.042E-2	2.026E-2

Note: Unc. = uncoupled, Inc. = Incompressible fluid, Com. = Compressible fluid.

Table 10.4 Natural frequencies of spherically isotropic hollow sphere ($n = 0$).

Material	Material (2)			Material (3)		
Case	Unc.	Inc.	Com. ($\overline{\Omega}$)	Unc.	Inc.	Com. ($\overline{\Omega}$)
1	7.488740	3.990765	7.477556	2.935567	1.537433	2.902140
2	7.565788	5.801045	7.550058	2.943735	2.224719	2.935913
3	7.488740	3.063116	7.468760	2.935567	1.178097	2.869154
4	7.488740	3.063116	7.457198	2.935567	1.178097	2.829444

Table 10.4 presents the frequencies of two spherically isotropic hollow spheres made of Materials (2) and (3) defined in Table 10.1. Four different cases are considered: (1) $\rho_{0o} = 0.127$, $c_{0o} = 0.279$ and $t_R = 0.05$; (2) $\rho_{0o} = 0.127$, $c_{0o} = 0.279$ and $t_R = 0.20$; (3) $\rho_{0o} = 0.250$, $c_{0o} = 0.279$ and $t_R = 0.05$; and (4) $\rho_{0o} = 0.250$, $c_{0o} = 0.350$ and $t_R = 0.05$.

Clearly, the effect of the fluid incompressibility on the natural frequencies is considerable. The effect of the involved fluid on the natural frequency becomes less significant for thick shell than for thin shell, as one would expect. With the compressible fluid, $\overline{\Omega}$ decreases with c_{0o} and ρ_{0o} . This is reasonable since a larger density of the fluid indicates a more significant added-mass-effect and a larger sound of velocity implies that the fluid tends to be incompressible (Note that for an incompressible fluid, we have $c_f \to \infty$).

② Non-breathing mode ($n > 0$)

Equation (10.3.24) is a complicated complex transcendental equation and generally has complex roots. When $n = 1$, however, there exists a trivial solution $\Omega = 0$, corresponding to a rigid-body translation. We will consider $\Omega = 0$ as the lowest natural frequency for $n = 1$ in the following case studies:

With Material (2),

(A) $\rho_{0o} = 0.12$, $c_{0o} = 0.80$, $t_R = 0.10$;

(B) $\rho_{0o} = 0.12$, $c_{0o} = 0.80$, $t_R = 0.25$;

(C) $\rho_{0o} = 0.12$, $c_{0o} = 0.80$, $t_R = 2/3$;

(D) $\rho_{0o} = 0.20$, $c_{0o} = 0.80$, $t_R = 0.25$;

(E) $\rho_{0o} = 0.05$, $c_{0o} = 0.80$, $t_R = 0.25$;

(F) $\rho_{0o} = 0.12$, $c_{0o} = 0.60$, $t_R = 0.25$;

(G) $\rho_{0o} = 0.12$, $c_{0o} = 0.40$, $t_R = 0.25$;

and with Material (3),

(H) $\rho_{0o} = 0.12$, $c_{0o} = 0.80$, $t_R = 0.10$;

(I) $\rho_{0o} = 0.12$, $c_{0o} = 0.80$, $t_R = 0.25$; and

(J) $\rho_{0o} = 0.12$, $c_{0o} = 0.80$, $t_R = 2/3$.

In Tables 10.5 and 10.6, the corresponding real and damping components of the complex frequency of breathing mode are also listed for comparison.

It is easy to observe that the non-breathing mode frequencies (of the lower order ones) are usually lower than the corresponding breathing ones and the natural frequency increases with the mode number, although this trend is not so for the damping component. In all the cases, the natural frequency increases with the thickness-to-mean radius ratio. The non-dimensional fluid parameters have a stronger influence on the damping component. In addition, the material constants show a significant effect on the frequencies, which indicates that if the dynamic behaviour of a structure is a concern, its material needs to be carefully selected in design. Furthermore, if only the lowest frequency is of interest, as Table 10.5 demonstrates, one can focus on the analysis with $n = 2$.

It is emphasized that the results presented in Tables 10.5 and 10.6 are calculated based on the three-dimensional theory of elasticity, and in this respect, they can serve as a benchmark to clarify any two-dimensional shell theories or numerical methods. Of course, since the final frequency equation is solved using an approximate numerical method as described earlier, these results, especially the damping components, still need to be refined using a more accurate or even exact algorithm. Nevertheless, the method of small damping coefficient can provide satisfactory results from the view point of engineering (Ding and Chen, 1995; Ding and Chen, 1998).

Now let us consider the effect of the fluid compressibility on the frequencies of the non-breathing mode. Table 10.7 shows that the compressibility contributes differently to the non-breathing and breathing modes (Refer to Table 10.4). Associated with the former, the natural frequency of the sphere in incompressible fluid is usually higher except $n=2$, but in the latter, it is lower. In general, fluid affects vibration in a very complicated manner since the fluid pressure, the most influential factor, is dependent on various parameters such as density ratio, acoustic velocity, mode number, radius of the interface and even the circular frequency, as Eq. (10.3.14) shows.

Table 10.5 Real parts of complex frequencies of submerged hollow sphere ($\overline{\Omega}$).

n	0	2	3	4	5	6	7
Case A	7.468006	2.235557	2.715750	3.059974	3.575541	4.208886	4.926897
Case B	7.599575	2.479684	3.366141	4.194064	5.069111	5.985878	6.929021
Case C	7.732083	2.653631	3.730959	4.649549	5.509923	6.332937	7.128944
Case D	7.601787	2.433691	3.325275	4.163264	5.051962	5.982736	6.939577
Case E	7.602692	2.520424	3.405901	4.230268	5.098805	6.008208	6.943875
Case F	7.603157	2.519821	3.409696	4.238612	5.109599	6.020234	6.956630
Case G	7.605469	2.538708	3.427503	4.253339	5.121512	6.029433	6.963390
Case H	2.811506	1.024316	1.295404	1.545883	1.844209	2.236060	2.720179
Case I	2.901939	1.153718	1.623537	2.258252	2.912457	3.710925	4.572337
Case J	2.998997	1.404193	2.331940	3.316970	4.273039	5.189961	6.058445

Table 10.6 The corresponding damping components of complex frequencies of submerged hollow sphere ($\underline{\Omega}$).

n	0	2	3	4	5	6	7
Case A	0.991015	0.409412	0.471553	0.285791	0.210927	0.197920	0.312054
Case B	0.208124	0.180252	0.248399	0.290710	0.321812	0.347146	0.369602
Case C	0.159778	0.088334	0.123903	0.163077	0.210403	0.264742	0.324575
Case D	0.432849	0.337294	0.499099	0.607659	0.691439	0.761497	0.825321
Case E	0.738890	0.068286	0.090000	0.102931	0.112130	0.120310	0.126016
Case F	0.156527	0.128717	0.180428	0.199127	0.212409	0.222123	0.231648
Case G	0.104565	0.092740	0.114065	0.123553	0.128742	0.133581	0.138293
Case H	1.039176	0.014797	0.004555	0.001068	0.000314	0.000157	0.000126
Case I	0.264346	0.021501	0.019113	0.022054	0.033804	0.059741	0.107882
Case J	0.107574	0.052275	0.112683	0.155707	0.191940	0.233136	0.280684

The classical thin shell theory (Lou and Su, 1978) is quite suitable for predicting the dynamic behaviour of thin shells, as shown in Table 10.8 for the free vibration of a steel hollow sphere submerged in water, where the parameters used are $\nu=0.3$, $t_R=0.03$, $\rho_{0o}=0.127$, and $c_{fo}/c_s=0.279$, with $c_s^2=E/[\rho(1-\nu^2)]$. The results of the classical thin shell theory agree well with the three-dimensional calculations. However, when the thickness-to-mean radius ratio becomes larger, the classical thin shell theory becomes invalid as is seen from Fig. 10.6.

Table 10.7 Effect of fluid on the frequencies of the non-breathing mode of Case A.

n	2	3	4	5	6	7	8
Com. ($\overline{\Omega}$)	2.235557	2.715750	3.059974	3.575541	4.208886	4.926897	5.706012
Inc.	2.169552	2.764602	3.264429	3.835696	4.510762	5.276902	6.111749
Unc.	2.469229	3.264429	3.618109	4.193775	4.877448	5.655621	6.504196

Table 10.8 Non-breathing mode frequencies of a submerged steel hollow sphere.

Theory	3-D exact			Classical thin shell theory		
n	Unc.	Inc.	Com. ($\overline{\Omega}$)	Unc.	Inc.	Com. ($\overline{\Omega}$)
2	1.178202	0.825925	0.750631	1.185847	0.839852	0.755658
3	1.422408	1.023740	0.945724	1.409633	1.022484	0.947819
4	1.509535	1.125318	1.072121	1.510164	1.134534	1.082488
5	1.579488	1.216529	1.178935	1.582316	1.226582	1.190035
6	1.653525	1.307740	1.282974	1.658028	1.319259	1.291195
7	1.745574	1.412041	1.388375	1.753323	1.425550	1.401883
8	1.866211	1.538438	1.516342	1.877835	1.554669	1.532888

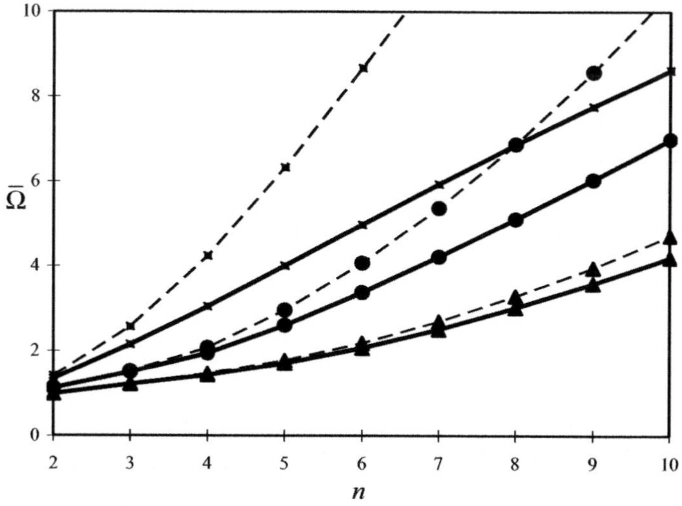

■ $t_R = 0.5$; ● $t_R = 0.2$; ▲ $t_R = 0.1$

Fig. 10.6 The classic thin shell theory (dotted line) *vs* the 3-D theory (solid line).

(*b*) Fluid-filled hollow sphere

① Breathing mode ($n = 0$)

As pointed out earlier, when the filled fluid is incompressible, the dynamic pressure of the fluid becomes infinite and the shell cannot vibrate purely radially. Table 10.9 shows the influence of compressible fluid under a variety of case conditions of (1)

$\rho_{0i}=0.2$, $c_{0i}=0.2$, $t_R=0.01$; (2) $\rho_{0i}=0.2$, $c_{0i}=0.2$, $t_R=0.05$; (3) $\rho_{0i}=0.2$, $c_{0i}=0.2$, $t_R=0.10$; (4) $\rho_{0i}=0.2$, $c_{0i}=0.2$, $t_R=0.20$; (5) $\rho_{0i}=0.2$, $c_{0i}=0.2$, $t_R=0.50$; (6) $\rho_{0i}=0.2$, $c_{0i}=0.2$, $t_R=1.00$; (7) $\rho_{0i}=0.2$, $c_{0i}=0.4$, $t_R=0.20$; (8) $\rho_{0i}=0.2$, $c_{0i}=0.6$, $t_R=0.20$; (9) $\rho_{0i}=0.3$, $c_{0i}=0.2$, $t_R=0.20$; and (10) $\rho_{0i}=0.4$, $c_{0i}=0.2$, $t_R=0.20$. The fluid has a clear effect on the vibration of breathing mode and tends to lower the frequency significantly. It is also interesting to note that the velocity ratio, $c_{0i}=c_f/v_2$, alters the natural frequency considerably, which is not so pronounced when the shell is submerged in fluid.

Table 10.9 Frequencies (Ω) of a hollow sphere of Material (2).

Case	1	2	3	4	5
Empty	7.483459	7.488766	7.504841	7.565791	7.811508
Fluid-filled	0.890927	0.918804	0.944275	0.997329	1.186463
Case	6	7	8	9	10
Empty	7.015444	7.565791	7.565791	7.565791	7.565791
Fluid-filled	3.086910	1.987081	2.959784	0.996729	0.996129

② Non-breathing mode ($n>0$)

The lowest natural frequencies of a spherically isotropic hollow sphere [Material (2), $t_R=0.50$] filled with fluid ($\rho_{0i}=0.2$ and $c_{0i}=0.2$), are listed in Table 10.10, which shows that the fluid effect on the non-breathing mode is very small and that incompressible fluid plays a more influential role[4].

Table 10.10 Non-breathing mode frequencies (Ω) of a hollow sphere of Material (2).

n	2	3	4	5	6	7	8
Empty	2.645278	3.689713	4.620881	5.534704	6.440711	7.334035	8.210779
Inc.	2.561958	3.587627	4.517806	5.439261	6.357950	7.266746	8.159327
Com.	2.645038	3.688544	4.620424	5.534089	6.440645	7.333917	8.210708

10.4 VIBRATION COUPLED WITH THE SURROUNDING ELASTIC MEDIUM

In this section, we will study the free vibration of a spherically isotropic hollow sphere embedded in an infinite elastic medium that is modelled as a two-parameter elastic

[4] A more detailed discussion on the free vibration of fluid-filled spherically isotropic spherical shells can be found in Chen and Ding (1999*a*). Chen, Wang and Ding (1999) studied the effect of the radial inhomogeneity and corrected the numerical errors in Chen and Ding (1999*a*).

foundation — the *Pasternak foundation model.* The thickness effect of the shear layer, which has been neglected in related previous studies (Kerr, 1964; Paliwal and Bhalla, 1993; Paliwal and Pandey, 2001), will be taken into account.

10.4.1 Pasternak Model of Elastic Foundation

The Pasternak elastic foundation model (Pasternak, 1954) is an extension of the well known *Winkler model* (Winkler, 1867). Since the Winkler model, as shown in Fig. 10.7, only considers the normal deflection, it treats the elastic foundation as a set of closely-packed, independent linear springs and thus the load (*P*) -deflection (*w*) relationship can be simply described as

$$P = kw, \tag{10.4.1}$$

where k is the foundation modulus. However, an elastic foundation is a continuum and its deformation under surface indentation must contain shearing distortion, as illustrated in Fig. 10.8, thus the Winkler model does not describe sufficiently accurately the foundation deformation, particularly when shear is considerable. Pasternak (1954) therefore modified the Winkler model by introducing shear interactions among the Winkler springs. This was done by connecting the ends of the springs to a beam or plate called a shear layer, which deforms only by transverse shear, as shown in Fig. 10.9. The thickness of the shear layer is often considered a unit (Kerr, 1964), but the reason for this is vague. In the following, we will use the *principle of minimum potential energy* to derive a $P-w$ relationship of a modified Pasternak model involving a shear layer thickness.

Assume that the thickness of the shear layer is ε. We consider the layer as a thin spherical shell with inner radius b and outer radius $b+\varepsilon$. According to the definition of the shear layer, only shear deformation will contribute to the strain energy and hence the effect of the displacements in θ and α directions on the shear deformation is neglected. This gives rise to

$$\gamma_{R\theta} = \frac{w_\theta}{R}, \quad \gamma_{\alpha R} = \frac{w_\alpha}{R\sin\theta}, \tag{10.4.2}$$

where w is the radial displacement in the shear layer, assumed to be independent of r, and

$$w_{\theta} = \frac{\partial w}{\partial \theta}, \quad w_{\alpha} = \frac{\partial w}{\partial \alpha}. \tag{10.4.3}$$

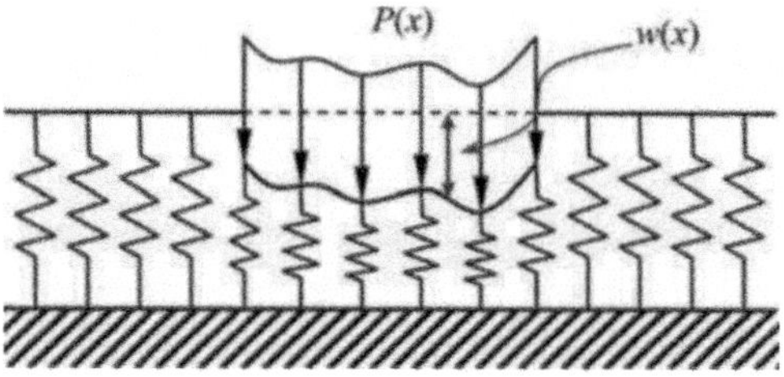

Fig. 10.7 Winkler model of elastic foundation.

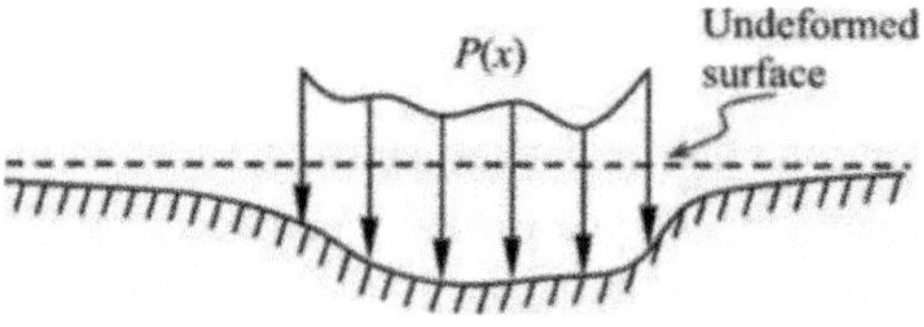

Fig. 10.8 Actual deformation of elastic foundation.

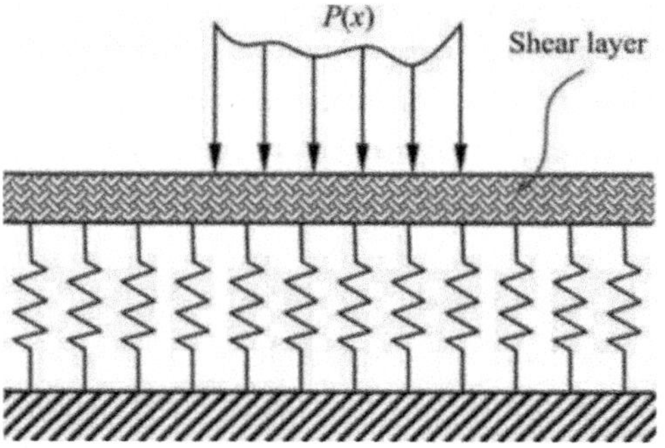

Fig. 10.9 Pasternak model of elastic foundation.

Under pressure $P(\theta, \alpha)$, the *total potential energy* of the foundation is

$$\begin{aligned}\Pi = &\int_b^{b+\varepsilon} \int_0^{\pi} \int_0^{2\pi} \frac{\widetilde{\mu}}{2} (\gamma_{R\theta}^2 + \gamma_{\alpha R}^2) R^2 \sin\theta \,\mathrm{d}R \,\mathrm{d}\theta \,\mathrm{d}\alpha \\ &- \int_0^{\pi} \int_0^{2\pi} P w b^2 \sin\theta \,\mathrm{d}\theta \,\mathrm{d}\alpha + \int_0^{\pi} \int_0^{2\pi} \frac{1}{2} \widetilde{k} w^2 (b+\varepsilon)^2 \sin\theta \,\mathrm{d}\theta \,\mathrm{d}\alpha,\end{aligned} \tag{10.4.4}$$

where $\tilde{\mu}$ is the shear modulus of the elastic medium and $\tilde{k}$ is the spring constant. By substituting Eq. (10.4.2) into Eq. (10.4.4) we get

$$\Pi = \int_0^{\pi}\int_0^{2\pi} F(\theta,\alpha,w,w_\theta,w_\alpha)\mathrm{d}\theta\,\mathrm{d}\alpha\,, \tag{10.4.5}$$

where

$$F = \frac{1}{2}\tilde{\mu}\varepsilon[\sin\theta w_\theta^2 + \csc\theta w_\alpha^2] - b^2 Pw\sin\theta + \frac{1}{2}(b+\varepsilon)^2\tilde{k}w^2\sin\theta\,. \tag{10.4.6}$$

The *Euler equation*[5] of this functional is known as

$$\frac{\partial F}{\partial w} - \frac{\partial}{\partial\theta}\left(\frac{\partial F}{\partial w_\theta}\right) - \frac{\partial}{\partial\alpha}\left(\frac{\partial F}{\partial w_\alpha}\right) = 0\,. \tag{10.4.7}$$

With Eq. (10.4.6), this equation gives

$$P = kw - \frac{K}{b^2}\nabla_1^2 w\,, \tag{10.4.8}$$

where

$$k = \left(1+\frac{\varepsilon}{b}\right)^2\tilde{k}\,, \quad K = \tilde{\mu}\varepsilon\,. \tag{10.4.9}$$

Equation (10.4.8) reduces to the Winkler model when K is zero or when the problem under consideration is spherically symmetrical so that w becomes independent of θ and α. Therefore, k in Eq. (10.4.9) must be the same as that in Eq. (10.4.1). Both k and K can be determined by experiment[6].

In the following analysis, we assume that the bonding between the spherical shell and elastic medium is perfect and frictionless (Men and Yuan, 1990; Nath and Jain, 1983). In other words, the vibration of the shell is affected by the elastic medium only through

[5] The condition to render Π a stationary value, i.e. $\delta\Pi=0$, see Hu (1984).
[6] A detailed discussion on various foundation models as well as the determination of model parameters can be found in book by Selvadurai (1979).

their normal interaction, *i.e.*, the pressure P. Of course, the displacement continuity condition $w = u_R$ applies at $R = b$.

10.4.2 Frequency Equations

Consider the free vibration of an embedded hollow sphere. According to the discussion above, we know that the boundary conditions on the spherical surfaces are

$$\begin{aligned} &\tau_{\xi\theta} = \tau_{\alpha\xi} = 0, \quad (\xi = 1 \pm t_R/2), \\ &\sigma_\xi = 0, \quad (\xi = 1 - t_R/2), \\ &\sigma_\xi = -P/c_{13}, \quad (\xi = 1 + t_R/2). \end{aligned} \tag{10.4.10}$$

As in the analyses in **Sections 10.1** and **10.3**, we have two independent classes of vibration. The first class is the same as the uncoupled vibration. The second class will be affected by the surrounding elastic medium.

(1) Breathing mode ($n = 0$)

The corresponding frequency equation is

$$\left|E_{ij}^{11}\right| = 0, \quad (i, j = 1, 2), \tag{10.4.11}$$

where

$$\begin{aligned} E_{11}^{11} &= [2 + (\zeta - 1/2)(f_4/f_3)]J_\zeta(\gamma t_1) - (f_4/f_3)\gamma t_1 J_{\zeta+1}(\gamma t_1), \\ E_{12}^{11} &= [2 + (\zeta - 1/2)(f_4/f_3)]Y_\zeta(\gamma t_1) - (f_4/f_3)\gamma t_1 Y_{\zeta+1}(\gamma t_1), \\ E_{21}^{11} &= [2 + (\zeta - 1/2)(f_4/f_3)]J_\zeta(\gamma t_2) - (f_4/f_3)\gamma t_2 J_{\zeta+1}(\gamma t_2) + (f_5/f_3)J_\zeta(\gamma t_2), \\ E_{22}^{11} &= [2 + (\zeta - 1/2)(f_4/f_3)]Y_\zeta(\gamma t_2) - (f_4/f_3)\gamma t_2 Y_{\zeta+1}(\gamma t_2) + (f_5/f_3)Y_\zeta(\gamma t_2), \end{aligned} \tag{10.4.12}$$

in which f_5 is the dimensionless foundation parameter defined by $f_5 = kb/c_{44}$.

(2) Non-breathing mode ($n > 0$)

The frequency equation is

$$\left|E_{ij}^{12}\right| = 0, \quad (i, j = 1, 2, 3, 4), \tag{10.4.13}$$

where

$$
\begin{aligned}
E_{1i}^{12} &= 2W_{ni}(t_1)/t_1 + n(n+1)V_{ni}(t_1)/t_1 + (f_4/f_3)W'_{ni}(t_1),\\
E_{2i}^{12} &= W_{ni}(t_1)/t_1 + V_{ni}(t_1)/t_1 - V'_{ni}(t_1),\\
E_{3i}^{12} &= W_{ni}(t_2)/t_2 + V_{ni}(t_2)/t_2 - V'_{ni}(t_2),\\
E_{4i}^{12} &= 2W_{ni}(t_2)/t_2 + n(n+1)V_{ni}(t_2)/t_2 + (f_4/f_3)W'_{ni}(t_2)\\
&\quad + \frac{f_5 t_2 + n(n+1)f_6 t_R}{f_3 t_2^2} W_{ni}(t_2),
\end{aligned}
\tag{10.4.14}
$$

and $f_6 = K/(c_{44}h)$.

Equations (10.4.11) and (10.4.13) will reduce to those of the uncoupled vibration, Eqs. (10.2.5) and (10.2.7), respectively, when $k = K = 0$. As Eqs. (10.4.11) and (10.4.12) show, under the breathing mode, the frequency equation is independent of K, indicating that, for spherical symmetry, the Pasternak and Winkler models will bring about the same result. Furthermore, as in the case of uncoupled vibration, integer m does not appear in these frequency equations.

10.4.3 Numerical Results

(1) Breathing mode ($n = 0$)

Figures 10.10 and 10.11 show the spectra of the lowest dimensionless frequencies (Ω), of the spheres of Materials (2) and (3) when the dimensionless elastic foundation constant f_5 and the thickness-to-mean radius ratio change. It can be seen that the Ω of the breathing mode increases with f_5. This is natural since f_5 corresponds to the spring constant of the elastic foundation. A comparison of the figures shows that the frequency spectra vary greatly with material properties.

(2) Non-breathing mode ($n > 0$)

In the case of the non-breathing mode, the natural frequency is affected by both the foundation parameters f_5 and f_6, as shown in Figs. 10.12 to 10.14. Because the Pasternak model reduces to the Winkler model when $f_6 = 0$, Fig. 10.12 actually demonstrates the behaviour with the Winkler model. If $f_5 = 0$, the vibration is uncoupled and when $n = 1$ there exists a rigid-body translation with $\Omega = 0$. When f_5 is big enough, say greater than 40, the frequency stays almost constant. Figures 10.13 and 10.14 also show that the frequency generally does not vary with f_5 when f_6 is large. As is seen from Eq. (10.4.14), the two foundation parameters play a similar role in the free vibration analysis, and with the increase of them, the foundation becomes stiffer and stiffer. Thus all curves presented in Figs. 10.12 to 10.14 as well as those in Figs. 10.10 and 10.11 for the breathing mode will gradually approach their limit values which correspond to the shell contacting smoothly with a surrounding rigid base.

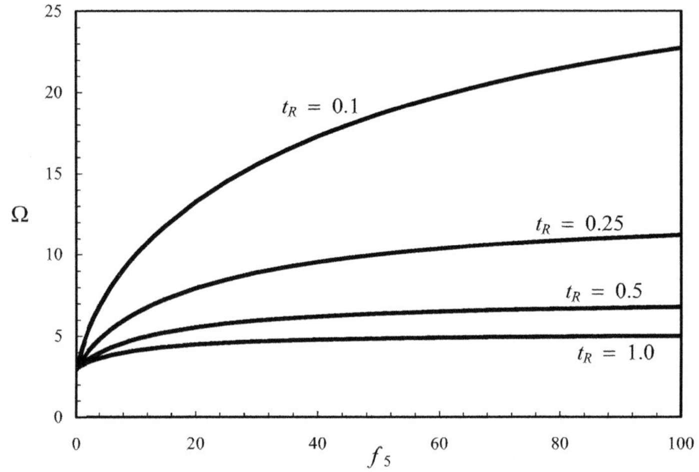

Fig. 10.10 Spectra of Ω *vs* f_5 [Material (3), breathing mode].

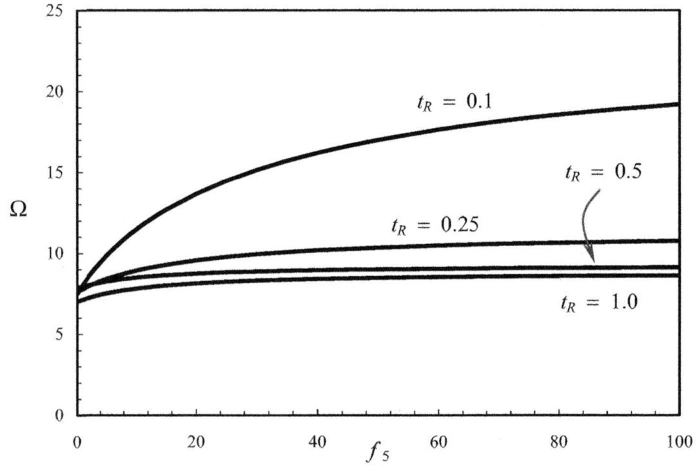

Fig. 10.11 Spectra of Ω *vs* f_5 [Material (2), breathing mode].

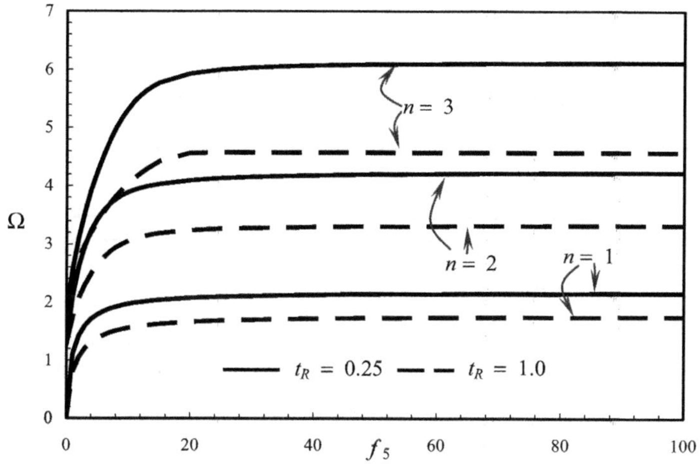

Fig. 10.12 Spectra of Ω *vs* f_5 [Material (3), $f_6 = 0$].

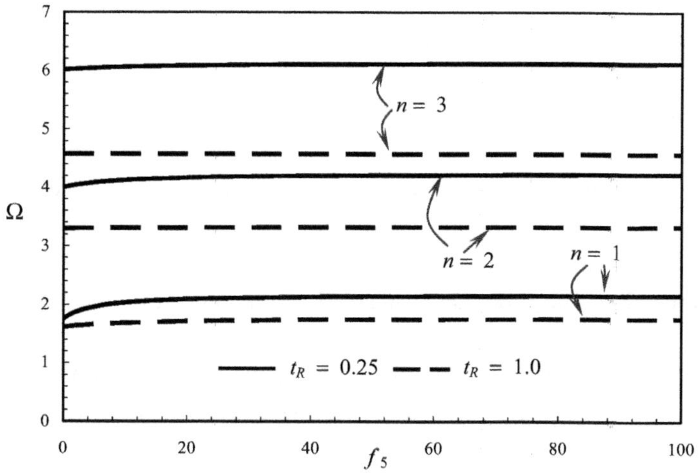

Fig. 10.13 Spectra of Ω *vs* f_5 [Material (3), $f_6 = 10$].

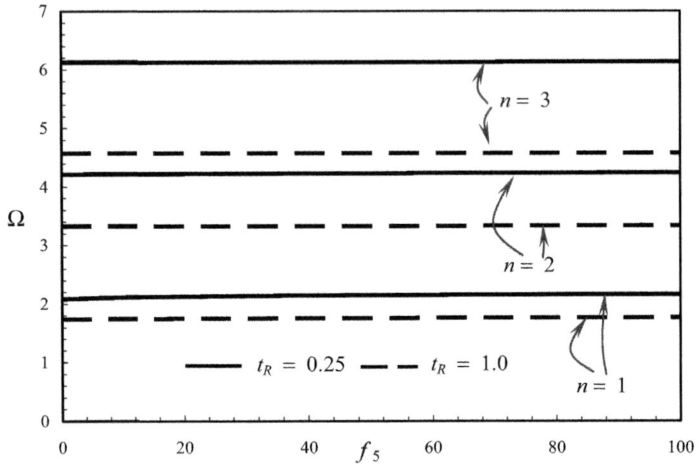

Fig. 10.14 Spectra of Ω *vs* f_5 [Material (3), $f_6 = 50$].

Further comparisons of the three-dimensional theory with the shell theories can be found in Chen (1995) and Ding and Chen (1996*a*).

10.5 LAMINATED SPHERICAL SHELLS

The three-dimensional method for a laminated spherical shell described in **Section 10.1** will face numerical problems when the matrix order of the equations increases. In the following, we will present a state-space formulation to overcome the difficulty so that an effective numerical solution will become possible.

10.5.1 State-Space Formulations for Spherically Isotropic Elasticity

The three-dimensional state-space formulation for generally anisotropic elasticity in Cartesian coordinates has been presented in **Section 2.1**. In spherical coordinates, however, it can only be done for spherically isotropic materials (Chen and Ding, 2001*a*,*b*).

Assume that the center of the spherical anisotropy is identical with the origin of spherical coordinates (R, θ, α). The linear constitutive relations in Eq. (1.2.40*a*) can be rewritten as

$$
\begin{aligned}
\Sigma_\theta &= R\sigma_\theta = c_{11}S_\theta + c_{12}S_\alpha + c_{13}S_R, \\
\Sigma_\alpha &= R\sigma_\alpha = c_{12}S_\theta + c_{11}S_\alpha + c_{13}S_R, \\
\Sigma_R &= R\sigma_R = c_{13}S_\theta + c_{13}S_\alpha + c_{33}S_R, \\
\Sigma_{\alpha R} &= R\tau_{\alpha R} = c_{44}S_{\alpha R}, \\
\Sigma_{R\theta} &= R\tau_{R\theta} = c_{44}S_{R\theta}, \\
\Sigma_{\theta\alpha} &= R\tau_{\theta\alpha} = c_{66}S_{\theta\alpha},
\end{aligned} \tag{10.5.1}
$$

where S_{ij} are defined by

$$
\begin{aligned}
S_R &= R\varepsilon_R = \nabla_2 u_R, \quad S_\theta = R\varepsilon_\theta = \frac{\partial u_\theta}{\partial\theta} + u_R, \\
S_\alpha &= R\varepsilon_\alpha = \frac{1}{\sin\theta}\frac{\partial u_\alpha}{\partial\alpha} + u_R + u_\theta\cot\theta, \\
S_{\alpha R} &= R\gamma_{\alpha R} = \frac{1}{\sin\theta}\frac{\partial u_R}{\partial\alpha} + \nabla_2 u_\alpha - u_\alpha, \\
S_{\theta\alpha} &= R\gamma_{\theta\alpha} = \frac{1}{\sin\theta}\frac{\partial u_\theta}{\partial\alpha} + \frac{\partial u_\alpha}{\partial\theta} - u_\alpha\cot\theta, \\
S_{R\theta} &= R\gamma_{R\theta} = \frac{\partial u_R}{\partial\theta} + \nabla_2 u_\theta - u_\theta,
\end{aligned} \tag{10.5.2}
$$

where $\nabla_2 = R\partial/\partial R$. In absence of body forces, the equations of motion in Eq. (1.2.12) can be rewritten as

$$
\begin{aligned}
&\nabla_2\Sigma_R + \frac{\partial\Sigma_{R\theta}}{\partial\theta} + \csc\theta\frac{\partial\Sigma_{\alpha R}}{\partial\alpha} + \Sigma_R - \Sigma_\theta - \Sigma_\alpha + \Sigma_{R\theta}\cot\theta = \rho R^2\frac{\partial^2 u_R}{\partial t^2}, \\
&\nabla_2\Sigma_{R\theta} + \frac{\partial\Sigma_\theta}{\partial\theta} + \csc\theta\frac{\partial\Sigma_{\theta\alpha}}{\partial\alpha} + 2\Sigma_{R\theta} + (\Sigma_\theta - \Sigma_\alpha)\cot\theta = \rho R^2\frac{\partial^2 u_\theta}{\partial t^2}, \\
&\nabla_2\Sigma_{\alpha R} + \frac{\partial\Sigma_{\theta\alpha}}{\partial\theta} + \csc\theta\frac{\partial\Sigma_\alpha}{\partial\alpha} + 2\Sigma_{\alpha R} + 2\Sigma_{\theta\alpha}\cot\theta = \rho R^2\frac{\partial^2 u_\alpha}{\partial t^2}.
\end{aligned} \tag{10.5.3}
$$

To establish the state-space equation, in addition to the displacement separation formulae given in Eq. (2.4.3), we need to make use of (Shul'ga *et al.*, 1988)

$$
\Sigma_{R\theta} = -\frac{1}{\sin\theta}\frac{\partial\Sigma_1}{\partial\alpha} - \frac{\partial\Sigma_2}{\partial\theta}, \quad \Sigma_{\alpha R} = \frac{\partial\Sigma_1}{\partial\theta} - \frac{1}{\sin\theta}\frac{\partial\Sigma_2}{\partial\alpha}, \tag{10.5.4}
$$

where Σ_1 and Σ_2 are the stress functions.

By virtue of Eq. (2.4.3) and (10.5.4), Eqs. (10.5.1) to (10.5.3) lead to

$$\nabla_2 \begin{Bmatrix} \Sigma_1 \\ \psi \end{Bmatrix} = \begin{bmatrix} -2 & -c_{66}(\nabla_1^2+2) + R^2\rho\partial^2/\partial t^2 \\ \dfrac{1}{c_{44}} & 1 \end{bmatrix} \begin{Bmatrix} \Sigma_1 \\ \psi \end{Bmatrix}, \tag{10.5.5}$$

$$\nabla_2 \begin{Bmatrix} \Sigma_R \\ \Sigma_2 \\ G \\ w \end{Bmatrix} = \begin{bmatrix} 2\beta-1 & \nabla_1^2 & k_1\nabla_1^2 & -2k_1 + R^2\rho\dfrac{\partial^2}{\partial t^2} \\ \beta & -2 & k_2\nabla_1^2 - 2c_{66} + R^2\rho\dfrac{\partial^2}{\partial t^2} & -k_1 \\ 0 & \dfrac{1}{c_{44}} & 1 & 1 \\ \dfrac{1}{c_{33}} & 0 & \beta\nabla_1^2 & -2\beta \end{bmatrix} \begin{Bmatrix} \Sigma_R \\ \Sigma_2 \\ G \\ w \end{Bmatrix}, \tag{10.5.6}$$

where ∇_1^2 is the two-dimensional Laplace operator on spherical surfaces, as given in Eq. (2.4.2), and

$$\beta = c_{13}/c_{33}, \quad k_1 = 2c_{13}\beta - (c_{11}+c_{12}), \quad k_2 = k_1/2 - c_{66}.$$

For the free vibration of a closed laminated spherical shell (hollow sphere), it can be assumed that

$$\Sigma_1 = ac_{44}^{(1)}\sum_{m=0}^{n}\sum_{n=0}^{\infty}\Sigma_{1n}(\xi)S_n^m(\theta,\alpha)\exp(i\omega t), \quad \psi = a\sum_{m=0}^{n}\sum_{n=0}^{\infty}\psi_n(\xi)S_n^m(\theta,\alpha)\exp(i\omega t),$$
$$\Sigma_R = ac_{44}^{(1)}\sum_{m=0}^{n}\sum_{n=0}^{\infty}\Sigma_{Rn}(\xi)S_n^m(\theta,\alpha)\exp(i\omega t), \quad \Sigma_2 = ac_{44}^{(1)}\sum_{m=0}^{n}\sum_{n=0}^{\infty}\Sigma_{2n}(\xi)S_n^m(\theta,\alpha)\exp(i\omega t),$$
$$G = a\sum_{m=0}^{n}\sum_{n=0}^{\infty}G_n(\xi)S_n^m(\theta,\alpha)\exp(i\omega t), \quad w = a\sum_{m=0}^{n}\sum_{n=0}^{\infty}w_n(\xi)S_n^m(\theta,\alpha)\exp(i\omega t), \tag{10.5.7}$$

where $S_n^m(\theta,\alpha)$ are spherical harmonics, n and m are integers, ω is the circular frequency, $\xi = R/a$ is the dimensionless radial variable, a is the inner radius of the spherical shell defined in Fig. 10.1, and $c_{44}^{(1)}$ represents the value of c_{44} in the first layer and so on. Substituting the above equation into Eqs. (10.5.5) and (10.5.6), we obtain for an arbitrary couple of (m,n),

$$\frac{\mathrm{d}}{\mathrm{d}\xi}\mathbf{V}_i(\xi)=\mathbf{M}_i(\xi)\mathbf{V}_i(\xi),\quad (i=1,2),\tag{10.5.8}$$

where $\mathbf{V}_1=[\Sigma_{1n},\psi_n]^{\mathrm{T}}$, $\mathbf{V}_2=[\Sigma_{rn},\Sigma_{2n},G_n,w_n]^{\mathrm{T}}$, and

$$\mathbf{M}_1=\frac{1}{\xi}\begin{bmatrix} -2 & \dfrac{(l-2)c_{66}}{c_{44}^{(1)}}-\dfrac{\xi^2\Omega^2\rho}{\rho^{(1)}} \\ \dfrac{c_{44}^{(1)}}{c_{44}} & 1 \end{bmatrix},\tag{10.5.9}$$

$$\mathbf{M}_2=\frac{1}{\xi}\begin{bmatrix} 2\beta-1 & -l & -\dfrac{k_1 l}{c_{44}^{(1)}} & -\dfrac{2k_1}{c_{44}^{(1)}}-\dfrac{\Omega^2\xi^2\rho}{\rho^{(1)}} \\ \beta & -2 & \dfrac{-k_2 l-2c_{66}}{c_{44}^{(1)}}-\dfrac{\Omega^2\xi^2\rho}{\rho^{(1)}} & -\dfrac{k_1}{c_{44}^{(1)}} \\ 0 & \dfrac{c_{44}^{(1)}}{c_{44}} & 1 & 1 \\ \dfrac{c_{44}^{(1)}}{c_{33}} & 0 & -\beta l & -2\beta \end{bmatrix},\tag{10.5.10}$$

where $l=n(n+1)$, and $\Omega=a\omega\sqrt{\rho^{(1)}/c_{44}^{(1)}}$ is the non-dimensional frequency. As one can see the variable ξ appears in the coefficient matrix $\mathbf{M}_k$, *i.e.* we have established two separated three-dimensional state equations with variable coefficients.

10.5.2 Layerwise Method and State Vector Solution

As in **Section 9.3**, we will use the layerwise method (Pestel and Leckie, 1963; Fan and Zhang, 1992) to transform Eq. (10.5.8) to the ones with constant coefficients. To do so, divide the jth layer of the spherical shell into j_q uniform sub-layers, each with a very small thickness $h_j/j_q=(b_j-a_j)/j_q$. Thus in every sub-layer the variable ξ can be considered constant so that its value can be taken at the mid plane. For example, in the j_k^{th} ($j_k=1,2,\cdots,j_q$) sub-layer, we have $\xi=\xi_{j0}+(2j_k-1)(\xi_{j1}-\xi_{j0})/(2j_q)=\xi_{jj_k}$, where $\xi_{j0}=a_j/a$ and $\xi_{j1}=b_j/a$ are the dimensionless inner and outer radii of the jth

layer. In this way, the coefficient matrix $\mathbf{M}_i$ in Eq. (10.5.8) becomes constant in the j_k th sub-layer of the j th layer, denoted as $\mathbf{M}_i^{jj_k}$. Thus the solution to Eqs. (10.5.8) is (Bellman, 1970)

$$\mathbf{V}_i(\xi) = \exp[(\xi - \xi_{jj_k}^0)\mathbf{M}_i^{jj_k}]\mathbf{V}_i(\xi_{jj_k}^0), \quad (\xi_{jj_k}^0 \le \xi \le \xi_{jj_k}^1, \ j_k = 1, 2, \cdots, j_q; \ j = 1, 2, \cdots, p), \tag{10.5.11}$$

where $\xi_{jj_k}^0 = \xi_{j0} + (j_k - 1)(\xi_{j1} - \xi_{j0}) / j_q$ and $\xi_{jj_k}^1 = \xi_{j0} + j_k(\xi_{j1} - \xi_{j0}) / j_q$. The exponential matrix $\exp[(\xi - \xi_{jj_k}^0)\mathbf{M}_i^{jj_k}]$, known as the transfer matrix, can be expressed in terms of a polynomial of matrix $\mathbf{M}_i^{jj_k}$ by using the Cayley-Hamilton theorem described in **Section 8.2**.

The continuity conditions at the interfaces of two adjacent sub-layers need to be satisfied, which leads to

$$\mathbf{V}_i(\xi_{j1}) = \mathbf{T}_{in}^j \mathbf{V}_i(\xi_{j0}), \quad (j = 1, 2, \cdots, p; \ n = 1, 2, 3, \cdots), \tag{10.5.12}$$

where $\mathbf{T}_{1n}^j = \prod_{j_k = j_q}^{1} \exp[(\xi_{j1} - \xi_{j0})\mathbf{M}_1^{jj_k} / j_q]$ and $\mathbf{T}_{2n}^j = \prod_{j_k = j_q}^{1} \exp[(\xi_{j1} - \xi_{j0})\mathbf{M}_2^{jj_k} / j_q]$ are the second-order and fourth-order square matrices, respectively.

Further using the continuity conditions between two adjacent natural layers, we obtain from Eq. (10.5.12)

$$\mathbf{V}_i(\xi_1) = \mathbf{T}_{in} \mathbf{V}_i(1), \quad (i = 1, 2; \ n = 1, 2, 3, \cdots), \tag{10.5.13}$$

where $\mathbf{T}_{1n} = \prod_{j=p}^{1} \mathbf{T}_{1n}^j$ and $\mathbf{T}_{2n} = \prod_{j=p}^{1} \mathbf{T}_{2n}^j$ are the second-order and fourth-order square matrices, respectively, and $\xi_1 = b / a$ is the outer radius-to-inner radius ratio.

Σ_{10}, ψ_0, Σ_{20} and G_0 will not appear in the expressions of displacements and stresses, according to Eqs. (2.4.3) and (10.5.4). Thus they can be simply set to zero for convenience of formulation. Hence when $n = 0$ Eq. (10.5.8) reduces to

$$\frac{\mathrm{d}}{\mathrm{d}\xi}\begin{Bmatrix}\Sigma_{R0}\\ w_0\end{Bmatrix}=\frac{1}{\xi}\begin{bmatrix}2\beta-1 & \dfrac{-2k_1}{c_{44}^{(1)}}-\dfrac{\xi^2\Omega^2\rho}{\rho^{(1)}}\\ \dfrac{c_{44}^{(1)}}{c_{33}} & -2\beta\end{bmatrix}\begin{Bmatrix}\Sigma_{R0}\\ w_0\end{Bmatrix}. \tag{10.5.14}$$

Hence,

$$\begin{Bmatrix}\Sigma_{R0}(\xi_1)\\ w_0(\xi_1)\end{Bmatrix}=\mathbf{T}_{20}\begin{Bmatrix}\Sigma_{R0}(1)\\ w_0(1)\end{Bmatrix},\quad (n=0), \tag{10.5.15}$$

where $\mathbf{T}_{20}$ is a second-order square matrix.

10.5.3 Frequency Equations

For free vibration, the boundary conditions are

$$\sigma_R=\tau_{R\theta}=\tau_{\alpha R}=0,\quad (R=a,b). \tag{10.5.16}$$

From Eq. (10.5.16), we have

$$\Sigma_R=\Sigma_1=\Sigma_2=0,\quad (\xi=1,\xi_1). \tag{10.5.17}$$

In the present case, both the governing equations and the boundary and continuity conditions can be separated into two independent classes. The first involves two state variables Σ_1 and ψ only and the second contains the other four state variables. Using Eqs. (10.5.17), Eqs. (10.5.13) and (10.5.15), we get

$$\begin{Bmatrix}0\\ \psi_n(\xi_1)\end{Bmatrix}=\mathbf{T}_{1n}\begin{Bmatrix}0\\ \psi_n(1)\end{Bmatrix},\quad (n=1,2,3,\cdots), \tag{10.5.18}$$

$$\begin{Bmatrix}0\\ 0\\ G_n(\xi_1)\\ w_n(\xi_1)\end{Bmatrix}=\mathbf{T}_{2n}\begin{Bmatrix}0\\ 0\\ G_n(1)\\ w_n(1)\end{Bmatrix},\quad (n=1,2,3,\cdots), \tag{10.5.19}$$

$$\begin{Bmatrix} 0 \\ w_0(\xi_1) \end{Bmatrix} = \mathbf{T}_{20} \begin{Bmatrix} 0 \\ w_0(1) \end{Bmatrix}, \quad (n = 0). \tag{10.5.20}$$

For nontrivial solutions, these equations give the frequency equations of the two independent classes of vibrations, respectively,

$$T_{1n12} = 0, \quad (n = 1, 2, 3, \cdots), \tag{10.5.21}$$

for the first class of vibration, and

$$T_{2012} = 0, \quad (n = 0), \tag{10.5.22}$$

$$\begin{vmatrix} T_{2n13} & T_{2n14} \\ T_{2n23} & T_{2n24} \end{vmatrix} = 0, \quad (n = 1, 2, 3, \cdots), \tag{10.5.23}$$

for the second class of vibration. In Eqs. (10.5.21) to (10.5.23), T_{knij} represents the element on the ith row and j^{th} column of the matrix $\mathbf{T}_{kn}$.

For calculating the mode shapes, once the frequency is obtained, the eigen-state vectors at the inner and/or outer spherical surfaces can be solved from Eqs. (10.5.18) to (10.5.20). The state vectors at any interior point can then be calculated by the following formula:

$$\mathbf{V}_i(\xi) = \exp[(\xi - \xi_{jj_k}^0)\mathbf{M}_i^{jj_k}] \prod_{j_l = j_k - 1}^{1} \exp[(\xi_{j1} - \xi_{j0})\mathbf{M}_i^{jj_l} / j_q] \prod_{r = j-1}^{1} \mathbf{T}_{in}^r \mathbf{V}_i(1),$$
$$(i = 1, 2;\ \xi_{jj_k}^0 \le \xi \le \xi_{jj_k}^1;\ j = 1, 2, \cdots, p). \tag{10.5.24}$$

The induced variables are determined by

$$\begin{aligned}
\Sigma_\theta - \Sigma_\alpha &= 2c_{66}\left(\nabla_1^2 G - 2\frac{\partial^2 G}{\partial \theta^2} + 2\cot\theta\csc\theta\frac{\partial \psi}{\partial \alpha} - 2\csc\theta\frac{\partial^2 \psi}{\partial \theta \partial \alpha} \right), \\
\Sigma_\theta + \Sigma_\alpha &= 2\beta\Sigma_R + k_1\nabla_1^2 G - 2k_1 w, \\
\Sigma_{\theta\alpha} &= -c_{66}\left(\nabla_1^2 \psi - 2\frac{\partial^2 \psi}{\partial \theta^2} - 2\cot\theta\csc\theta\frac{\partial G}{\partial \alpha} + 2\csc\theta\frac{\partial^2 G}{\partial \theta \partial \alpha} \right).
\end{aligned} \tag{10.5.25}$$

10.5.4 Numerical Results

Consider a thick homogeneous hollow sphere made of Material (2) defined in Table 10.1. Its inner radius-to-outer radius ratio is $a:b=1:2$. Table 10.11 compares the numerical results calculated by various methods, the one proposed in this section, that based on the Talyor's expansion theorem (Chen and Ding, 2001*b*) and that described in **Section 10.1**. When using the present method, two schemes are considered, *i.e.* the sphere is divided equally into $p=30$ layers and $p=35$ layers, respectively. It can be seen that the results obtained by the three methods mentioned above are very close to each other.

Table 10.11 The lowest non-dimensional frequency $\Omega=\omega a\sqrt{\rho/c_{44}}$ for a homogeneous spherically isotropic hollow sphere.

	n	0	1	2	3	4	5
Present ($p=30$)	The first class	...	3.55565	2.39224	3.69722	4.87407	5.99857
	The second class	5.15158	3.39325	1.78222	2.49624	3.10452	3.67347
Present ($p=35$)	The first class	...	3.55569	2.39216	3.69706	4.87384	5.99826
	The second class	5.15154	3.39329	1.78223	2.49625	3.10451	3.67343
Taylor's expansion	The first class	...	3.55579	2.39193	3.69664	4.87320	5.99740
	The second class	5.15141	3.39334	1.78225	2.49629	3.10472	3.67441
Section 10.1	The first class	...	3.55579	2.39193	3.69664	4.87320	5.99739
	The second class	5.15141	3.39334	1.78224	2.49626	3.10448	3.67332

Table 10.12 The lowest non-dimensional frequencies of a three-layered hollow sphere. ($\Omega=\omega a\sqrt{\rho^{(1)}/c_{44}^{(1)}}$)

	n	0	1	2	3	4	5
$q=10$	The first class	...	2.62215	2.76048	4.23163	5.47461	6.58646
	The second class	5.05042	2.51475	2.00311	2.93600	3.78071	4.54906
$q=15$	The first class	...	2.62305	2.76009	4.23104	5.47385	6.58554
	The second class	5.04985	2.51580	2.00327	2.93665	3.78228	4.55153

For the free vibration of a three-layered hollow sphere with $b:b_2:b_1:a=2.0:1.7:1.3:1.0$, assume that the inner and outer layers are both made of Material (2) but the middle layer is Material (1) that is isotropic. The density ratio between Material (1) and Material (2) is $\rho^{(1)}:\rho^{(2)}=1:1.2$. Table 10.12 presents a comparison between results for two division schemes, *i.e.* each layer in the laminate is further divided equally into 10 sub-layers and 15 sub-layers. It can be seen that the error between the two results is less than 1%, indicating a rapid convergence rate of the layerwise method.

Appendix A

ADDITIONAL NOTES AND BIBLIOGRAPHY TO CHAPTERS

This appendix gives some additional notes and bibliography to the subjects treated in respective chapters.

Ch. 1

The mathematical details of the basic equations of elasticity theory can be found in a variety of textbooks and monographs, *e.g.* Love (1927), Green and Zerna (1954), Sokolnikoff (1956), Chien and Yeh (1956), Boley and Weiner (1960), Hermon (1961), Lur'e (1964), Fung (1965), Timoshenko and Goodier (1970), Flügge (1972), Eringen and Suhubi (1975), Nowinski (1978), Kupradze *et al.* (1979), Wu and Wang (1980), Lekhnitskii (1981), Hu (1984), Nowacki (1986), Du *et al.* (1986), Xie *et al.* (1988), Barber (1992), Luo and Li (1994), Ting (1996), Ding *et al.* (1997), Noda *et al.* (2000), Zhang (2001), and Wang (2002).

Ch. 2

Section 2.1

The state-space formulations are also known as the Hamiltonian formulae (or system) in some papers or monographs, see Tang (1991), Ouyang (1993), Wang and Tang (1995), Zhong (1995), Yao and Zhong (2002), to name a few.

Section 2.2

Early in 1940, Lekhnitskii obtained a general solution for axisymmetric problems of transversely isotropic bodies (Lekhnitskii, 1940, 1981); the completeness of the solution was proved later by Eubanks and Sternberg (1953) and Wang, Xu and Wang (1994).

Note that Wang and Xu (1995) also studied the general solutions of axisymmetric problems. Hu (1953, 1956) and Nowacki (1954) made a generalization of Lekhnitskii's solution to the three-dimensional case. Elliott (1948), Hu (1954*a*) and Lodge (1955) further presented the general solution in a simpler form. Wang and Wang (1995) and Wang and Shi (1998) obtained the general solution in other forms and proved the completeness of Hu-Nowacki's solution and Elliott-Hu-Lodge's solution. Chen (1966*b*) made a generalization of the solution to the problem for a uniformly moving load by employing the Galilean transformation. Some other interesting works on the general solutions of transversely isotropic elastic media are also available, see Bickford (1977) and Okumura and Dohba (1989).

Section 2.4

As early as 1865, Saint-Venant discussed the solution of a spherical shell of a spherically isotropic material subjected to a uniform external pressure. Saint-Venant's work was well cited in the two famous textbooks on elasticity, *i.e.* Love (1927) and Lekhnitskii (1981). Because of the complicated equations involved, the general solution of this class of problems was not properly addressed until the work by Hu in 1954, who opened a new solution avenue (Hu, 1954*b*). Based on Hu's method, a series of further studies became possible, including a detailed discussion on some basic solutions (Chen, 1966*a*), a special solution for axisymmetric deformation (Xia and He, 1988), thermal stress analysis (Ding and Ren, 1990), equilibrium problems (Ding and Ren, 1991), analysis of vibration (Ding, Ren, and Zou *et al.*, 1994), and coupled problems (Ding and Chen, 1996*a*,*b*). Puro (1980) found that the variable separation method was also valid for a spherically isotropic medium with a radial inhomogeneity.

Ch. 3

Section 3.1

Elliott (1948) first obtained the solution of an infinite transversely isotropic solid subjected to a concentrated force perpendicular to the material's isotropic plane. Kröner (1953) and Hu (1956) developed independently the solution for the case under a concentrated force in any direction. Hu overcame the difficulty with $s_1 = s_2$ by taking a limit to cover an isotropic material. Pan and Chou (1976) tried to establish a unified solution, but the expressions they derived for isotropic and transversely isotropic materials were basically different and not unified; an immediate discussion of this work was given by Chen (1977). Also, two errors were found in Pan and Chou's work by Koguchi *et al.* (1990) and Liang (1994). Pan (1989*b*) also presented an exact solution of a transversely isotropic infinite space subjected to a concentrated force. Ding, Liang and Chen (1997) derived a really unified solution of an infinite transversely isotropic body. Fabrikant (2004) reported a similar solution, with a further discussion on annular loading. Note that a related solution was obtained by Chen (1964) to the elasticity problem where the discontinuity is in the displacement component parallel to the plane area inside a transversely isotropic medium. Zhu (1992) reported a systematic study of obtaining the time-harmonic Green's functions in a transversely isotropic body.

Section 3.2

There exist some relevant studies for both isotropic materials (Rongved, 1955; Dundurs and Hetényi, 1965) and transversely isotropic materials (Sveklo, 1969; Pan and Chou, 1979*a*). Their methods, however, were not so straightforward and the applications of the solutions are limited to some special cases. For example, Sveklo's solution was only for materials with $s_1 \neq s_2$ under a point force perpendicular to an isotropic plane. Pan and Chou's study contained some errors as pointed out by Koguchi *et al.* (1990). Huang and Wang (1991) obtained fundamental solution for a bi-material elastic space. Hanson and Wang (1997), Ding and Chi (1999), and Ding and Liang (2001) studied the problem of an infinite and semi-infinite transversely isotropic body subjected to a concentrated ring force, respectively. Recently, Liew *et al.* (2001) investigated the elastic field of a bi-material space.

Section 3.3

The solution of a spheroidal inclusion or a cavity in a transversely isotropic material was first obtained by Chen (1968), who later investigated the axisymmetric stress field perturbed by a spheroidal inhomogeneity (Chen, 1971). Somaratna and Ting (1986) studied the stress singularities at conical notches and inclusions in a transversely isotropic material. Zureick (1988) obtained an exact solution of a rigid spheroidal inclusion in a transversely isotropic medium under an axial pull. Zureick and Eubanks (1988) and Zureick (1989) considered the problems of a spheroidal cavity with prescribed asymmetric stress and displacement in a three-dimensional transversely isotropic body. Heinrich and Wang (1989) investigated the stress interaction of two spheroidal cavities in a transversely isotropic elastic body. Podil'chuk and Sokolovskii

(1991*a*,*b*) studied the stress state of a transversely isotropic medium containing an anisotropic elastic prolate spheroid subjected to uniform and linear stress fields, respectively at infinity. Podil'chuk (1997) also derived an exact solution of the problem of a rigid elliptical inclusion in a transversely isotropic body. All the above-mentioned studies are based on displacement functions using a trial-and-error method, as illustrated in this section.

Ch. 4

The solution to the stresses and displacements in a half-space or a layered solid of transverse isotropy is fundamental to the development of the theory of elasticity and is of primary importance to many engineering applications such as the design of high strength foundations and underground structures. Over the last century, many scientists made important contributions to this field. For example, Michell (1900) obtained the stresses and displacements in an anisotropic half-space subjected to a normal point force on the surface. Mindlin (1936) developed the solution for an isotropic half-space under an interior point force. He also found that the solution consists of two parts, the singular fundamental solution of an infinite body and the non-singular solution of a half-space. (The reader can see in **Section 4.2** of this book that this is also true for a transversely isotropic half-space.) Soon after, Lekhnitskii (1940) solved the problem of an anisotropic half space subjected to a normal point force. His result was used by Dahan and Zarka (1977), with the aid of Hankel transform, to develop the solution of contact stresses between an isotropic sphere and a transversely isotropic half-space. In 1954, Hu developed the expressions of stresses and displacements of a transversely isotropic half-space ($s_1 \neq s_2$) under a normal and a tangential point force. He also discussed the applications of the solution to the cases with various boundary conditions (Hu, 1954*a*). Shield (1951), Sveklo (1969), Pan and Chou (1979*b*), and Sneddon (1992) also carried out a series of relevant studies.

Ding and Xu (1982) studied the equilibrium of a transversely isotropic elastic layer using the general solution. Ding, Liang and Wang (1996) obtained an analytical solution of a transversely isotropic layer under the action of a concentrated force. The stress analysis of multi-layered elastic media has attracted much interest. Relative studies on isotropic materials include Burmister (1945*a*,*b*), Bufler (1971), Small and Booker (1984, 1986), Benitez and Rosakis (1987), Kausel and Roësset (1981), and Kausel and Seale (1987), to name a few. Studies on anisotropic materials are also available, for example,

Qiu and Wang (1988), Shi and Zeng (1989), Pan (1989*a*,*c*), Choi and Thangjitham (1991), Huang (1994), and Jin *et al.* (1996), etc.

Ch. 5

The problems considered in **Sections 5.2-5.5** are very typical in the theory of elasticity, and also practically useful. The solutions presented can be easily degenerated to those for isotropic materials that can be found in the textbook of Timoshenko and Goodier (1970).

The stress and deformation problem of elastic cones is a classic mechanics problem. There are some relevant investigations on cone problems. Fore example, Hu (1953) studied the bending of a transversely isotropic cone, Lekhnitskii (1981) obtained the solution of a cone under axial compression, Chen (1965) discussed both the bending and compression and extended its discussion on conical shells. Recently, Podil'chuk (1991) carried out a research on a cone subjected to general surface loading. Ding, Zou and Ren (1995) developed the solution of a spherically isotropic cone under a tip load. Some special cases, such as the effect of the self-weight, have also been addressed (Zou, 1992).

Ch. 6

Section 6.1

The thermal stresses of transversely isotropic materials have been studied to a certain extent (*e.g.*, Sharma, 1958; Shaldyrvan, 1980; Tsai, 1983*a*,*b*; Korolev, 1992; Birman, 1995). Similar to method presented in **Section 2.2**, Podil'chuk (1987, 1988), Podil'chuk and Sokolovskii (1994), Podil'chuk *et al.* (1994) employed three displacement functions to represent the displacements. Borodachev (1988) presented several classes of thermoelastic general solutions of transversely isotropic materials. The method presented in this section is more convenient for solving practical problems. It is noted that Noda *et al.* (Noda and Ashida, 1987, 1988; Noda *et al.*, 1989, 1994; Ashida *et al.*, 1993) presented some analytical solutions for dynamic thermoelastic problems of transversely isotropic materials.

Section 6.2

Employing the Hankel transform method, Mehta (1966) investigated the thermal stress around a crack in an elastic solid of transversely isotropic material, but most final expressions were given in integral form. Tsai (1983*a*,*b*) used a similar method to study the thermal stress in a transversely isotropic medium containing a penny-shaped crack; the expressions for normal stress and axial displacement at the plane of crack surface were explicitly derived. Singh *et al.* (1987) examined the steady state thermoelastic behaviour of an external circular crack subjected to a symmetric temperature distribution at the crack surfaces. Podil'chuk and Sokolovskii (1994) employed a general solution for transversely isotropic elasticity to solve the steady state problem of an infinite medium with an internal elliptical crack by the trial-and-error method. Tsai (1998, 2000) extended his earlier works (Tsai, 1983*a*,*b*) to the flat toroidal crack case using the techniques of triple integral equations and multiplying factors. Chen and Ding (2003) proposed a general solution expressed in terms four quasi-harmonic functions only. Using this solution, Chen *et al.* (2004) obtained a fundamental solution for a point temperature loaded penny-shaped crack in a transversely isotropic solid.

Section 6.3

In the literature, Nowinski (1959) and Eason (1964) studied the spherically symmetric thermoelastic problem of spherically isotropic hollow spheres. Raju (1975) performed a finite element thermal stress analysis of a spherically isotropic shallow shell. Vasilenko *et al.* (1976) studied the thermal stresses in a multi-layered nonhomogeneous spherical shell using a transfer matrix method along with a harmonic functions expansion technique. Ding and Ren (1990) investigated the thermal stresses using the harmonic functions expansion technique. Misra *et al.* (1994) presented a transient analysis of an infinite spherically isotropic medium containing a cavity.

Ch. 7

Since the pioneering work of Hertz on the elastic contact of two isotropic bodies in the 1880s, numerous studies have been carried out considering more and more factors, such as friction, surface roughness and anisotropy of materials. Various solution methods have also been developed to overcome the mathematical difficulties encountered. The complex function method by Muskhelishvili (1953) and the integral transform method by Sneddon (1951) are the two distinct examples of development, which have played an important role in the effective solution of both the contact problems and general elasticity. In the early stage of studying contact mechanics, emphasis was mainly placed on isotropic materials, as summarized by Gladwell (1980) and Johnson (1985). Other relevant treatments based on the general theory of elasticity have also been discussed in many texts on solid mechanics, such as those by Love (1927), Muskhelishvili (1953), Green and Zerna (1954), Lur'e (1964), Xie *et al.* (1988) and Zhang (2001).

The technological development and applications of high strength foundations and surface coatings promoted the research on the contact mechanics of anisotropic materials. Elliott (1949) was the first to use two harmonic functions (Elliott, 1948) to investigate the axisymmetric deformation of a transversely isotropic half-space when it was in contact with a rigid cone, sphere or cylindrical punch. Since then, a series of investigations were carried out to understand the effect of anisotropy on elastic deformation, develop efficient solution methods and explore the similarity and difference of stress and deformation with the changes of material properties and geometry of contacting bodies. Some interesting studies can be briefly summarized in chronological order as follows:

- The compliance of elastic bodies in contact (Mindlin, 1949).
- The contact with an elliptic punch (Shield, 1951).
- The plane contact and the similarity of isotropic deformation to anisotropy (Green and Zerna, 1954; Galin, 1961).
- The application of Fourier transform, which converted a Hertz contact problem to a corresponding contour integral so that some closed form solutions for a transversely isotropic half-plane became feasible (Willis, 1966, 1967).
- The axisymmetric deformation of a transversely isotropic half-space subjected to the indentation of a rigid sphere, with certain friction considerations (Conway, *et al.* 1967; Conway and Farnham, 1967).
- The contact of plane deformation of anisotropic materials and three-dimensional deformation of transversely isotropic materials (Chen, 1969). The stress field considering the frictional indentation was also developed. The drawbacks in Green and Zerna (1954) and Galin (1961) were discussed and it was pointed out that under some special conditions, the stress functions for both anisotropic and isotropic plane contact problems became the same. Furthermore, it was found that the boundary value problems of isotropy and transverse isotropy were identical, which in turn made it possible to use the formulae of Hamilton and Goodman (1966) to produce the analytical solutions of displacements and stresses.
- Thermal contact analysis (Grilitskii and Shelestovskii, 1970).

- The application of Hankel transform in solving frictionless Hertz contact problems involving transversely isotropic materials (Dahan and Zarka, 1977)[1].
- The polynomial solution of an anisotropic half-space under the indentation of an elliptic punch (Gladwell, 1978).
- The contact of transversely isotropic materials with adhesive sliding friction (Keer and Mowry, 1979).
- The punch problem of a transversely isotropic layer elastically supported at edges (Rogowski, 1986).
- Symmetrical contact of a thick transversely isotropic plate (Tsai, 1986).
- The application of the finite difference method and potential theory for the thermal contact between a rigid ring and a transversely isotropic cylinder (Noda and Ashida, 1988).
- The Hertzian and non-Hertzian contact problems of transversely isotropic materials (Lin, *et al.*, 1991), using the potential function method (Green and Zerna, 1954; Pan and Chou, 1976).
- The circular punch problem of a transversely isotropic layer on a rigid foundation (Sakamoto *et al.*, 1991)
- The application of Hankel transform (Kuo and Keer, 1992). A good example of which was the spherical indentation of a transversely isotropic layer adhered with a transversely isotropic half-space.
- The frictionless contact of a layered anisotropic half-plane (Pindera and Lane, 1993), using the Fourier transform technique.
- The frictional Hertzian contact of inelastic anisotropic bodies using a similarity approach (Borodich, 1993).
- Two-dimensional contact on an anisotropic elastic half-space (Fan and Keer, 1994).
- The development of the potential theories and their applications to a large class of contact problems, with a variety of indentor profiles (Fabrikant, 1989, 1991). The method converted a boundary value problem to an integral equation without using any transform technique and in some cases the solutions could be obtained in terms of elementary functions.
- A unified and more direct method to achieve the solutions of four fundamental frictional contact problems (Hanson, 1992*a*,*b*, 1993, 1994; Hanson and Johnson, 1993), with the aid of the Green's function for a transversely isotropic half-space.

---

[1] The results were in a different from compared to those obtained by Chen (1969).

Ch. 8

Readers who are interested in plate theories such as the classical thin plate theory, Reissner plate theory, Mindlin plate theory *etc.* are referred to the pertinent papers and monographs, *e.g.* Mindlin (1951*a*), Timoshenko and Woinowsky-Krieger (1959), Reissner (1945, 1975), Ashton and Whitney (1970), and Reddy (1999).

Section 8.1

Pagano (1969, 1970, 1972), Srinivas *et al.* (Srinivas and Rao 1970; Srinivas *et al.*, 1970), and Sundara Raja Iyengar and Raman (1977) developed a systematic three-dimensional method for analysing the bending, vibration and stability of orthotropic rectangular plates, which naturally can be applied to transversely isotropic materials. Wang (1986) and Wang and He (1986) studied the free vibration and stability of transversely isotropic rectangular plates, using the general solution suggested by Hu (1956). An analysis similar to their work has been described in **Section 8.1** for a simply-supported rectangular plate and further extended to a circular plate with particular boundary conditions. The out-of-plane vibration of circular plate was considered by Ding and Xu (2000*a*), where the separation of vibration modes was not performed. Readers are also referred to two interesting papers by Pan (1992) and Ding, Xu, and Guo (1999), respectively for additional discussions. It is noted that Xu (1999) solved the bending and free vibration of sectorial plates and sectorial annular plates using the general solution presented in **Chapter 2**.

Section 8.2

Fan and Ye (1990*a*,*b*) first presented the state-space formulations for orthotropic elasticity and analysed the static and dynamic behaviour of thick or laminated rectangular plates with simply supported conditions. Pan (1992) suggested a propagator matrix method for the vibration of a transversely isotropic simply supported laminated rectangular plate. Chen, Ding and Yu (1996) further considered the coupled vibration of an orthotropic laminated rectangular plate on a Pasternak foundation. Wang, Tarn and Hsu (2000) considered the stress decay in laminated anisotropic plates using a state-space approach. A new state space formulation for transversely isotropic materials was recently derived by Ding, Chen and Xu (2001) that was introduced in **Section 8.2**.

Ch. 9

Readers who are interested in shell theories are referred to the pertinent textbooks and review papers, *e.g.* Timoshenko and Woinowsky-Krieger (1959), Ambartsumian (1964), Leissa (1973), Calladine (1983), Niordson (1985), Flügge (1990), Noor and Burton (1990), Mirza (1991), and Soldatos (1994). A large number of papers on anisotropic (laminated) cylindrical shells can be also found in literature, *e.g.* Mirsky (1964), Dong (1968), Dong and Tso (1972), Whitney and Sun (1974), Hyer and Cooper (1986), Upadhyay and Mishra (1988), Spencer *et al.* (1992), Fan and Ding (1995), Tessler *et al.* (1995), Ren (1995), Chen, Ding, and Guo *et al.* (1997), Jones (1998), Tang *et al.* (1998), Shakeri *et al.* (1999), Paliwal and Singh (1999), Ramasamy and Ganesan (1999), Lam *et al.* (2000), Xi *et al.* (2000), Ding, Wang and Chen (2001), Xia and Ding (2001), Paliwal and Pandey (2001), Bhaskar and Ganapathysaran (2002), and Li and Chen (2002), to name a few.

As regards transversely isotropic materials, Hu (1956) first proposed a displacement decomposition formula in cylindrical coordinates for vibration analysis of shells of revolution. Buchwald (1961) suggested a simpler displacement decomposition formula and obtained a Bessel function solution for wave motions in a transversely isotropic cylindrical shell. Mirsky (1965) also got a Bessel function solution of free vibration of a simply supported transversely isotropic cylindrical shell. Vendhan and Archer (1978) obtained an analytical solution of axisymmetric problems of a transversely isotropic finite cylinder. White and Tongtaow (1981) studied the cylindrical wave in a transversely isotropic medium. Chan and Tsang (1983) considered the propagation of acoustic waves in a fluid-filled borehole surrounded by concentrically layered, transversely isotropic media. Tsai (1991) gave an exact analysis of the longitudinal motion of a thick transversely isotropic hollow cylinder. Chau (1994) employed a method similar to that of Mirsky (1965) to study the free vibration problems of transversely isotropic cylinders. Zhang *et al.* (1994) studied the acoustic field due to multipole sources in a fluid-filled borehole surrounded by a transversely isotropic medium. Honarvar and Sinclair (1996) investigated the acoustic wave scattering from transversely isotropic cylinders by employing a decomposition formula, which is the same as that for isotropic materials (Morse and Feshbach, 1953). Chen *et al.* (Chen, Cai, and Ye *et al.*, 1998; Chen, Ying and Yang, 1998; Chen, Ding and Xu, 1998; Cai *et al.*, 1999; Chen and Ding, 1999*b*) presented a Bessel function solution with complex arguments and studied several coupled free vibration problems, where a displacement decomposition formula was employed that is very similar to that of Buchwald (1961). Rahman and Ahmad (1998) discussed the difference between two displacement decomposition formulae employed by Buchwald (1961) and Honarvar and Sinclair (1996) respectively, and, they reconsidered the acoustic scattering problem (Ahmad and Rahman, 2000).

Ch. 10

The study of anisotropic spherical shells has been a long history. Hoppmann II and Baker (1961) studied the extensional vibration of anisotropic spheres, using the momentless theory. Employing the finite Hankel transform, Mukherjee (1968) discussed the radial and rotary vibrations of a spherically isotropic spherical shell and presented analytical expressions for stresses. Ramakrishnan and Shah (1970) studied the free vibrations of spherically isotropic spherical shells by introducing displacement functions, where the effects of transverse shear deformation, rotatory inertia and transverse normal strain were included; two different classes of vibrations were found. Based on the three-dimensional elasticity equations of spherically isotropic media, Cohen *et al.* (1972) introduced two auxiliary functions to study the free vibration problem. However, it was very difficult to express the displacements and stresses by the two functions. Roy (1977) gave an analytical solution of a nonhomogeneous spherically isotropic elastic sphere with a concentric spherical inclusion. Chao and Chern (1988) analyzed the free vibration of complete orthotropic spherical shells by virtue of the shell theory. Narasimhan (1992) studied the dynamic responses of a laminated orthotropic spherical shell subjected to axisymmetric dynamic loads. Wu and Ma (1992) derived the governing equations of motion of orthotropic spherical shell, taking the displacements and rotary angles as the independent variables, and calculated the responses under impact load using the finite difference method. Wang (1994) investigated the dynamic response of anisotropic spheres by dividing the response into two parts, *i.e.* the quasi-static part that was obtained using the method suggested by Lekhnitskii (1981) and the dynamic part solved through the finite Hankel transform. Shul'ga (1980) and Ramskaya (1983) discussed the axisymmetric vibration of transversely isotropic hollow spheres by expanding the displacements in terms of Legendre functions along with the power series method. Shul'ga *et al.* (1988) further presented a non-axisymmetric free vibration analysis of a transversely isotropic hollow sphere with radial inhomogeneity. Recently, Ding, Wang and Chen (2002) obtained an analytical thermo-elastodynamic solution of an inhomogeneous spherically isotropic hollow sphere.

Section 10.3

Spherical shells filled with or submerged in fluid media are often encountered in engineering such as oil tanks in petrochemical industry, vessels in ocean exploitation and exploration, underwater sonar, and rocket fuel bins in aerospace structures. Even in the area of geophysics, Montagner and Anderson (1989) reported a mantel model of a multi-layered spherically isotropic sphere containing a fluid nucleus. In many technological applications of shell-structured elements, static, dynamic and environmental factors are often coupled and need to be considered (Maiti, 1975; Zou, 1992; Chen, 1995; Ding, Zou and Ding, 1996). However, a coupled vibration is so difficult to solve such that only the simpler shell theories, such as the momentless and classical thin bending theories, are often employed for analysis (Su and Lou, 1975; Su, 1981*a,b*, 1982; Junger and Feit, 1986) except some special cases such as the purely torsional and radial vibrations. The application of new composite materials and high-molecular compounds, which are mostly anisotropic, added more complexity to the solution of the problems.

A quite systematic three-dimensional method for analysing the small vibration of rigid and isotropic elastic bodies submerged in or filled with viscous, compressible fluid was presented by Guz' (1980*a*, 1980*b*, 1981, 1993*a,b*). The linearized Navier-Stokes equations were used to describe the fluid with a small perturbation and the displacement separation method was employed to simplify the basic equations. However, the coupled vibration of spherically isotropic hollow spheres was not addressed.

Section 10.4

The vibration of plates and shells coupled with elastic media or foundations is one of the problems that are frequently met in civil engineering. To simplify the analysis, several important foundation models have been proposed such as Winkler model, Kerr model, Pasternak model, viscoelastic model and high-order models etc. (Kerr, 1964, Bharatha and Levinson, 1980). The mathematical expressions for these models are different from each other because of the distinct assumptions employed. Among them, the simplest one is Winkler model, which however, has been used widely in civil engineering. Nowinski

(1959) considered the thermoelastic problem of a transversely isotropic hollow sphere in an infinite elastic medium, which is treated as this model. With the development of science and technology, especially the computer science, the higher order elastic foundation models have attracted more attention. Nath and Jain (1983) analyzed the nonlinear responses of a thin spherical shell on Pasternak foundation. Nath *et al.* (1987) investigated the nonlinear dynamic response of doubly curved shallow shells resting on elastic foundations. Paliwal and Bhalla (1993) studied the free vibration of thin spherical shells on Pasternak foundation, that the effect of large deformation was considered. Paliwal and Srivastava (1994) studied the vibration of a thin spherical shell on Kerr foundation. The study of anisotropic spherical shell coupled with elastic medium is much less.

Section 10.5

Shul'ga *et al.* (1988) first established a state equation for spherically isotropic media using the separation formulae for stresses and displacements. However, they just regarded it as a first-order ordinary differential equation set that was solved by a numerical method. Ye and Soldatos (1996) first applied the state-space approach to the axisymmetric vibration of laminated hollow spheres composed of transversely isotropic layers and employed a layerwise method for accomplishing the analysis. Chen and Ding (2001*a*) developed an exact three-dimensional stress analysis of a spherically isotropic spherical shell based on state-space formulations. Employing the Taylor's expansion method, they further studied the three-dimensional free vibration of a spherically isotropic hollow sphere (Chen and Ding, 2001*b*).

Appendix B

SPECIAL FUNCTIONS

This appendix presents an introductory description of the special functions associated with the three-dimensional *Helmholtz equation* in cylindrical and spherical coordinates, which are used extensively throughout the book. More details of the properties of these functions can be found in many sources, *e.g.* Watson (1922), Courant and Hilbert (1953), McLachlan (1954), Guo (1991), Wu (1999), and Wang and Guo (2000).

B. 1 HELMHOLTZ EQUATION AND SEPARATION OF VARIABLES

The Helmholtz equation is

$$\nabla^2 u + k^2 u = 0\,, \tag{B.1.1}$$

where k is a constant, known as the wave number that is equal to a circular frequency divided by a wave speed for certain dynamic problems considered in the text. ∇^2 is the three-dimensional Laplace operator defined by

$$\nabla^2 = \frac{1}{r}\frac{\partial}{\partial r}\left(r\frac{\partial}{\partial r}\right) + \frac{1}{r^2}\frac{\partial^2}{\partial \alpha^2} + \frac{\partial^2}{\partial z^2} \tag{B.1.2}$$

in cylindrical coordinates and

$$\nabla^2 = \frac{1}{R}\frac{\partial}{\partial R}\left(R^2\frac{\partial}{\partial R}\right) + \frac{1}{R^2 \sin\theta}\frac{\partial}{\partial \theta}\left(\sin\theta\frac{\partial}{\partial \theta}\right) + \frac{1}{R^2 \sin^2\theta}\frac{\partial^2}{\partial \alpha^2} \tag{B.1.3}$$

in spherical coordinates.

Consider first the Helmholtz equation in cylindrical coordinates. By separation of variables

$$u(r,\alpha,z)=U(r,\alpha)Z(z), \tag{B.1.4}$$

the Helmholtz equation is transformed to the following two equations

$$\frac{1}{r}\frac{\partial}{\partial r}\left(r\frac{\partial U}{\partial r}\right)+\frac{1}{r^2}\frac{\partial^2 U}{\partial \alpha^2}+(k^2-\beta)U=0, \tag{B.1.5}$$

$$\frac{\mathrm{d}^2 Z}{\mathrm{d}z^2}+\beta Z=0, \tag{B.1.6}$$

where β is a constant to be determined from the particular problem. Again by separation of variables, $U=S(r)T(\alpha)$, Eq. (B.1.5) yields

$$\frac{1}{r}\frac{\mathrm{d}}{\mathrm{d}r}\left(r\frac{\mathrm{d}S}{\mathrm{d}r}\right)+\left(k^2-\beta-\frac{\gamma}{r^2}\right)S=0, \tag{B.1.7}$$

$$\frac{\mathrm{d}^2 T}{\mathrm{d}\alpha^2}+\gamma T=0, \tag{B.1.8}$$

where γ is another constant to be determined. Thus, by separation of variables, the solution of Helmholtz equation can be expressed as the product of three solutions to Eqs. (B.1.7), (B.1.8) and (B.1.6). Both Eqs. (B.1.6) and (B.1.8) are second-order ordinary differential equation (ODE) with constant coefficients; their solutions are easy to obtain. Equation (B.1.7) is a second-order ODE with coefficients being r-dependent, and is widely known as the Bessel's equation.

Similarly, the Helmholtz equation in spherical coordinates, with the solution in the form of $U(R,\theta,\alpha)=S(R)Z(\theta)T(\alpha)$, can be transformed to the following three equations:

$$\frac{1}{R^2}\frac{\mathrm{d}}{\mathrm{d}R}\left(R^2\frac{\mathrm{d}S}{\mathrm{d}R}\right)+\left(k^2-\frac{\beta}{R^2}\right)S=0, \tag{B.1.9}$$

$$\frac{1}{\sin\theta}\frac{\mathrm{d}}{\mathrm{d}\theta}\left(\sin\theta\frac{\mathrm{d}Z}{\mathrm{d}\theta}\right)+\left(\beta-\frac{\gamma}{\sin^2\theta}\right)Z=0, \tag{B.1.10}$$

$$\frac{d^2 T}{d\alpha^2} + \gamma T = 0. \tag{B.1.11}$$

Equation (B.1.11) is exactly the same as Eq. (B.1.8), Eq. (B.1.9) is known as the *spherical Bessel's equation*, and Eq. (B.1.10) is the *associated Legendre's equation*.

In most problems considered in the book, u and hence T must be periodic functions of α with period 2π. We then obtain the solution to Eq. (B.1.8) or Eq. (B.1.11) as $T = A\cos(m\alpha) + B\sin(m\alpha)$ or, in complex form, $T = C\,e^{im\alpha}$, here $m = \sqrt{\gamma}$ is any non-negative integer, A and B are two arbitrary real constants, and C is an arbitrary complex constant. In solving Legendre's equation and Bessel's equation, Frobenius power series method is generally adopted to get solutions in a series form.

B. 2 LEGENDRE FUNCTION AND ASSOCIATED LEGENDRE FUNCTION

Taking the transform $x = \cos\theta$ and $y(x) = Z(\theta)$, and writing $\beta = \nu(\nu+1)$, we get from Eq. (B.1.10)

$$(1-x^2)y'' - 2xy' + \left[\nu(\nu+1) - \frac{m^2}{1-x^2}\right]y = 0, \tag{B.2.1}$$

where the prime denotes differentiation with respect to x. When $m = 0$, Eq. (B.2.1) is known as the Legendre's equation, whose general solution is

$$y(x) = c_1 P_\nu(x) + c_2 Q_\nu(x), \tag{B.2.2}$$

where c_1 and c_2 are arbitrary, and $P_\nu(x)$ and $Q_\nu(x)$ are Legendre functions of the first and second kinds, respectively. For most problems considered in the book, the solution should be regular at points $x = \pm 1$; it follows that $c_2 = 0$ and $\nu = n$, n is a non-negative integer. Thus, the solution becomes

$$y(x) = c_1 P_n(x), \quad (n = 0, 1, 2, \cdots), \tag{B.2.3}$$

where $P_n(x)$ are Legendre polynomials:

$$P_n(x) = \sum_{i=0}^{n} \frac{1}{(i!)^2} \frac{(n+i)!}{(n-i)!} \left(\frac{x-1}{2} \right)^i, \quad (n = 0, 1, 2, \cdots). \tag{B.2.4}$$

The general solution to the associated Legendre's equation (B.2.1) $(m \neq 0)$ is

$$y(x) = c_1 P_\nu^m(x) + c_2 Q_\nu^m(x), \tag{B.2.5}$$

where $P_\nu^m(x)$ and $Q_\nu^m(x)$ are the associated Legendre functions of the first and second kinds, respectively. The solution that is regular at points $x = \pm 1$ is

$$y(x) = c_1 P_n^m(x), \ (n = m, m+1, m+2, \cdots). \tag{B.2.6}$$

The following orthogonality relation holds for $P_n^m(x)$

$$\begin{aligned} \int_{-1}^{1} P_n^m(x) P_l^m(x)\,\mathrm{d}x &= \int_0^{\pi} P_n^m(\cos\theta) P_l^m(\cos\theta) \sin\theta\,\mathrm{d}\theta \\ &= \begin{cases} 0, & l \neq n, \\ \dfrac{2(n+m)!}{(2n+1)(n-m)!}, & l = n. \end{cases} \end{aligned} \tag{B.2.7}$$

We also mention the following two recurrence formulae

$$\begin{aligned} &(n-m+1)P_{n+1}^m(x) - (2n+1)xP_n^m(x) + (n+m)P_{n-1}^m(x) = 0, \\ &(1-x^2)\frac{\mathrm{d}P_n^m(x)}{\mathrm{d}x} = (n+1)xP_n^m(x) - (n-m+1)P_{n+1}^m(x). \end{aligned} \tag{B.2.8}$$

B. 3 SPHERICAL HARMONICS

Spherical harmonics $S_n^m(\theta, \alpha)$ satisfy the following equation:

$$\frac{1}{\sin\theta} \frac{\partial}{\partial\theta} \left(\sin\theta \frac{\partial S_n^m}{\partial\theta} \right) + \frac{1}{\sin^2\theta} \frac{\partial^2 S_n^m}{\partial\alpha^2} + n(n+1)S_n^m = 0. \tag{B.3.1}$$

The reader should be very careful since various forms of spherical harmonics have been proposed in literature. Here we adopt the following definition (Guo, 1991):

$$S_n^m(\theta,\alpha) = P_n^m(\cos\theta)\mathrm{e}^{\mathrm{i}m\alpha}, \quad (n = 0, 1, 2, \cdots; m = 0, \pm 1, \pm 2, \cdots, \pm n). \tag{B.3.2}$$

The orthogonality relation for spherical harmonics is

$$\int_0^\pi \int_0^{2\pi} S_n^m(\theta,\alpha)\bar{S}_k^l(\theta,\alpha)\sin\theta\,\mathrm{d}\theta\,\mathrm{d}\alpha = \frac{(n+m)!}{(n-m)!}\frac{4\pi}{2n+1}\delta_{nk}\delta_{ml}, \tag{B.3.3}$$

where δ_{nk} is the Kronecker delta, and an overbar denotes the complex conjugate. The term *surface spherical harmonics* denotes any linear combination of real spherical harmonics

$$A_n P_n(\cos\theta) + \sum_{m=1}^{n}[A_n^m \cos(m\alpha) + B_n^m \sin(m\alpha)]P_n^m(\cos\theta) \tag{B.3.4}$$

for n fixed. The *solid harmonics*, satisfying the Laplace's equation, are equal to the surface spherical harmonics of order n multiplied by r^n or $r^{-(n+1)}$.

B. 4 BESSEL FUNCTIONS

If $k^2 - \beta \neq 0$, by taking the transform $x = \sqrt{k^2 - \beta}r$ and $y(x) = S(r)$, we can rewrite Eq. (B.1.7) as

$$x^2 y'' + xy' + (x^2 - \nu^2)y = 0, \tag{B.4.1}$$

where $\nu = \sqrt{\gamma}$. This is the standard form of Bessel's equation, of order ν, with the general solution

$$y(x) = c_1 J_\nu(x) + c_2 Y_\nu(x), \tag{B.4.2}$$

where $J_\nu(x)$ and $Y_\nu(x)$ are Bessel functions, of order ν, of the first and second kinds, respectively. Both functions belong to the *cylinder functions* $C_\nu(x)$ that satisfy

$$\begin{aligned}
&\frac{\mathrm{d}}{\mathrm{d}x}[x^\nu C_\nu(x)] = x^\nu C_{\nu-1}(x), \\
&\frac{\mathrm{d}}{\mathrm{d}x}[x^{-\nu} C_\nu(x)] = -x^{-\nu} C_{\nu+1}(x) \\
&C_{\nu-1}(x) + C_{\nu+1}(x) = \frac{2\nu}{x} C_\nu(x), \\
&C_{\nu-1}(x) - C_{\nu+1}(x) = 2C_\nu'(x).
\end{aligned} \tag{B.4.3}$$

For x near zero, we have the following asymptotic formulae:

$$\begin{aligned}
&J_\nu(x) \sim \frac{1}{\Gamma(\nu+1)}\left(\frac{x}{2}\right)^\nu, \quad x \to 0^+, \\
&Y_0(x) \sim \frac{2}{\pi}\ln\left(\frac{x}{2}\right), \quad x \to 0^+, \\
&Y_\nu(x) \sim \frac{\Gamma(\nu)}{\pi}\left(\frac{x}{2}\right)^{-\nu}, \quad x \to 0^+ \quad (\mathrm{Re}\,\nu > 0).
\end{aligned} \tag{B.4.4}$$

This indicates that, for $\mathrm{Re}\,\nu \geq 0$, $J_\nu(x)$ is finite while $Y_\nu(x)$ is singular at $x = 0$. For large values of $|x|$, we have

$$\begin{aligned}
J_\nu(x) \sim \sqrt{\frac{2}{\pi x}}\Bigg[&\cos(x - \tfrac{1}{2}\nu\pi - \tfrac{1}{4}\pi)\sum_{m=0}^{\infty}\frac{(-1)^m \cdot (\nu, 2m)}{(2x)^{2m}} \\
&- \sin(x - \tfrac{1}{2}\nu\pi - \tfrac{1}{4}\pi)\sum_{m=0}^{\infty}\frac{(-1)^m \cdot (\nu, 2m+1)}{(2x)^{2m+1}}\Bigg], \\
Y_\nu(x) \sim \sqrt{\frac{2}{\pi x}}\Bigg[&\sin(x - \tfrac{1}{2}\nu\pi - \tfrac{1}{4}\pi)\sum_{m=0}^{\infty}\frac{(-1)^m \cdot (\nu, 2m)}{(2x)^{2m}} \\
&+ \cos(x - \tfrac{1}{2}\nu\pi - \tfrac{1}{4}\pi)\sum_{m=0}^{\infty}\frac{(-1)^m \cdot (\nu, 2m+1)}{(2x)^{2m+1}}\Bigg],
\end{aligned} \tag{B.4.5}$$

where $|\arg(x)| < \pi$, and

$$(\nu, m) = \frac{(4\nu^2 - 1^2)(4\nu^2 - 3^2)\cdots[4\nu^2 - (2m-1)^2]}{2^{2m} \cdot m!}. \tag{B.4.6}$$

We have the orthogonality property of Bessel function of the first kind as

$$\int_0^a J_\nu(k_m x) J_\nu(k_n x) x \,\mathrm{d}x = \begin{cases} 0, & m \neq n, \\ \frac{a^2}{2}[J_{\nu+1}(k_n a)]^2, & m = n, \end{cases} \tag{B.4.7}$$

where k_m and k_n are roots of $J_\nu(ka) = 0$. With this property, a piecewise continuous function $f(x)$ defined on $0 \le x \le a$ can be expanded into the *Fourier-Bessel series*

$$f(x) = \sum_{n=1}^{\infty} c_n J_\nu(k_n x), \quad (\nu > -1/2), \tag{B.4.8}$$

where k_1, k_2, k_3 … are the positive roots of $J_\nu(ka) = 0$ arranged in ascending order of magnitude, and the coefficients are given by

$$c_n = \frac{2}{a^2[J_{\nu+1}(k_n b)]^2} \int_0^a f(x) J_\nu(k_n x) x \,\mathrm{d}x. \tag{B.4.9}$$

B. 5 BESSEL FUNCTIONS OF OTHER KINDS

B.5.1 Modified Bessel Functions

If $k^2 - \beta < 0$, by taking the transform $x = \sqrt{\beta - k^2}\, r$ and $y(x) = S(r)$, we can rewrite Eq. (B.1.7) as

$$x^2 y'' + xy' - (x^2 + \nu^2) y = 0, \tag{B.5.1}$$

which differs from Bessel's equation only in the coefficient of y and is known as the *modified Bessel's equation*. The solution to this equation can still be expressed by Eq.

(B.4.2) with the argument x replaced with $\mathrm{i}x$. In practice, we usually write the solution in the following form

$$y(x) = c_1 I_\nu(x) + c_2 K_\nu(x), \tag{B.5.2}$$

where $I_\nu(x)$ and $K_\nu(x)$ are the modified Bessel functions of the first and second kinds, respectively; they are defined as

$$\begin{aligned} I_\nu(x) &= \mathrm{e}^{-\mathrm{i}\nu\pi/2} J_\nu(x\mathrm{e}^{\mathrm{i}\pi/2}), \\ K_\nu(x) &= \frac{\pi}{2\sin(\nu\pi)}[I_{-\nu}(x) - I_\nu(x)]. \end{aligned} \tag{B.5.3}$$

The recurrence formulae for $I_\nu(x)$ and $K_\nu(x)$ are

$$\begin{aligned} &\frac{\mathrm{d}}{\mathrm{d}x}[x^\nu I_\nu(x)] = x^\nu I_{\nu-1}(x), && \frac{\mathrm{d}}{\mathrm{d}x}[x^{-\nu} I_\nu(x)] = x^{-\nu} I_{\nu+1}(x), \\ &I_{\nu-1}(x) + I_{\nu+1}(x) = 2I'_\nu(x), && I_{\nu-1}(x) - I_{\nu+1}(x) = \frac{2\nu}{x} I_\nu(x), \\ &\frac{\mathrm{d}}{\mathrm{d}x}[x^\nu K_\nu(x)] = -x^\nu K_{\nu-1}(x), && \frac{\mathrm{d}}{\mathrm{d}x}[x^{-\nu} K_\nu(x)] = -x^{-\nu} K_{\nu+1}(x), \\ &K_{\nu-1}(x) + K_{\nu+1}(x) = -2K'_\nu(x), && K_{\nu-1}(x) - K_{\nu+1}(x) = -\frac{2\nu}{x} K_\nu(x). \end{aligned} \tag{B.5.4}$$

Just as $J_\nu(x)$ and $Y_\nu(x)$ respectively, $I_\nu(x)$ are finite while $K_\nu(x)$ are singular at $x = 0$ when $\mathrm{Re}\,\nu \geq 0$. On the other hand, $K_\nu(x)$ decay exponentially and $I_\nu(x)$ grow exponentially as $x \to \infty$. These asymptotic properties are important in determining the proper form of solution to the modified Bessel's equation for a practical boundary value problem.

B.5.2 Hankel Functions

The Hankel functions, or Bessel functions of the third kind, are defined as

$$H_\nu^{(1)}(x) = J_\nu(x) + \mathrm{i}Y_\nu(x), \tag{B.5.5}$$

$$H_\nu^{(2)}(x) = J_\nu(x) - \mathrm{i}\,Y_\nu(x),$$

where $H_\nu^{(1)}(x)$ and $H_\nu^{(2)}(x)$ are Hankel functions of the first and second kinds, respectively. Since these functions are linear combinations of Bessel functions of the first and second kinds, they satisfy Bessel's equation (B.4.1) identically. The recurrence formulae for $H_\nu^{(1)}(x)$ and $H_\nu^{(2)}(x)$ are also given by Eq. (B.4.3).

One of the most important applications of Hankel functions is related to wave propagation analysis in cylindrical coordinates. Since for $|x| \to \infty$,

$$\begin{aligned} H_\nu^{(1)}(x) &\sim \sqrt{\frac{2}{\pi x}}\exp\left[\mathrm{i}\left(x - \frac{\nu\pi}{2} - \frac{\pi}{4}\right)\right], \\ H_\nu^{(2)}(x) &\sim \sqrt{\frac{2}{\pi x}}\exp\left[-\mathrm{i}\left(x - \frac{\nu\pi}{2} - \frac{\pi}{4}\right)\right], \end{aligned} \tag{B.5.6}$$

if combined with the time factor $\mathrm{e}^{i\omega t}$ (ω is the circular frequency), $H_\nu^{(1)}(x)$ represents the wave propagating backward along x, and $H_\nu^{(2)}(x)$ the wave propagating forward along x.

B.5.3 Sperical Bessel Functions

By taking the transform $x = kR$ and $y(x) = S(R)$, we can rewrite Eq. (B.1.9) as

$$x^2 y'' + 2xy' + [x^2 - \nu(\nu+1)]y = 0, \tag{B.5.7}$$

Further by the transform $y(x) = x^{-1/2}u(x)$, we find $u(x)$ satisfying Bessel's equation of order $\nu + 1/2$. Thus the general solution to Eq. (B.5.7) is

$$y(x) = x^{-1/2}[c_1 J_{\nu+1/2}(x) + c_2 Y_{\nu+1/2}(x)]. \tag{B.5.8}$$

When $\nu = n$ is an integer, we define

$$j_n(x) = \sqrt{\frac{2}{\pi x}} J_{n+1/2}(x), \quad y_n(x) = \sqrt{\frac{2}{\pi x}} Y_{n+1/2}(x). \tag{B.5.9}$$

where $j_n(x)$ and $y_n(x)$ are spherical Bessel functions of the first and second kinds, respectively, both satisfying Eq. (B.5.7) for $\nu = n$. These functions can be expressed in terms of elementary functions, *e.g.*

$$\begin{aligned} j_0(x) &= \frac{\sin x}{x}, \quad j_1(x) = \frac{1}{x^2}(\sin x - x\cos x), \\ j_2(x) &= \frac{1}{x^3}[(3-x^2)\sin x - 3x\cos x], \quad y_0(x) = -\frac{\cos x}{x}, \\ y_1(x) &= -\frac{1}{x^2}(\cos x + x\sin x), \quad y_2(x) = -\frac{1}{x^3}[(3-x^2)\cos x + 3x\sin x]. \end{aligned} \tag{B.5.10}$$

The recurrence formulae for spherical Bessel functions can be readily derived from those of the standard Bessel functions in Eq. (B.4.3) as

$$\begin{aligned} &\frac{\mathrm{d}}{\mathrm{d}x}[x^{n+1}c_n(x)] = x^{n+1}c_{n-1}(x), \qquad \frac{\mathrm{d}}{\mathrm{d}x}[x^{-n}c_n(x)] = -x^{-n}c_{n+1}(x) \\ &c_{n-1}(x) + c_{n+1}(x) = \frac{2n+1}{x}c_n(x), \quad nc_{n-1}(x) - (n+1)c_{n+1}(x) = 2(n+1)c_n'(x), \end{aligned} \tag{B.5.11}$$

where $c_n(x)$ represents either $j_n(x)$ or $y_n(x)$. We also can define the modified spherical Bessel functions of the first and second kinds, $i_n(x)$ and $k_n(x)$, as

$$i_n(x) = \sqrt{\frac{2}{\pi x}} I_{n+1/2}(x), \quad k_n(x) = \sqrt{\frac{2}{\pi x}} K_{n+1/2}(x). \tag{B.5.12}$$

Typically we have $i_0(x) = \sinh x / x$ and $k_0(x) = \mathrm{e}^{-x}/x$. Similarly, the spherical Hankel functions of the first and second kinds are given by

$$\begin{aligned} h_n^{(1)}(x) &= j_n(x) + \mathrm{i}\, y_n(x), \\ h_n^{(2)}(x) &= j_n(x) - \mathrm{i}\, y_n(x). \end{aligned} \tag{B.5.13}$$

These functions satisfy Eq. (B.5.7) and have important applications in wave propagation analysis in spherical coordinates.

Appendix C

NOMENCLATURE

In the following list, a number in brackets {} denotes the page number on which a symbol is first introduced.

a, b	Length of semi-major and semi-minor axes of an ellipse or a spheroid; inner and outer radii of a cylindrical shell, or spherical shell {93, 210, 284, 331}
c	$=\omega/k$, wave velocity {322}
c_0	$=c_f/v_2$, ratio between the sound velocity in fluid and shear wave velocity {316}
c_1	$=\sqrt{(\lambda+2\mu)/\rho_m}$, longitudinal wave velocity {321}
c_2	$=\sqrt{\mu/\rho_m}$, transverse wave velocity {321}
c_f	Velocity of sound in fluid {316}
c_{ij}	Elastic stiffnesses (or moduli) {10}
c_{ijkl}	Components of the elastic stiffness tensor of rank four {8}
$\overline{c}_1$, $\overline{c}_2$	$=c/c_1, c/c_2$, respectively, the dimensionless velocities {323}
$\overline{c}_{13}$	$=\sqrt{c_{11}c_{33}}$ {45}
D, L, M, N	Dimensionless material parameters defined in Eq. (2.4.23) {66}
$D(e)$	$=[K(e)-E(e)]/e^2$ {211}
D_p, D_T	Material parameters defined in Eqs. (4.2.9) and (4.2.15), respectively {120, 121}
e	$=(s_{11}+s_{12})/s_{11}$, or eccentricity of an ellipse {57, 210}
E, v, G	Young's modulus, Poisson's ratio and shear modulus of an isotropic material, respectively {21}
$E(e)$	Complete elliptical integral of the second kind {211}
$E(\nabla)$	Operator matrix defined in Eq. (1.2.6) {6}
E_i	Elastic (Young's) modulus in the i-direction {13}
$f, \overline{f}$	$=f_x+\mathrm{i}f_y$, $f_x-\mathrm{i}f_y$, respectively, the complex frictional coefficients

	{222}
f_1, f_2, f_3, f_4	$=c_{11}/c_{44},\ c_{12}/c_{44}, c_{13}/c_{44},\ c_{33}/c_{44}$, respectively, the dimensionless elastic constants {322}
$f_5,\ f_6$	$=kb/c_{44},\ K/(c_{44}h)$, respectively, the dimensionless parameters of the Pasternak foundation model {354, 355}
$f_x,\ f_y$	Frictional coefficients in x- and y- directions {219}
$\overline{f}(\alpha,\beta)$	Double Fourier transform of function $f(x,y)$ {124}
F, H	Stress functions for problems of axisymmetric deformation {55, 57}
F_i	Components of body force per unit volume in Cartesian coordinates ($i = x, y, z$), cylindrical coordinates ($i = r, \alpha, z$), or spherical coordinates ($i = R, \theta, \alpha$) {7}
G_{ij}	Shear modulus in the i-j plane {13}
h	The z-coordinate of the point where a concentrated load acts, or the thickness of a plate, a cylindrical shell or a spherical shell {80, 148, 247, 284, 331}
h_i	Thickness of the i-th layer of a laminated plate or shell {272, 314, 367}
$h_i\ (i=1,2,3)$	$=s_i h$ {80}
$h_n^{(1)}(x),\ h_n^{(2)}(x)$	Spherical Hankel functions of the first and second kinds, respectively, of order n {345}
H	Thickness of an elastic layer, or a material parameter defined in Eq. (7.1.37) {134, 215}
H_i	Thickness of the i-th layer of a multi-layered elastic medium {142}
$H_\nu^{(1)}(x),\ H_\nu^{(2)}(x)$	Hankel functions of the first and second kinds, respectively, of order ν {394, 395}
$i_n(x),\ k_n(x)$	Modified spherical Bessel functions of the first and second kinds, respectively, of order n {396}
$I_n(x),\ K_n(x)$	Modified Bessel functions of the first and second kinds respectively, of order n {303}
$j_n(x),\ y_n(x)$	Spherical Bessel functions of the first and second kinds, respectively, of order n {345}
$J_n(x),\ Y_n(x)$	Bessel functions of the first and second kinds respectively, of order n {189, 286}
k	Wave number, elastic constant of the Winkler foundation model, or the first parameter of the Pasternak foundation model {322, 357, 359}
k_4, k_5, k_6	Dimensionless material parameters defined in Eq. (2.2.67) {50}
$k_{ji}\ (j=1,2,3; i=1,2)$	Dimensionless material parameters defined in Eq. (2.2.64) {49}
$k_x,\ k_z$	Coefficients of thermal conduction in x- and z-directions,

	respectively {186}
$\tilde{k}$	Spring constant in the elastic foundation model {359}
K	The second parameter of the Pasternak foundation model {359}
$K(e)$	Complete elliptical integral of the first kind {211}
$K_{ij}\ (i,j=1,2)$	Principal curvatures of a curved surface {206}
K_{I}	Stress intensity factor of type I {199}
$l_{p'i}\ (p',i=1,2,3)$	Direction cosines between one axis of the (x,y,z)-coordinates and one axis of the (x',y',z')-coordinates {3}
L	Length of a cylindrical shell {284}
L_0, L_s	Operators defined in Eq. (2.2.26) and (2.2.41), respectively {42, 45}
L_1, L_2, L_3, L_4	Operators defined in Eq. (2.4.8) {64}
L_5	$=2-\beta_3/\beta_1-(\beta_3/\beta_1)\nabla_2$ {200}
L_6	$=L_4\nabla_1^2+L_5L_2$ {200}
L_7	$=L_3+L_5L_1$ {200}
L^*	Operator defined in Eq. (2.4.22) {66}
m_1, m_2	$=c_{44}/c_{66}$, c_{12}/c_{11}, respectively {289, 293}
$\mathbf{M}$	$=M_x\mathbf{i}+M_y\mathbf{j}+M_z\mathbf{k}$, concentrated moment applied at the apex of a cone (conical shell) {167}
$\vec{n}$	Unit external normal vector {4}
$N(z)$	Step function, $=0$ when $z<0$; $=1$ when $z\geq 0$ {129}
p_x, p_y, p_z	Components of surface traction in Cartesian coordinates {4}
P	Concentrated force in the z-direction {71}
P, Q	$=D-3M$, $8(3N-L)$, respectively, dimensionless material parameters {103}
$P(n)$	$=[2D-Mn(n+1)]/2$ {203}
$P_n(x)$	Legendre polynomial, of order n {100}
$P_v(x)$, $Q_v(x)$	Legendre functions of the first and second kinds, respectively, of order v {389}
$P_v^m(x)$, $Q_v^m(x)$	Associated Legendre functions of the first and second kinds, respectively, of order v {384}
$\mathbf{P}$	$=p_x\mathbf{i}+p_y\mathbf{j}+p_z\mathbf{k}$, concentrated force applied at the apex of a cone (conical shell) {167}
$Q(n)$	$=(n+2)(n-1)[Nn(n+1)-2L]$ {203}
r^2	$=x^2+y^2$ {71}

r_0, r_1	Inner and outer radii of an annular plate {148}
R, R_0	$=(a+b)/2$, mean radii of a cylindrical shell and spherical shell, respectively {284, 331}
R_i $(i=1,2,3)$	$=\sqrt{r^2+z_i^2}$ {71}
R_{ij} $(i,j=1,2,3)$	$=\sqrt{x^2+y^2+z_{ij}^2}$ {80}
R_i^* $(i=1,2,3)$	$=R_i+s_i\|z\|$ {71}
R'_{ij} $(i,j=1,2,3)$	$=\sqrt{x^2+y^2+z'^2_{ij}}$ {80}
$\overline{R}_{ij}$ $(i,j=1,2,3)$	$=\sqrt{x^2+y^2+\overline{z}_{ij}^2}$ {80}
s_1, s_2, s_3	Eigenvalues of a material, given in Eqs. (2.2.44), (2.2.45) and (2.2.59) {45, 48}
s_{ij}	Elastic compliances {10}
s_{ijkl}	Components of the elastic compliance tensor of rank four {21}
$\mathrm{sign}(z)$	Sign function: $=1$ when $z\geq 0$; $=-1$ when $z<0$ {71}
$S_n^m(\theta,\alpha)$	Spherical harmonics {166}
t	Time variable {8}
t_1, t_2	$=1\mp t_r/2$, inner (outer) radius-to-mean radius ratio of a cylindrical shell; $=1\mp t_R/2$, that of a spherical shell {289, 334}
t_L	$=m\pi R/L$ or $m\pi a/L$ {289, 313}
t_r	$=h/R$, thickness-to-mean radius ratio of a cylindrical shell {289}
t_R	$=h/R_0$, thickness-to-mean radius ratio of a spherical shell {331}
t_α	$=n\pi/\alpha_0$ {313}
T	Temperature, or concentrated force in the x direction {25, 71}
u, v, w	Displacement components in Cartesian coordinates {1}
u_1, u_2, u_3	Components of displacement vector. {1}
u_r, u_α, w	Displacement components in Cylindrical coordinates {1}
u_R, u_θ, u_α	Displacement components in Spherical coordinates {1}
v_1, v_2, v_3	$=\sqrt{c_{11}/\rho}$, $\sqrt{c_{44}/\rho}$, $\sqrt{c_{66}/\rho}$, respectively, velocities of various waves {292, 297, 286}
v_f	Fluid velocity in the radial direction {317}
v_r, v_s	$=\partial u_r/\partial t$, $\partial u_R/\partial t$, respectively, radial velocity of a material point in a cylindrical shell and spherical shell {317, 347}

w, ψ, G	Displacement functions for transversely (spherically) isotropic bodies {39}
W	Strain energy density {10}
$z_i \quad (i=1,2,3)$	$=s_i z$ {47}
$z_{ij} \quad (i,j=1,2,3)$	$=z_i+h_j$ {80}
$z_i' \quad (i=1,2,3)$	$=s_i' z$ {80}
$z_{ij}' \quad (i,j=1,2,3)$	$=z_i'-h_j$ {80}
$\bar{z}_{ij} \quad (i,j=1,2,3)$	$=z_i-h_j$ {80}
α_0	Central angle of a cylindrical panel {309}
$\alpha_1, \alpha_2, \alpha_3=-\alpha_4$	Dimensionless material parameters defined in Eqs. (2.2.57) and (2.2.62) {47, 48, 74}
α_{ij} (or α)	Thermal expansion coefficient of an anisotropic (or isotropic) body {25, 27}
β_0, β_1	External and internal conical angles of a hollow cone {159}
β_{ij} (or β_j or β)	Thermal modulus of an anisotropic (or orthotropic or isotropic) body {25, 27}
$\gamma_{ij}=2\varepsilon_{ij} \quad (i \neq j)$	Engineering shear strains {2}
δ_{ij}	Kronecker delta, *i.e.* $=1$ for $i=j$, and $=0$ for $i \neq j$ {197}
$\Delta, \bar{\Delta}$	$=\partial/\partial x+\mathrm{i}\partial/\partial y$, $\partial/\partial x-\mathrm{i}\partial/\partial y$, respectively {50}
ε	Damping factor {350}
ϑ_4	$=c_{33}s_1(\alpha_1+\alpha_4)$ {74}
$\vartheta_i \quad (i=1,2)$	$=c_{33}\alpha_i s_i-c_{13}$ {72}
$\varphi_1, \varphi_2, \varphi_3$	Displacement functions {47}
λ, μ	Laméconstants {20}
Λ	$=\frac{\partial^2}{\partial x^2}+\frac{\partial^2}{\partial y^2}=\frac{\partial^2}{\partial r^2}+\frac{1}{r}\frac{\partial}{\partial r}+\frac{1}{r^2}\frac{\partial^2}{\partial \alpha^2}$, two-dimensional harmonic (Laplace) operator {40, 52}
$\tilde{\mu}$	Shear modulus of the elastic medium {359}
ν_{ij}	Poisson's ratio, *i.e.* the negative of the transverse strain in the j-direction over the strain in the i-direction when stress is applied in the i-direction {13}
$\xi_i \quad (i=1,2)$	$=c_{13}\alpha_i s_i-c_{12}$ {72}
ρ	Density of material {7}
ρ_0	$=\rho_f/\rho$, density ratio between fluid and solid media {317}

ρ_f Density of fluid {317}

ρ_m, ν_m Density and Poisson's ratio of the surrounding elastic medium, respectively {321, 324}

σ_i, ε_i Normal stresses and strains in Cartesian coordinates for $i = x, y, z$, in cylindrical coordinates for $i = r, \alpha, z$, or in spherical coordinates for $i = R, \theta, \alpha$ {2}

σ_{ij}, ε_{ij} Components of stress and strain tensor, respectively {3}

τ_{ij}, γ_{ij} $(i \neq j)$ Shear stresses and strains in Cartesian coordinates for $i, j = x, y, z$, in cylindrical coordinates for $i, j = r, \alpha, z$, or in spherical coordinates for $i, j = R, \theta, \alpha$ {2}

ω Angular velocity or circular frequency {154, 249}

ω_i $(i = 1,2)$ $= c_{44}(s_i + \alpha_i)$ {72}

ω_3 $= c_{44}s_3$ {76}

ω_4 $= c_{44}(s_1 + \alpha_4)$ {84}

ω_s $= v_2 / R$ {302}

Ω Dimensionless frequency {249}

$\bar{\Omega}$ Real part of the dimensionless complex frequency {350}

∇^2 Three-dimensional harmonic (Laplace) operator:

$$= \frac{\partial^2}{\partial x^2} + \frac{\partial^2}{\partial y^2} + \frac{\partial^2}{\partial z^2} \quad \text{(Cartesian coordinates)}$$

$$= \frac{1}{r}\frac{\partial}{\partial r}\left(r\frac{\partial}{\partial r}\right) + \frac{1}{r^2}\frac{\partial^2}{\partial \alpha^2} + \frac{\partial^2}{\partial z^2} \quad \text{(cylindrical coordinates)}$$

$$= \frac{1}{R}\frac{\partial}{\partial R}\left(R^2\frac{\partial}{\partial R}\right) + \frac{1}{R^2 \sin\theta}\left(\sin\theta\frac{\partial}{\partial \theta}\right) + \frac{1}{R^2 \sin^2\theta}\frac{\partial^2}{\partial \alpha^2}$$

(spherical coordinates)

{316, 345}

∇_2 $= R\frac{\partial}{\partial R}$ {33}

∇_2^2 $= R\frac{\partial}{\partial R}R\frac{\partial}{\partial R}$ {33}

∇_3^2 $= \nabla_2^2 + \nabla_2$ {33}

∇_1^2 $= \frac{\partial^2}{\partial \theta^2} + \cot\theta\frac{\partial}{\partial \theta} + \frac{1}{\sin^2\theta}\frac{\partial^2}{\partial \alpha^2}$, surface spherical harmonic operator {33}

(r, α, z) Cylindrical coordinates {1}

(R, θ, α)	Spherical coordinates {1}
(x, y, z)	Cartesian coordinates {1}

REFERENCES[1]

Ahmad, F. and Khan, A. (2001). Effect of rotation on wave propagation in transversely isotropic medium, *Mathematical Problems in Engineering*, **7**, 147-154. *22*

Ahmad, F. and Rahman, A. (2000). Acoustic scattering by transversely isotropic cylinders, *International Journal of Engineering Science*, **38**, 325-335. *383*

Ahmad, F., Kiyani, N., Yousaf, F. and Shams, M. (2002). Guided waves in a fluid-loaded transversely isotropic plate, *Mathematical Problems in Engineering*, **8**, 151-159. *22*

Almansi, E. (1899). Sull' integrazione dell' equazione differenziale $\nabla^{2n}=0$, *Annali di Matematica Pura ed Applicata, Ser. 3*, **2**, 1-51. *46*

Ambartsumian, S. A. (1964). *Theory of Anisotropic Shells*. NASA Technical Translation, TTF-118. *382*

Ashida, F., Noda, N. and Okumura, I. (1993). General solution technique for transient thermoelasticity of transversely isotropic solids in cylindrical coordinates, *Acta Mechanica*, **101**, 215-230. *377*

Ashton, J. E. and Whitney, J. M. (1970). *Theory of Laminated Plates*. Stamford: Technomic. *381*

Bahar, L. Y. (1972). Transfer matrix approach to layered systems, *Journal of Engineering Mechanics*, **98**, 1159-1172. *33*

Bahar, L. Y. (1977). Discussion on "A mixed method in elasticity", *Journal of Applied Mechanics*, **44**, 790-791. *33*

Baltrukonis, J. H. and Gottenberg, W. G. (1959). Thickness-shear vibrations of circular bars, *Journal of the Acoustical Society of America*, **31**, 734-739. *296*

Bank, C. G. (2002). *Seismic Measurements Using Ultrasound on Gneiss Samples*. Ph.D. Dissertation, Vancouver: University of British Columbia. *22*

Barber, J. R. (1992). *Elasticity*. Dordrecht: Kluwer Academic Publishers. *373*

Behrens, E. (1971). Elastic constants of fiber-reinforced composite with transversely isotropic constituents, *Journal of Applied Mechanics*, **38**, 1062-1065. *22*

Bellman, R. (1970). *Introduction to Matrix Analysis*. New York: McGraw-Hill. *272, 314, 368*

[1] The italic numbers attached to each reference represent the pages on which the reference was cited.

Benitez, F. G. and Rosakis, A. J. (1987). Three-dimensional elastostatics of a layer and a layered medium, *Journal of Elasticity*, **18**, 3-50. *139, 376*

Bharatha, S. and Levinson, M. (1980). A theory of elastic foundations, *Archive for Rational Mechanics and Analysis*, **74**, 249-266. *384*

Bhaskar, K. and Ganapathysaran, N. (2002). Elasticity solution for a laminated orthotropic cylindrical shell subjected to a localized longitudinal shear force, *Journal of Applied Mechanics*, **69**, 700-703. *382*

Bickford, W. B. (1977). A generalization of the Boussinesq-Papkovich functions to an anisotropic media, *Zeitschrift für angewandte Mathematik und Physik*, **28**, 723-728. *374*

Birman, V. (1995). On three-dimensional state of thermal stresses in a transversely isotropic plate with a circular hole, *International Journal of Engineering Science*, **33**, 95-103. *377*

Boley, B. A. and Weiner, J. H. (1960). *Theory of Thermal Stresses*. New York: John Wiley. *373*

Borodachev, A. N. (1988). Parametric class of solutions of thermoelasticity equations for transversally isotropic bodies, *Soviet Applied Mechanics*, **24**, 678-682. *377*

Borodich, F. M. (1993). The Hertz frictional contact between nonlinear elastic anisotropic bodies (the similarity approach), *International Journal of Solids and Structures*, **30**, 1513-1526. *380*

Buchwald, V. T. (1961). Rayleigh waves in transversely isotropic media, *Quarterly Journal of Mechanics and Applied Mathematics*, **14**, 293-317. *382, 383*

Bufler, H. (1971). Theory of elasticity of a multilayered medium, *Journal of Elasticity*, **1**, 125-143. *376*

Bufler, H. (1998). The arbitrarily and the periodically laminated elastic hollow sphere: Exact solutions and homogenization, *Archive of Applied Mechanics*, **68**, 579-588. *19*

Burmister, D. M. (1945*a*). The general theory of stresses and displacements in layered systems. I, *Journal of Applied Physics*, **16**, 89-93. *376*

Burmister, D. M. (1945*b*). The general theory of stresses and displacements in layered soil systems. II, *Journal of Applied Physics*, **16**, 126-127. *376*

Calladine, C. R. (1983). *Theory of Shell Structures*. Cambridge: Cambridge University Press. *382*

Cai, J. B., Chen, W. Q., Ye, G. R. and Ding, H. J. (1999). On natural frequencies of a transversely isotropic cylindrical panel on a Kerr foundation, *Journal of Sound and Vibration*, **232**, 997-1004. *383*

Celep, Z. (1978). On the axially symmetric vibration of thick circular plates,

Ingenieur-Archive, **47**, 411-420. *275*
Celep, Z. (1980). Free vibration of some circular plates of arbitrary thickness,*Journal of Sound and Vibration*, **70**, 379-388. *275*
Chan, A. K. and Tsang, L. (1983). Propagation of acoustic waves in a fluid-filled borehole surrounded by a concentrically layered transversely isotropic formation, *Journal of the Acoustical Society of America*, **74**, 1605-1616. *382*
Chao, C. C. and Chern, Y. C. (1988). Axisymmetric free vibration of orthotropic complete spherical shells, *Journal of Composite Materials*, **22**, 1116-1130. *383*
Chau, K. T. (1994). Vibrations of transversely isotropic finite circular cylinders, *Journal of Applied Mechanics*, **61**, 964-970. *285, 303, 304, 382*
Chen, W. Q. (1995). *Coupled Free Vibrations of Spherically Isotropic Hollow Spheres*. Ph.D. Dissertation, Hangzhou: Zhejiang University. (in Chinese) *64, 330, 344, 364, 384*
Chen, W. Q., Cai, J. B., Ye, G. R. and Ding, H. J. (1998). 3-D exact free vibration analysis of transversely isotropic cylindrical panels, *Journal of Vibration and Acoustics*, **120**, 982-986. *310, 382*
Chen, W. Q. and Ding, H. J. (1999*a*). Natural frequencies of a fluid-filled anisotropic spherical shell, *Journal of the Acoustical Society of America*, **105**, 174-182. *356*
Chen, W. Q. and Ding, H. J. (1999*b*). Natural frequencies of fluid-filled transversely isotropic cylindrical shells, *International Journal of Mechanical Sciences*, **41**, 677-684. *320, 383*
Chen, W. Q. and Ding, H. J. (2001*a*). A state-space based stress analysis of multilayered spherical shell with spherical isotropy, *Journal of Applied Mechanics*, **68**, 109-114. *364, 385*
Chen, W. Q. and Ding, H. J. (2001*b*). Free vibration of multi-layered spherically isotropic hollow spheres, *International Journal of Mechanical Sciences*, **43**, 667-680. *313, 364, 371, 385*
Chen, W. Q. and Ding, H. J. (2003). Three-dimensional general solution of transversely isotropic thermoelasticity and the potential theory method, *Acta Mechanica Sinica*, **35**, 578-583. (in Chinese) *189, 378*
Chen, W. Q., Ding, H. J., Guo, Y. M. and Yang, Q. D. (1997). Free vibrations of fluid-filled orthotropic cylindrical shells, *Journal of Engineering Mechanics*, **123**, 1130-1133. *382*
Chen, W. Q., Ding, H. J. and Ling, D. S. (2004). Thermoelastic field of a transversely isotropic elastic medium containing a penny-shaped crack: Exact fundamental solution, *International Journal of Solids and Structures*, **41**, 69-83. *378*
Chen, W. Q., Ding, H. J. and Xu, R. Q. (1998). On exact analysis of free vibrations of

embedded transversely isotropic cylindrical shells, *International Journal of Pressure Vessels and Piping*, **75**, 961-966. *383*

Chen, W. Q., Ding, H. J. and Yu, R. J. (1996). Exact analysis of laminated orthotropic plates on Pasternak foundation, In: *Proceedings from the International Conference on Advances in Steel Structures* (edited by Chan, S. L. and Teng, J. G.), Vol. **2**, pp. 1092-1097. New York: Elsevier Ltd. *273, 381*

Chen, W. Q., Wang, X. and Ding, H. J. (1999). Free vibration of a fluid-filled hollow sphere of a functionally graded material with spherical isotropy, *Journal of the Acoustical Society of America*, **106**, 2588-2594. *356*

Chen, W. Q., Ying, J. and Yang, Q. D. (1998). Free vibrations of transversely isotropic cylinders and cylindrical shells, *Journal of Pressure Vessel Technology*, **120**, 321-324. *306, 383*

Chen, W. T. (1964). Displacement discontinuity over a transversely isotropic elastic half-space, *IBM Journal*, **8**, 435-442. *375*

Chen, W. T. (1965). Stresses in a transversely isotropic elastic cone under an asymmetric force at its vertex, *Zeitschrift für angewandte Mathematik und Physik*, **16**, 337-344. *377*

Chen, W. T. (1966*a*). On some problems in spherically isotropic elastic materials, *Journal of Applied Mechanics*, **33**, 539-546. *99, 374*

Chen, W. T. (1966*b*). On some problems in transversely isotropic elastic materials, *Journal of Applied Mechanics*, **33**, 347-355. *374*

Chen, W. T. (1968). Axisymmetric stress field around spheroidal inclusions and cavities in a transversely isotropic material, *Journal of Applied Mechanics*, **35**, 770-773. *93, 98, 375*

Chen, W. T. (1969). Stresses in some anisotropic materials due to indentation and sliding, *International Journal of Solids and Structures*, **5**, 191-294. *224, 379, 380*

Chen, W. T. (1971). Elastic analysis of an axisymmetric stress field perturbed by a spheroidal inhomogeneity, *Quarterly of Applied Mathematics*, **28**, 517-525. *375*

Chen, W. T. (1977). Discussion on "Point force solution for an infinite transversely isotropic solid", *Journal of Applied Mechanics*, **44**, 514-515. *375*

Chien, W. Z. and Yeh, K. Y. (1956). *Elasticity*. Beijing: Science Press. (in Chinese) *373*

Choi, H. J. and Thangjitham, S. (1991). Stress analysis of multilayered anisotropic elastic media, *Journal of Applied Mechanics*, **58**, 382-387. *377*

Chree, C. (1889). The equations of an isotropic elastic solid in polar and cylindrical coordinates, their solutions and applications, *Transactions of the Cambridge Philosophical Society*, **14**, 250-369. *301*

Cohen, H., Shah, A. H. and Ramakrishnan, C. V. (1972). Free vibrations of a spherically isotropic hollow sphere, *Acustica*, **26**, 329-333. *329, 337, 383*

Conway, H. D. and Farnham, K. A. (1967). Shear stresses induced on an axis of symmetry by circular sliding contact on a transversely isotropic body, *Journal of Applied Mechanics*, **34**, 756-757. *379*

Conway, H. D., Farnham, K. A. and Ku, T. C. (1967). The indentation of a transversely isotropic half space by a rigid sphere, *Journal of Applied Mechanics*, **34**, 491-492. *379*

Courant, R. and Hilbert, D. (1953). *Methods of Mathematical Physics*. New York: John Wiley. *387*

Dahan, M. and Zarka, J. (1977). Elastic contact between a sphere and a semi infinite transversely isotropic body, *International Journal of Solids and Structures*, **13**, 229-238. *376, 380*

Ding, H. J. (1987). On the stress function of axisymmetric deformation in a solid of revolution, *Shanghai Journal of Mechanics*, **8**(1), 42-48. (in Chinese) *54*

Ding, H. J., Chen, B. and Liang, J. (1996). General solutions for coupled equations for piezoelectric media, *International Journal of Solids and Structures*, **33**, 2283-2298. *39, 41, 46*

Ding, H. J. and Chen, W. Q. (1995). Moderately thick shell theory analysis of free vibrations of submerged spherically isotropic spherical shells, *Shanghai Journal of Mechanics*, **16**, 1-12. (in Chinese) *344, 353*

Ding, H. J. and Chen, W. Q. (1996*a*). Nonaxisymmetric free vibrations of a spherically isotropic spherical shell embedded in an elastic medium, *International Journal of Solids and Structures*, **33**, 2575-2590. *344, 364, 374*

Ding, H. J. and Chen, W. Q. (1996*b*). Natural frequencies of an elastic spherically isotropic hollow sphere submerged in a compressible fluid medium, *Journal of Sound and Vibration*, **192**, 173-198. *351, 374*

Ding, H. J. and Chen, W. Q. (1998). Exact shell theory analysis of free vibrations of submerged thin spherical shells, *International Journal of Solids and Structures*, **35**, 4381-4389. *350, 353*

Ding, H. J., Chen, W. Q. and Liu, Z. (1995). Solutions to equations of vibrations of spherical and cylindrical shells, *Applied Mathematics and Mechanics*, **16**, 1-15. *330*

Ding, H. J., Chen, W. Q. and Xu, R. Q. (2001). On the bending, vibration and stability of laminated rectangular plates with transversely isotropic layers, *Applied Mathematics and Mechanics*, **22**, 17-24. *33, 268, 275, 382*

Ding, H. J. and Chi, Y. W. (1999). Fundamental solutions for axisymmetric problems of transversely isotropic materials, *Chinese Journal of Applied Mechanics*, **16**(4), 114-118. (in Chinese) *375*

Ding, H. J. *et al.* (1997). *Transversely Isotropic Elasticity*. Hangzhou: Zhejiang University Press. (in Chinese) *373*

Ding, H. J. and Liang, J. (2001). Fundamental solutions for axisymmetric problems of semi-infinite transversely isotropic media, *Journal of Zhejiang University*, **35**, 233-237. (in Chinese) *375*

Ding, H. J., Liang, J. and Chen, B. (1997). The united point force solution for both isotropic and transversely isotropic media, *Communications in Numerical Methods in Engineering*, **13**, 95-102. *78, 375*

Ding, H. J., Liang, J. and Wang, Y. (1996). Point force solution for a transversely isotropic elastic layer, *Applied Mathematics and Mechanics*, **17**, 309-318. *376*

Ding, H. J. and Ren, Y. J. (1990). Thermal stress problem of spherically isotropic bodies, *Journal of Shanghai Jiaotong University*, **24**(5-6), 1-6. (in Chinese) *374, 378*

Ding, H. J. and Ren, Y. J. (1991). Equilibrium problems of spherically isotropic bodies, *Applied Mathematics and Mechanics*, **12**, 155-162. *374*

Ding, H. J., Ren, Y. J., Zou, D. Q. and Chen, W. Q. (1994). Displacement method of elasticity problems in spherically isotropic media, *Acta Mechanica Sinica*, **26**, 186-197. (in Chinese) *374*

Ding, H. J., Wang, H. M. and Chen, W. Q. (2001). A theoretical solution of cylindrically isotropic cylindrical tube for axisymmetric plane strain dynamic thermoelastic problem, *Acta Mechanica Solida Sinica*, **14**, 357-363. *382*

Ding, H. J., Wang, H. M. and Chen, W. Q. (2002). Analytical thermo-elastodynamic solutions for a nonhomogeneous transversely isotropic hollow sphere, *Archive of Applied Mechanics*, **72**, 545-553. *384*

Ding, H. J. and Xu, B. H. (1988). General solutions of axisymmetric problems in transversely isotropic body, *Applied Mathematics and Mechanics*, **9**, 143-151. *54, 62*

Ding, H. J. and Xu, R. Q. (2000*a*). Exact solutions for free vibrations of transversely isotropic circular plates, *Acta Mechanica Solida Sinica*, **13**, 105-111. *381*

Ding, H. J. and Xu, R. Q. (2000*b*). Free axisymmetric vibration of laminated transversely isotropic annular plates, *Journal of Sound and Vibration*, **230**, 1031-1044. *280*

Ding, H. J., Xu, R. Q. and Guo, F. L. (1999). Exact axisymmetric solutions for laminated transversely isotropic piezoelectric circular plate (II) — Exact solutions for elastic circular plates and numerical results, *Science in China (E)*, **42**, 471-478. *261, 381*

Ding, H. J., Xu, R. Q., Chen, W. Q. and Chi, Y.W.(1998). Free axisymmetric vibration of transversely isotropic laminated circular plates, *Acta Mechanica Solida Sinica*, **11**, 209-215. *275*

Ding, H. J. and Xu, X. (1982). The equilibrium of transversely isotropic elastic layers, *Journal of Zhejiang University*, **16**(2), 141-154. (in Chinese) *376*

Ding, H. J., Zou, D. Q. and Ren, Y. J. (1995). Elasticity solutions of spherically isotropic cones under concentrated loads at apex, *Applied Mathematics and Mechanics*, **16**, 1025-1038. *377*

Ding, Z. Y., Zou, D. Q. and Ding, H. J. (1996). A study of effects of the Earth's radial anisotropy on the tidal stress field, *Tectonophysics*, **258**, 103-114. *15, 384*

Dong, S. B. (1968). Free vibration of laminated orthotropic cylindrical shells,*Journal of the Acoustical Society of America*, **44**, 1628-1635. *382*

Dong, S. B. and Tso, F. K. W. (1972). On a laminated orthotropic shell theory including transverse shear deformation, *Journal of Applied Mechanics*, **39**, 1091-1097. *382*

Du, Q. H., Yu, S. W. and Yao, Z. H. (1986). *Theory of Elasticity*. Beijing: Science Press. (in Chinese) *373*

Dundurs, J. and Hetényi, M. (1965). Transmission of force between two semi-infinite solids, *Journal of Applied Mechanics*, **32**, 671-674. *375*

Eason, G. (1964). Transient thermal stresses in anisotropic bodeis with spherical symmetry, *Applied Scientific Research*, **A13**, 1-15. *378*

Elliott, H. A. (1948). Three-dimensional stress distributions in hexagonal aeolotropic crystals, *Proceedings of the Cambridge Philosophical Society*, **44**, 522-553. *48, 374, 379*

Elliott, H. A. (1949). Axial symmetric stress distributions in aeolotropic hexagonal crystals: the problem of the plane and related problems, *Proceedings of the Cambridge Philosophical Society*, **45**, 621-630. *226, 231, 379*

Eringen, A. C. and Suhubi, E. S. (1975). *Elastodynamics, Vol. 2: Linear Theory*. New York: Academic Press. *25, 373*

Eshelby, J. D. (1961). Elastic inclusions and inhomogeneities, In: *Progress in Solid Mechanics* (edited by Sneddon, I. N. and Hill, T.), Vol. **2**, pp. 89-140. Amsterdam: North Holland. *94*

Eubanks, R. A. and Sternberg, E. (1953). On the axisymmetric problem of elasticity theory for a medium with transverse isotropy, *Journal of Rational Mechanics and Analysis*, **3**, 89-101. *46, 61, 62, 373*

Fabrikant, V. I. (1988). Elastic field around a circular punch, *Journal of Applied Mechanics*, **55**, 604-610. *218, 226-228, 236, 237*

Fabrikant, V. I. (1989). *Applications of Potential Theory in Mechanics: A Selection of New Results*. Dordrecht: Kluwer Academic Publishers. *195, 197, 216, 219, 380*

Fabrikant, V. I. (1991). *Mixed Boundary Value Problems of Potential Theory and Their Applications in Engineering*. Dordrecht: Kluwer Academic Publishers. *216, 219, 380*

Fabrikant, V. I. (2004). A new form of the Green function for a transversely isotropic body, *Acta Mechanica*, **167**, 101-111. *78, 375*

Fan, H. and Keer, L. M. (1994). Two-dimensional contact on an anisotropic elastic half-space, *Journal of Applied Mechanics*, **61**, 250-255. *380*

Fan, J. R. and Ding, K. W. (1995). The exact solution for thick laminated cantilever cylindrical shell, *Applied Mathematics and Mechanics*, **16**, 143-159. *382*

Fan, J. R. and Ye, J. Q. (1990*a*). An exact solution for the statics and dynamics of laminated thick plates with orthotropic layers, *International Journal of Solids and Structures*, **26**, 655-662. *33, 268, 314, 381*

Fan, J. R. and Ye, J. Q. (1990*b*). A series solution of the exact equation for thick orthotropic plates, *International Journal of Solids and Structures*, **26**, 773-778. *33, 268, 314, 381*

Fan, J. R. and Ye, J. Q. (1990*c*). Exact solutions for axisymmetric vibration of laminated circular plates, *Journal of Engineering Mechanics*, **116**, 920-927. *275*

Fan, J. R. and Zhang, J. Y. (1992). Exact solutions for thick laminated shells, *Science in China (A)*, **35**, 1343-1355. *310, 313, 367*

Feng, K. and Qin, M. Z. (1991). Hamiltonian algorithms for Hamiltonian dynamical systems, *Progress in Natural Science*, **1**, 105-116. *33*

Flügge, W. (1972). *Tensor Analysis and Continuum Mechanics*. Berlin: Springer-Verlag. *373*

Flügge, W. (1990). *Stresses in Shells* (2nd ed.). Berlin: Springer-Verlag. *382*

Fung, Y. C. (1965). *Foundations of Solid Mechanics*. New York: Prentice-Hall. *373*

Galin, L. A. (1961). *Contact Problems in Theory of Elasticity*. North Carolina State College. *379*

Gardos, M. N. (1990). On the elastic constants of thin solid lubricant films, In: *Mechanics of Coatings* (edited by Dowson, D., Taylor, C.M. and Godet, M.), pp. 3-13. Amsterdam: Elsevier. *22*

Gladwell, G. M. L. (1978). Polynomial solutions for an ellipse on an anisotropic elastic half-space, *Quarterly Journal of Mechanics and Applied Mathematics*, **31**, 251-260. *380*

Gladwell, G. M. L. (1980). *Contact Problems in the Classical Theory of Elasticity*. Alphen aan den Rijn: Sijthoff and Noordhoff. *216, 379*

Gonçalves, P. B. and Batista, R. C. (1988). Non-linear vibration analysis of fluid-filled cylindrical shell, *Journal of Sound and Vibration*, **127**, 133-143. *326*

Goodman, L. E. and Keer, L. M. (1975). Influence of an elastic layer on the tangential compliance of bodies in contact, In: *The Mechanics of the Contact between Deformable Bodies* (edited by De Pater, A. D. and Kalker, J. J.), pp. 127-151. Delft: Delft University Press. *224*

Green, A. E. and Zerna, W. (1954). *Theoretical Elasticity*. Oxford: Clarendon Press. *373, 379, 380*

Grilitskii, D. V. and Shelestovskii, B. G. (1970). The axisymmetric contact problem of thermoelasticity for a transversely isotropic half-space, *Prikladnaya Mekhanika*, **6**(8), 3-8. (in Russian) *379*

Guo, D. R. (1991). *Methods of Mathematical Physics*. Beijing: Higer Education Press. (in Chinese) *387, 391*

Guz', A. N. (1980*a*). Wave propagation in a cylindrical shell containing a viscous compressible liquid, *Soviet Applied Mechanics*, **16**, 842-850. *384*

Guz', A. N. (1980*b*). Hydroelasticity problems for viscous compressible liquids in a spherical coordinate system, *Soviet Applied Mechanics*, **16**, 937-943. *384*

Guz', A. N. (1981). Dynamics of rigid bodies in a compressible viscous fluid (quiescent liquid), *Soviet Applied Mechanics*, **17**, 207-223. *384*

Guz', A. N. (1993*a*). Limiting cases of the dynamics of rigid and elastic bodies in a compressible viscous liquid, *International Applied Mechanics*, **29**, 569-573. *384*

Guz', A N. (1993*b*). Vibrations of a sphere in the flow of a compressible viscous fluid, *International Applied Mechanics*, **29**, 959-968. *384*

Hamilton, G. M. and Goodman, L. E. (1966). The stress field created by a circular sliding contact, *Journal of Applied Mechanics*, **33**, 371-376. *379*

Hanson, M. T. (1992*a*). The elastic field for conical indentation including sliding friction for transverse isotropy, *Journal of Applied Mechanics*, **59**, S123-S130. *216, 231-234, 380*

Hanson, M. T. (1992*b*). The elastic field for spherical Hertzian contact including sliding friction for transverse isotropy, *Journal of Tribology*, **114**, 606-611. *216, 222, 224, 380*

Hanson, M. T. (1993). The elastic field for a sliding conical punch on an isotropic half-space, *Journal of Applied Mechanics*, **60**, 557-559. *216, 380*

Hanson, M. T. (1994). The elastic field for an upright or tilted sliding circular flat punch on a transversely isotropic half space, *International Journal of Solids and Structures*, 1994, **31**, 567-586. *216, 220, 229, 239, 242, 380*

Hanson, M. T. and Johnson, T. (1993). The elastic field for spherical Hertzian contact of isotropic bodies revisited: Some alternative expressions, *Journal of Tribology*, **115**, 327-332. *216, 380*

Hanson, M. T. and Wang, Y. (1997). Concentrated ring loadings in a full space or half space: Solutions for transverse isotropy and isotropy, *International Journal of Solids and Structures*, **34**, 1379-1418. *375*

Hart, D. J. (2000). *Laboratory Measurements of Poroelastic Constants and Flow Parameters and Some Associated Phenomena*. Ph.D. Dissertation, Madison: University of Wisconsin-Madison. *22*

Heinrich, S. M. and Wang, J. Y. (1989). Stress interference in a transversely isotropic body under axisymmetric loading, *Journal of Engineering Mechanics*, **115**, 555-577. *375*

Hermon, R. F. S. (1961). *An Introduction to Applied Anisotropic Elasticity*. Oxford: Oxford University Press. *373*

Hertz, H. (1895). Über die Berührung fester elastischer Körper, *Gesammelte Werke*, Bd. **1**, Leipzig. *210*

Hoff, N. J. and Soong, T. C. (1965). Buckling of circular cylindrical shells in axial compression, *International Journal of Mechanical Sciences*, **7**, 489-520. *297*

Hoff, N. J. and Rehfield, L. W. (1965). Buckling of axially compressed circular cylindrical shells at stresses smaller than the classical critical value, *Journal of Applied Mechanics*, **32**, 542-546. *297*

Honarvar, F. and Sinclair, A. N. (1996). Acoustic wave scattering from transversely isotropic cylinders, *Journal of the Acoustical Society of America*, **100**, 57-63. *382, 383*

Hoppmann II, W. H. and Baker, W. E. (1961). Extensional vibrations of elastic orthotropic spherical shells, *Journal of Applied Mechanics*, **28**, 229-237. *383*

Hu, H. C. (1953). On the three-dimensional problems of the theory of elasticity of a transversely isotropic body, *Scientia Sinica*, **2**, 145-151. [First published in Chinese in *Acta Physica Sinica*, **9**, 130-147 (1953)] *44, 374, 377*

Hu, H. C. (1954*a*). On the equilibrium of a transversely isotropic elastic half space, *Scientia Sinica*, **3**, 463-479. [First published in Chinese in *Acta Physica Sinica*, **10**, 239-258 (1954)] *48, 374, 376*

Hu, H. C. (1954*b*). On the general theory of elasticity for a spherically isotropic medium, *Scientia Sinica*, **3**, 247-260. [First published in Chinese in *Acta Physica Sinica*, **10**, 57-70 (1954)] *64, 374*

Hu, H. C. (1956). On the equilibrium and vibration of a transversely isotropic elastic body, *Scientia Sinica*, **5**, 1-18. [First published in Chinese in *Acta Physica Sinica*, **11**, 219-238 (1955)] *44, 374, 381, 382*

Hu, H. C. (1984). *Variational Principles of Theory of Elasticity with Applications*. Beijing: Science Press. *359, 373*

Huang, K. F. and Wang, M. Z. (1991). Fundamental solution of bi-material elastic space, *Science in China (A)*, **34**, 309-315. *375*

Huang, X. M. (1994). Stress analysis of sandwich pavement structure on soft-soil foundation, *Chinese Journal of Geotechnical Engineering*, **16**(1), 69-78. (in Chinese) *377*

Huntington, H. B. (1958). *The Elastic Constants of Crystals*. New York: Academic Press. *98*

Hyer, M. W. and Cooper, D. E. (1986). Stresses and deformation in composite tubes due to a circumferential temperature gradient, *Journal of Applied Mechanics*, **53**, 757-764. *382*

Jain, R. K. (1974). Vibration of fluid-filled, orthotropic cylindrical shells, *Journal of Sound and Vibration*, **37**, 379-384. *23, 307, 308, 319, 320*

Jin, B., Tang, J. C. and Sun, B. N. (1996). The Mindlin's solution for axially symmetric problem of multilayer subsoils, *Computational Structural Mechanics and Applications*, **13**, 187-192. (in Chinese) *377*

Johnson, K. L. (1985). *Contact Mechanics*. Cambridge: Cambridge University Press. *216, 224, 241, 242, 379*

Jones, I. A. (1998). Flügge shell theory and solution for orthotropic cylindrical shells under pinching loads, *Composite Structures*, **42**, 53-72. *382*

Junger, M. C. and Feit, D. (1986). *Sound, Structures, and Their Interaction*. Cambridge: MIT Press. *351, 384*

Kameswara Rao, N. S. V. and Das, Y. C. (1977). A mixed method in elasticity, *Journal of Applied Mechanics*, **44**, 51-56. *33*

Kassir, M. K. and Sih, G. C. (1975). *Three Dimensional Crack Problems, Mechanics of Fracture*, 2. Leyden: Noordhoff International Publishing. *228*

Kausel, E. and Roësset, J. M. (1981). Stiffness matrices for layered soils, *Bulletin of the Seismological Society of America*, **71**, 1743-1761. *376*

Kausel, E. and Seale, S. H. (1987). Static loads in layered halfspaces, *Journal of Applied Mechanics*, **54**, 403-408. *376*

Keer, L. M. and Mowry, D. B. (1979). The stress field created by a circular sliding contact on transversely isotropic spheres, *International Journal of Solids and Structures*, **15**, 33-39. *380*

Kelvin, L. (1848). Note on the integration of the equations of equilibrium of an elastic solid, *Cambridge and Dublin Mathematics Journal*, **3**, 87-89. *74, 78*

Kerr, A. D. (1964). Elastic and viscoelastic foundation models, *Journal of Applied Mechanics*, **31**, 491-498. *357, 384*

Khoma, I. Y. (1998). Stress state of a transversely isotropic spherical composite shell with a circular hole, *International Applied Mechanics*, **34**, 471-480. *19*

Koguchi, H., Ohshima, M. and Yada, T. (1990). Fundamental solution for two-phase transversely isotropic materials, *Transaction of the Japan Society of Mechanical Engineers*, **A56**, 84-92. (in Japanese) *375*

Korolev, V. V. (1992). Stress state of a transversally isotropic half space subjected to local heating, *International Applied Mechanics*, **28**, 347-353. *377*

Kriz, R. D. and Stinchcomb, W. W. (1979). Elastic moduli of transversely isotropic graphite fibers and their composites, *Experimental Mechanics*, **19**, 41-49. *22*

Kröner, E. (1953). Das Fundamentalintegral der anisotropen elastischen Differentialgleichungen, *Zeitschrift für Physik,* **136**, 402-410. *374*

Kuo, C. H. and Keer, L. M. (1992). Contact stress analysis of a layered transversely isotropic half-space, *Journal of Tribology*, **114**, 253-262. *22, 380*

Kupradze, V. D., Gegelia, T. G., Basheleishvili, M. O. and Burchuladze, T. V. (1979) *Three-Dimensional Problems of the Mathematically Theory of Elasticity and Thermoelasticity*. Amsterdam: North-Holland. *373*

Lam, K. Y., Ng, T. Y. and Wu, Q. (2000). Vibration analysis of thick laminated composite cylindrical shells, *AIAA Journal*, **38**, 1102-1107. *382*

Lamb, H. (1882). On the vibrations of an elastic sphere, *Proceedings of the London Mathematical Society*, **13**, 189-212. *255*

Lees, S. and Rollins, F. R., Jr. (1972). Anisotropy in hard dental tissues, *Journal of Biomechanics*, **5**, 557-566. *23*

Leissa, A. W. (1973). *Vibration of Shells*. NASA SP-288. *382*

Lekhnitskii, S. G. (1940). Symmetrical deformation and torsion of a body of revolution with a special kind of anisotropy, *PPM*, **4**(3), 43-60. (in Russian) *61, 373, 376*

Lekhnitskii, S. G. (1981). *Theory of Elasticity of an Anisotropic Elastic Body*. Moscow: Mir Publishers. *19, 54, 61, 373, 374, 377, 383*

Levy, M., Bass, H. E. and Stern, R. R. (2001). *Handbook of Elastic Properties of Solids, Liquids, and Gases*. San Diego: Academic Press. *23*

Li, X. and Chen, Y. (2002). Transient dynamic response analysis of orthotropic circular cylindrical shell under external hydrostatic pressure, *Journal of Sound and Vibration*, **257**, 967-976. *382*

Liang, J. (1994). *Fundamental Solutions of Transversely Isotropic Elasticity and Boundary Element Methods*. Ph. D. dissertation, Hangzhou: Zhejiang University. (in Chinese) *39, 41, 375*

Liew, K. M., Liang, J., Ding, H. J. and Lu, C. (2001). Elastic fields in two-joined half-spaces subjected to point force and uniform ring loads, *Computer Methods in Applied Mechanics and Engineering*, **190**, 3749-3769. *375*

Lin, H. C. (2002). *Three-Dimensional Finite Element Modeling and Stress Analysis of Mandible*. Master's thesis, Tainan: National Cheung Kong University. *22*

Lin, W., Kuo, C. H. and Keer, L. M. (1991). Analysis of a transversely isotropic half space under normal and tangential loadings, *Journal of Tribology*, **113**, 335-338. *380*

Lodge, A. S. (1955). The transformation to isotropic form of the equilibrium equations for a class of anisotropic elastic solids, *Quarterly Journal of Mechanics and Applied Mathematics*, **8**, 211-225. *48, 374*

Lou, Y. K. and Su, T. C. (1978). Free oscillations of submerged spherical shells, *Journal of the Acoustical Society of America*, **63**, 1402-1408. *350, 354*

Love, A. E. H. (1927). *A Treatise on the Mathematical Theory of Elasticity*. Cambridge: Cambridge University Press. *19, 61, 292, 301, 331, 337, 338, 351, 373, 374, 379*

Luo, Z. D. and Li, S. J. (1994). *Mechanics of Anisotropic Materials*. Shanghai: Shanghai Jiaotong University. (in Chinese) *373*

Lur'e, A. I. (1964). *Three-Dimensional Problems of the Theory of Elasticity*. New York: Interscience Publishers. *373, 379*

Lusher, C. P. and Hardy, W. N. (1988). Axisymmetric free vibrations of a transversely isotropic finite cylindrical rod, *Journal of Applied Mechanics*, **55**, 855-862. *285, 306*

Maiti, M. (1975). Stresses in anisotropic nonhomogeneous sphere, *Journal of Engineering Mechanics*, **101**, 101-108. *384*

Martin, R. M. (1972). Relation between elastic tensors of wurtzite and zinc-blende structure materials, *Physical Review B*, **6**, 4546-4553. *23*

McLachlan, N. W. (1954). *Bessel Functions for Engineers* (2nd ed.). Oxford: Clarendon Press. *387*

Mehta, Y. K. (1966). The distribution of thermal stresses around a crack in a semi-infinite elastic solid of transversely isotropic material, *Archiwum Mechaniki Stosowanej*, **18**, 749-757. *378*

Melo, J. D. D. and Radford, D. W. (2002). Determination of the elastic constants of a transversely isotropic lamina using laminate coefficients of thermal expansion, *Journal of Composite Materials*, **36**, 1321-1329. *23*

Men, F. L. and Yuan, X. M. (1990). Vibration of fluid-filled pipeline buried in soil, *Journal of Pressure Vessel Technology*, **112**, 386-391. *326, 359*

Messina, A. and Soldatos, K. P. (2002). A general vibration model of angle-ply laminated plates that accounts for the continuity of interlaminar stresses, *International Journal of Solids and Structures*, **39**, 617-635. *256*

Michell, J. H. (1900). The stress in an aeolotropic elastic solid with an infinite plane boundary, *Proceedings of the London Mathematical Society*, **32**, 247-259. *376*

Mindlin, R. D. (1936). Force at a point in the interior of a semi-infinite solid, *Physics*, **7**, 195-202. *132, 376*

Mindlin, R. D. (1949). Compliance of elastic bodies in contact, *Journal of Applied Mechanics*, **16**, 259-268. *379*

Mindlin, R. D. (1951*a*). Influence of rotatory inertia and shear on flexural motions of isotropic elastic plates, *Journal of Applied Mechanics*, **18**, 31-38. *381*

Mindlin, R. D. (1951*b*). Thickness-shear and flexural vibrations of crystal plates, *Journal of Applied Physics*, **22**, 316-323. *254*

Mindlin, R. D. and Deresiewicz, H. (1955). Timoshenko's shear coefficient for flexural vibrations of beams, *Proceedings of the Second U.S. National Congress on Applied Mechanics*, pp. 175-178. *296*

Mirsky, I. (1964). Vibrations of orthotropic, thick, cylindrical shells, *Journal of the Acoustical Society of America*, **36**, 41-51. *307, 382*

Mirsky, I. (1965). Wave propagation in transversely isotropic circular cylinders (Part I and Part II), *Journal of the Acoustical Society of America*, **37**, 1016-1026. *303, 304, 382*

Mirza, S. (1991). Recent research in vibration of layered shells, *Journal of Pressure Vessel Technology*, **113**, 321-325. *382*

Misra, J. C., Chattopadhyay, N. C. and Samanta, S. C. (1994). Thermoelastic stress wave in a spherically aeolotropic medium with a spherical cavity, induced by a distributed heat source within the medium, *International Journal of Engineering Science*, **32**, 1769-1789. *378*

Montagner, J. P. and Anderson, D. L. (1989). Constrained reference mantle model, *Physics of the Earth and Planetary Interiors*, **58**, 205-227. *15, 384*

Morse, P. M. and Feshbach, H. (1953). *Methods of Theoretical Physics*. New York: McGraw-Hill. *317, 382*

Mukherjee, J. (1968). Stresses due to vibration of spherical shells of aeolotropic materials, *AIAA Journal*, **6**, 1563-1565. *383*

Muskhelishvili, N. I. (1953). *Some Basic Problems of the Mathematical Theory of Elasticity*. Groningen: Noordhoff. *378, 379*

Narasimhan, M. C. (1992). Dynamic response of laminated orthotropic spherical shell, *Journal of the Acoustical Society of America*, **91**, 2714-2720. *383*

Nath, Y. and Jain, R. K. (1983). Non-linear dynamic analysis of shallow spherical shells on elastic foundations, *International Journal of Mechanical Sciences*, **25**, 409-419. *359, 385*

Nath, Y., Mahrenholtz, O. and Varma, K. K. (1987). Non-linear dynamic response of a doubly curved shallow shell on an elastic foundation, *Journal of Sound and Vibration*, **112**, 53-61. *385*

Niordson, F. I. (1985). *Shell Theory*. Amsterdam: North-Holland. *382*

Noda, N. and Ashida, F. (1987). Transient thermoelastic fields in a transversely isotropic infinite solid with a penny-shaped crack, *Journal of Applied Mechanics*, **54**, 854-860. *377*

Noda, N. and Ashida, F. (1988). Transient thermoelastic contact problem of a transversely isotropic cylinder, *Ingenieur-Archiv*, **58**, 241-251. *377, 380*

Noda, N., Ashida, F. and Matsunaga, Y. (1994). Stress intensity factor for external and penny-shaped cracks in a transversely isotropic cylinder subjected to thermal shock, *Archive of Applied Mechanics*, **64**, 383-394. *377*

Noda, N., Ashida, F. and Tsuji, T. (1989). An inverse transient thermoelastic problem for a transversely-isotropic body, *Journal of Applied Mechanics*, **56**, 791-797. *377*

Noda, N., Hetnarski, R. B. and Tanigawa, Y. (2000). *Thermal Stresses*. New York: Lastran Corporation Rochester. *373*

Noor, A. K. and Burton, W. S. (1990). Assessment of computational models for multilayered composite shells, *Applied Mechanics Reviews*, **43**, 67-97. *382*

Nosier, A., Kapania, R. K. and Reddy, J. N. (1993). Free vibration analysis of laminated plates using a layerwise theory, *AIAA Journal*, **31**, 2335-2346. *252*

Nowacki, W. (1954). The stress function in three-dimensional problems concerning an elastic body characterized by transverse isotropy, *Bulletin de l'Academie Polonaise Sciences, Cl. IV*, **2**, 21-25. *374*

Nowacki, W. (1986). *Thermoelasticity* (2nd ed.). Warszawa: PWN-Polish Scientific Publishers. *25, 373*

Nowinski, J. L. (1959). Note on a thermoelastic problem for a transversely isotropic hollow sphere embedded in an elastic medium, *Journal of Applied Mechanics*, **26**, 649-650. *378, 384-385*

Nowinski, J. L. (1978). *Theory of Thermoelasticity with Applications*. Alphen aan den Rijn: Sijhoff & Noordhoff International Publishers. *373*

Okumura, I. A. and Dohba, H. (1989). Generalization of Elliott's solution to transversely isotropic elasticity problems in Cartesian coordinates, *JSME International Journal, Series I*, **32**, 331-336. *374*

Ouyang, H. J. (1993). Hamiltonian formulation of anisotropic elasticity, *Acta Mechanica Solida Sinica*, **6**, 453-458. *373*

Pagano, N. J. (1969). Exact solutions for composite laminates in cylindrical bending, *Journal of Composite Materials*, **3**, 398-411. *256, 381*

Pagano, N. J. (1970). Exact solutions for rectangular bidirectional composites and sandwich plates, *Journal of Composite Materials*, **4**, 20-34. *256, 381*

Pagano, N. J. (1972). Dynamic characteristics of composite laminates, *Journal of Sound and Vibration*, **23**, 127-143. *256, 381*

Paliwal, D. N. and Bhalla, V. (1993). Large amplitude free vibration of shallow spherical shell on a Pasternak foundation, *Journal of Vibration and Acoustics*, **115**, 70-74. *357, 385*

Paliwal, D. N. and Pandey, R. (2001). Free vibration of an orthotropic cylindrical shell on a Pasternak foundation, *AIAA Journal*, **39**, 2188-2191. *357, 382*

Paliwal, D. N. and Singh, S. (1999). Free vibrations of orthotropic cylindrical shell on elastic foundation, *AIAA Journal*, **37**, 1135-1139. *382*

Paliwal, D. N. and Srivastava, R. (1994). Vibrations of a shallow spherical shell on a Kerr foundation, *Journal of Vibration and Acoustics*, **116**, 47-52. *385*

Pan, E. N. (1989*a*). Elastic dislocations in a transversely isotropic and layered half-space, *Acta Geophysica Sinica*, **32**, 516-527. (in Chinese) *377*

Pan, E. N. (1989*b*). Concentrated force in an infinite space of transversely isotropic material, *Acta Mechanica*, **80**, 127-135. *375*

Pan, E. N. (1989*c*). Static response of a transversely isotropic and layered half-space to general surface loads, *Physics of the Earth and Planetary Interiors*, **54**, 353-363. *377*

Pan, E. N. (1992). Vibration of a transversely isotropic, simply supported and layered rectangular plate, *Journal of Elasticity*, **27**, 167-181. *381*

Pan, Y. C. and Chou, T. W. (1976). Point force solution for an infinite transversely isotropic solid, *Journal of Applied Mechanics*, **43**, 608-612. *374, 380*

Pan, Y. C. and Chou, T. W. (1979*a*). Green's functions for two-phase transversely isotropic materials, *Journal of Applied Mechanics*, **46**, 551-556. *375*

Pan, Y. C. and Chou, T. W. (1979*b*). Green's function solutions for semi-infinite transversely isotropic materials, *International Journal of Engineering Science*, **17**, 545-551. *376*

Panet, M., Cheng, C., Deschamps, M., Poncelet, O. and Audoin, B. (2002). Microconcrete ageing ultrasonic identification, *Cement and Concrete Research*, **32**, 1831-1838. *22*

Pao, Y. H. and Mow, C. C. (1971). *Diffraction of Elastic Waves and Dynamic Stress Concentrations*. New York: The Rand Corporation. *321*

Pasternak, P. L. (1954). On a new method of analysis of an elastic foundation by means of two foundation constants, *Gosudarstrennoe Izdatelstvo Literaturi po Stroitelstvu I Arkhitekture*, Moscow. (in Russian) *357*

Payton, R. G. (1983). *Elastic Wave Propagation in Transversely Isotropic Media*. The Hague: Martinus Nijhoff Publishers. *23, 295*

Pestel, E. C. and Leckie, F. A. (1963). *Matrix Methods in Elastomechanics*. New York: McGraw-Hill. *313, 367*

Pindera, M. J. and Lane, M. S. (1993). Frictionless contact of layered half-planes, Part I and Part II, *Journal of Applied Mechanics*, **60**, 633-645. *380*

Pochhammer, J. (1876). Ueber die Fortpflanzungsgeschwindigkeiten kleiner Schwingungen in einem unbegrenzten isotropen Kreiscylinder, *J. fur reine und angewandte Mathematik (Crelle)*, **81**, 324-336. *301*

Podil'chuk, Y. N. (1987). Thermoelastic deformation of a transversally isotropic prolate spheroid, *Soviet Applied Mechanics*, **23**, 1126-1133. *377*

Podil'chuk, Y. N. (1988). Thermoelastic deformation of transversally isotropic two-cavity hyperboloid of revolution, *Soviet Applied Mechanics*, **24**, 833-839. *377*

Podil'chuk, Y. N. (1991). General problem of the theory of elasticity for a transversely isotropic cone, *Soviet Applied Mechanics*, **27**, 1153-1157. *377*

Podil'chuk, Y. N. (1997). Stress state of a transversely-isotropic body with elliptical inclusion, *International Applied Mechanics*, **33**, 881-887. *376*

Podil'chuk, Y. N. and Sokolovskii, Y. I. (1991*a*). Stress state of a transversely isotropic medium with an anisotropic spheroidal inclusion: Arbitrary uniform stress field at infinity, *Soviet Applied Mechanics*, **27**, 551-558. *375-376*

Podil'chuk, Y. N. and Sokolovskii, Y. I. (1991*b*). Stress state of a transversely isotropic medium with an anisotropic inclusion: Arbitrary linear force field at infinity, *Soviet Applied Mechanics*, **27**, 644-653. *375-376*

Podil'chuk, Y. N. and Sokolovskii, Y. I. (1994). Thermostress in an infinite transversely isotropic medium with an internal elliptical crack, *International Applied Mechanics*, **30**, 834-840. *377, 378*

Podil'chuk, Y. N., Kucheryarvy, V. A. and Sokolovskii, Y. I. (1994). Thermal stressed state of a transversely isotropic medium with spheroidal inclusion, *International Applied Mechanics*, **30**, 403-412. *377*

Prasad, C. and Jain, R. K. (1966). Vibrations of transversely isotropic cylindrical shells of finite length, *Journal of the Acoustical Society of America*, **38**, 1006-1009. *304*

Puro, A. E. (1980). Variable separation in elasticity-theory equations for spherically transversely isotropic inhomogeneous bodies, *Soviet Applied Mechanics*, **16**,117-120. *374*

Qiu, Z. Q. and Wang, R. (1988). Static solution for transversely isotropic elastic *N*-layered systems, *Acta Scientiarum Naturalium Universitatis Pekinensis*, **24**, 202-211. (in Chinese) *377*

Rahman, A. and Ahmad, F. (1998). Representation of the displacement in terms of scalar functions for use in transversely isotropic materials, *Journal of the Acoustical Society of America*, **104**, 3675-3676. *383*

Raju, P. P. (1975). On shallow shells of transversely isotropic materials, *Journal of Pressure Vessel Technology*, **97**,185-191. *378*

Ramakrishnan, C. V. and Shah, A. H. (1970). Vibration of an aeolotropic spherical shell, *Journal of the Acoustical Society of America*, **47**, 1366-1374. *338, 344, 383*

Ramasamy, R. and Ganesan, N. (1999) Vibration and damping analysis of fluid-filled orthotropic cylindrical shells with constrained viscoelastic damping, *Computers and Structures*, **70**, 363-376. *382*

Ramskaya, E. I. (1983). The analysis of the proper frequencies and forms of axisymmetric oscillations of a transversely isotropic hollow sphere, *Prikladnaya Mekhanika*, **19**(7), 103-107. *383*

Reddy, J. N. (1999). *Theory and Analysis of Elastic Plates*. Philadelphia: Taylor & Francis. *381*

Reilly, D. T. and Burstein, A. H. (1975). The elastic and ultimate properties of compact bone tissue, *Journal of Biomechanics*, **8**, 391-405. *22, 23*

Reissner, E. (1945). The effect of transverse shear deformation on the bending of elastic plates, *Journal of Applied Mechanics*, **12**, A69-A77. *381*

Reissner, E. (1975). On transverse bending of plates, including the effect of transverse shear deformation, *International Journal of Solids and Structures*, **11**, 569-573. *381*

Ren, J. G. (1995). Analysis of laminated circular cylindrical shells under axisymmetric loading, *Composite Structures*, **30**, 271-280. *382*

Ritter, M. A. (1992). Timber bridges design, construction, inspection and maintenance, *U.S. Forest Service Document*, EM 7700-8. *23*

Rogowski, B. (1986). Inclusion, punch and crack problems in an elastically supported transversely isotropic layer, *Solid Mechanics Archives*, **11**, 65-102. *380*

Rongved, L. (1955). Force interior to one of two joined semi-infinite solids, In: *Proceedings of the Second Midwestern Conference on Solid Mechanics*, pp. 1-13. Lafayette: Purdue University. *375*

Rosler, S., Lucklum, R., Borngraber, R., Hartmann, J. and Hauptmann, P. (1998). Sensor system for the detection of organic pollutants in water by thickness shear mode resonators, *Sensors and Actuators*, **B48**, 415-424. *284*

Roy, S. C. (1977). Radial deformation of a nonhomogeneous spherically isotropic elastic sphere with a concentric spherical inclusion, *Proceedings of the Indian Academy of Sciences*, **A85**, 338-350. *383*

Royer, D. and Dieulesaint, E. (2000). *Elastic Waves in Solids I*. Berlin: Springer-Verlag. *22, 23*

Sáez, A. and Domínguez, J. (2000). Far field dynamic Green's functions for BEM in transversely isotropic solids, *Wave Motion*, **32**, 113-123. *22*

Sakamoto, M., Hara, T., Shibuya, T. and Koizumi, T. (1991). Indentation by a circular rigid punch of a transversely isotropic layer on a rigid foundation, *JSME International Journal, Series I*, **34**,130-134. *380*

Sayir, M. (1980). Flexural vibrations of strongly anisotropic beams, *Ingenieur-Archiv*, **49**, 309-323. *256*

Selvadurai, A. P. S. (1979). *Elastic Analysis of Soil-Foundation Interaction*. Amsterdam: Elsevier. *359*

Shakeri, M., Elsami, M. R. and Yas, M. H. (1999). Elasticity solution and free vibration analysis of laminated anisotropic cylindrical shells, *Structural Engineering and Mechanics*, **7**, 181-202. *382*

Shaldyrvan, V. A. (1980). Three-dimensional thermoelasticity problems for transversally isotropic plates, *Soviet Applied Mechanics*, **16**, 370-375. *377*

Sharma, B. (1958). Thermal stresses in transversely isotropic semi-infinite elastic solid, *Journal of Applied Mechanics*, **25**, 86-88. *377*

Shenderov, E. L. (1985). Vibrational impedances of a transversely isotropic spherical layer, *Soviet Physics Acoustics*, **31**, 386-389. *19*

Shi, Z. and Ramalingam, S. (2001). Thermal and mechanical stresses in transversely isotropic coatings, *Surface and Coatings Technology*, **138**, 173-184. *22*

Shi, Z. Y. and Zeng, G. X. (1989). Determination of displacements and stresses in multilayered anisotropic soil system by initial function method, *Journal of Zhejiang University*, **23**, 222-232. (in Chinese) *377*

Shield, R. T. (1951). Notes on problems in hexagonal aeolotropic materials, *Proceedings of the Cambridge Philosophical Society*, **47**, 401-409. *376, 379*

Shul'ga, N. A. (1980). The proper oscillations of a transversally isotropic hollow sphere, *Prikladnaya Mekhanika*, **16**(12), 108-111. *383*

Shul'ga, N. A., Grigorenko, A. Y. and Efimova, T. L. (1988). Free non-axisymmetric oscillations of a thick-walled, nonhomogeneous, transversally isotropic, hollow sphere, *Soviet Applied Mechanics*, **24**, 439-444. *365, 383, 385*

Silbiger, A. (1962). Non-axisymmetric modes of vibrations of thin spherical shell, *Journal of the Acoustical Society of America*, **34**, 862. *336*

Singh, B. M., Danyluk, H. T. and Selvadurai, A. P. S. (1987). Thermal stresses in a transversely isotropic elastic solid weakened by an external circular crack, *International Journal of Solids and Structures*, **23**, 403-412. *378*

Sittig, E. K. and Coquin, G. A. (1970). Visualization of plane-strain vibration modes of a long cylinder capable of producing sound radiation, *Journal of the Acoustical Society of America*, **48**, 1150-1159. *284*

Small, J. C. and Booker, J. R. (1984). Finite layer analysis of layered elastic materials using flexibility approach. Part 1 — strip loadings, *International Journal for Numerical Methods in Engineering*, **20**, 1025-1037. *376*

Small, J. C. and Booker, J. R. (1986) Finite layer analysis of layered elastic materials using flexibility approach, Part 2 — circular and rectangular loadings, *International Journal for Numerical Methods in Engineering*, **23**, 959-978. *376*

Sneddon, I. N. (1948). Boussinesq's problem for a rigid cone, *Proceedings of the Cambridge Philosophical Society*, **44**, 492-507. *231, 233*

Sneddon, I. N. (1951). *Fourier Transforms*. New York: McGraw-Hill. *277, 280, 378*

Sneddon, I. N. (1992). Fourier-transform solution of a Boussinesq problem for a hexagonally aeolotropic elastic half-space, *Quarterly Journal of Mechanics and Applied Mathematics*, **45**, 607-616. *376*

Sokolnikoff, I. S. (1956). *Mathematical Theory of Elasticity* (2nd ed.). New York: McGraw-Hill. *74, 373*

Soldatos, K. P. (1994). Review of three dimensional dynamic analyses of circular cylinders and cylindrical shells, *Applied Mechanics Reviews*, **47**, 501-516. *382*

Soldatos, K. P. and Hadjigeorgiou, V. P. (1990). Three-dimensional solution of the free vibration problem of homogeneous isotropic cylindrical shells and panels, *Journal of Sound and Vibration*, **137**, 369-384. *309*

Somaratna, N. and Ting, T. C. T. (1986). Three-dimensional stress singularities at conical notches and inclusions in transversely isotropic materials, *Journal of Applied Mechanics*, **53**, 89-96. *375*

Spencer, A. J. M., Watson, P. and Rogers, T. G. (1992). Thermoelastic distortions in laminated anisotropic tubes and channel sections, *Journal of Thermal Stresses*, **15**, 129-141. *33, 382*

Srinivas, S., Joga Rao, C. V. and Rao, A. K. (1970). An exact analysis for vibration of simply-supported homogeneous and laminated thick rectangular plates, *Journal of Sound and Vibration*, **12**, 187-199. *256, 381*

Srinivas, S. and Rao, A. K. (1970). Bending, vibration and buckling of simply supported thick orthotropic rectangular plates and laminates, *International Journal of Solids and Structures*, **6**, 1463-1481. *256, 267, 381*

Steele, C. R. and Kim, Y. Y. (1992). Modified mixed variational principle and the state-vector equation for elastic bodies and shells of revolution, *Journal of Applied Mechanics*, **59**, 587-595. *33*

Su, T. C. (1981*a*). The effect of viscosity on free oscillations of fluid-filled spherical shells, *Journal of Sound and Vibration*, **74**, 205-220. *384*

Su, T. C. (1981*b*). The effect of viscosity on free vibrations of submerged fluid-filled spherical shells, *Journal of Sound and Vibration*, **77**, 101-125. *384*

Su, T. C. (1982). Natural frequencies of an elastic spherical shell submerged in a compressible viscous fluid medium, *Journal of Sound and Vibration*, **83**, 163-169. *384*

Su, T. C. and Lou, Y. K. (1975). The effect of viscosity on the dynamics of a submerged spherical shell, *Journal of Engineering for Industry*, **97**, 1338-1344. *384*

Sundara Raja Iyengar, K. T. and Raman, P. V. (1977). Free vibrations of rectangular plates of arbitrary thickness, *Journal of Sound and Vibration*, **54**, 229-236. *33, 381*

Sveklo, V. A. (1969). Concentrated force in a transversally isotropic half-space and in a composite space, *PMM: Journal of Applied Mathematics and Mechanics*, **33**, 517-523. *375, 376*

Tang, L. M. (1991). Mixed formulation and Hamilton canonical equations of theory of elasticity, *Computational Structural Mechanics and Applications*, **8**, 343-350. (in Chinese) *373*

Tang, W. Y., Zhang, S, K. and Chen, T. Y. (1998). Dynamic structure-fluid interaction response of a composite cylindrical shell submerged in water, *China Ocean Engineering*, **12**, 297-307. *382*

Tarn, J. Q. (2002). A state space formalism for anisotropic elasticity. Part I: Rectilinear anisotropy, *International Journal of Solids and Structures*, **39**, 5143-5155. *33*

Tessler, A., Tsui, T. and Saether, E. (1995) A {1,2}-order theory for elastodynamic analysis of thick orthotropic shells, *International Journal of Solids and Structures*, **32**, 3237-3260. *382*

Thurston, R. N. (1992). Elastic waves in rods and optical fibers, *Journal of Sound and Vibration*, **159**, 441-467. *296*

Timoshenko, S. P. and Goodier, J. N. (1970). *Theory of Elasticity* (3rd ed.). New York: McGraw-Hill. *157, 205, 373, 377*

Timoshenko, S. P. and Woinowsky-Krieger, S. (1959). *Theory of Plates and Shells* (2nd ed.). New York: McGraw-Hill. *381, 382*

Ting, T. C. T. (1996). *Anisotropic Elasticity: Theory and Applications*. Oxford: Oxford University Press. *373*

Tsai, Y. M. (1983*a*). Thermal stress in a transversely isotropic medium containing a penny-shaped crack, *Journal of Applied Mechanics*, **50**, 24-28. *199, 377, 378*

Tsai, Y. M. (1983*b*). Transversely isotropic thermoelastic problem of uniform heat flow disturbed by a penny-shaped crack, *Journal of Thermal Stresses*, **6**, 379-389. *377, 378*

Tsai, Y. M. (1986). Symmetrical contact problem of thick transversely isotropic plate, *Journal of Elasticity*, **16**, 179-188. *380*

Tsai, Y. M. (1991). Longitudinal motion of a thick transversely isotropic hollow cylinder, *Journal of Pressure Vessel Technology*, **113**, 585-589. *382*

Tsai, Y. M. (1998). Thermoelastic behavior of a transversely isotropic material containing a flat toroidal crack, *Journal of Thermal Stresses* **21**, 881-895. *378*

Tsai, Y. M., (2000). Thermoelastic problem of uniform heat flow disturbed by a flat toroidal crack in a transversely isotropic medium, *Journal of Thermal Stresses*, **23**, 217-231. *378*

Upadhyay, P. C. and Mishra, B. K. (1988). Non-axisymmetric dynamic response of buried orthotropic cylindrical shells, *Journal of Sound and Vibration*, **121**, 149-160. *382*

Uyaner, M., Akdemir, A., Erim, S. and Avci, A. (2000). Plastic zones in a transversely isotropic solid cylinder containing a ring-shaped crack, *International Journal of Fracture*, **106**, 161-175. *22*

Vasilenko, A. T., Grigorenko, Y M. and Pankratova, N. D. (1976). Stress state of transversely isotropic nonhomogeneous thickwall spherical shells, *Mechanics of Solids*, **11**(1), 56-61. *378*

Vendhan, C. P. and Archer, R. R. (1978). Axisymmetric stresses in transversely isotropic finite cylinders, *International Journal of Solids and Structures*, **14**, 305-318. *382*

Vlasov, V. Z. and Leontev, N. N. (1966). *Beams, Plates and Shells on Elastic Foundations*. NASA: TTF-357, TT65-50135. *33*

Wachtman, J. B., Jr., Tefft, W. E., Lam, Jr., D. G. and Stinchfield, R. P. (1960). Elastic constants of synthetic single crystal Corundum at room temperature, *Journal of Research of the National Bureau of Standards — A. Physics and Chemistry*, **64A**, 213-228. *22*

Wang, F. Y. (1986). Exact analysis of stability problems for transversely isotropic plates, *Journal of Zhejiang University*, **20**, 91-96. (in Chinese) *256, 381*

Wang, F. Y. and He, F. B. (1986). A study of free vibration problem based on the theory of elasticity for simply supported transversely isotropic thick plates, *Journal of Shock and Vibration*, **5**(1), 1-11. (in Chinese) *252, 255, 256, 381*

Wang, L. T. (2004). Space of degeneracy in the Stroh eigensystem and surface waves in transversely isotropic elastic media, *Wave Motion*, **40**, 173-190. *23*

Wang, M. Z. (2002). *Advanced Elasticity*. Beijing: Peking University Press. (in Chinese) *373*

Wang, M. Z. and Wang, L. N. (1989). Derivation of some special stress function from

Beltrami-Schaefer stress function, *Applied Mathematics and Mechanics*, **10**, 665-673. *54*

Wang, M. Z. and Wang, W. (1995). Completeness and nonuniqueness of general solutions of transversely isotropic elasticity, *International Journal of Solids and Structures*, **32**, 501-513. *374*

Wang, W. and Shi, M. X. (1998). On the general solutions of transversely isotropic elasticity, *International Journal of Solids and Structures*, **35**, 3283-3297. *374*

Wang, W. and Xu, X. S. (1995). On the general solutions of axisymmetric problems of a transversely isotropic body, *Acta Scientiarum Naturalium Universitatis Pekinensis*, **31**, 303-308. (in Chinese) *374*

Wang, W., Xu, X. S. and Wang, M. Z. (1994). Completeness of general solutions to axisymmetric problems of transversely isotropic body, *Science in China (A)*, **37**, 580-596. *61, 373*

Wang, X. (1994). An elastodynamic solution for an anisotropic hollow sphere, *International Journal of Solids and Structures*, **31**, 903-911. *383*

Wang, Y. M., Tarn, J. Q. and Hsu, C. K. (2000). State space approach for stress decay in laminates, *International Journal of Solids and Structures*, **37**, 3535-3553. *381*

Wang, Z. G. and Tang, L. M. (1995). Hamiltonian systems in elasticity and their variational principles, *Applied Mathematics and Mechanics*, **16**, 125-131. *373*

Wang, Z. X. and Guo, D. R. (2000). *Introduction to Special Function*. Beijing: Peking University Press. (in Chinese) *330, 387*

Watson, G. N. (1922). *A Treatise on the Theory of Bessel Functions*. Cambridge: Cambridge University Press. *261, 336, 387*

White, J. E. and Tongtaow, C. (1981). Cylindrical waves in transversely isotropic media, *Journal of the Acoustical Society of America*, **70**, 1147-1155. *382*

Whitney, J. M. and Sun, C. T. (1974). A refined theory for laminated anisotropic, cylindrical shells, *Journal of Applied Mechanics*, **41**, 471-476. *382*

Willis, J. R. (1966). Hertzian contact of anisotropic bodies, *Journal of the Mechanics and Physics of Solids*, **14**, 163-176. *379*

Willis, J. R. (1967). Boussinesq problems for an anisotropic half-space, *Journal of the Mechanics and Physics of Solids*, **15**, 331-339. *379*

Winkler, E. (1867). *Die Lehre von der Elasticitaet und Festigkeit*. Prag: Dominicus. *357*

Wu, B. and Ma, H. Z. (1992). Elastic wave propagation of orthotropic spherical shell subjected to a shock load, *Acta Materiae Compositae Sinica*, **9**(1), 49-58. (in Chinese) *383*

Wu, C. S. (1999). *Methods of Mathematical Physics*. Beijing: Peking University Press. (in Chinese) *387*

Wu, J. K. and Wang, M. Z. (1980). *An Introduction to Elasticity*. Beijing: Peking University Press. (in Chinese) *373*

Xi, Z. C., Liu, G. R., Lam, K. Y. and Shang, H. M. (2000). A strip-element method for analyzing wave scattering by a crack in a fluid-filled composite cylindrical shell, *Composite Science and Technology*, **60**, 1985-1996. *382*

Xia, L. P. and Ding, K. W. (2001). Three-dimensional thermoelastic solution for laminated cantilever cylindrical shell, *Aerospace Science Technology*, **5**, 339-346. *382*

Xia, T. and He, F. B. (1988). Three-dimensional elastic solution of the axisymmetrically bending problem for the transversely isotropic spherical shells, *Chinese Journal of Applied Mechanics*, **5**(3), 55-66. (in Chinese) *374*

Xie, Y. Q., Lin, Z. X. and Ding, H. J. (1988). *Elasticity*. Hangzhou: Zhejiang University Press. (in Chinese) *210, 373, 379*

Xu, R. Q. (1999). *Exact Solutions of Free Vibration and Equilibrium Problems of Circular, Annular and Sectorial Plates*. Ph.D. Dissertation, Hangzhou: Zhejiang University. (in Chinese) *264, 266, 275, 280, 381*

Yao, W. A. and Zhong, W. X. (2002). *Symplectric Elasticity*. Beijing: Higher Education Press. (in Chinese) *373*

Ye, J. Q. and Soldatos, K. P. (1996). Axisymmetric vibration analysis of laminated hollow spheres composed of transversely isotropic layers, In: *Proceedings of Engineering Systems Design and Analysis Conference*, **9**, 161-169. Montpellier, France. *385*

Zhang, B., Dong, H. and Wang, K. (1994). Multipole sources in a fluid-filled borehole surrounded by a transversely isotropic elastic solid, *Journal of the Acoustical Society of America*, **96**, 2546-2555. *382*

Zhang, L. C. (2001) *Solid Mechanics for Engineers*. Basingstoke: Macmillan Press. *216, 373, 379*

Zheng, Y. and Zhang, H. Q. (1996). The canonical Hamiltonian representations in solid mechanics, *Acta Mechanica Sinica*, **28**(1), 119-125. (in Chinese) *38*

Zhong, W. X. (1995). *A New Systematic Methodology for Theory of Elasticity*. Dalian: Dalian University of Technology Press. (in Chinese) *33, 373*

Zhu, H. (1992). A method to evaluate three-dimensional time-harmonic elastodynamic Green's functions in transversely isotropic media, *Journal of Applied Mechanics*, **59**, S96-S101. *375*

Zou, D. Q. (1992). *Elasticity of Spherically Isotropic Materials and Its Preliminary Application in Geoscience*. Ph. D. dissertation, Hangzhou: Zhejiang University. (in Chinese) *377, 384*

Zou, D. Q., Liang, J. and Ding, H. J. (1994). General solution of transversely isotropic elasticity problem, *Journal of Zhejiang University*, **28**, 273-282. (in Chinese) *39, 41, 64, 374*

Zureick, A. H. (1988). Transversely isotropic medium with a rigid spheroidal inclusion under an axial pull, *Journal of Applied Mechanics*, **55**, 495-497. *375*

Zureick, A. H. (1989). Spheroidal cavity with prescribed asymmetric displacements in three-dimensional transverse isotropy, *Journal of Engineering Mechanics*, **115**, 2427-2439. *375*

Zureick, A. H. and Eubanks, R. A. (1988). Spheroidal cavity with prescribed asymmetric tractions in three-dimensional transverse isotropy, *Journal of Engineering Mechanics*, **114**, 24-48. *375*

INDEX

Mechanics

SOLID MECHANICS AND ITS APPLICATIONS

Series Editor: G.M.L. Gladwell

Aims and Scope of the Series

The fundamental questions arising in mechanics are: *Why?*, *How?*, and *How much?* The aim of this series is to provide lucid accounts written by authoritative researchers giving vision and insight in answering these questions on the subject of mechanics as it relates to solids. The scope of the series covers the entire spectrum of solid mechanics. Thus it includes the foundation of mechanics; variational formulations; computational mechanics; statics, kinematics and dynamics of rigid and elastic bodies; vibrations of solids and structures; dynamical systems and chaos; the theories of elasticity, plasticity and viscoelasticity; composite materials; rods, beams, shells and membranes; structural control and stability; soils, rocks and geomechanics; fracture; tribology; experimental mechanics; biomechanics and machine design.

1. R.T. Haftka, Z. Gürdal and M.P. Kamat: *Elements of Structural Optimization.* 2nd rev.ed., 1990 ISBN 0-7923-0608-2
2. J.J. Kalker: *Three-Dimensional Elastic Bodies in Rolling Contact.* 1990 ISBN 0-7923-0712-7
3. P. Karasudhi: *Foundations of Solid Mechanics.* 1991 ISBN 0-7923-0772-0
4. *Not published*
5. *Not published.*
6. J.F. Doyle: *Static and Dynamic Analysis of Structures.* With an Emphasis on Mechanics and Computer Matrix Methods. 1991 ISBN 0-7923-1124-8; Pb 0-7923-1208-2
7. O.O. Ochoa and J.N. Reddy: *Finite Element Analysis of Composite Laminates.* ISBN 0-7923-1125-6
8. M.H. Aliabadi and D.P. Rooke: *Numerical Fracture Mechanics.* ISBN 0-7923-1175-2
9. J. Angeles and C.S. López-Cajún: *Optimization of Cam Mechanisms.* 1991 ISBN 0-7923-1355-0
10. D.E. Grierson, A. Franchi and P. Riva (eds.): *Progress in Structural Engineering.* 1991 ISBN 0-7923-1396-8
11. R.T. Haftka and Z. Gürdal: *Elements of Structural Optimization.* 3rd rev. and exp. ed. 1992 ISBN 0-7923-1504-9; Pb 0-7923-1505-7
12. J.R. Barber: *Elasticity.* 1992 ISBN 0-7923-1609-6; Pb 0-7923-1610-X
13. H.S. Tzou and G.L. Anderson (eds.): *Intelligent Structural Systems.* 1992 ISBN 0-7923-1920-6
14. E.E. Gdoutos: *Fracture Mechanics.* An Introduction. 1993 ISBN 0-7923-1932-X
15. J.P. Ward: *Solid Mechanics.* An Introduction. 1992 ISBN 0-7923-1949-4
16. M. Farshad: *Design and Analysis of Shell Structures.* 1992 ISBN 0-7923-1950-8
17. H.S. Tzou and T. Fukuda (eds.): *Precision Sensors, Actuators and Systems.* 1992 ISBN 0-7923-2015-8
18. J.R. Vinson: *The Behavior of Shells Composed of Isotropic and Composite Materials.* 1993 ISBN 0-7923-2113-8
19. H.S. Tzou: *Piezoelectric Shells.* Distributed Sensing and Control of Continua. 1993 ISBN 0-7923-2186-3
20. W. Schiehlen (ed.): *Advanced Multibody System Dynamics.* Simulation and Software Tools. 1993 ISBN 0-7923-2192-8
21. C.-W. Lee: *Vibration Analysis of Rotors.* 1993 ISBN 0-7923-2300-9
22. D.R. Smith: *An Introduction to Continuum Mechanics.* 1993 ISBN 0-7923-2454-4
23. G.M.L. Gladwell: *Inverse Problems in Scattering.* An Introduction. 1993 ISBN 0-7923-2478-1

Mechanics

SOLID MECHANICS AND ITS APPLICATIONS

Series Editor: G.M.L. Gladwell

24. G. Prathap: *The Finite Element Method in Structural Mechanics*. 1993 ISBN 0-7923-2492-7
25. J. Herskovits (ed.): *Advances in Structural Optimization*. 1995 ISBN 0-7923-2510-9
26. M.A. González-Palacios and J. Angeles: *Cam Synthesis*. 1993 ISBN 0-7923-2536-2
27. W.S. Hall: *The Boundary Element Method*. 1993 ISBN 0-7923-2580-X
28. J. Angeles, G. Hommel and P. Kovács (eds.): *Computational Kinematics*. 1993 ISBN 0-7923-2585-0
29. A. Curnier: *Computational Methods in Solid Mechanics*. 1994 ISBN 0-7923-2761-6
30. D.A. Hills and D. Nowell: *Mechanics of Fretting Fatigue*. 1994 ISBN 0-7923-2866-3
31. B. Tabarrok and F.P.J. Rimrott: *Variational Methods and Complementary Formulations in Dynamics*. 1994 ISBN 0-7923-2923-6
32. E.H. Dowell (ed.), E.F. Crawley, H.C. Curtiss Jr., D.A. Peters, R. H. Scanlan and F. Sisto: *A Modern Course in Aeroelasticity*. Third Revised and Enlarged Edition. 1995 ISBN 0-7923-2788-8; Pb: 0-7923-2789-6
33. A. Preumont: *Random Vibration and Spectral Analysis*. 1994 ISBN 0-7923-3036-6
34. J.N. Reddy (ed.): *Mechanics of Composite Materials*. Selected works of Nicholas J. Pagano. 1994 ISBN 0-7923-3041-2
35. A.P.S. Selvadurai (ed.): *Mechanics of Poroelastic Media*. 1996 ISBN 0-7923-3329-2
36. Z. Mróz, D. Weichert, S. Dorosz (eds.): *Inelastic Behaviour of Structures under Variable Loads*. 1995 ISBN 0-7923-3397-7
37. R. Pyrz (ed.): *IUTAM Symposium on Microstructure-Property Interactions in Composite Materials*. Proceedings of the IUTAM Symposium held in Aalborg, Denmark. 1995 ISBN 0-7923-3427-2
38. M.I. Friswell and J.E. Mottershead: *Finite Element Model Updating in Structural Dynamics*. 1995 ISBN 0-7923-3431-0
39. D.F. Parker and A.H. England (eds.): *IUTAM Symposium on Anisotropy, Inhomogeneity and Nonlinearity in Solid Mechanics*. Proceedings of the IUTAM Symposium held in Nottingham, U.K. 1995 ISBN 0-7923-3594-5
40. J.-P. Merlet and B. Ravani (eds.): *Computational Kinematics '95*. 1995 ISBN 0-7923-3673-9
41. L.P. Lebedev, I.I. Vorovich and G.M.L. Gladwell: *Functional Analysis*. Applications in Mechanics and Inverse Problems. 1996 ISBN 0-7923-3849-9
42. J. Menčik: *Mechanics of Components with Treated or Coated Surfaces*. 1996 ISBN 0-7923-3700-X
43. D. Bestle and W. Schiehlen (eds.): *IUTAM Symposium on Optimization of Mechanical Systems*. Proceedings of the IUTAM Symposium held in Stuttgart, Germany. 1996 ISBN 0-7923-3830-8
44. D.A. Hills, P.A. Kelly, D.N. Dai and A.M. Korsunsky: *Solution of Crack Problems*. The Distributed Dislocation Technique. 1996 ISBN 0-7923-3848-0
45. V.A. Squire, R.J. Hosking, A.D. Kerr and P.J. Langhorne: *Moving Loads on Ice Plates*. 1996 ISBN 0-7923-3953-3
46. A. Pineau and A. Zaoui (eds.): *IUTAM Symposium on Micromechanics of Plasticity and Damage of Multiphase Materials*. Proceedings of the IUTAM Symposium held in Sèvres, Paris, France. 1996 ISBN 0-7923-4188-0
47. A. Naess and S. Krenk (eds.): *IUTAM Symposium on Advances in Nonlinear Stochastic Mechanics*. Proceedings of the IUTAM Symposium held in Trondheim, Norway. 1996 ISBN 0-7923-4193-7
48. D. Ieşan and A. Scalia: *Thermoelastic Deformations*. 1996 ISBN 0-7923-4230-5

Mechanics

SOLID MECHANICS AND ITS APPLICATIONS

Series Editor: G.M.L. Gladwell

49. J.R. Willis (ed.): *IUTAM Symposium on Nonlinear Analysis of Fracture*. Proceedings of the IUTAM Symposium held in Cambridge, U.K. 1997 ISBN 0-7923-4378-6
50. A. Preumont: *Vibration Control of Active Structures*. An Introduction. 1997 ISBN 0-7923-4392-1
51. G.P. Cherepanov: *Methods of Fracture Mechanics: Solid Matter Physics*. 1997 ISBN 0-7923-4408-1
52. D.H. van Campen (ed.): *IUTAM Symposium on Interaction between Dynamics and Control in Advanced Mechanical Systems*. Proceedings of the IUTAM Symposium held in Eindhoven, The Netherlands. 1997 ISBN 0-7923-4429-4
53. N.A. Fleck and A.C.F. Cocks (eds.): *IUTAM Symposium on Mechanics of Granular and Porous Materials*. Proceedings of the IUTAM Symposium held in Cambridge, U.K. 1997 ISBN 0-7923-4553-3
54. J. Roorda and N.K. Srivastava (eds.): *Trends in Structural Mechanics*. Theory, Practice, Education. 1997 ISBN 0-7923-4603-3
55. Yu.A. Mitropolskii and N. Van Dao: *Applied Asymptotic Methods in Nonlinear Oscillations*. 1997 ISBN 0-7923-4605-X
56. C. Guedes Soares (ed.): *Probabilistic Methods for Structural Design*. 1997 ISBN 0-7923-4670-X
57. D. François, A. Pineau and A. Zaoui: *Mechanical Behaviour of Materials*. Volume I: Elasticity and Plasticity. 1998 ISBN 0-7923-4894-X
58. D. François, A. Pineau and A. Zaoui: *Mechanical Behaviour of Materials*. Volume II: Viscoplasticity, Damage, Fracture and Contact Mechanics. 1998 ISBN 0-7923-4895-8
59. L.T. Tenek and J. Argyris: *Finite Element Analysis for Composite Structures*. 1998 ISBN 0-7923-4899-0
60. Y.A. Bahei-El-Din and G.J. Dvorak (eds.): *IUTAM Symposium on Transformation Problems in Composite and Active Materials*. Proceedings of the IUTAM Symposium held in Cairo, Egypt. 1998 ISBN 0-7923-5122-3
61. I.G. Goryacheva: *Contact Mechanics in Tribology*. 1998 ISBN 0-7923-5257-2
62. O.T. Bruhns and E. Stein (eds.): *IUTAM Symposium on Micro- and Macrostructural Aspects of Thermoplasticity*. Proceedings of the IUTAM Symposium held in Bochum, Germany. 1999 ISBN 0-7923-5265-3
63. F.C. Moon: *IUTAM Symposium on New Applications of Nonlinear and Chaotic Dynamics in Mechanics*. Proceedings of the IUTAM Symposium held in Ithaca, NY, USA. 1998 ISBN 0-7923-5276-9
64. R. Wang: *IUTAM Symposium on Rheology of Bodies with Defects*. Proceedings of the IUTAM Symposium held in Beijing, China. 1999 ISBN 0-7923-5297-1
65. Yu.I. Dimitrienko: *Thermomechanics of Composites under High Temperatures*. 1999 ISBN 0-7923-4899-0
66. P. Argoul, M. Frémond and Q.S. Nguyen (eds.): *IUTAM Symposium on Variations of Domains and Free-Boundary Problems in Solid Mechanics*. Proceedings of the IUTAM Symposium held in Paris, France. 1999 ISBN 0-7923-5450-8
67. F.J. Fahy and W.G. Price (eds.): *IUTAM Symposium on Statistical Energy Analysis*. Proceedings of the IUTAM Symposium held in Southampton, U.K. 1999 ISBN 0-7923-5457-5
68. H.A. Mang and F.G. Rammerstorfer (eds.): *IUTAM Symposium on Discretization Methods in Structural Mechanics*. Proceedings of the IUTAM Symposium held in Vienna, Austria. 1999 ISBN 0-7923-5591-1

Mechanics

SOLID MECHANICS AND ITS APPLICATIONS

Series Editor: G.M.L. Gladwell

69. P. Pedersen and M.P. Bendsøe (eds.): *IUTAM Symposium on Synthesis in Bio Solid Mechanics*. Proceedings of the IUTAM Symposium held in Copenhagen, Denmark. 1999 ISBN 0-7923-5615-2
70. S.K. Agrawal and B.C. Fabien: *Optimization of Dynamic Systems*. 1999 ISBN 0-7923-5681-0
71. A. Carpinteri: *Nonlinear Crack Models for Nonmetallic Materials*. 1999 ISBN 0-7923-5750-7
72. F. Pfeifer (ed.): *IUTAM Symposium on Unilateral Multibody Contacts*. Proceedings of the IUTAM Symposium held in Munich, Germany. 1999 ISBN 0-7923-6030-3
73. E. Lavendelis and M. Zakrzhevsky (eds.): *IUTAM/IFToMM Symposium on Synthesis of Nonlinear Dynamical Systems*. Proceedings of the IUTAM/IFToMM Symposium held in Riga, Latvia. 2000 ISBN 0-7923-6106-7
74. J.-P. Merlet: *Parallel Robots*. 2000 ISBN 0-7923-6308-6
75. J.T. Pindera: *Techniques of Tomographic Isodyne Stress Analysis*. 2000 ISBN 0-7923-6388-4
76. G.A. Maugin, R. Drouot and F. Sidoroff (eds.): *Continuum Thermomechanics*. The Art and Science of Modelling Material Behaviour. 2000 ISBN 0-7923-6407-4
77. N. Van Dao and E.J. Kreuzer (eds.): *IUTAM Symposium on Recent Developments in Non-linear Oscillations of Mechanical Systems*. 2000 ISBN 0-7923-6470-8
78. S.D. Akbarov and A.N. Guz: *Mechanics of Curved Composites*. 2000 ISBN 0-7923-6477-5
79. M.B. Rubin: *Cosserat Theories: Shells, Rods and Points*. 2000 ISBN 0-7923-6489-9
80. S. Pellegrino and S.D. Guest (eds.): *IUTAM-IASS Symposium on Deployable Structures: Theory and Applications*. Proceedings of the IUTAM-IASS Symposium held in Cambridge, U.K., 6–9 September 1998. 2000 ISBN 0-7923-6516-X
81. A.D. Rosato and D.L. Blackmore (eds.): *IUTAM Symposium on Segregation in Granular Flows*. Proceedings of the IUTAM Symposium held in Cape May, NJ, U.S.A., June 5–10, 1999. 2000 ISBN 0-7923-6547-X
82. A. Lagarde (ed.): *IUTAM Symposium on Advanced Optical Methods and Applications in Solid Mechanics*. Proceedings of the IUTAM Symposium held in Futuroscope, Poitiers, France, August 31–September 4, 1998. 2000 ISBN 0-7923-6604-2
83. D. Weichert and G. Maier (eds.): *Inelastic Analysis of Structures under Variable Loads*. Theory and Engineering Applications. 2000 ISBN 0-7923-6645-X
84. T.-J. Chuang and J.W. Rudnicki (eds.): *Multiscale Deformation and Fracture in Materials and Structures*. The James R. Rice 60th Anniversary Volume. 2001 ISBN 0-7923-6718-9
85. S. Narayanan and R.N. Iyengar (eds.): *IUTAM Symposium on Nonlinearity and Stochastic Structural Dynamics*. Proceedings of the IUTAM Symposium held in Madras, Chennai, India, 4–8 January 1999 ISBN 0-7923-6733-2
86. S. Murakami and N. Ohno (eds.): *IUTAM Symposium on Creep in Structures*. Proceedings of the IUTAM Symposium held in Nagoya, Japan, 3-7 April 2000. 2001 ISBN 0-7923-6737-5
87. W. Ehlers (ed.): *IUTAM Symposium on Theoretical and Numerical Methods in Continuum Mechanics of Porous Materials*. Proceedings of the IUTAM Symposium held at the University of Stuttgart, Germany, September 5-10, 1999. 2001 ISBN 0-7923-6766-9
88. D. Durban, D. Givoli and J.G. Simmonds (eds.): *Advances in the Mechanis of Plates and Shells The Avinoam Libai Anniversary Volume*. 2001 ISBN 0-7923-6785-5
89. U. Gabbert and H.-S. Tzou (eds.): *IUTAM Symposium on Smart Structures and Structonic Systems*. Proceedings of the IUTAM Symposium held in Magdeburg, Germany, 26–29 September 2000. 2001 ISBN 0-7923-6968-8

Mechanics

SOLID MECHANICS AND ITS APPLICATIONS

Series Editor: G.M.L. Gladwell

90. Y. Ivanov, V. Cheshkov and M. Natova: *Polymer Composite Materials – Interface Phenomena & Processes*. 2001 ISBN 0-7923-7008-2
91. R.C. McPhedran, L.C. Botten and N.A. Nicorovici (eds.): *IUTAM Symposium on Mechanical and Electromagnetic Waves in Structured Media*. Proceedings of the IUTAM Symposium held in Sydney, NSW, Australia, 18-22 Januari 1999. 2001 ISBN 0-7923-7038-4
92. D.A. Sotiropoulos (ed.): *IUTAM Symposium on Mechanical Waves for Composite Structures Characterization*. Proceedings of the IUTAM Symposium held in Chania, Crete, Greece, June 14-17, 2000. 2001 ISBN 0-7923-7164-X
93. V.M. Alexandrov and D.A. Pozharskii: *Three-Dimensional Contact Problems*. 2001 ISBN 0-7923-7165-8
94. J.P. Dempsey and H.H. Shen (eds.): *IUTAM Symposium on Scaling Laws in Ice Mechanics and Ice Dynamics*. Proceedings of the IUTAM Symposium held in Fairbanks, Alaska, U.S.A., 13-16 June 2000. 2001 ISBN 1-4020-0171-1
95. U. Kirsch: *Design-Oriented Analysis of Structures*. A Unified Approach. 2002 ISBN 1-4020-0443-5
96. A. Preumont: *Vibration Control of Active Structures*. An Introduction (2nd Edition). 2002 ISBN 1-4020-0496-6
97. B.L. Karihaloo (ed.): *IUTAM Symposium on Analytical and Computational Fracture Mechanics of Non-Homogeneous Materials*. Proceedings of the IUTAM Symposium held in Cardiff, U.K., 18-22 June 2001. 2002 ISBN 1-4020-0510-5
98. S.M. Han and H. Benaroya: *Nonlinear and Stochastic Dynamics of Compliant Offshore Structures*. 2002 ISBN 1-4020-0573-3
99. A.M. Linkov: *Boundary Integral Equations in Elasticity Theory*. 2002 ISBN 1-4020-0574-1
100. L.P. Lebedev, I.I. Vorovich and G.M.L. Gladwell: *Functional Analysis*. Applications in Mechanics and Inverse Problems (2nd Edition). 2002 ISBN 1-4020-0667-5; Pb: 1-4020-0756-6
101. Q.P. Sun (ed.): *IUTAM Symposium on Mechanics of Martensitic Phase Transformation in Solids*. Proceedings of the IUTAM Symposium held in Hong Kong, China, 11-15 June 2001. 2002 ISBN 1-4020-0741-8
102. M.L. Munjal (ed.): *IUTAM Symposium on Designing for Quietness*. Proceedings of the IUTAM Symposium held in Bangkok, India, 12-14 December 2000. 2002 ISBN 1-4020-0765-5
103. J.A.C. Martins and M.D.P. Monteiro Marques (eds.): *Contact Mechanics*. Proceedings of the 3rd Contact Mechanics International Symposium, Praia da Consolação, Peniche, Portugal, 17-21 June 2001. 2002 ISBN 1-4020-0811-2
104. H.R. Drew and S. Pellegrino (eds.): *New Approaches to Structural Mechanics, Shells and Biological Structures*. 2002 ISBN 1-4020-0862-7
105. J.R. Vinson and R.L. Sierakowski: *The Behavior of Structures Composed of Composite Materials*. Second Edition. 2002 ISBN 1-4020-0904-6
106. Not yet published.
107. J.R. Barber: *Elasticity*. Second Edition. 2002 ISBN Hb 1-4020-0964-X; Pb 1-4020-0966-6
108. C. Miehe (ed.): *IUTAM Symposium on Computational Mechanics of Solid Materials at Large Strains*. Proceedings of the IUTAM Symposium held in Stuttgart, Germany, 20-24 August 2001. 2003 ISBN 1-4020-1170-9

Mechanics

SOLID MECHANICS AND ITS APPLICATIONS

Series Editor: G.M.L. Gladwell

109. P. Ståhle and K.G. Sundin (eds.): *IUTAM Symposium on Field Analyses for Determination of Material Parameters – Experimental and Numerical Aspects.* Proceedings of the IUTAM Symposium held in Abisko National Park, Kiruna, Sweden, July 31 – August 4, 2000. 2003 ISBN 1-4020-1283-7
110. N. Sri Namachchivaya and Y.K. Lin (eds.): *IUTAM Symposium on Nonlinear Stochastic Dynamics.* Proceedings of the IUTAM Symposium held in Monticello, IL, USA, 26 – 30 August, 2000. 2003 ISBN 1-4020-1471-6
111. H. Sobieckzky (ed.): *IUTAM Symposium Transsonicum IV.* Proceedings of the IUTAM Symposium held in Göttingen, Germany, 2–6 September 2002, 2003 ISBN 1-4020-1608-5
112. J.-C. Samin and P. Fisette: *Symbolic Modeling of Multibody Systems.* 2003 ISBN 1-4020-1629-8
113. A.B. Movchan (ed.): *IUTAM Symposium on Asymptotics, Singularities and Homogenisation in Problems of Mechanics.* Proceedings of the IUTAM Symposium held in Liverpool, United Kingdom, 8-11 July 2002. 2003 ISBN 1-4020-1780-4
114. S. Ahzi, M. Cherkaoui, M.A. Khaleel, H.M. Zbib, M.A. Zikry and B. LaMatina (eds.): *IUTAM Symposium on Multiscale Modeling and Characterization of Elastic-Inelastic Behavior of Engineering Materials.* Proceedings of the IUTAM Symposium held in Marrakech, Morocco, 20-25 October 2002. 2004 ISBN 1-4020-1861-4
115. H. Kitagawa and Y. Shibutani (eds.): *IUTAM Symposium on Mesoscopic Dynamics of Fracture Process and Materials Strength.* Proceedings of the IUTAM Symposium held in Osaka, Japan, 6-11 July 2003. Volume in celebration of Professor Kitagawa's retirement. 2004 ISBN 1-4020-2037-6
116. E.H. Dowell, R.L. Clark, D. Cox, H.C. Curtiss, Jr., K.C. Hall, D.A. Peters, R.H. Scanlan, E. Simiu, F. Sisto and D. Tang: *A Modern Course in Aeroelasticity.* 4th Edition, 2004 ISBN 1-4020-2039-2
117. T. Burczyński and A. Osyczka (eds.): *IUTAM Symposium on Evolutionary Methods in Mechanics.* Proceedings of the IUTAM Symposium held in Cracow, Poland, 24-27 September 2002. 2004 ISBN 1-4020-2266-2
118. D. Ieşan: *Thermoelastic Models of Continua.* 2004 ISBN 1-4020-2309-X
119. G.M.L. Gladwell: *Inverse Problems in Vibration.* Second Edition. 2004 ISBN 1-4020-2670-6
120. J.R. Vinson: *Plate and Panel Structures of Isotropic, Composite and Piezoelectric Materials, Including Sandwich Construction.* 2005 ISBN 1-4020-3110-6
121. *Forthcoming*
122. G. Rega and F. Vestroni (eds.): *IUTAM Symposium on Chaotic Dynamics and Control of Systems and Processes in Mechanics.* Proceedings of the IUTAM Symposium held in Rome, Italy, 8–13 June 2003. 2005 ISBN 1-4020-3267-6
123. E.E. Gdoutos: *Fracture Mechanics. An Introduction. 2nd edition.* 2005 ISBN 1-4020-3267-6
124. M.D. Gilchrist (ed.): *IUTAM Symposium on Impact Biomechanics from Fundamental Insights to Applications.* 2005 ISBN 1-4020-3795-3
125. J.M. Huyghe, P.A.C. Raats and S.C. Cowin (eds.): *IUTAM Symposium on Physicochemical and Electromechanical Interactions in Porous Media.* 2005 ISBN 1-4020-3864-X
126. H. Ding and W. Chen: *Elasticity of Transversely Isotropic Materials.* 2006 ISBN 1-4020-4033-4
127. W. Yang (ed): *IUTAM Symposium on Mechanics and Reliability of Actuating Materials.* Proceedings of the IUTAM Symposium held in Beijing, China, 1–3 September 2004. 2006 ISBN 1-4020-4131-6

Mechanics

SOLID MECHANICS AND ITS APPLICATIONS

Series Editor: G.M.L. Gladwell

128. J.-P. Merlet: *Parallel Robots.* 2006 ISBN 1-4020-4132-2
129. G.E.A. Meier and K.R. Sreenivasan (eds.): *IUTAM Symposium on One Hundred Years of Boundary Layer Research.* Proceedings of the IUTAM Symposium held at DLR-Göttingen, Germany, August 12–14, 2004. 2006 ISBN 1-4020-4149-7
130. H. Ulbrich and W. Günthner (eds.): *IUTAM Symposium on Vibration Control of Nonlinear Mechanisms and Structures.* 2006 ISBN 1-4020-4160-8
131. L. Librescu and O. Song: *Thin-Walled Composite Beams.* Theory and Application. 2006 ISBN 1-4020-3457-1
132. G. Ben-Dor, A. Dubinsky and T. Elperin: *Applied High-Speed Plate Penetration Dynamics.* 2006 ISBN 1-4020-3452-0
133. X. Markenscoff and A. Gupta (eds.): *Collected Works of J. D. Eshelby.* Mechanics and Defects and Heterogeneities. 2006 ISBN 1-4020-4416-X
134. R.W. Snidle and H.P. Evans (eds.): *IUTAM Symposium on Elastohydrodynamics and Micro-elastohydrodynamics.* Proceedings of the IUTAM Symposium held in Cardiff, UK, 1–3 September, 2004. 2006 ISBN 1-4020-4532-8
135. T. Sadowski (ed.): *IUTAM Symposium on Multiscale Modelling of Damage and Fracture Processes in Composite Materials.* Proceedings of the IUTAM Symposium held in Kazimierz Dolny, Poland, 23–27 May 2005. 2006 ISBN 1-4020-4565-4
136. A. Preumont: *Mechatronics.* Dynamics of Electromechanical and Piezoelectric Systems. 2006 ISBN 1-4020-4695-2
137. M.P. Bendsoe, N. Olhoff and O. Sigmund (eds.): *IUTAM Symposium on Topological Design Optimization of Structures, Machines and Materials.* Status and Perspectives. 2006 ISBN 1-4020-4729-0

GPSR Compliance

The European Union's (EU) General Product Safety Regulation (GPSR) is a set of rules that requires consumer products to be safe and our obligations to ensure this.

If you have any concerns about our products, you can contact us on ProductSafety@springernature.com

In case Publisher is established outside the EU, the EU authorized representative is:

Springer Nature Customer Service Center GmbH
Europaplatz 3
69115 Heidelberg, Germany

Batch number: 10755284

Printed by Printforce, the Netherlands